INTERNATIONAL UNION OF CRYSTALLOGRAPHY BOOK SERIES

IUCr BOOK SERIES COMMITTEE

J. Bernstein, *Israel*
P. Colman, *Australia*
J. R. Helliwell, *UK*
K. A. Kantardjieff, *USA*
T. Mak, *China*
P. Müller, *USA*
Y. Ohashi, *Japan*
P. Paufler, *Germany*
H. Schenk, *The Netherlands*
D. Viterbo (*Chairman*), *Italy*

IUCr Monographs on Crystallography

1   *Accurate molecular structures*
    A. Domenicano, I. Hargittai, editors
2   *P.P. Ewald and his dynamical theory of X-ray diffraction*
    D.W.J. Cruickshank, H.J. Juretschke, N. Kato, editors
3   *Electron diffraction techniques, Vol. 1*
    J.M. Cowley, editor
4   *Electron diffraction techniques, Vol. 2*
    J.M. Cowley, editor
5   *The Rietveld method*
    R.A. Young, editor
6   *Introduction to crystallographic statistics*
    U. Shmueli, G.H. Weiss
7   *Crystallographic instrumentation*
    L.A. Aslanov, G.V. Fetisov, J.A.K. Howard
8   *Direct phasing in crystallography*
    C. Giacovazzo
9   *The weak hydrogen bond*
    G.R. Desiraju, T. Steiner
10 *Defect and microstructure analysis by diffraction*
    R.L. Snyder, J. Fiala, H.J. Bunge
11 *Dynamical theory of X-ray diffraction*
    A. Authier
12 *The chemical bond in inorganic chemistry*
    I.D. Brown
13 *Structure determination from powder diffraction data*
    W.I.F. David, K. Shankland, L.B. McCusker, Ch. Baerlocher, editors
14 *Polymorphism in molecular crystals*
    J. Bernstein
15 *Crystallography of modular materials*
    G. Ferraris, E. Makovicky, S. Merlino
16 *Diffuse X-ray scattering and models of disorder*
    T.R. Welberry
17 *Crystallography of the polymethylene chain: an inquiry into the structure of waxes*
    D.L. Dorset

18 *Crystalline molecular complexes and compounds*: structure and principles
    F.H. Herbstein
19 *Molecular aggregation*: structure analysis and molecular simulation of crystals and liquids
    A. Gavezzotti
20 *Aperiodic crystals*: from modulated phases to quasicrystals
    T. Janssen, G. Chapuis, M. de Boissieu
21 *Incommensurate crystallography*
    S. van Smaalen
22 *Structural crystallography of inorganic oxysalts*
    S.V. Krivovichev
23 *The nature of the hydrogen bond*: outline of a comprehensive hydrogen bond theory
    G. Gilli, P. Gilli
24 *Macromolecular crystallization and crystal perfection*
    N.E. Chayen, J.R. Helliwell, E.H. Snell
25 *Neutron protein crystallography*: hydrogen, protons, and hydration in bio-macromolecules
    N. Niimura, A. Podjarny

IUCr Texts on Crystallography

 1 *The solid state*
    A. Guinier, R. Julien
 4 *X-ray charge densities and chemical bonding*
    P. Coppens
 8 *Crystal structure refinement*: a crystallographer's guide to SHELXL
    P. Müller, editor
 9 *Theories and techniques of crystal structure determination*
    U. Shmueli
10 *Advanced structural inorganic chemistry*
    Wai-Kee Li, Gong-Du Zhou, Thomas Mak
11 *Diffuse scattering and defect structure simulations*: a cook book using the program DISCUS
    R.B. Neder, T. Proffen
12 *The basics of crystallography and diffraction, third edition*
    C. Hammond
13 *Crystal structure analysis: principles and practice, second edition*
    W. Clegg, editor
14 *Crystal structure analysis: a primer, third edition*
    J.P. Glusker, K.N. Trueblood
15 *Fundamentals of crystallography, third edition*
    C. Giacovazzo, editor
16 *Electron crystallography*: electron microscopy and electron diffraction
    X. Zou, S. Hovmöller, P. Oleynikov
17 *Symmetry in crystallography*: understanding the International Tables
    P.G. Radaelli
18 *Symmetry relationships between crystal structures*: applications of crystallographic group theory in crystal chemistry
    U. Müller
19 *Small angle X-ray and neutron scattering from biomacromolecular solutions*
    D.I. Svergun, M.H.J. Koch, P.A. Timmins, R.P. May
20 *Phasing in crystallography*: a modern perspective
    C. Giacovazzo

Phasing in Crystallography

A Modern Perspective

CARMELO GIACOVAZZO

Professor of Crystallography,
University of Bari, Italy
Institute of Crystallography, CNR, Bari, Italy

Great Clarendon Street, Oxford, OX2 6DP,
United Kingdom

Oxford University Press is a department of the University of Oxford.
It furthers the University's objective of excellence in research, scholarship,
and education by publishing worldwide. Oxford is a registered trade mark of
Oxford University Press in the UK and in certain other countries

© Carmelo Giacovazzo 2014

The moral rights of the author have been asserted

First Edition published in 2014

Impression: 1

All rights reserved. No part of this publication may be reproduced, stored in
a retrieval system, or transmitted, in any form or by any means, without the
prior permission in writing of Oxford University Press, or as expressly permitted
by law, by licence or under terms agreed with the appropriate reprographics
rights organization. Enquiries concerning reproduction outside the scope of the
above should be sent to the Rights Department, Oxford University Press, at the
address above

You must not circulate this work in any other form
and you must impose this same condition on any acquirer

Published in the United States of America by Oxford University Press
198 Madison Avenue, New York, NY 10016, United States of America

British Library Cataloguing in Publication Data

Data available

Library of Congress Control Number: 2013943731

ISBN 978–0–19–968699–5

Printed in Great Britain by
Butler Tanner & Dennis Ltd

Links to third party websites are provided by Oxford in good faith and for information
only. Oxford disclaims any responsibility for the materials contained in any third party
website referenced in this work.

Dedication

To my mother,
to my wife Angela,
my sons Giuseppe and Stefania,
to my grandchildren Agostino, Stefano and Andrea Morris

Acknowledgements

I acknowledge the following colleagues and friends for their generous help:

Caterina Chiarella, for general secretarial management of the book and for her assistance with the drawings;

Angela Altomare, Benedetta Carrozzini, Corrado Cuocci, Giovanni Luca Cascarano, Annamaria Mazzone, Anna Grazia Moliterni, and Rosanna Rizzi for their kind support, helpful discussions, and critical reading of the manuscript. Corrado Cuocci also took care of the cover figure.

Facilities provided by the Istituto di Cristallografia, CNR, Bari, are gratefully acknowledged.

Preface

A short analysis of the historical evolution of phasing methods may be a useful introduction to this book because it will allow us to better understand efforts and results, the birth and death of scientific paradigms, and it will also explain the general organization of this volume. This analysis is very personal, and arises through the author's direct interactions with colleagues active in the field; readers interested in such aspects may find a more extensive exposition in *Rend. Fis. Acc. Lincei* (2013), **24**(1), pp. 71–76.

In a historical sense, crystallographic phasing methods may be subdivided into two main streams: the small and medium-sized molecule stream, and the macro-molecule stream; these were substantially independent from each other up until the 1990s. Let us briefly consider their achievements and the results of their subsequent confluence.

Small and medium-sized molecule stream

The *Patterson* (1934) *function* was the first general phasing tool, particularly effective for heavy-atom structures (e.g. this property met the requirements of the earth sciences, the first users of early crystallography). Even though subsequently computerized, it was soon relegated to a niche by *direct methods*, since these were also able to solve light-atom structures (a relevant property towards the development of organic chemistry).

Direct methods were introduced, in their modern probabilistic guise, by Hauptman and Karle (1953) and Cochran (1955); corresponding phasing procedures were automated by Woolfson and co-workers, making the crystal structure solution of small molecules more straightforward. Efforts were carried out exclusively in reciprocal space (*first paradigm of direct methods*); the paradigm was systematized by the neighbourhood (Hauptman, 1975) and representation theories (Giacovazzo, 1977, 1980). Structures up to 150 non-hydrogen (non-H) atoms in the asymmetric unit were routinely able to be solved.

The complete success of this stream may be deduced from the huge numbers of structures deposited in appropriate data banks. Consequently, western national research agencies no longer supported any further research in the small to medium-sized molecule area (the work was done!); research groups working on methods moved instead to powder crystallography, electron crystallography, or to proteins, all areas of technological interest for which phasing was still a challenge. Direct space approaches were soon developed, which enhanced our capacity to solve structures, even from low quality diffraction data.

The macromolecule stream

Since the 1950s, efforts were confined to *isomorphous replacement* (SIR, MIR; Green et al., 1954), *molecular replacement* (MR; Rossmann and Blow, 1962), and *anomalous dispersion techniques* (SAD-MAD; Okaya and Pepinsky, 1956; Hoppe and Jakubowski, 1975). *Ab initio approaches*, the main techniques of interest for the small and medium-sized molecule streams, were neglected as being unrealistic; indeed, they are less demanding in terms of prior information but are very demanding in terms of data resolution.

The popularity of protein phasing techniques changed dramatically over the years. At the very beginning, SIR-MIR was the most popular method, but soon MR started to play a more major role as good structural models became progressively more readily available. About 75% of structures today are solved using MR. The simultaneous technological progress in synchrotron radiation and its wide availability have increased the appeal of SAD-MAD techniques.

The achievements obtained within the macromolecular stream have been impressive. A huge number of protein structures has been deposited in the Protein Data Bank, and the solution of protein structures is no longer confined to just an elite group of scientists, it is performed in many laboratories spread over four continents, often by young scientists. Crucial to this has been the role of the *CCP4* project, for the coordination of new methods and new computer programs.

The synergy of the two streams

It is the opinion of the author that synergy between the two streams originated due to a common interest in *EDM* (*electron density modification*) techniques. This approach, first proposed by Hoppe and Gassman (1968) for small molecules, was later extensively modified to be useful for both streams. Confluence of the two streams began in the 1990s (even if contacts were begun in the 1980s), when EDM techniques were used to improve the efficiency of direct methods. That was the beautiful innovation of *shake and bake* (Weeks et al., 1994); both direct and reciprocal space were explored to increase phasing efficiency (this was the *second paradigm of direct methods*). It was soon possible to solve ab initio structures with up to 2000 non-hydrogen atoms in the asymmetric unit, provided data at atomic or quasi-atomic resolution are available. As a consequence, the ab initio approach for proteins started to attract greater attention. A secondary effect of the EDM procedures was the recent discovery of new ab initio techniques, such as *charge flipping* and *VLD* (*vive la différence*), and the newly formulated Patterson techniques.

The real revolution in the macromolecular area occurred when probabilistic methods, already widely used in small and medium-sized molecules, erupted into the protein field. Joint probability distributions and maximum likelihood approaches were tailored to deal with large structures, imperfect isomorphism, and errors in experimental data; and they were applied to SAD-MAD, MR, and SIR-MIR cases. For example, protein substructures with around 200 atoms in the asymmetric unit, an impossible challenge for traditional techniques, could easily be solved by the new approaches.

High-throughput crystallography is now a reality: protein structures, 50 years ago solvable only over months or years, can now be solved in hours or days; also due to technological advances in computer sciences.

The above considerations have been the basic reason for reconsidering the material and the general guidelines given in my textbook *Direct Phasing in Crystallography*, originally published in 1998. This was essentially a description of the mathematical bases of direct methods and of their historical evolution, with some references to applicative aspects and ancillary techniques.

The above described explosion in new phasing techniques and the improved efficiency of the revisited old methods made impellent the need for a new textbook, mainly addressing the phasing approaches which are alive today, that is those which are applicable to today's routine work. On the other hand, the wide variety of new methods and their intricate relationship with the old methods requires a new rational classification: methods similar regarding the type of prior information exploited, mathematical technique, or simply their mission, are didactically correlated, in such a way as to offer an organized overview of the current and of the old approaches. This is the main aim of this volume, which should not therefore simply be considered as the second edition of *Direct Phasing in Crystallography*, but as a new book with different guidelines, different treated material, and a different purpose.

Attention will be focused on both the theoretical and the applicative aspects, in order to provide a friendly companion for our daily work. To emphasize the new design the title has been changed to *Phasing in Crystallography,* with the subtitle, *A Modern Perspective*. In order to make the volume more useful, historical developments of phasing approaches that are not in use today, are simply skipped, and readers interested in these are referred to *Direct Phasing in Crystallography*.

This volume also aims at being a tool to inspire new approaches. On the one hand, we have tried to give, in the main text, descriptions of the various methods that are as simple as possible, so that undergraduate and graduate students may understand their general purpose and their applicative aspects. On the other hand, we did not shrink from providing the interested reader with mathematical details and/or demonstrations (these are necessary for any book dealing specifically with methods). These are confined in suitable appendices to the various chapters, and aimed at the trained crystallographer. At the end of the book, we have collected together mathematical appendices of a general character, appendices denoted by the letter M for mathematics and devoted to the bases of the methods (e.g. probability theory, basic crystallography, concepts of analysis and linear algebra, specific mathematical techniques, etc.), thus offering material of interest for professional crystallographers.

A necessary condition for an understanding of the content of the book is a knowledge of the fundamentals of crystallography. Thus, in Chapter 1 we have synthesized the essential elements of the general crystallography and we have also formulated the *basic postulate of structural crystallography*; the entire book is based on its validity.

In Chapter 2, the statistics of structure factors is described simply: it will be the elementary basis of most of the methods described throughout the volume.

Chapter 3 is a simplified description of the concepts of structure invariant and seminvariant, and of the related origin problem.

In Chapter 4, we have synthesized the methods of joint probability distributions and neighbourhoods–representation theories. The application of these methods to three-phase and four-phase structure invariants are described in Chapter 5. The probabilistic estimation of structure seminvariants has been skipped owing to their marginal role in modern phasing techniques. In Chapter 6, we discuss direct methods and the most traditional phasing approaches.

Chapter 7 is dedicated to joint probability distribution functions when a model is available, with specific attention to two- and to three-phase invariants. The most popular Fourier syntheses are described in the same chapter and their potential discussed in relation with the above probability distributions.

Chapter 8 is dedicated to phase improvement and extension via electron density modification techniques, Chapter 9 to two new phasing approaches, *charge flipping* and *VLD* (*vive la difference*), and Chapter 10, to *Patterson techniques*. Their recent revision has made them one of the most powerful techniques for ab initio phasing and particularly useful for proteins.

X-rays are not always the most suitable radiation for performing a diffraction experiment. Indeed, neutron diffraction may provide information complementary to that provided by X-ray data, electron diffraction becoming necessary when only nanocrystals are available. In Chapter 11 phasing procedures useful for this new scenario are described.

Often single crystals of sufficient size and quality are not available, but microcrystals can be grown. In this case powder data are collected; diffraction techniques imply a loss of experimental information, and therefore phasing via such data requires significant modifications to the standard methods. These are described in Chapter 12.

Chapters 13 to 15 are dedicated to the most effective and popular methods used in macromolecular crystallography: the non-ab initio methods, *Molecular Replacement (MR)*, *Isomorphous Replacement (SIR-MIR)*, and *Anomalous Dispersion (SAD-MAD)* techniques.

The reader should not think that the book has been partitioned into two parts, the first devoted to small and medium-sized molecules, the second to macromolecules. Indeed in the first twelve chapters, most of the mathematical tools necessary to face the challenges of macromolecular crystallography are described, together with the main algorithms used in this area and the fundamentals of the probabilistic approaches employed in macromolecular phasing. This design allows us to provide, in the last three chapters, simpler descriptions of MR, SIR-MIR, and SAD-MAD approaches.

Contents

Symbols and notation   xvii

1 Fundamentals of crystallography   1

   1.1 Introduction   1
   1.2 Crystals and crystallographic symmetry in direct space   1
   1.3 The reciprocal space   5
   1.4 The structure factor   11
   1.5 Symmetry in reciprocal space   12
      1.5.1 Friedel law   12
      1.5.2 Effects of symmetry operators in reciprocal space   12
      1.5.3 Determination of reflections with restricted phase values   13
      1.5.4 Systematic absences   15
   1.6 The basic postulate of structural crystallography   17
   1.7 The legacy of crystallography   24

2 Wilson statistics   27

   2.1 Introduction   27
   2.2 Statistics of the structure factor: general considerations   28
   2.3 Structure factor statistics in $P1$ and $P\bar{1}$ 29
   2.4 The $P(z)$ distributions   35
   2.5 Cumulative distributions   35
   2.6 Space group identification   36
   2.7 The centric or acentric nature of crystals: Wilson statistical analysis   42
   2.8 Absolute scaling of intensities: the Wilson plot   43
   2.9 Shape of the Wilson plot   47
   2.10 Unit cell content   49
   Appendix 2.A Statistical calculations in $P1$ and $P\bar{1}$ 50
   2.A.1 Structure factor statistics in $P1$ 50
   2.A.2 Structure factor statistics in $P\bar{1}$ 52
   Appendix 2.B Statistical calculations in any space group   53
   2.B.1 The algebraic form of the structure factor   53
   2.B.2 Structure factor statistics for centric and acentric space groups   55
   Appendix 2.C The Debye formula   58

3 The origin problem, invariants, and seminvariants   60

   3.1 Introduction   60
   3.2 Origin, phases, and symmetry operators   61

3.3	The concept of structure invariant	63
3.4	Allowed or permissible origins in primitive space groups	65
3.5	The concept of structure seminvariant	69
3.6	Allowed or permissible origins in centred cells	76
3.7	Origin definition by phase assignment	81

4 The method of joint probability distribution functions, neighbourhoods, and representations   83

4.1	Introduction	83
4.2	Neighbourhoods and representations	87
4.3	Representations of structure seminvariants	89
4.4	Representation theory for structure invariants extended to isomorphous data	91

Appendix 4.A The method of structure factor joint probability distribution functions   93

4.A.1	Introduction	93
4.A.2	Multivariate distributions in centrosymmetric structures: the case of independent random variables	94
4.A.3	Multivariate distributions in non-centrosymmetric structures: the case of independent random variables	97
4.A.4	Simplified joint probability density functions in the absence of prior information	99
4.A.5	The joint probability density function when some prior information is available	102
4.A.6	The calculation of $P(E)$ in the absence of prior information	103

5 The probabilistic estimation of triplet and quartet invariants   104

5.1	Introduction	104
5.2	Estimation of the triplet structure invariant via its first representation: the P1 and the P$\bar{1}$ case	104
5.3	About triplet invariant reliability	108
5.4	The estimation of triplet phases via their second representation	110
5.5	Introduction to quartets	112
5.6	The estimation of quartet invariants in P1 and P$\bar{1}$ via their first representation: Hauptman approach	112
5.7	The estimation of quartet invariants in P1 and P$\bar{1}$ via their first representation: Giacovazzo approach	115
5.8	About quartet reliability	116

Appendix 5.A	The probabilistic estimation of the triplet invariants in P1	117
Appendix 5.B	Symmetry inconsistent triplets	120
Appendix 5.C	The P_{10} formula	121
Appendix 5.D	The use of symmetry in quartet estimation	123

6 Traditional direct phasing procedures — 125

- 6.1 Introduction — 125
- 6.2 The tangent formula — 128
- 6.3 Procedure for phase determination via traditional direct methods — 130
  - 6.3.1 Set-up of phase relationships — 131
  - 6.3.2 Assignment of starting phases — 134
  - 6.3.3 Phase determination — 136
  - 6.3.4 Finding the correct solution — 137
  - 6.3.5 E-map interpretation — 138
  - 6.3.6 Phase extension and refinement: reciprocal space techniques — 140
  - 6.3.7 The limits of the tangent formula — 141
- 6.4 Third generation direct methods programs — 144
  - 6.4.1 The shake and bake approach — 144
  - 6.4.2 The half-bake approach — 147
  - 6.4.3 The SIR2000-N approach — 148
- Appendix 6.A Finding quartets — 149

7 Joint probability distribution functions when a model is available: Fourier syntheses — 151

- 7.1 Introduction — 151
- 7.2 Estimation of the two-phase structure invariant $(\phi_\mathbf{h} - \phi_{p\mathbf{h}})$ — 152
- 7.3 Electron density maps — 155
  - 7.3.1 The ideal Fourier synthesis and its properties — 156
  - 7.3.2 The observed Fourier synthesis — 162
  - 7.3.3 The difference Fourier synthesis — 164
  - 7.3.4 Hybrid Fourier syntheses — 166
- 7.4 Variance and covariance for electron density maps — 168
- 7.5 Triplet phase estimate when a model is available — 170
- Appendix 7.A Estimation of σ_A — 173
- Appendix 7.B Variance and covariance expressions for electron density maps — 174
- Appendix 7.C Some marginal and conditional probabilities of $P(R, R_p, \phi, \phi_p)$ — 176

8 Phase improvement and extension — 177

- 8.1 Introduction — 177
- 8.2 Phase extension and refinement via direct space procedures: EDM techniques — 177
- 8.3 Automatic model building — 184
- 8.4 Applications — 188
- Appendix 8.A Solvent content, envelope definition, and solvent modelling — 190
  - 8.A.1 Solvent content according to Matthews — 190
  - 8.A.2 Envelope definition — 191
  - 8.A.3 Models for the bulk solvent — 192
- Appendix 8.B Histogram matching — 193
- Appendix 8.C A brief outline of the *ARP/wARP* procedure — 196

9 Charge flipping and VLD (vive la difference) — 198

- 9.1 Introduction — 198
- 9.2 The charge flipping algorithm — 199
- 9.3 The *VLD* phasing method — 201
  - 9.3.1 The algorithm — 201
  - 9.3.2 *VLD* and hybrid Fourier syntheses — 205
  - 9.3.3 *VLD* applications to ab initio phasing — 205
- **Appendix 9.A** About *VLD* joint probability distributions — 206
- 9.A.1 The *VLD* algorithm based on difference Fourier synthesis — 206
- 9.A.2 The *VLD* algorithm based on hybrid Fourier syntheses — 211
- **Appendix 9.B** The *RELAX* algorithm — 212

10 Patterson methods and direct space properties — 214

- 10.1 Introduction — 214
- 10.2 The Patterson function — 215
  - 10.2.1 Mathematical background — 215
  - 10.2.2 About interatomic vectors — 216
  - 10.2.3 About Patterson symmetry — 217
- 10.3 Deconvolution of Patterson functions — 218
  - 10.3.1 The traditional heavy-atom method — 219
  - 10.3.2 Heavy-atom search by translation functions — 220
  - 10.3.3 The method of implication transformations — 221
  - 10.3.4 Patterson superposition methods — 223
  - 10.3.5 The C-map and superposition methods — 225
- 10.4 Applications of Patterson techniques — 227
- **Appendix 10.A** Electron density and phase relationships — 230
- **Appendix 10.B** Patterson features and phase relationships — 232

11 Phasing via electron and neutron diffraction data — 234

- 11.1 Introduction — 234
- 11.2 Electron scattering — 235
- 11.3 Electron diffraction amplitudes — 236
- 11.4 Non-kinematical character of electron diffraction amplitudes — 237
- 11.5 A traditional experimental procedure for electron diffraction studies — 239
- 11.6 Electron microscopy, image processing, and phasing methods — 241
- 11.7 New experimental approaches: precession and rotation cameras — 244
- 11.8 Neutron scattering — 245
- 11.9 Violation of the positivity postulate — 247
- **Appendix 11.A** About the elastic scattering of electrons: the kinematical approximation — 249

12 Phasing methods for powder data — 252

- 12.1 Introduction — 252
- 12.2 About the diffraction pattern: peak overlapping — 253

12.3	Modelling the diffraction pattern	258
12.4	Recovering $\|F_{hkl}\|^2$ from powder patterns	260
12.5	The amount of information in a powder diagram	263
12.6	Indexing of diffraction patterns	264
12.7	Space group identification	266
12.8	Ab initio phasing methods	267
12.9	Non-ab initio phasing methods	270
Appendix 12.A	Minimizing texture effects	272

13 Molecular replacement 275

13.1	Introduction	275
13.2	About the search model	277
13.3	About the six-dimensional search	279
13.4	The algebraic bases of vector search techniques	280
13.5	Rotation functions	282
13.6	Practical aspects of the rotation function	284
13.7	The translation functions	286
13.8	About stochastic approaches to MR	289
13.9	Combining MR with 'trivial' prior information: the *ARCIMBOLDO* approach	289
13.10	Applications	291
Appendix 13.A	Calculation of the rotation function in orthogonalized crystal axes	294
13.A.1	The orthogonalization matrix	294
13.A.2	Rotation in Cartesian space	295
13.A.3	Conversion to fractional coordinates	297
13.A.4	Symmetry and the rotation function	299
Appendix 13.B	Non-crystallographic symmetry	304
13.B.1	NCS symmetry operators	304
13.B.2	Finding NCS operators	305
13.B.3	The translational NCS	308
Appendix 13.C	Algebraic forms for the rotation and translation functions	311

14 Isomorphous replacement techniques 314

14.1	Introduction	314
14.2	Protein soaking and co-crystallization	315
14.3	The algebraic bases of SIR techniques	317
14.4	The algebraic bases of MIR techniques	320
14.5	Scaling of experimental data	322
14.6	The probabilistic approach for the SIR case	323
14.7	The probabilistic approach for the MIR case	327
14.8	Applications	329
Appendix 14.A	The SIR case for centric reflections	330
Appendix 14.B	The SIR case: the one-step procedure	331
Appendix 14.C	About methods for estimating the scattering power of the heavy-atom substructure	333

15 Anomalous dispersion techniques — 335

- 15.1 Introduction — 335
- 15.2 Violation of the Friedel law as basis of the phasing method — 337
- 15.3 Selection of dispersive atoms and wavelengths — 340
- 15.4 Phasing via SAD techniques: the algebraic approach — 344
- 15.5 The SIRAS algebraic bases — 347
- 15.6 The MAD algebraic bases — 352
- 15.7 The probabilistic approach for the SAD-MAD case — 354
- 15.8 The probabilistic approach for the SIRAS-MIRAS case — 360
- 15.9 Anomalous dispersion and powder crystallography — 363
- 15.10 Applications — 364
- **Appendix 15.A** A probabilistic formula for the SAD case — 365
- **Appendix 15.B** Structure refinement for MAD data — 366
- **Appendix 15.C** About protein phase estimation in the SIRAS case — 368

Appendices — 370

- **Appendix M.A** Some basic results in probability theory — 370
  - M.A.1 Probability distribution functions — 370
  - M.A.2 Moments of a distribution — 371
  - M.A.3 The characteristic function — 371
  - M.A.4 Cumulants of a distribution — 373
  - M.A.5 The normal or Gaussian distribution — 374
  - M.A.6 The central limit theorem — 375
  - M.A.7 Multivariate distributions — 375
  - M.A.8 Evaluation of the moments in structure factor distributions — 377
  - M.A.9 Joint probability distributions of the signs of the structure factors — 379
  - M.A.10 Some measures of location and dispersion in the statistics of directional data — 380
- **Appendix M.B** Moments of the P(Z) distributions — 382
- **Appendix M.C** The gamma function — 382
- **Appendix M.D** The Hermite and Laguerre polynomials — 383
- **Appendix M.E** Some results in the theory of Bessel functions — 385
  - M.E.1 Bessel functions — 385
  - M.E.2 Generalized hypergeometric functions — 389
- **Appendix M.F** Some definite integrals and formulas of frequent application — 390

References — 394
Index — 412

Symbols and notation

The following symbols and conventions will be used throughout the full text. The **bold** character is used for denoting vectors and matrices.

$\mathbf{h} \cdot \mathbf{r}$	the dot indicates the scalar product of the two vectors $\mathbf{h}$ and $\mathbf{r}$
$\mathbf{a} \wedge \mathbf{b}$	cross-product of the two vectors $\mathbf{a}$ and $\mathbf{b}$
$\bar{\mathbf{A}}$	the bar indicates the transpose of the matrix $\mathbf{A}$
s.f.	structure factor
n.s.f.	normalized structure factor
s.i.	structure invariant
s.s.	structure seminvariant
cs.	centrosymmetric
n.cs.	non-centrosymmetric
RES	experimental data resolution (in Å)
CORR	correlation between the electron density map of the target structure (the one we want to solve) and that of a model map
$R_{cryst} = \frac{\sum_\mathbf{h} ||F_{obs}| - |F_{calc}||}{\sum_\mathbf{h} |F_{obs}|}$	crystallographic residual
SIR-MIR	single–multiple isomorphous replacement
SAD-MAD	single–multiple anomalous dispersion
MR	molecular replacement

Fundamentals of crystallography

1.1 Introduction

In this chapter we summarize the basic concepts, formulas and tables which constitute the essence of general crystallography. In Sections 1.2 to 1.5 we recall, without examples, definitions for unit cells, lattices, crystals, space groups, diffraction conditions, etc. and their main properties: reading these may constitute a useful reminder and support for daily work. In Section 1.6 we establish and discuss the *basic postulate of structural crystallography*: this was never formulated, but during any practical phasing process it is simply assumed to be true by default. We will also consider the consequences of such a postulate and the caution necessary in its use.

1.2 Crystals and crystallographic symmetry in direct space

We recall the main concepts and definitions concerning crystals and crystallographic symmetry.

Crystal. This is the periodic repetition of a motif (e.g. a collection of molecules, see Fig. 1.1). An equivalent mathematical definition is: the crystal is the convolution between a lattice and the unit cell content (for this definition see (1.4) below in this section).

Unit cell. This is the parallelepiped containing the motif periodically repeated in the crystal. It is defined by the unit vectors $\mathbf{a}, \mathbf{b}, \mathbf{c}$, or, by the six scalar parameters $a, b, c, \alpha, \beta, \gamma$ (see Fig. 1.1). The generic point into the unit cell is defined by the vector

$$\mathbf{r} = x\mathbf{a} + y\mathbf{b} + z\mathbf{c},$$

where x, y, z are fractional coordinates (dimensionless and lying between 0 and 1). The volume of the unit cell is given by (see Fig. 1.2)

$$V = \mathbf{a} \wedge \mathbf{b} \cdot \mathbf{c} = \mathbf{b} \wedge \mathbf{c} \cdot \mathbf{a} = \mathbf{c} \wedge \mathbf{a} \cdot \mathbf{b}. \tag{1.1}$$

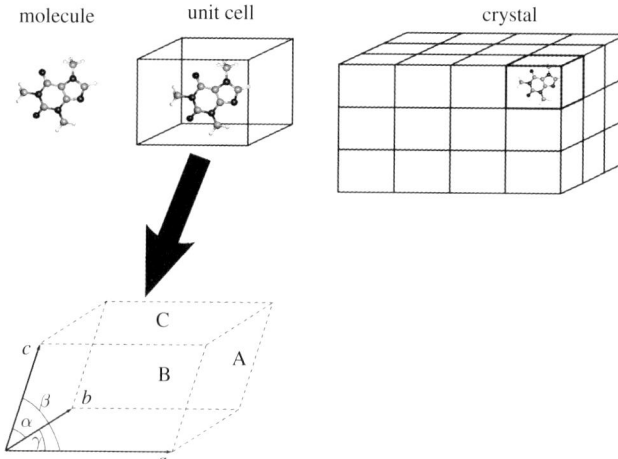

Fig. 1.1
The motif, the unit cell, the crystal.

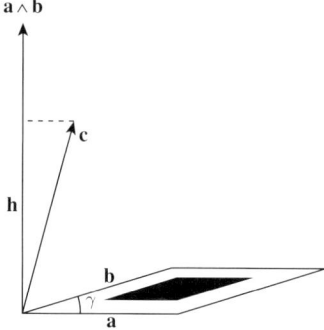

Fig. 1.2
The vector $\mathbf{a} \wedge \mathbf{b}$ is perpendicular to the plane $(\mathbf{a}, \mathbf{b})$: its modulus $|ab \sin \gamma|$ is equal to the shaded area on the base. The volume of the unit cell is the product of the base area and $\mathbf{h}$, the projection of $\mathbf{c}$ over the direction perpendicular to the plane $(\mathbf{a}, \mathbf{b})$. Accordingly, $V = (\mathbf{a} \wedge \mathbf{b}) \cdot \mathbf{c}$.

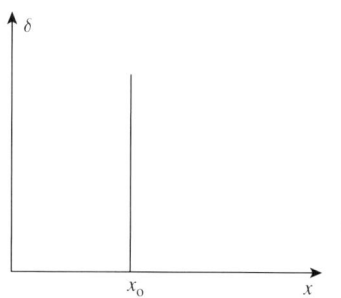

Fig. 1.3
Schematic representation of the Dirac function $\delta(x - x_0)$.

Dirac delta function. In a three-dimensional space the Dirac delta function $\delta(\mathbf{r} - \mathbf{r}_0)$ is defined by the following properties:

$$\delta = 0 \quad \text{for} \quad (\mathbf{r} \neq \mathbf{r}_0), \quad \delta = \infty \quad \text{for} \quad (\mathbf{r} = \mathbf{r}_0), \quad \int_S \delta(\mathbf{r} - \mathbf{r}_0) d\mathbf{r} = 1,$$

where S is the full $\mathbf{r}$ space. The function δ is highly discontinuous and is qualitatively represented in Fig. 1.3 as a straight line.

Crystal lattice. This describes the repetition geometry of the unit cell (see Fig. 1.4). An equivalent mathematical definition is the following: a crystal lattice is represented by the lattice function $L(\mathbf{r})$, where

$$L(\mathbf{r}) = \sum_{u,v,w=-\infty}^{+\infty} \partial(\mathbf{r} - \mathbf{r}_{u,v,w}); \tag{1.2}$$

where $\partial(\mathbf{r} - \mathbf{r}_{u,v,w})$ is the Dirac delta function centred on $\mathbf{r}_{u,v,w} = u\mathbf{a} + v\mathbf{b} + w\mathbf{c}$ and u, v, w are integer numbers.

Convolution. The convolution of two functions $\rho(\mathbf{r})$ and $g(\mathbf{r})$ (this will be denoted as $\rho(\mathbf{r}) \otimes g(\mathbf{r})$) is the integral

$$C(\mathbf{u}) = \rho(\mathbf{r}) \otimes g(\mathbf{r}) = \int_S \rho(\mathbf{r}) g(\mathbf{u} - \mathbf{r}) d\mathbf{r}. \tag{1.3}$$

The reader will notice that the function g is translated by the vector $\mathbf{u}$ and inverted before being integrated.

The convolution of the function $\rho(\mathbf{r})$, describing the unit cell content, with a lattice function centred in $\mathbf{r}_0$, is equivalent to shifting $\rho(\mathbf{r})$ by the vector $\mathbf{r}_0$. Indeed

$$\delta(\mathbf{r} - \mathbf{r}_0) \otimes \rho(\mathbf{r}) = \rho(\mathbf{r} - \mathbf{r}_0).$$

Accordingly, the convolution of $\rho(\mathbf{r})$ with the lattice function $L(\mathbf{r})$ describes the periodic repetition of the unit cell content, and therefore describes the crystal (see Fig. 1.5):

$$L(\mathbf{r}) \otimes \rho(\mathbf{r}) = \sum_{u,v,w=-\infty}^{+\infty} \partial(\mathbf{r} - \mathbf{r}_{u,v,w}) \otimes \rho(\mathbf{r}) = \sum_{u,v,w=-\infty}^{+\infty} \rho(\mathbf{r} - \mathbf{r}_{u,v,w}). \tag{1.4}$$

Crystals and crystallographic symmetry in direct space

Primitive and centred cells. A cell is primitive if it contains only one lattice point and centered if it contains more lattice points. The cells useful in crystallography are listed in Table 1.1: for each cell the multiplicity, that is the number of lattice points belonging to the unit cell, and their positions are emphasized.

Symmetry operators. These relate symmetry equivalent positions. Two positions **r** and **r**′ are symmetry equivalent if they are related by the symmetry operator **C** = (**R**, **T**), where **R** is the rotational component and **T** the translational component. More explicitly,

$$\begin{vmatrix} x' \\ y' \\ z' \end{vmatrix} = \begin{vmatrix} R_{11} & R_{12} & R_{13} \\ R_{21} & R_{22} & R_{23} \\ R_{31} & R_{32} & R_{33} \end{vmatrix} \begin{vmatrix} x \\ y \\ z \end{vmatrix} + \begin{vmatrix} T_1 \\ T_2 \\ T_3 \end{vmatrix}, \qquad (1.5)$$

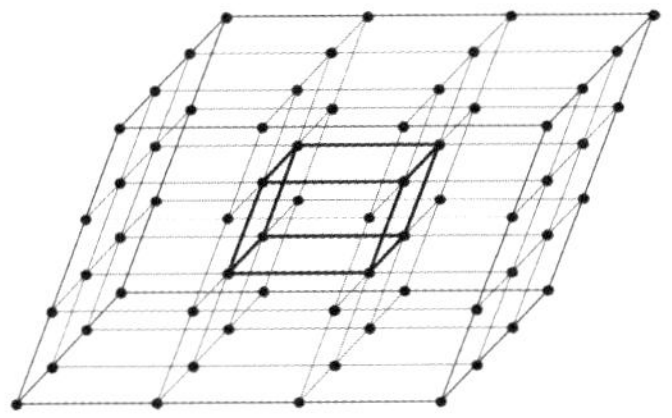

Fig. 1.4
The unit cell (bold lines) and the corresponding lattice.

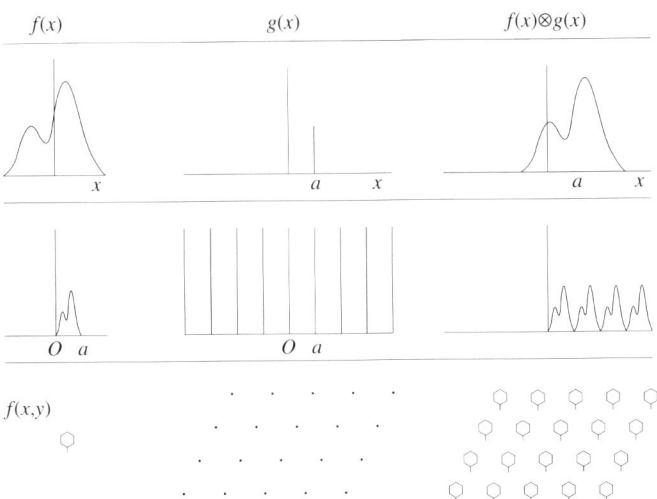

Fig. 1.5
The convolution of the motif f with a delta function is represented in the first line. In the second line f is still the motif, g is a one-dimensional lattice, $f(x) \otimes g(x)$ is a one-dimensional crystal. In the third line, a two-dimensional motif and lattice are used.

Table 1.1 The conventional types of unit cell and corresponding lattice multiplicity

Symbol	Type	Positions of additional lattice points	Number of lattice points per cell
P	Primitive	—	1
I	body-centred	(1/2, 1/2, 1,2)	2
A	A-face centred	(0, 1/2, 1/2)	2
B	B-face centred	(1/2, 0, 1/2)	2
C	C-face centred	(1/2, 1/2, 0)	2
F	All faces centred	(1/2, 1/2, 0), (1/2, 0, 1/2) (0, 1/2, 1/2)	4
R	Rhombohedrally centred (description with 'hexagonal axes')	(1/3, 2/3, 2/3), (2/3, 1/3, 1/3)	3

where (x',y',z') and (x,y,z) are the coordinates of $\mathbf{r}'$ and $\mathbf{r}$ respectively. In a vectorial form,

$$\mathbf{r}' = \mathbf{R}\mathbf{r} + \mathbf{T}.$$

If the determinant $|\mathbf{R}| = 1$ the symmetry operator is proper and refers to objects directly congruent; if $|\mathbf{R}| = -1$ the symmetry operator is improper and refers to enantiomorph objects. The type of symmetry operator may be identified according to Table 1.2:

Table 1.2 Trace and determinant of the rotation matrix for crystallographic symmetry operators

Element	1	2	3	4	6	$\bar{1}$	$\bar{2}$	$\bar{3}$	$\bar{4}$	$\bar{6}$
trace	3	$\bar{1}$	0	1	2	$\bar{3}$	1	0	$\bar{1}$	$\bar{2}$
determinant	1	1	1	1	1	$\bar{1}$	$\bar{1}$	$\bar{1}$	$\bar{1}$	$\bar{1}$

Point group symmetry. This is a compatible combination of symmetry operators, proper or improper, without translational components, and intersecting at one point. The number of crystallographic point groups is 32 and their symbols are shown in Table 1.3. Most of the physical properties depend on the point group symmetry of the crystal (they show a symmetry equal to or larger than the point group symmetry: *Neumann principle*).

Crystal systems. Crystals belonging to point groups with common features can be described by unit cells of the same type. For example, crystals with only three twofold axes, no matter if proper or improper, can be described by an orthogonal cell. These crystals then belong to the same crystal system, the orthorhombic system. The relations between crystal system-point groups are shown in Table 1.4. For each system the allowed Bravais lattices, the characterizing symmetry, and the type of unit cell parameters are reported.

Table 1.3 List of the 32 crystal point groups, Laue groups, and lattice point groups

Crystal systems	Point groups			Laue classes	Lattice point groups
	Non-centrosymmetric		Centrosymmetric		
Triclinic	1		$\bar{1}$	$\bar{1}$	$\bar{1}$
Monoclinic	2	m	$2/m$	$2/m$	$2/m$
Orthorhombic	222	$mm2$	mmm	mmm	mmm
Tetragonal	$\begin{bmatrix} 4 \\ 422 \end{bmatrix}$	$\bar{4}$ $4mm, \bar{4}2m$	$4/m$ $4/mmm$	$4/m$ $4/mmm$	$\Big]4/mmm$
Trigonal	$\begin{bmatrix} 3 \\ 32 \end{bmatrix}$	$3m$	$\bar{3}$ $\bar{3}m$	$\bar{3}$ $\bar{3}m$	$\Big]\bar{3}m$
Hexagonal	$\begin{bmatrix} 6 \\ 622 \end{bmatrix}$	$\bar{6}$ $6mm, \bar{6}2m$	$6/m$ $6/mmm$	$6/m$ $6/mmm$	$\Big]6/mmm$
Cubic	$\begin{bmatrix} 23 \\ 432 \end{bmatrix}$	$\bar{4}3m$	$m\bar{3}$ $m\bar{3}m$	$m\bar{3}$ $m\bar{3}m$	$\Big]m\bar{3}m$

Table 1.4 Crystal systems, characterizing symmetry and unit cell parameters

Crystal system	Bravais type(s)	Characterizing symmetry	Unit cell properties
Triclinic	P	None	a, b, c, α, β, γ
Monoclinic	P, C	Only one 2-fold axis	a, b, c, 90°, β, 90°
Orthorhombic	P, I, F	Only three perpendicular 2-fold axes	a, b, c, 90°, 90°, 90°
Tetragonal	P, I	Only one 4-fold axis	a, a, c, 90°, 90°, 90°
Trigonal	P, R	Only one 3-fold axis	a, a, c, 90°, 90°, 120°
Hexagonal	P	Only one 6-fold axis	a, a, c, 90°, 90°, 120°
Cubic	P, F, I	Four 3-fold axes	a, a, a, 90°, 90°, 90°

Space groups. Three-dimensional crystals show a symmetry belonging to one of the 230 space groups reported in Table 1.5. The space group is a set of symmetry operators which take a three- dimensional periodic object (say a crystal) into itself. In other words, the crystal is invariant under the symmetry operators of the space group.

The space group symmetry defines the *asymmetric unit*: this is the smallest part of the unit cell applying to which the symmetry operators, the full content of the unit cell, and then the full crystal, are obtained. This last statement implies that the space group also contains the information on the repetition geometry (this is the first letter in the space group symbol, and describes the type of unit cell).

1.3 The reciprocal space

We recall the main concepts and definitions concerning crystal reciprocal space.

Reciprocal space. In a scattering experiment, the amplitude of the wave (say $F(\mathbf{r}^*)$, in Thomson units) scattered by an object represented by the function $\rho(\mathbf{r})$, is the Fourier transform of $\rho(\mathbf{r})$:

$$F(\mathbf{r}^*) = T[\rho(\mathbf{r})] = \int_S \rho(\mathbf{r}) \exp(2\pi i \mathbf{r}^* \cdot \mathbf{r}) d\mathbf{r}, \qquad (1.6)$$

where T is the symbol of the Fourier transform, S is the full space where the scattering object is immersed, $\mathbf{r}^* = \mathbf{s} - \mathbf{s}_0$ is the difference between the unit vector $\mathbf{s}$, oriented along the direction in which we observe the radiation, and the unit vector $\mathbf{s}_0$ along which the incident radiation comes (see Fig. 1.6). We recall that $|\mathbf{r}^*| = 2\sin\theta/\lambda$, where 2θ is the angle between the direction of incident radiation and the direction along which the scattered radiation is observed, and λ is the wavelength. We will refer to $\mathbf{r}^*$ as to the generic point of the reciprocal space S^*, the space of the Fourier transform.

$F(\mathbf{r}^*)$ is a complex function, say $F(\mathbf{r}^*) = A(\mathbf{r}^*) + iB(\mathbf{r}^*)$. It may be shown that, for two enantiomorphous objects, the corresponding $F(\mathbf{r}^*)$ are the complex conjugates of each other: they therefore have the same modulus $|F(\mathbf{r}^*)|$. As a consequence, for a centrosymmetrical object, $F(\mathbf{r}^*)$ is real.

Fundamentals of crystallography

Table 1.5 The 230 three-dimensional space groups arranged by crystal systems and point groups. Point groups not containing improper symmetry operators are in a square box (the corresponding space groups are the only ones in which proteins may crystallize). Space groups (and enantiomorphous pairs) that are uniquely determinable from the symmetry of the diffraction pattern and from systematic absences (see Section 1.5) are shown in bold type

Crystal system	Point group	Space groups
Triclinic	$\boxed{1}$	P1
	$\bar{1}$	P$\bar{1}$
Monoclinic	$\boxed{2}$	P2, P2_1, C2
	m	Pm, Pc, Cm, Cc
	$2/m$	P$2/m$, P$2_1/m$, C$2/m$, P$2/c$, **P$2_1/c$**, C$2/c$
Orthorhombic	$\boxed{222}$	P222, **P222_1**, **P2_12_12**, **P$2_12_12_1$**, **C222_1**, C222, F222, I222, I$2_12_12_1$
	$mm2$	P$mm2$, P$mc2_1$, P$cc2$, P$ma2_1$, P$ca2_1$, P$nc2_1$, P$mn2_1$, P$ba2$, P$na2_1$, P$nn2$, C$mm2$, C$mc2_1$, C$cc2$, A$mm2$, A$bm2$, A$ma2$, A$ba2$, F$mm2$, **F$dd2$**, I$mm2$, I$ba2$, I$ma2$
	mmm	Pmmm, **Pnnn**, Pccm, **Pban**, Pmma, **Pnna**, Pmna, **Pcca**, Pbam, **Pccn**, Pbcm, Pnnm, Pmmn, **Pbcn**, **Pbca**, Pnma, Cmcm, Cmca, Cmmm, Cccm, Cmma, Ccca, Fmmm, **Fddd**, Immm, Ibam, **Ibca**, Imma
Tetragonal	$\boxed{4}$	P4, **P4_1**, P4_2, **P4_3**, I4, **I4_1**
	$\bar{4}$	P$\bar{4}$, I$\bar{4}$
	$4/m$	P$4/m$, P$4_2/m$, **P$4/n$**, **P$4_2/n$**, I$4/m$, **I$4_1/a$**
	$\boxed{422}$	P422, P42_12, P4_122, **P4_12_12**, **P4_222**, **P4_22_12**, **P4_322**, **P4_32_12**, I422, **I4_122**
	$4mm$	P$4mm$, P$4bm$, P4_2cm, P4_2nm, P$4cc$, P$4nc$, P4_2mc, P4_2bc, I$4mm$, I$4cm$, I4_1md, **I4_1cd**
	$\bar{4}m$	P$\bar{4}2m$, P$\bar{4}2c$, P$\bar{4}2_1m$, **P$\bar{4}2_1c$**, P$\bar{4}m2$, P$\bar{4}c2$, P$\bar{4}b2$, P$\bar{4}n2$, I$\bar{4}m2$, I$\bar{4}c2$, I$\bar{4}2m$, I$\bar{4}2d$
	$4/mmm$	P$4/mmm$, P$4/mcc$, **P$4/nbm$**, **P$4/nnc$**, P$4/mbm$, P$4/mnc$, **P$4/nmm$**, **P$4/ncc$**, P$4_2/mmc$, P$4_2/mcm$, **P$4_2/nbc$**, **P$4_2/nnm$**, P$4_2/mbc$, P$4_2/mnm$, **P$4_2/nmc$**, **P$4_2/ncm$**, I$4/mmm$, I$4/mcm$, **I$4_1/amd$**, **I$4_1/acd$**
Trigonal–hexagonal	$\boxed{3}$	P3, **P3_1**, **P3_2**, R3
	$\bar{3}$	P$\bar{3}$, R$\bar{3}$
	$\boxed{32}$	P312, P321, **P3_112**, **P3_121**, **P3_212**, **P3_221**, R32
	$3m$	P$3m1$, P$31m$, P$3c1$, P$31c$, R$3m$, R$3c$
	$\bar{3}m$	P$\bar{3}1m$, P$\bar{3}1c$, P$\bar{3}m1$, P$\bar{3}c1$, R$\bar{3}m$, R$\bar{3}c$
	$\boxed{6}$	P6, **P6_1**, **P6_5**, P6_3, **P6_2**, **P6_4**
	$\bar{6}$	P$\bar{6}$
	$6/m$	P$6/m$, P$6_3/m$
	$\boxed{622}$	P622, **P6_122**, **P6_522**, **P6_222**, **P6_422**, P6_322
	$6mm$	P$6mm$, P$6cc$, P6_3cm, P6_3mc
	$\bar{6}m$	P$\bar{6}m2$, P$\bar{6}c2$, P$\bar{6}2m$, P$\bar{6}2c$
	$6/mmm$	P$6/mmm$, P$6/mcc$, P$6_3/mcm$, P$6_3/mmc$
Cubic	$\boxed{23}$	P23, F23, I23, **P2_13**, I2_13
	$m\bar{3}$	P$m\bar{3}$, **P$n\bar{3}$**, F$m\bar{3}$, **F$d\bar{3}$**, I$m\bar{3}$, **P$a\bar{3}$**, **I$a\bar{3}$**
	$\boxed{432}$	P432, **P4_232**, F432, **F4_132**, I432, **P4_332**, **P4_132**, **I4_132**
	$\bar{4}3m$	P$\bar{4}3m$, F$\bar{4}3m$, I$\bar{4}3m$, P$\bar{4}3n$, F$\bar{4}3c$, **I$\bar{4}3d$**
	$m\bar{3}m$	P$m\bar{3}m$, **P$n\bar{3}n$**, P$m\bar{3}n$, **P$n\bar{3}m$**, F$m\bar{3}m$, F$m\bar{3}c$, **F$d\bar{3}m$**, **F$d\bar{3}c$**, I$m\bar{3}m$, **I$a\bar{3}d$**

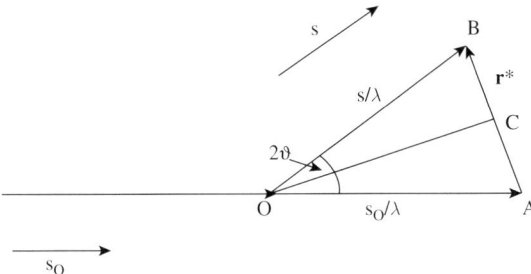

Fig. 1.6
The scatterer is at O, $\mathbf{s}_0$ and $\mathbf{s}$ are unit vectors, the first along the incident X-ray radiation, the second along the direction in which the scattered intensity is observed. To calculate $|\mathbf{r}^*|$ it is sufficient to notice that the triangle AOB is isosceles and that point C divides AB into two equal parts.

$\rho(\mathbf{r})$ may be recovered via the inverse Fourier transform of $F(\mathbf{r}^*)$:

$$\rho(\mathbf{r}) = T^{-1}[F(\mathbf{r}^*)] = \int_{S^*} F(\mathbf{r}^*) \exp(-2\pi i \mathbf{r}^* \cdot \mathbf{r}) d\mathbf{r}^*. \quad (1.7)$$

The reciprocal lattice. It is usual in crystallography to take, as a reference system for the reciprocal space, the reciprocal vectors $\mathbf{a}^*, \mathbf{b}^*, \mathbf{c}^*$, defined below. Given a direct lattice, with unit vectors $\mathbf{a}, \mathbf{b}, \mathbf{c}$, its reciprocal lattice is identified by the vectors $\mathbf{a}^*, \mathbf{b}^*, \mathbf{c}^*$ satisfying the following two conditions:

1. $\mathbf{a}^* \wedge \mathbf{b} = \mathbf{a}^* \wedge \mathbf{c} = \mathbf{b}^* \wedge \mathbf{a} = \mathbf{b}^* \wedge \mathbf{c} = \mathbf{c}^* \wedge \mathbf{a} = \mathbf{c}^* \wedge \mathbf{b} = 0$
2. $\mathbf{a}^* \cdot \mathbf{a} = \mathbf{b}^* \cdot \mathbf{b} = \mathbf{c}^* \cdot \mathbf{c} = 1$

Condition 1 defines the orientation of the reciprocal basis vectors (e.g. $\mathbf{a}^*$ is perpendicular to $\mathbf{b}$ and $\mathbf{c}$, etc.), whereas condition 2 fixes their modulus. From the above conditions the following relations arise:

(i) $\quad \mathbf{a}^* = \dfrac{1}{V}\mathbf{b} \wedge \mathbf{c}, \ \mathbf{b}^* = \dfrac{1}{V}\mathbf{c} \wedge \mathbf{a}, \ \mathbf{c}^* = \dfrac{1}{V}\mathbf{a} \wedge \mathbf{b}, \ V^* = V^{-1}, \quad (1.8)$

(ii) the scalar product of the two vectors $\mathbf{r} = x\mathbf{a} + y\mathbf{b} + z\mathbf{c}$ and $\mathbf{r}^* = x^*\mathbf{a}^* + y^*\mathbf{b}^* + z^*\mathbf{c}^*$, one defined in direct and the other in reciprocal space, reduces to the sum of the products of the corresponding coordinates:

$$\mathbf{r} \cdot \mathbf{r}^* = x^*x + y^*y + z^*z = \bar{\mathbf{X}}^*\mathbf{X} = |x^* y^* z^*| \begin{vmatrix} x \\ y \\ z \end{vmatrix}; \quad (1.9)$$

(iii) the generic reciprocal lattice point is defined by the vector $\mathbf{r}^*_{hkl} = h\mathbf{a}^* + k\mathbf{b}^* + l\mathbf{c}^*$, with integer values of h, k, l. We will also denote it by $\mathbf{r}^*_\mathbf{H}$ or $\mathbf{r}^*_\mathbf{h}$, where $\mathbf{H}$ or $\mathbf{h}$ represent the triple h,k,l.

(iv) $\mathbf{r}^*_{hkl}$ represents the family (in direct space) of lattice planes with Miller indices (hkl). Indeed $\mathbf{r}^*_{hkl}$ is perpendicular to the planes of the family (hkl) and its modulus is equal to the spacing of the planes (hkl): i.e.

$$\mathbf{r}^*_{hkl} \perp (hkl), \quad \text{and} \quad |\mathbf{r}^*_{hkl}| = 1/d_{hkl}. \quad (1.10)$$

(v) the reciprocal lattice may be represented by the reciprocal lattice function

$$L(\mathbf{r}^*) = \dfrac{1}{V} \sum_{h,k,l=-\infty}^{+\infty} \partial\left(\mathbf{r}^* - \mathbf{r}^*_\mathbf{H}\right); \quad (1.11)$$

$L(\mathbf{r}^*)$ is the Fourier transform of the direct lattice:

$$L(\mathbf{r}^*) = T[L(\mathbf{r})] = T\left[\sum_{u,v,w=-\infty}^{+\infty} \partial(\mathbf{r} - \mathbf{r}_{u,v,w})\right] = \frac{1}{V}\sum_{\mathbf{H}} \partial(\mathbf{r}^* - \mathbf{r}_{\mathbf{H}}^*). \quad (1.12)$$

Atomic scattering factor $f(\mathbf{r}^*)$. This is the amplitude, in Thomson units, of the wave scattered by the atom and observed at the reciprocal space point $\mathbf{r}^*$. $f(\mathbf{r}^*)$ is the Fourier transform of the atomic electron density ρ_a:

$$f(\mathbf{r}^*) = T[\rho_a(\mathbf{r})] = \int_S \rho_a(\mathbf{r}) \exp(2\pi i \mathbf{r}^* \cdot \mathbf{r}) d\mathbf{r}. \quad (1.13)$$

Usually $\rho_a(\mathbf{r})$ includes thermal displacement: accordingly, under the isotropic scattering assumption,

$$f(r^*) = f_0(r^*)\exp\left(-Br^{*2}/4\right) = f_0(r^*)\exp(-B\sin^2\theta/\lambda^2), \quad (1.14)$$

where $f_0(r^*)$ is the scattering factor of the atom at rest, and B is the isotropic temperature factor. At $r^* = 0$, $f(r^*)$ is maximum (then $f(r^*) = Z$, where Z is the atomic number). The decay with r^* is sharper for high B values (see Fig. 1.7).

Molecular scattering factor $F_M(\mathbf{r}^*)$. This is the amplitude, in Thomson units, of the wave scattered by a molecule, observed at the reciprocal space point $\mathbf{r}^*$. It is the Fourier transform of the electron density of the molecule:

$$F_M(\mathbf{r}^*) = T[\rho_M(\mathbf{r})] = \int_S \sum_{j=1}^{N} \rho_{aj}(\mathbf{r} - \mathbf{r}_j)\exp(2\pi i \mathbf{r}^* \cdot \mathbf{r})d\mathbf{r}$$
$$= \sum_{j=1}^{N} f_j \exp(2\pi i \mathbf{r}^* \cdot \mathbf{r}_j), \quad (1.15)$$

where $\rho_M(\mathbf{r})$ is the electron density of the molecule and N is the corresponding number of atoms. $F_M(\mathbf{r}^*)$ is a continuous function of $\mathbf{r}^*$.

Structure factor $F_M(\mathbf{r}^*)$ *of a unit cell*. This is the amplitude, in Thomson units, of the wave scattered by all the molecules contained in the unit cell and observed at the reciprocal space point $\mathbf{r}^*$. $F_M(\mathbf{r}^*)$ is the Fourier transform of the electron density of the unit cell:

$$F_M(\mathbf{r}^*) = T[\rho_M(\mathbf{r})] = \int_S \sum_{j=1}^{N} \rho_{aj}(\mathbf{r} - \mathbf{r}_j)\exp\left(2\pi i \mathbf{r}^* \cdot \mathbf{r}\right) d\mathbf{r}$$
$$= \sum_{j=1}^{N} f_j \exp\left(2\pi i \mathbf{r}^* \cdot \mathbf{r}_j\right). \quad (1.16)$$

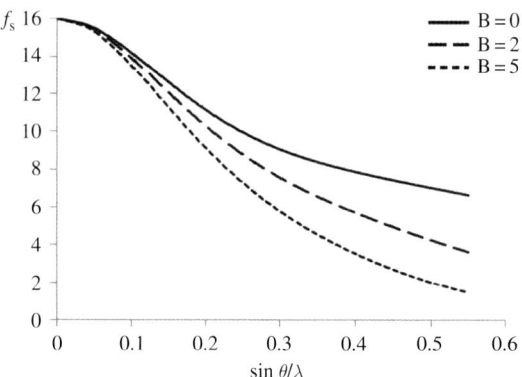

Fig. 1.7
Scattering factor of sulphur for different values of the temperature factor.

$\rho_M(\mathbf{r})$ is now the electron density in the unit cell, N is the corresponding number of atoms, and $F_M(\mathbf{r}^*)$ is a continuous function of $\mathbf{r}^*$. The reader will certainly have noted that we have used for the unit cell the same notation employed for describing the scattering from a molecule: indeed, from a physical point of view, the unit cell content may be considered to be a collection of molecules.

Structure factor $F(\mathbf{r}^*)$ for a crystal. This is the amplitude, in Thomson units, of the wave scattered by the crystal as observed at the reciprocal space point $\mathbf{r}^*$. It is the Fourier transform of the electron density of the crystal. In accordance with equation 1.4

$$F(\mathbf{r}^*) = T[\rho_{cr}(\mathbf{r})] = T[\rho_M(\mathbf{r}) \otimes L(\mathbf{r})]$$

and, owing to the convolution theorem,

$$F(\mathbf{r}^*) = T[\rho_M(\mathbf{r})] \cdot T[L(\mathbf{r})] = F_M(\mathbf{r}^*) \cdot \frac{1}{V} \sum_{\mathbf{H}} \partial(\mathbf{r}^* - \mathbf{r}^*_{\mathbf{H}}). \qquad (1.17)$$

$F(\mathbf{r}^*)$ is now a highly discontinuous function which is different from zero only at the reciprocal lattice points defined by the vectors $\mathbf{r}^*_{\mathbf{H}}$. From now on, $F_M(\mathbf{r}^*_{\mathbf{H}})$ will be written as $F_{\mathbf{H}}$ and will simply be called the *structure factor*. The study of $F_{\mathbf{H}}$ and of its statistical properties is basic for phasing methods.

Limits of a diffraction experiment. Diffraction occurs when $\mathbf{r}^*_{\mathbf{H}}$ meet the Ewald sphere (see Fig. 1.8). A diffraction experiment only allows measurement of reflections with $\mathbf{r}^*_{\mathbf{H}}$ contained within the *limiting sphere* (again, see Fig. 1.8). Data resolution is usually described in terms of the maximum measurable value of $|\mathbf{r}^*_{\mathbf{H}}|$ (say $|\mathbf{r}^*_{\mathbf{H}}|_{max}$): in this case the resolution is expressed in Å^{-1}. More frequently, because of equation (1.10), in terms of the minimum measurable value of $d_{\mathbf{H}}$ (say $(d_{\mathbf{H}})_{min}$): in this case data resolution is expressed in Å. Accordingly, stating that data resolution is 2 Å is equivalent to saying that only reflections with $d_{\mathbf{H}} > 2$ Å were measured. Severe resolution limits are frequent for proteins: often reflections inside and close to the limiting sphere cannot be measured because of the poor quality of the crystal. Usually, better

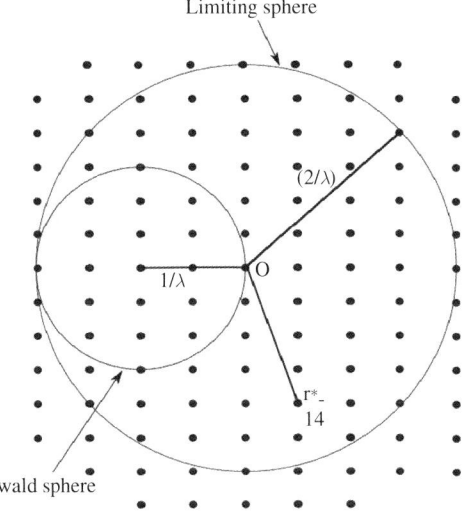

Fig. 1.8
Ewald and limiting spheres.

data can be collected, not by diminishing λ, but by performing the experiment in cryo-conditions, to fight decay of the scattering factor due to thermal displacement.

Electron density calculations. According to equation (1.17), the electron density in a point **r** having fractional coordinates (x,y,z) may be estimated via the Fourier series

$$\rho(\mathbf{r}) = \frac{1}{V} \sum_{h=-\infty}^{\infty} \sum_{k=-\infty}^{\infty} \sum_{l=-\infty}^{\infty} F_{hkl} \exp[-2\pi i(hx + ky + lz)]$$

$$= \frac{2}{V} \sum_{h=0}^{\infty} \sum_{k=-\infty}^{\infty} \sum_{l=-\infty}^{\infty} |F_{hkl}| \cos[\phi_{hkl} - 2\pi(hx + ky + lz)].$$

(1.18)

The last term is obtained by applying the Friedel law, and shows that the electron density is a real function. As previously recalled, there are limitations to the number of measurable reflections: accordingly, series (1.18) will show truncation effects which are more and more severe as soon as the resolution becomes worse (see Fig. 1.9 and Section 7.3.1).

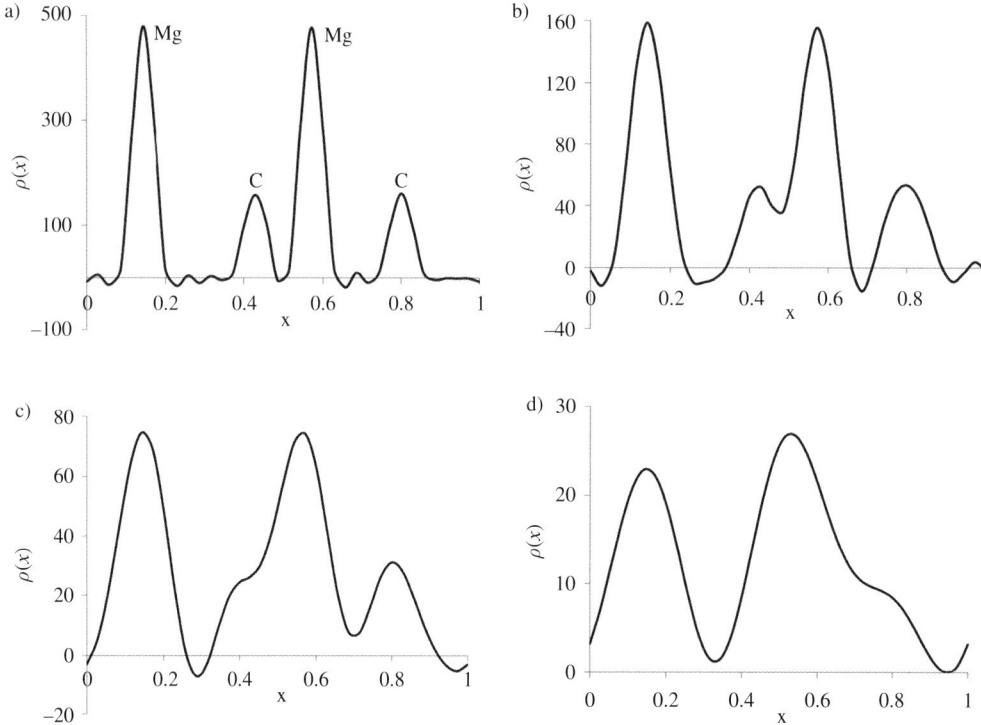

Fig. 1.9
Electron density maps of a (non-realistic) four-atom one-dimensional structure. Data up to: (a) 0.9 Å; (b) 1.5 Å; (c) 2 Å; (d) 3 Å. In all cases true phases have been used: the differences between the maps are only due to truncation effects. Changes in peak intensity and positions are clearly visible.

1.4 The structure factor

The structure factor $F_\mathbf{h}$ plays a central role in phasing methods: its simple geometrical interpretation is therefore mandatory. Let N be the number of atoms in the unit cell, f_j the scattering factor of the jth atom, and x_j, y_j, z_j its fractional coordinates: then

$$F_\mathbf{h} = \sum_{j=1}^{N} f_j \exp\left(2\pi i \mathbf{h} \cdot \mathbf{r}_j\right) = \sum_{j=1}^{N} f_j \exp\left[2\pi i(hx_j + ky_j + lz_j)\right]. \quad (1.19)$$

f_j includes the thermal displacement and must be calculated at the $\sin\theta/\lambda$ corresponding to the reflection $\mathbf{h}$: to do that, firstly, the modulus of the vector $\mathbf{r}^*_{hkl} = h\mathbf{a}^* + k\mathbf{b}^* + l\mathbf{c}^*$ should be calculated and then, by using the equation $|\mathbf{r}^*| = 2\sin\theta/\lambda$, the searched f value may be obtained.

Let us rewrite (1.19) in the form

$$F_\mathbf{h} = \sum_{j=1}^{N} f_j \exp(i\alpha_j) = |F_\mathbf{h}|\exp(i\phi_\mathbf{h}) = A_\mathbf{h} + iB_\mathbf{h}, \quad (1.20)$$

where

$$\alpha_j = 2\pi \mathbf{h} \cdot \mathbf{r}_j, \quad A_\mathbf{h} = \sum_{j=1}^{N} f_j \cos(2\pi \mathbf{h} \cdot \mathbf{r}_j),$$

$$B_\mathbf{h} = \sum_{j=1}^{N} f_j \sin(2\pi \mathbf{h} \cdot \mathbf{r}_j).$$

On representing $F_\mathbf{h}$ in an Argand diagram (Fig. 1.10), we obtain a vectorial diagram with N vectors each characterized by a modulus f_j and an angle α_j with the real axis: the value

$$\phi_\mathbf{h} = \tan^{-1}(B_\mathbf{h}/A_\mathbf{h}) \quad (1.21)$$

depends on the moduli and on the mutual orientation of the vectors $\mathbf{f}_j$ and is said to be the *phase* of $F_\mathbf{h}$.

In a space group with symmetry higher than P1, with point group symmetry of order m, for each atomic position $\mathbf{r}_j$, located in the asymmetric unit, there are m symmetry equivalent positions

$$\mathbf{r}_{js} = \mathbf{R}_s \mathbf{r}_j + \mathbf{T}_s.$$

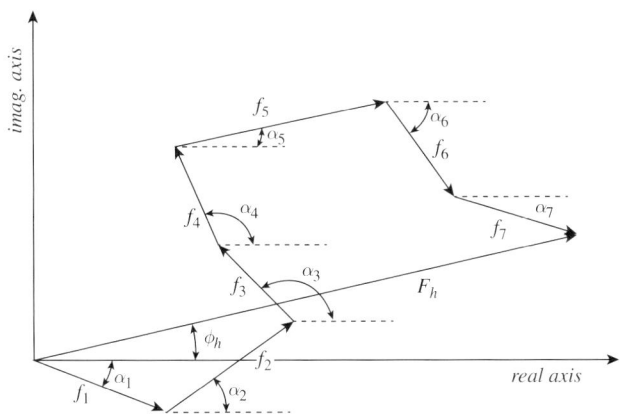

Fig. 1.10
The structure factor $F_\mathbf{h}$ is represented in the Argand plane as the sum of $N = 7$ $\mathbf{f}_j$ vectors, with modulus f_j and phase angle α_j.

Then the structure factor takes the form

$$F_{\mathbf{h}} = \sum_{j=1}^{t} f_j \sum_{s=1}^{m} \exp 2\pi i\mathbf{h}(\mathbf{R}_s\mathbf{r}_j + \mathbf{T}_s)$$

where t is the number of atoms in the asymmetric unit.

1.5 Symmetry in reciprocal space

A diffraction experiment allows us to *see* the reciprocal space: it is then very important to understand which symmetry relations can be discovered there as a consequence of the symmetry present in direct space. Here we summarize the main effects.

1.5.1 Friedel law

In accordance with equation (1.20) we write $F_h = A_h + iB_h$. Then it will follow that $F_{-h} = A_h - iB_h$, and consequently

$$\phi_{-h} = -\phi_h. \tag{1.22}$$

The value of ϕ_{-h} is the opposite of the value of ϕ_h, see Fig. 1.11. Since

$$I_h = (A_h - iB_h)(A_h + iB_h) = A_h^2 + B_h^2,$$

$$I_{-h} = (A_h + iB_h)(A_h - iB_h) = A_h^2 + B_h^2,$$

we deduce the Friedel law, according to which the diffraction intensities associated with the vectors $\mathbf{h}$ and $-\mathbf{h}$ of reciprocal space are equal. Since these intensities appear to be related by a centre of symmetry, usually, although imperfectly, it is said that the diffraction by itself introduces a centre of symmetry.

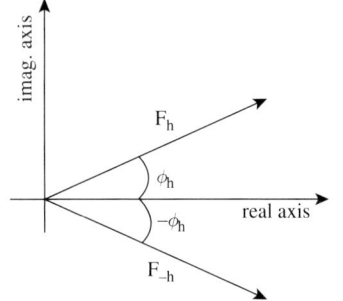

Fig. 1.11
The Friedel law.

1.5.2 Effects of symmetry operators in reciprocal space

Let us suppose that the symmetry operator $C = (\mathbf{R}, \mathbf{T})$ exists in direct space. We wonder what kind of relationships the presence of the operator C brings in reciprocal space.

Since

$$F_{\bar{h}\mathbf{R}} \exp\left(2\pi i\bar{h}\mathbf{T}\right) = \sum_{j=1}^{N} f_j \exp\left(2\pi i\bar{h}\mathbf{R}\mathbf{X}_j\right) \cdot \exp\left(2\pi i\bar{h}\mathbf{T}\right)$$

$$= \sum_{j=1}^{N} f_j \exp\left[2\pi i\bar{h}(\mathbf{R}\mathbf{X}_j + \mathbf{T})\right] = F_h,$$

we can write

$$F_{\bar{h}\mathbf{R}} = F_h \exp(-2\pi i\bar{h}\mathbf{T}). \tag{1.23}$$

Sometimes it is convenient to split equation (1.23) into two relations, the first involving moduli and the second the phases

$$|F_{\bar{h}\mathbf{R}}| = |F_h|, \tag{1.24}$$

$$\phi_{\bar{h}\mathbf{R}} = \phi_h - 2\pi\bar{h}\mathbf{T}. \tag{1.25}$$

From (1.23) it is concluded that intensities I_h and $I_{\bar{h}R}$ are equal, while their phases are related by equation (1.25).

Reflections related by (1.24) and by the Friedel law are said to be *symmetry equivalent reflections*. For example, in P2 the set of symmetry equivalent reflections is

$$|F_{hkl}| = |F_{\bar{h}k\bar{l}}| = |F_{\bar{h}\bar{k}\bar{l}}| = |F_{hk\bar{l}}|. \tag{1.26}$$

The reader will easily verify that space groups P4, P$\bar{4}$, and P4/m show the following set of symmetry equivalent reflections:

$$|F_{hkl}| = |F_{\bar{h}\bar{k}l}| = |F_{\bar{k}hl}| = |F_{k\bar{h}l}| = |F_{\bar{h}\bar{k}\bar{l}}| = |F_{hk\bar{l}}|, = |F_{k\bar{h}\bar{l}}| = |F_{\bar{k}h\bar{l}}|.$$

1.5.3 Determination of reflections with restricted phase values

Let us suppose that for a given set of reflections the relationship $\bar{h}\mathbf{R} = -\bar{h}$ is satisfied. If we apply (1.25) to this set we will obtain $2\phi_h = 2\pi\bar{h}\mathbf{T} + 2n\pi$, from which

$$\phi_h = \pi\bar{h}\mathbf{T} + n\pi. \tag{1.27}$$

Equation (1.27) restricts the phase ϕ_h to two values, $\pi\bar{h}\mathbf{T}$ or $\pi(\bar{h}\mathbf{T} + 1)$. These reflections are called reflections with restricted phase values, or less correctly, 'centrosymmetric'.

If the space group is centrosymmetric (cs.) the inversion operator

$$\mathbf{R} = \begin{vmatrix} \bar{1} & 0 & 0 \\ 0 & \bar{1} & 0 \\ 0 & 0 & \bar{1} \end{vmatrix}, \quad \mathbf{T} = \begin{vmatrix} T_1 \\ T_2 \\ T_3 \end{vmatrix}$$

will exist. In this case every reflection is a restricted phase reflection and will assume the values $\pi\bar{h}\mathbf{T}$ or $\pi(\bar{h}\mathbf{T} + 1)$. If the origin is assumed to be the centre of symmetry then $\mathbf{T} = 0$ and the permitted phase values are 0 and π. Then F_h will be a real positive number when ϕ_h is equal to 0, and a negative one when ϕ_h is equal to π. For this reason we usually talk in cs. space groups about the sign of the structure factor rather than about the phase.

In Fig. 1.12, F_h is represented as an Argand diagram for a centrosymmetric structure of six atoms. Since for each atom at r_j another symmetry equivalent atom exists at $-r_j$, the contribution of every couple to F_h will have to be real.

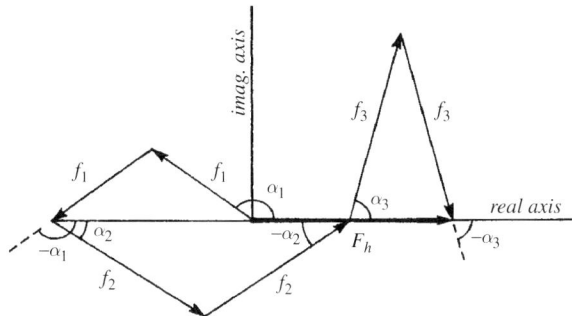

Fig. 1.12

$F_\mathbf{h}$ is represented in the Argand plane for a cs. crystal structure with $N = 6$. It is $\alpha_j = 2\pi\bar{\mathbf{H}}\mathbf{X}_j$.

Table 1.6 Restricted phase reflections for the 32 point groups

Point group	Sets of restricted phase reflections
1	None
$\bar{1}$	All
m	(0, k, 0)
2	(h, 0, l)
2/m	All
mm2	[(h, k, 0) masks (h, 0, 0), (0, k, 0)]
222	Three principal zones only
mmm	All
4	(h, k, 0)
$\bar{4}$	(h, k, 0); (0, 0, l)
4/m	All
422	(h, k, 0); {h, 0, l}; {h, h, l}
$\bar{4}2m$	[(h, k, 0), {h, h, 0}]; [{h, 0, l}, (0, 0, l)]
4mm	[(h, k, 0), {h, 0, 0}, {h, h, 0}]
4/mmm	All
3	None
$\bar{3}$	All
3m	{h, 0, $\bar{h}$, 0}
32	{h, 0, $\bar{h}$, l}
$\bar{3}m$	All
6	(h, k, 0)
$\bar{6}$	(0, 0, l)
6/m	All
$\bar{6}m2$	[{h, h, l}, {h, h, 0}, (0, 0, l)]
6mm	[(h, k, 0), {h, h, 0}, {h, 0, 0}]
62	(h, k, 0); (h, 0, l); (h, h, l)
6/mmm	All
23	{h, k, 0}
$m\bar{3}$	All
$\bar{4}3m$	[{h, k, 0}, {h, h, 0}]
432	{h, k, 0}; {h, h, l}
$m\bar{3}m$	All

Table 1.7 If $\bar{h}R = -h$ the allowed phase values ϕ_a of F_h are $\pi\bar{h}T$ and $\pi\bar{h}T + \pi$. Allowed phases are multiples of 15° and are associated, in direct methods programs, with a symmetry code (SCODE). For general reflections SCODE = 1

$\phi_a^{(0)}$	SCODE
Any	1
(30, 210)	3
(45, 225)	4
(60, 240)	5
(90, 270)	7
(120, 300)	9
(135, 315)	10
(150, 330)	11
(180, 360)	13

As an example of a non-centrosymmetric (n.cs.) space group let us examine $P2_12_12_1$, $[(x, y, z), (\frac{1}{2} - x, \bar{y}, \frac{1}{2} + z), (\frac{1}{2} + x, \frac{1}{2} - y, \bar{z}), (\bar{x}, \frac{1}{2} + y, \frac{1}{2} - z)]$, where the reflections (hk0), (0kl), (h0l) satisfy the relation $\bar{h}R = -h$ for $R = R_2, R_3, R_4$ respectively. By introducing $T = T_2$ in equation (1.27) we obtain $\phi_{hk0} = (\pi h/2) + n\pi$. Thus ϕ_{hk0} will have phase 0 or π if h is even and phase $\pm \pi/2$ if h is odd. By introducing $T = T_3$ in equation (1.27) we obtain $\phi_{0kl} = (\pi k/2) + n\pi$: i.e. ϕ_{0kl} will have phase 0 or π if k is even and $\pm \pi/2$ if k is odd. In the same way, by introducing $T = T_4$ in equation (1.27) we obtain $\phi_{h0l} = (\pi l/2) + n\pi$: i.e. ϕ_{h0l} will have phase 0 or π if l is even and $\pm \pi/2$ if l is odd. In Table 1.6 the sets of restricted phase reflections are given for the 32 point groups.

The allowed values of restricted phases depend on the translational component of the symmetry element and on its location with respect to the cell origin. For conventional three-dimensional space groups the allowed phase values are multiples of 15°. In Table 1.7 the different types of phase restriction are shown: in the second column the characteristic codes associated in direct methods programs with the various restrictions are quoted. It should not be forgotten that symmetry equivalent reflections can have different allowed phase values. For example, in the space group $P4_12_12$ $\left[(x, y, z); (\bar{x}, \bar{y}, z + \frac{1}{2}); (\bar{y} + \frac{1}{2}, x + \frac{1}{2}, z + \frac{1}{4}); (y + \frac{1}{2}, \bar{x} + \frac{1}{2}, z + \frac{3}{4}); (\bar{x} + \frac{1}{2}, y + \frac{1}{2}, \bar{z} + \frac{1}{4}); (x + \frac{1}{2}, \bar{y} + \frac{1}{2}, \bar{z} + \frac{3}{4}); (y, x, \bar{z}); (\bar{y}, \bar{x}, \bar{z} + \frac{1}{2})\right]$, the reflection (061) has phase values restricted to $(-(\pi/4), 3\pi/4)$. Its equivalent reflections are also symmetry restricted, but the allowed phase values may be different from $(-(\pi/4), 3\pi/4)$. On the assumption that $\phi_{061} = 3\pi/4$, the reader will find for the equivalent reflections the phase restrictions shown in Fig. 1.13.

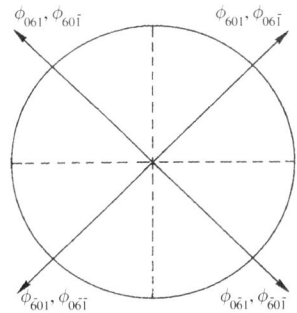

Fig. 1.13
Phase restrictions for the reflection (061) and its symmetry equivalents.

1.5.4 Systematic absences

Let us look for the class of reflections for which $\bar{h}R = \bar{h}$ and apply equation (1.23) to them. This relation would be violated for those reflections for which $\bar{h}T$ is not an integer number unless $|F_h| = 0$. From this fact the rule follows: reflections for which $\bar{h}R = \bar{h}$ and $\bar{h}T$ is not an integer will have diffraction intensity zero or, as is usually stated, will be systematically absent or extinct. Let us give a few examples.

In the space group $P2_1$ $\left[(x, y, z), (\bar{x}, y + \frac{1}{2}, \bar{z})\right]$, the reflections $(0k0)$ satisfy the condition $\bar{h}R_2 = \bar{h}$. If k is odd, $\bar{h}T_2$ is semi-integer. Thus, the reflections $(0k0)$ with $k \neq 2n$ are systematically absent.

In the space group $P4_1$ $\left[(x, y, z), (\bar{x}, \bar{y}, \frac{1}{2} + z), (\bar{y}, x, \frac{1}{4} + z), (y, \bar{x}, \frac{3}{4} + z)\right]$, only the reflections $(00l)$ satisfy the condition $hR_j = h$ for $j = 2,3,4$. Since $\bar{h}T_2 = l/2, \bar{h}T_3 = l/4, \bar{h}T_4 = 3l/4$, the only condition for systematic absence is $l \neq 4n$, with n integer.

In the space group Pc $\left[(x, y, z), (x, \bar{y}, z + \frac{1}{2})\right]$, the reflections $(h0l)$ satisfy the condition $\bar{h}R_2 = \bar{h}$. Since $\bar{h}R_2 = l/2$, the reflections $(h0l)$ with $l \neq 2n$ will be systematically absent.

Note that the presence of a slide plane imposes conditions for systematic absences to bidimensional reflections. In particular, slide planes opposite to a, b, and c impose conditions to classes $(0kl)$, $(h0l)$, and $(hk0)$ respectively. The condition will be $h = 2n$, $k = 2n$, $l = 2n$ for the slide planes of type a, b, or c respectively.

Let us now apply the same considerations to the symmetry operators centring the cell. If the cell is of type A, B, C, I, symmetry operators will exist whose rotational matrix is always the identity, while the translational matrices are

$$\mathbf{T}_A = \begin{bmatrix} 0 \\ \frac{1}{2} \\ \frac{1}{2} \end{bmatrix} \quad \mathbf{T}_B = \begin{bmatrix} \frac{1}{2} \\ 0 \\ \frac{1}{2} \end{bmatrix} \quad \mathbf{T}_C = \begin{bmatrix} \frac{1}{2} \\ \frac{1}{2} \\ 0 \end{bmatrix} \quad \mathbf{T}_I = \begin{bmatrix} \frac{1}{2} \\ \frac{1}{2} \\ \frac{1}{2} \end{bmatrix}$$

respectively. If we use these operators in equation (1.24), we find that (1) the relation $\bar{h}R = \bar{h}$ is satisfied for any reflection and (2) the systematic absences, of three-dimensional type, are $k + l = 2n, h + l = 2n, h + k = 2n, h + k + l = 2n$, respectively.

A cell of type F is simultaneously A-, B-, and C-centred, so the respective conditions for systematic absences must be simultaneously valid. Consequently, only the reflections for which h, k, and l are all even or all odd will be present.

The same criteria lead us to establish the conditions for systematic absences for rhombohedral lattices ($-h + k + l \neq 3n$ for obverse setting and $h - k + l \neq 3n$ for reverse setting). The list of systematic absences for any symmetry element is given in Table 1.8.

Table 1.8 Systematic absences

Symmetry elements		Set of reflections	Conditions
Lattice	P	hkl	None
	I		$h + k + l = 2n$
	C		$h + k = 2n$
	A		$k + l = 2n$
	B		$h + l = 2n$
	F		$\begin{cases} h + k = 2n \\ k + l = 2n \\ h + l = 2n \end{cases}$
	R_{obv}		$-h + k + l = 3n$
	R_{rev}		$h - k + l = 3n$
Glide plane ∥ (001)	a	$hk0$	$h = 2n$
	b		$k = 2n$
	n		$h + k = 2n$
	d		$h + k = 4n$
Glide plane ∥ (100)	b	$0kl$	$k = 2n$
	c		$l = 2n$
	n		$k + l = 2n$
	d		$k + l = 4n$
Glide plane ∥ (010)	a	$h0l$	$h = 2n$
	c		$l = 2n$
	n		$h + l = 2n$
	d		$h + l = 4n$
Glide plane ∥ (110)	c	hhl	$l = 2n$
	b		$h = 2n$
	n		$h + l = 2n$
	d		$2h + l = 4n$
Screw axis ∥ c	$2_1, 4_2, 6_3$	$00l$	$l = 2n$
	$3_1, 3_2, 6_2, 6_4$		$l = 3n$
	$4_1, 4_3$		$l = 4n$
	$6_1, 6_5$		$l = 6n$
Screw axis ∥ a	$2_1, 4_2$	$h00$	$h = 2n$
	$4_1, 4_3$		$h = 4n$
Screw axis ∥ b	$2_1, 4_2$	$0k0$	$k = 2n$
	$4_1, 4_3$		$k = 4n$
Screw axis ∥ [110]	2_1	$hh0$	$h = 2n$

1.6 The basic postulate of structural crystallography

In the preceding paragraphs we have summarized the basic relations of general crystallography: these can be found in more extended forms in any standard textbook. The reader is now ready to learn about the topic of phasing, one of the most intriguing problems in the history of crystallography. We will start by illustrating its logical aspects (rather than its mathematics) via a short list of questions.

Given a model structure, can we calculate the corresponding set (say $\{|F_\mathbf{h}|\}$) of structure factor moduli? The answer is trivial; indeed we have only to introduce the atomic positions and the corresponding scattering factors (including temperature displacements) into equation 1.19. As a result of these calculations, moduli and phases of the structure factors can be obtained. It may therefore be concluded that there is no logical or mathematical obstacle to the symbolic operation

$$\rho(\mathbf{r}) \Rightarrow \{|F_\mathbf{h}|\}.$$

A second question is: *given only the structure factor moduli, can we entertain the hope of recovering the crystal structure, or, on the contrary, is there some logical impediment to this (for example, an irrecoverable loss of information)?* In symbols, this question deals with the operation

$$\{|F_\mathbf{h}|\} \Rightarrow \rho(\mathbf{r}). \tag{1.28}$$

As an example, let us suppose that the diffraction experiment has provided 30 000 structure factor moduli and lost 30 000 phases. Can we recover the 30 000 phases given the moduli, and consequently determine the structure, or are the phases irretrievably lost?

A first superficial answer may be provided by our daily experience. To give a simple example, if we are looking for a friend in New York but we have lost his address, it would be very difficult to find him. This allegorical example is appropriate as in New York there are millions of addresses, similarly, millions of structural models may be conceived that are compatible with the experimental unit cell. The search for our friend would be much easier if some valuable information were still in our hands: e.g. he lives in a flat on the 130th floor. In this case we could discard most of the houses in New York. But where, in the diffraction experiment, is the information hidden which can allow us to discard millions of structural models and recover the full structure?

A considered answer to the problem of phase recovery should follow reference to modern structural databases (see Section 1.7). In Fig. 1.14 statistics are shown from the *Cambridge Structural Database*, where the growth in numbers of deposited structures per year is shown. Hundreds of thousands of crystal structures have been deposited, the large majority of these having been solved starting from the diffraction moduli only. In Fig. 1.15, similar statistics are shown for the *Protein Data Bank* (*PDB*).

Fig. 1.14
Cumulative growth per year of the structures deposited in the *Cambridge Structural Database* (*CSD*).

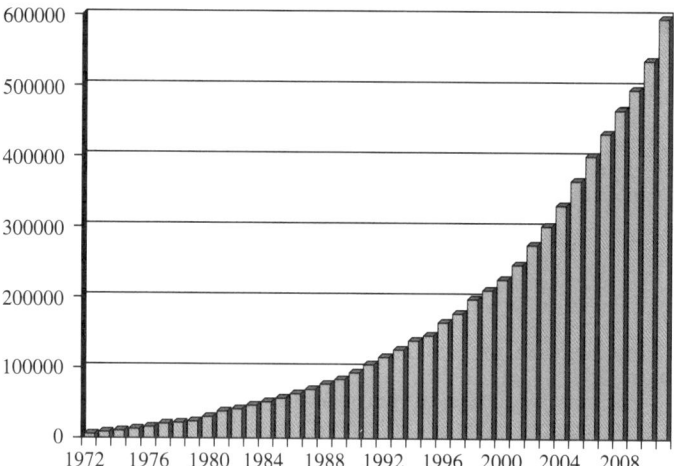

Fig. 1.15
Growth per year of the structures deposited in the *Protein Data Bank* (*PDB*).

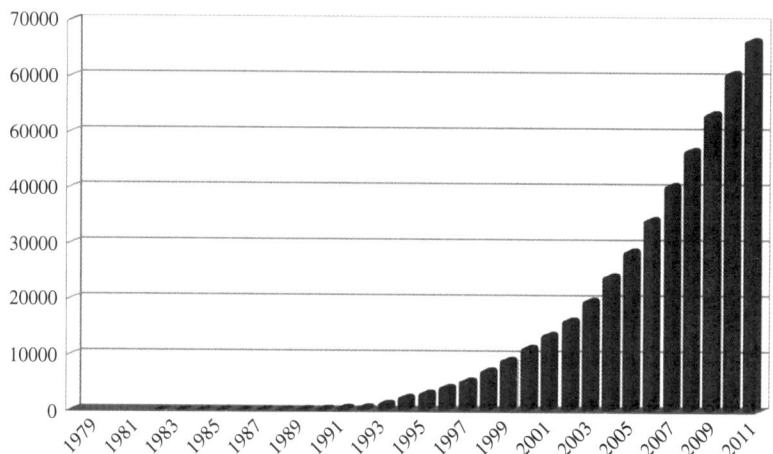

Such huge numbers of structures could not have been solved without valuable information provided by experiment and since X-ray experiments only provide diffraction amplitudes we have to conclude that the phase information is hidden in the amplitudes. But at the moment we do not know how this is codified.

Before dealing with the code problem, we should answer a preliminary question: *how can we decide (and accept) that such huge numbers of crystal structures are really (and correctly) solved?* Each deposited structure is usually accompanied by a *cif* file, where the main experimental conditions, the list of the collected experimental data, their treatment by crystallographic programs, and the structural model are all described. Usually residuals such as (Booth, 1945)

$$R_{cryst} = \frac{\sum_{\mathbf{h}} ||F_{obs}| - |F_{calc}||}{\sum_{\mathbf{h}} |F_{obs}|} \qquad (1.29)$$

are mentioned as mathematical proof of the correctness of a model: if R_{cryst} is smaller than a given threshold and no crystal chemical rule is violated by the proposed model, then the model is assumed to be correct. This assumption is universally accepted, and is the basic guideline for any structural crystallographer, even though it is not explicitly formulated and not demonstrated mathematically. But, how can we exclude two or more crystal structures which may exist, which do not violate well-established chemical rules, and fit the same experimental data? A postulate should therefore be evoked and legitimized, in order to allow us to accept that a crystal structure is definitively solved: this is what we call the basic postulate of structural crystallography.

The basic postulate of structural crystallography: *only one chemically sound crystal structure exists that is compatible with the experimental diffraction data.*

Before legitimizing such a postulate mathematically a premise is necessary: the postulate is valid for crystal structures, that is, for structures for which chemical (i.e. the basic chemical rules) and physical constraints hold. Among physical constraints we will mention atomicity (the electrons are not dispersed in the unit cell, but lie around the nucleus) and positivity (i.e. the electron density is non-negative everywhere). The latter two conditions are satisfied if X-ray data and, by extension, electron data (electrons are sensible to the potential field) are collected: the positivity condition does not hold for neutron diffraction, but we will see that the postulate may also be applied to neutron data.

Let us now check the postulate by using the non-realistic four-atom one-dimensional structure shown in Fig. 1.9a: we will suppose that the chosen interatomic distances comply with the chemistry (it is then a *feasible model*). In Fig. 1.16a–c three electron densities are shown at 0.9 Å resolution, obtained by using, as coefficients of the Fourier series (1.18), the amplitudes of the true structure combined with random phases. All three models, by construction, have the same diffraction amplitudes ($R_{cryst} = 0$ for such models), but only one, that shown in Fig. 1.9a, satisfies chemistry and positivity–atomicity postulates. All of the random models show positive peaks (say potential atoms) in random positions, there are always a number of negative peaks present, and the number of positive peaks may not coincide with the original structure. Any attempt to obtain other feasible models by changing the phases in a random way will not succeed: this agrees well with the postulate.

A more realistic example is the following (structure code *Teoh*, space group *I-4*, $C_{42} H_{40} O_6 Sn_2$). Let us suppose that the crystallographer has requested his phasing program to stop when a model structure is found for which $R_{cryst} < 0.18$ and that the program stops, providing the model depicted in Fig. 1.17a, for which $R_{cryst} = 0.16$. This model, even if it is further refinable up to smaller values of R_{cryst}, has to be rejected because it is chemically invalid, even if the crystallographic residual is sufficiently small. If the crystallographer asks the phasing program to stop only when a model is found for which $R_{cryst} < 0.10$, then the model shown in Fig. 1.17b is obtained, for which $R_{cryst} = 0.09$. This new model satisfies basic crystal chemical rules and may be further refined.

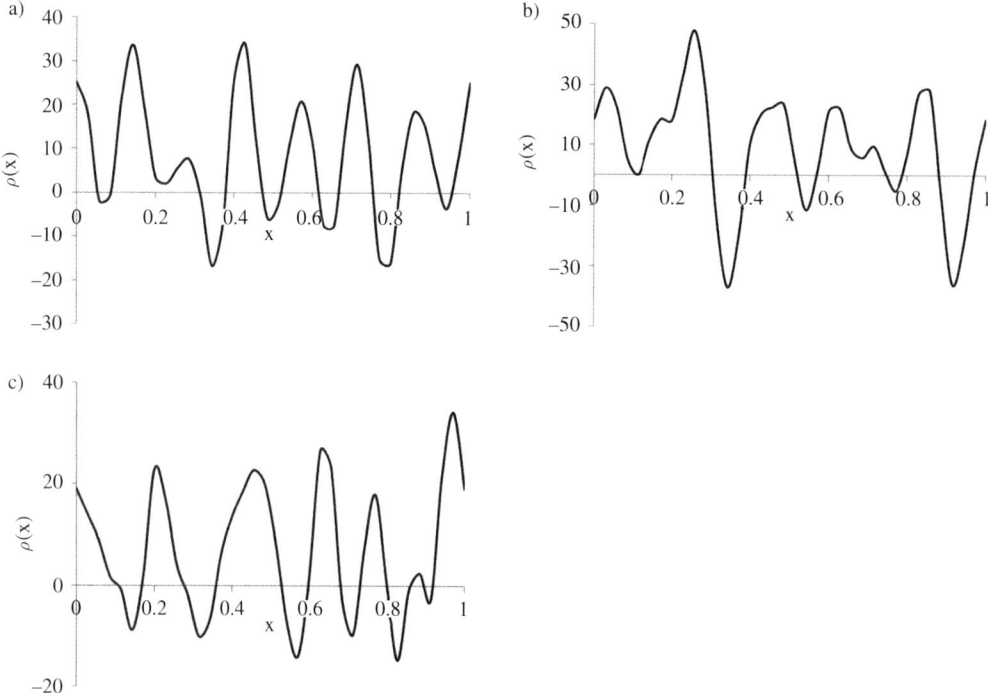

Fig. 1.16
For the four-atom one-dimensional structure shown in Fig. 1.9a, three models, obtained using random phases, are shown. Data resolution: 0.9 Å.

The above results lead to a practical consequence: even if experimental data are of high quality, and even if there is very good agreement between experiment and model (i.e. a small value of R_{cryst}), *structure validation* (i.e. the control that the basic crystal chemical rules are satisfied by the model) *is the necessary final check of the structure determination process*. Indeed it is an obligatory step in modern crystallography, a tool for a posteriori confirmation of the basic postulate of crystallography.

The basic postulate may be extended to neutron data, but now the positivity condition does not hold: it has to be replaced by the chemical control and validation of the model, but again, there should not exist two chemically sound crystal structures which both fit high quality experimental data.

In order to legitimize the basic postulate of structural crystallography mathematically, we now describe how the phase information is codified in the diffraction amplitudes. We observe that the modulus square of the structure factor, say

$$|F_\mathbf{h}|^2 = F_\mathbf{h} \cdot F_{-\mathbf{h}} = \sum_{j=1}^{N} f_j \exp\left(2\pi i \mathbf{h} \cdot \mathbf{r}_j\right) \cdot \sum_{j=1}^{N} f_j \exp\left(-2\pi i \mathbf{h} \cdot \mathbf{r}_j\right)$$

$$= \sum_{j1,j2=1}^{N} f_{j1} f_{j2} \exp\left[2\pi i \mathbf{h} \cdot (\mathbf{r}_{j1} - \mathbf{r}_{j2})\right]$$

(1.30)

depends on the interatomic distances: inversely, the set of interatomic distances defines the diffraction moduli. If one assumes that only a crystal structure exists with the given set of interatomic distances, the obvious conclusion should be that only one structure exists (except for the enantiomorph structure) which is compatible with the set of experimental data, and vice versa, only one set of diffraction data is compatible with a given structure. In symbols

$$\text{crystal structure} \Leftrightarrow \{\mathbf{r}_i - \mathbf{r}_j\} \Leftrightarrow \{|F_\mathbf{h}|\}. \quad (1.31)$$

This coincides exactly with the previously defined basic postulate.

The conclusion (1.31), however, must be combined with structure validation, as stated in the basic postulate. Indeed Pauling and Shapell (1930) made the observation that for the mineral bixbyite there are two different solutions, not chemically equivalent, with the same set of interatomic vectors. Chemistry (i.e. structure validation) was invoked to define the correct structure. Patterson (1939, 1944) defined these kinds of structure as *homometric* and investigated the likelihood of their occurrence. Hosemann and Bagchi (1954) gave formal definitions of different types of homometric structures. Further contributions were made by Buerger (1959, pp. 41–50), Bullough (1961, 1964), and Hoppe (1962a,b). In spite of the above considerations it is common practice for crystallographers to postulate, for structures of normal complexity, a biunique correspondence between the set of interatomic vectors and atomic arrangement. Indeed for almost the entire range of the published structures, two different *feasible* (this property being essential) structures with the same set of observed moduli has never been found.

Some care, however, is necessary when the diffraction data are not of high quality and/or some pseudosymmetry is present. Typical examples of structural ambiguity are:

(a) The low quality of the crystal (e.g. high mosaicity), or the disordered nature of the structure. In this case the quality of the diffraction data is depleted, and therefore the precision of the proposed model may be lower.
(b) The structure shows a symmetry higher than the real one. For example, the structure is very close to being centric but it is really acentric, or it shows a strong pseudo tetragonal symmetry but it is really orthorhombic. Deciding between the two alternatives may not be easy, particularly when the pseudosymmetry is very close to crystal symmetry and data quality is poor.
(c) Strong pseudotranslational symmetry is present. This occurs when a high percentage of electron density satisfies a translational vector $\boldsymbol{u}$ smaller than that allowed by the crystal periodicity: for example, if $\boldsymbol{u} = \boldsymbol{a}/3$ and 90% of the electron density is invariant under the pseudotranslation. In this case reflections with h = 3n are very strong, the others are very weak. If only substructure reflections are measured, the substructure only is defined (probably with a quite good R_{cryst} value), but the fine detail of the structure is lost.

In all of the cases a–c the final decision depends on the chemistry and on the fit between model and observations. To give a general view of what the fit means

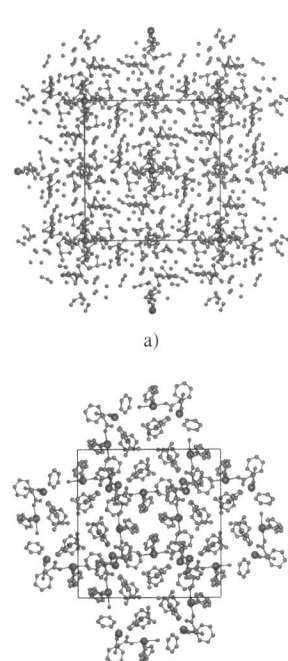

Fig. 1.17
Teoh: (a) false structural model with $R_{cryst} = 0.16$; (b) correct structural model with $R_{cryst} = 0.08$.

Table 1.9 Statistics on R_{cryst} for structures deposited in the *Cambridge Structural Database* up to 1 January 2012. For each range ΔR_{cryst} in which R_{cryst} lies, *Nstr* and % are the corresponding number of structures and percentage, respectively

ΔR_{cryst}	Nstr	%
0.01–0.03	62 774	10.5
0.03–0.04	122 706	20.6
0.04–0.05	135 525	22.7
0.05–0.07	163 269	27.4
0.07–0.09	60 651	10.2
0.09–0.10	13 370	2.2
0.10–0.15	18 353	3.1
0.15–....	3835	0.6

numerically today, we report in Table 1.9 statistics on the crystallographic residual R_{cryst} performed over the structures deposited in the *Cambridge Structural Database* up to January 2012. We see that the precision of the structural determination may vary over a wide range: indeed R_{cryst} values are found between 0.01 and more than 0.1, and this wide range is often due to the different quality of the crystals. For the large majority of structures, even those with a relatively high value of R_{cryst}, the structure is uniquely fixed in all details, eventually with limited precision in unit cell regions where structural disorder is present. These details, however, do not destroy the general validity of the basic postulate.

The basic postulate of structural crystallography should be considered by any rational crystallographer before initiating their daily structural work. This may be further summarized as follows: *in a diffraction experiment the phase information is not lost, it is only hidden within the diffraction amplitudes. Accordingly, any phasing approach is nothing else but a method for recovering the hidden phases from the set of diffraction amplitudes.*

Let us now suppose that the basic postulate is consciously considered by our young crystallographer. A further problem then arises: is the amount of information stored in the diffraction amplitudes sufficient to define the structure? For example, in the case of a low resolution diffraction experiment the crystallographic data may not be sufficient to define the short interatomic distances, making it impossible, therefore, to uniquely define the structure. This is a crucial problem for structural crystallography, since the crystal structure solution may depend on the amount of information provided by the diffraction experiment. *What then are the resolution limits for a useful diffraction experiment?*

Suppose we have a crystal with *P1* symmetry: let N be the number of non-hydrogen atoms in the unit cell, and $N_{sp} = 4N$ the number of structural parameters necessary for defining the structure (four parameters per atom, say x, y, z and the corresponding isotropic thermal factor). For a small- or medium-sized molecule, $V = k N$, where k is usually between 15.5 and 18.5; for a protein, owing to the presence of the solvent, k may be significantly larger, up to or even exceeding 40. According to equation (1.8), $V^* = V^{-1} = (kN)^{-1}$.

Let us suppose that a diffraction experiment provides data up to r^*_{max}, or, equivalently, up to d_{min}. The number of measurable reflections (say N_{ref}) may be calculated as follows. The reciprocal space measured volume may be parameterized as

$$\Phi^*_{meas} = \frac{4}{3}\pi \left(r^*_{max}\right)^3 = \frac{4\pi}{3d^3_{min}},$$

and

$$N_{ref} = \frac{\Phi^*_{meas}}{V^*} = \frac{4\pi}{3d^3_{min}} kN.$$

Let us now estimate the index,

$$R_{inf} = \text{ratio between the experimental information and the structural complexity.} \quad (1.32)$$

When no prior supplemental information is available besides experimental data, R_{inf} may be qualitatively approximated as follows:

$$R_{\text{inf}} = \text{number of measured symmetry independent reflections}/\text{number of structural parameters} \quad (1.33)$$

To compute R_{inf}, the Friedel law should be taken into account: thus we divide N_{ref} by 2 and then write the resulting expression for R_{inf}:

$$R_{\text{inf}} \approx \frac{\pi}{6 d_{\min}^3} k.$$

The numerical values of R_{inf} for specific values of k and d_{min} are shown in Table 1.10: larger values of R_{inf} correspond with cases in which the structure is overdetermined by the observations, while small values of R_{inf} do not uniquely fix the structure. Let us suppose, just as a rule of thumb, that a structure may be solved, from diffraction data only, if $R_{\text{inf}} \geq 3$: Table 1.10 suggests that $d_{\min} \approx 1.4$ Å is the resolution threshold below which a small molecule structure cannot be solved ab initio. The threshold moves to ≈ 1.6 Å for a protein with a small percentage of solvent, and to ≈ 1.8 Å for a protein with a larger solvent percentage.

The conclusion is that the solvent is a valuable source of information: the larger the solvent, the higher the threshold for the ab initio crystal structure solution (modern solvent flattening techniques are able to efficiently exploit this information). A special case occurs when one is interested in solving a substructure, for example the heavy-atom substructure in SIR-MIR cases and the anomalous scatterer substructure in SAD-MAD cases. If it is supposed that the structure factor amplitudes of such substructures are estimated with reasonable approximation, then the atoms belonging to the substructure are dispersed in a big empty space (i.e. the unit cell of the structure). In this case the estimated structure factor amplitudes of the substructure overdetermine it, and the substructure could be solved even at very low resolution (worse than 3.5 Å).

The above conclusions do not change significantly if the space group has symmetry higher than triclinic. Indeed in this case R_{inf} is the ratio between the number of unique reflections and the number of structural parameters corresponding to the symmetry independent atoms.

Additional difficulties with the phasing process arise when experimental data quality is poor. If there are errors in the diffraction amplitudes, since information on the phases is hidden within the amplitudes, such errors will inevitably cause a deterioration in the efficiency of any phasing procedure. This is particularly important in the case of powder data (see Chapter 12) and also electron data (see Chapter 11), but it is also important for proteins, because the presence of the solvent implies disordered regions in the unit cell and therefore limited data resolution.

So far we have answered the question: under what conditions is a structure univocally fixed from its diffraction data? We have skipped cases where some previous additional information is available; here, the number of measured symmetry independent reflections in the numerator of R_{inf} is only part of the total information available and therefore the conclusions drawn from

Table 1.10 R_{inf} in $P1$ is shown for some values of d_{min} and k. $k = 17$ is representative of the small- to medium-sized structures, $k = 25, 35$, of the proteins

d_{min}	$k = 17$	$k = 25$	$k = 35$
0.4	139	204	286
0.6	41	60.6	84.7
0.8	17.4	25.6	35.7
1.0	8.9	13.1	18.3
1.4	3.2	4.8	13.7
1.8	1.5	2.2	3.1
2.2	0.8	1.2	1.7
2.5	0.6	0.8	1.2
3.0	0.3	0.5	0.7
3.5	0.2	0.3	0.4
4.0	0.1	0.2	0.3

Table 1.10 must be corrected. In this book we will consider four cases in which additional information is present:

1. *Non-crystallographic symmetry*. This is an important source of information which permits a reduction in the number of structural parameters in equation 1.33. It occurs when there are more identical molecules in the asymmetric unit: in this case they may be defined in terms of one molecule by applying the local symmetry operators. *Non-crystallographic symmetry* allows the structural solution of large biological assemblies such as viruses.
2. *Molecular replacement*. A model molecule, similar geometrically to that under investigation, is available.
3. *Isomorphous derivatives*. Diffraction data for the target and one or more isomorphous structures are measured.
4. *Anomalous dispersion data*. Diffraction data with anomalous dispersion effects are collected (we will see that this case is similar to case 3).

Because of the additional experimental information available, the value of R_{inf} increases substantially which allows structure solution even at data resolutions larger than 4 Å.

1.7 The legacy of crystallography

Human beings periodically visit museums to enjoy the artistic masterpieces exhibited in witness of human sensitivity to beauty. Historical and technical museums are often consulted in relation to their acquaintance with the evolution of human civilization andwith man's capacity for improving human life through technical innovations. But, where can the products of crystallography be consulted, in witness of its immense legacy to chemistry, physics, mineralogy, and biology?

Over a period of about one century crystallographic phasing methods have solved a huge number of crystal structures, so enriching our understanding of the mineral world, of organic, metallorganic, and inorganic chemistry, and of the bio-molecules. This enormous mine of information is stored in dedicated databases, among which are the following.

1. The *Cambridge Structural Database* (*CSD*), <http://www.ccdc.cam.ac.uk/products/csd/>, where chemical and crystallographic information for organic molecules and metal–organic compounds determined by X-ray or neutron diffraction: powder diffraction studies are deposited.
2. *Inorganic Crystal Structure Database* (*ICSD*), <http://www.fiz-karlsruhe.de/icsd_content.html>, where structural data of pure elements, metals, minerals and intermetallic compounds are deposited. By January 2012 it contained more than 150 000 entries, 75.6% of them with a structure having been assigned.
3. *CRYSTMET*, <http://www.tothcanada.com/>, where structural information on metals and alloys are stored.

Table 1.11 *CSD* entries on 1 January 2012: Nr is the number of entries, % the corresponding percentage over the total

	Nr	%
Total number of structures	596 810	100
Number of compounds	544 565	
Organic compounds	254 475	42.6
Transition metal present	319 188	53.5
Neutron studies	1534	0.3
Powder diffraction studies	2354	0.4

4. *Protein Data Bank* (*PDB*), <http://www.rcsb.org/pdb/>, with about 75 056 entries up to January 2012.
5. *Nucleic Acid Database* (*NDB*), <http://ndbserver.rutgers.edu/>, oligonucleotide structures deposited up to April 2012.

For each structure deposited, an archive typically contains details of the structure solution, citation information, the list of atoms and their coordinates; the structure can be visualized and displayed on the user's computer. In this section we report some statistics on the entries in two of these databases, the *CSD* for small molecules and the *PDB* for macromolecules, in order to provide the reader with some essential information on some of the parameters to which phasing methods are sensitive. For example, which type of radiation is more useful in standard conditions, which are the most frequent space groups or crystal systems, how data resolution is distributed among the deposited structures, etc. This type of information is shown in Tables 1.11 to 1.16. It should be noted that:

(a) Tables 1.11 and 1.14 provide the numbers of deposited structures for small molecules and macromolecules, respectively. Also given is information on the type of radiation used for their solution. The tables justify the special attention we are giving to X-ray diffraction.
(b) Tables 1.13 to 1.16 suggest which are the most frequent space groups, for both small and large molecules. The reader should remember that these are expected to be very different for the two categories: indeed centric space groups and, in general, groups with inversion axes, are not allowed for proteins.

Table 1.12 *CSD*: crystallographic system statistics

System	%
Triclinic	24.7
Monoclinic	52.5
Orthorhombic	18.0
Tetragonal	2.2
Trigonal	1.7
Hexagonal	0.5
Cubic	0.5

Table 1.13 *CSD*: the 10 most frequent space groups

Space group	%
$P2_1/c$	34.9
$P\text{-}1$	23.8
$C2/c$	8.2
$P2_12_12_1$	7.6
$P2_1$	5.3
$Pbca$	3.5
$Pna2_1$	1.6
$Pnma$	1.2
Cc	1.1
$P1$	1.0

Table 1.14 *PDB*: entries for proteins, nucleic acids, and protein/NA complexes, according to experimental technique

	Proteins	Nucleic acids	Protein/NA complexes
Total	75056	2360	3609
X-ray	66381	1352	3298
NMR	8206	979	186
Electron microscopy	285	22	118

Table 1.15 *PDB*: distribution of data resolution for the deposited structures, in Å

Å	Nr	%
0.5–1	485	0.68
1.0–1.5	6134	8.63
1.5–2.0	27 385	38.53
2.0–2.5	22 144	31.16
2.5–3.0	11 210	15.77
3.0–3.5	2930	4.12
3.5–4.0	602	0.85

Table 1.16 *PDB*: the 10 most frequent space groups

Space group	Nr	%
$P2_12_12_1$	16 421	22.94
$P2_1$	10 948	15.30
C2	6777	9.47
$P2_12_12$	3924	5.48
$C222_1$	3511	4.91
P1	2975	4.16
$P4_32_12$	2702	3.78
$P3_221$	2547	3.56
$P3_121$	2373	3.32
$P4_12_12$	2365	3.13

(c) Table 1.15 does not have a counterpart for small molecules, where resolution is frequently atomic. The reason for this is very simple. Proteins have a large number of atoms in the unit cell: in only a few cases is the asymmetric unit composed of fewer than 500 non-H atoms, more often in excess of 8000. Thus the number of atoms in the unit cell, for high-symmetry space groups, is often more than 100 000. Protein molecules are irregular in shape and they pack together to form a crystal, but with gaps between them. The gaps are filled by another liquid (also called a *solvent*): such a disordered region ranges from 30–75% of the volume of the unit cell. Thus, while a classical molecular crystal may be described in terms of a regular lattice, in a protein, the crystalline array coexists with an extended disordered region. This will contribute to low-resolution diffraction, say up to 8–10 Å resolution as a rule of thumb. The disorder in the protein crystal and the high thermal motion within the protein do not permit diffraction intensities to be collected up to the resolution usually attainable for small molecules. According to Table 1.15, about 85% of the proteins show resolution between 1.5 and 3 Å and for about 8.6% of them, the resolution is better than 1.5 Å: in exceptional cases the resolution is higher than 1 Å.

The consequence is that data resolution is one of the most severe parameters in protein structural solvability.

Wilson statistics

2.1 Introduction

In a very traditional village game, popular over the period of Lent (usually on the pignata day, the first Sunday of Lent), a young player, suitably blindfolded and armed with a long cudgel, tries to hit a pot (the pignata) located some distance away, in order to win the sweetmeats contained inside. To break the pot they take random steps, and at each step they try to hit the pot with the cudgel. Is it possible to guess the distance of the player from their starting position after n random steps? Is it possible to guess the direction of the vectorial resultant of the n steps?

A very simple analysis of the problem suggests that the distance after n steps may be estimated but the direction of the resultant step cannot, because a preferred privileged orientation does not exist.

The situation is very similar to structure factor statistics. Each of the N atoms in the unit cell provides the vectorial contribution

$$\mathbf{f}_j = f_j \exp(2\pi i \mathbf{h} \cdot \mathbf{r}_j) = f_j \exp(i\theta_j)$$

to the structure factor; this is equivalent to a vectorial step of the pignata player. The modulus of the atomic contribution, like the amplitude of the step in the pignata game, is known (because the chemical composition of the molecules in the unit cell is supposed to be known), but the phase θ_j (corresponding to the direction of the step) remains unknown; indeed we do not know the position $\mathbf{r}_j$ of the jth atom.

The analogy with the pignata game suggests that some information on the moduli of the structure factors can be obtained via a suitable statistical approach, while no phase information can be obtained using this approach. This chapter deals just with this statistical approach and owing to the relevant contributions of A. J. C. Wilson, we call this chapter *Wilson statistics*. To allow greater fluency of reading, we have moved some of the mathematical proofs and formula derivations to the appendices at the end of the chapter, and we have completely omitted more specialized topics such as:

(i) non-ideal distributions correlated to special atomic positions (Howells et al., 1950; Hargreaves, 1955; Pradhan et al., 1985);
(ii) effects of measurement errors or data omission on the Wilson plot (Howells et al., 1950; Vicković and Viterbo, 1979; French and Wilson, 1978; Subramanian and Hall, 1982; Cascarano et al., 1991);
(iii) effects of pseudo- or hyper-symmetry on the structure factor distributions (Parthasarathy and Parthasarathi, 1976; Lipson and Woolfson, 1952; Rogers and Wilson, 1953; Cascarano et al., 1985a,b, 1987b, 1988a,b; Gramlich, 1984);
(iv) the Fourier series representation of the structure factor distributions as described by Shmueli et al. (1984) and Shmueli and Weiss (1995).

Readers interested in these topics are referred to *Phasing*, Chapter 1, or to the original papers.

2.2 Statistics of the structure factor: general considerations

The structure factors, when known in modulus and phase, contain all of the information about the crystal structure; therefore knowing about their statistical behaviour may be an aid towards success with the phasing problem. A statistical study may only be performed after defining which are the variables in the mathematical model and which are the parameters under study. The structure factor $F_\mathbf{h}$ is a function of the reciprocal vector $\mathbf{h}$ as well as of the atomic positional parameters $\mathbf{r}_j$: both may be assumed to be variables or parameters. We will consider three different ways of performing statistical analysis on $F_\mathbf{h}$:

(1) *$\mathbf{h}$ fixed, variable positions*. The distribution function of $F_\mathbf{h}$ is derived by taking the vectorial index $\mathbf{h}$ as fixed, and by assuming the atomic positions $\mathbf{r}_j$ as random variables. In the absence of any prior information, the $\mathbf{r}_j$s may be assumed to be uniformly distributed in the unit cell (in other words, any site of the unit cell has the same probability of accomodating an atomic position). In this case $2\pi \mathbf{h} \cdot \mathbf{r}_j$ is uniformly distributed over the trigonometric circle, or, equivalently, the fractional part of $\mathbf{h} \cdot \mathbf{r}_j$ is uniformly distributed between 0 and 1. This type of statistics may answer the following question: for fixed $\mathbf{h}$ what are the expected values of $|F_\mathbf{h}|$ and $\phi_\mathbf{h}$ when the structure varies in all possible ways?

Such statistics will only be useful in practice if crystals satisfy or nearly satisfy the above stated conditions. If the number of atoms per unit cell is very small (tentatively less than 8) the positions of the atoms are mainly controlled by physicochemical factors such as electrostatic interaction, packing, bond angles, etc. Under these conditions we have an ordered configuration for the crystal structure in the sense that, for a high percentage of atoms, the coordinates are not random and comply with relationships of the type

$$m_1 x_j + m_2 y_j + m_3 z_j = m_4, \tag{2.1}$$

where $m_i, i = 1, \ldots, 4$ are integers and not all zero (in a typical case, most of the atoms are in special positions). This is the case mentioned in

Section 2.1 and referred to as *non-ideal distributions correlated to special atomic positions*.

When N is sufficiently large, even though the above mentioned physicochemical interactions are still present, the atomic coordinates may be considered approximately as *random* variables. If the atoms are distributed at random in the unit cell, then any vector

$$\mathbf{f}_j = f_j \exp(2\pi i \mathbf{h} \cdot \mathbf{r}_j)$$

may be considered as a random step of a random walk. Thus the problem of evaluating the probability that the structure factor with index $\mathbf{h}$ takes on a value $F_\mathbf{h}$ appears similar to the *random walk problem* (Pearson, 1905; Kluyver, 1906), briefly schematized in Section 2.1; or to the problem of evaluating the probability, for a particle affected by Brownian motion, of occupying a position $\mathbf{r}$, if at time $t = 0$ it was at the origin of the reference system.

(2) *Fixed positions, $\mathbf{h}$ variable*. Let us assume that the structure is fixed (even if unknown) and that $\mathbf{h}$ varies within a sufficiently large subset of structure factors (for example the subset of reflections contained in a resolution shell or the full set of reflections up to *RES*). This type of statistics may answer the following question: for the actual fixed (even if unknown) structure, how are the $F_\mathbf{h}$ moduli and phases distributed when $\mathbf{h}$ varies freely within the chosen subset? From a mathematical point of view the conditions for the basis of the two statistics (1) and (2) look different, even though $\mathbf{h}$ and $\mathbf{r}_j$ are symmetrical factors of the product $\mathbf{h} \cdot \mathbf{r}_j$. In fact, with the former approach the $\mathbf{r}_j$ variables may take on rational values, while in the latter the $\mathbf{h}$ variables may take on only integer values. Nevertheless, Weyl (1915–16) showed that when an $\mathbf{r}_j$ vector has rationally independent x_j, y_j, z_j components (in the sense that there are no four non-zero integer numbers m_i, $i = 1, \ldots, 4$ satisfying (2.1)), then the fractional part of $\mathbf{h} \cdot \mathbf{r}_j$ is uniformly distributed within the interval (0,1) when $\mathbf{h}$ varies in the domain of the integer numbers.

Weyl's theorem proves that the two statistical approaches, the first with $\mathbf{h}$ fixed and $\mathbf{r}_j$ as random variables, and the second with fixed $\mathbf{r}_j$ and $\mathbf{h}$ varying over reciprocal space, although ideally distinct, are exactly alike. This is of fundamental importance in the subsequent steps of intensity statistics and also in the theory of assigning phases by algebraic or probabilistic methods.

(3) $\mathbf{h}$ is kept fixed, while atomic positions vary under suitable constraints. This situation may occur frequently during the phasing process. As an example, let us suppose that only part of the structure has been determined while the rest remains unknown. In this case the atomic positions of the fixed molecular fragment should be considered as fixed parameters, while the undetermined atomic positions are free variables. In various chapters of this book we will consider different cases in which this procedure may be applied.

2.3 Structure factor statistics in $P1$ and $P\bar{1}$

Wilson (1942, 1949) was the first to derive the structure factor probability distributions. Their expressions are identical to those derived by Rayleigh (1919a,b) for the distribution of the resultant obtained from the composition of N vibrations of equal amplitude but random phase. In this section we will deal with the distribution in $P1$ and $P\bar{1}$, under the assumption that **h** is fixed and that the $\mathbf{r}_j$ are uniformly varying in the unit cell, each atomic position independently of the others ((1) in Section 2.2). The same results, however, may be obtained using (2), where the structure is fixed and **h** variable.

Case P1: The statistical calculations described in Appendix 2.A.1 lead to the final distribution

$$_1P(|F|, \phi) = \frac{|F|}{\pi \Sigma} \exp\left(-\frac{|F|^2}{\Sigma}\right), \qquad (2.2)$$

where the prefixed subscript 1 indicates that the space group is $P1$,

$$\Sigma = \sum_{j=1}^{N} f_j^2 \qquad (2.3)$$

and N is the number of atoms in the unit cell. f_j includes the thermal factor, and is calculated at the $\sin\theta/\lambda$ corresponding to fixed **h**.

The reader should notice that (2.2) does not provide any information on the phase, even if $|F|$ is known. To explicitly check this feature we calculate the conditional distribution of ϕ when the amplitude $|F|$ is assumed to be known:

$$_1P(\phi||F|) = \frac{_1P(|F|, \phi)}{\int_0^{2\pi} {_1P(|F|, \phi)d\phi}} = \frac{1}{2\pi}.$$

The result is that ϕ is uniformly distributed between 0 and 2π: i.e. the phase value of a reflection cannot be determined by its diffraction modulus (see also Chapter 3, where the origin problem is discussed).

Therefore, in this chapter, we can forget the phase problem and concentrate our efforts on the amplitude distribution. The marginal distribution of the amplitudes (i.e. whatever the phase) is obtained by integrating ϕ in (2.2) over the interval $(0, 2\pi)$:

$$_1P(|F|) = \frac{2|F|}{\Sigma} \exp\left(-\frac{|F|^2}{\Sigma}\right) \qquad (2.4)$$

Equation 2.4 is known as the *acentric distribution of the structure factors*.

Case $P\bar{1}$: The statistical calculations described in Appendix 2.A.2 lead to the distribution

$$_{\bar{1}}P(|F|) = \left(\frac{2}{\pi \Sigma}\right)^{1/2} \exp\left(-\frac{|F|^2}{2\Sigma}\right). \qquad (2.5)$$

Equation 2.5 is known as the *centric distribution of the structure factors*.

Let us now discuss equations (2.4) and (2.5). Both distributions are structure and data resolution dependent: that is immediately evident if we calculate the expected value of $<|F|^2>$, say

$$<|F|^2> = \int_0^\infty |F|^2 P(|F|)d|F|.$$

For both distributions, $<|F|^2> = \Sigma$ and Σ depends on the number of atoms, on the atomic species present in the unit cell, and on the $\sin\theta/\lambda$ value at which the average is calculated.

Distributions (2.4) and (2.5) should be of more general usefulness if they could be transformed into universal (i.e. structure and data resolution independent) distributions. To obtain such a result we only need to normalize the variable F, by remembering that a variable y is normalized by dividing it by $<y^2>^{1/2}$. In our case we should replace F by the *normalized structure factor*,

$$E = F / \left(\sum\right)^{1/2}. \tag{2.6}$$

The corresponding distributions are (from now on $R = |E|$)

$$_1P(R) = 2R \exp(-R^2) \tag{2.7}$$

and

$$_{\bar{1}}P(R) = \left(\frac{2}{\pi}\right)^{1/2} \exp\left(-\frac{R^2}{2}\right). \tag{2.8}$$

Distributions (2.7) and (2.8) are known as *acentric and centric distributions of the normalized structure factors*, respectively and do not vary with structure complexity and data resolution: in some way they have a universal character, and they are represented in Fig. 2.1.

We now take a few lines to better describe the meaning of E and of the two curves in Fig. 2.1. In a more explicit form, definition (2.6) may be rewritten as

$$E = \frac{F}{\left(\sum_{j=1}^{N} f_j^2\right)^{1/2}} = \sum_{j=1}^{N} \nu_j \exp(2\pi i \mathbf{h} \cdot \mathbf{r}_j), \tag{2.9}$$

where $\nu_j = f_j / (\sum)^{1/2}$. Equation (2.9) suggests that the normalized scattering factor ν_j:

(i) does not change practically with $\sin\theta/\lambda$. Since atomic scattering factors decay in a similar way (provided they have a similar temperature factor; see Fig. 1.7), ν_j may be approximated as

$$\nu_j = f_j / \left(\sum\right)^{1/2} \approx Z_j / \left(\sum_{j=1}^{N} Z_j^2\right)^{1/2};$$

(ii) is perfectly constant if all the atoms in the unit cell are of one species and with the same thermal factor: then $\nu_j = 1/\sqrt{N}$.

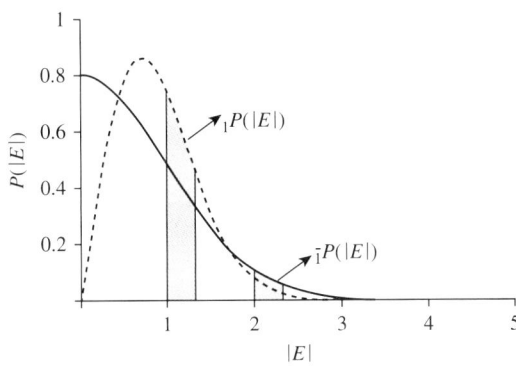

Fig. 2.1
$_1P(|E|)$ (dashed line) and $_{\bar{1}}P(|E|)$ (solid line).

The normalization is therefore equivalent to assuming the atoms to be point scatterers, the scattering amplitude of which no longer varies with $\sin\theta/\lambda$.

The plots in Fig. 2.1 may easily be interpreted by recalling that each distribution provides the expected percentage of reflections with R lying in a given interval (generally speaking, for good distributions such a percentage should correspond well with the experimental frequency). In Fig. 2.1 we emphasize the area below the distribution (2.8) in the R-interval (2.0–2.3): this area provides in $P\bar{1}$ the percentage of reflections which are expected to have R in the interval (2.0–2.3). We will see later in this section that this property also holds for all centric space groups. In the same figure we emphasize the area below the distribution (2.7) in the R-interval (1.0–1.3). Again the area represents in P1 the percentage of structure factors which are expected to have R in that interval. We will see later in this section that this property also holds for all acentric space groups. By generalizing the above statement, we say that $P(R)dR$ represents the percentage of normalized structure factors with modulus lying in a quite small interval around the chosen R value. Consequently, the full area below each distribution is equal to unity, by definition.

Distributions (2.7) and (2.8) show remarkable differences. Indeed: (i) at $R=0$, (2.7) attains a minimum and (2.8) a maximum; (ii) the maximum of (2.7) is at $R = 1/\sqrt{2} = 0.707$; (iii) for both distributions values of R larger than 3 are not very probable, but the expected percentage of structure factors with $R > 3$ is significantly larger for centric distributions.

Why are (2.7) and (2.8) so different? The main reason is as follows: (2.7) derives from (2.2), which is a two-dimensional distribution (indeed F is a complex variable, with real and imaginary components), while for centric groups F is a one-dimensional real variable. If this explanation is correct, we should expect that the distributions for acentric space groups should be more similar to (2.7) than to (2.8): the reverse should be true for centric space groups. This is what we derive mathematically in Appendix 2.B. In particular, the calculations indicate that the most general distribution for any acentric crystal is

$$_1P(|F|) = \frac{2|F|}{p\sum} \exp\left(-\frac{|F|^2}{p\sum}\right) \qquad (2.10)$$

and the most general distribution for any centric crystal is

$$_{\bar{1}}P(|F|) = \left(\frac{2}{\pi p \sum}\right)^{1/2} \exp\left(-\frac{|F|^2}{2p\sum}\right), \qquad (2.11)$$

where

$$p(\mathbf{h}) = \tau\varepsilon(\mathbf{h}) \qquad (2.12)$$

is the classical $\mathbf{h}$ dependent *Wilson weight*.

p is the product of two parameters: τ, which is the multiplicity of the lattice (or number of lattice points per unit cell; see Table 1.1) and $\varepsilon(\mathbf{h})$, which is the order of the point group which leaves $\mathbf{h}$ unchanged. That is to say, if m is the number of symmetry operators of the space group, $m/\varepsilon(\mathbf{h})$ is the number of distinct symmetry equivalent reflections; see Appendix 2.B). Table 2.1 shows the ε values for all of the point groups.

Structure factor statistics in $P1$ and $P\bar{1}$

Table 2.1 Values of ε for all point groups (for non-primitive unit cells, all figures must be multiplied by τ, the multiplicity of the lattice, to obtain the Wilson weight p)

(i) Triclinic, monoclinic, orthorhombic

Point group	hkl	$0kl$	$h0l$	$hk0$	$h00$	$0k0$	$00l$
1	1	1	1	1	1	1	1
$\bar{1}$	1	1	1	1	1	1	1
$2^\dagger$	1	1	1	1	1	2	1
$m^\dagger$	1	1	2	1	2	1	2
$2/m^\dagger$	1	1	2	1	2	2	2
222	1	1	1	1	2	2	2
$mm2^\ddagger$	1	2	2	1	2	2	4
mmm	1	2	2	2	4	4	4

(ii) Tetragonal

Point group	hkl	$h0l, 0kl$	$hhl, h\bar{h}l$	$hk0$	$hh0, h\bar{h}0$	$h00, 0k0$	$00l$
4	1	1	1	1	1	1	4
$\bar{4}$	1	1	1	1	1	1	2
$4/m$	1	1	1	2	2	2	4
422	1	1	1	1	2	2	4
$4mm$	1	2	2	1	2	2	8
$\bar{4}2m$	1	1	2	1	2	2	4
$\bar{4}m2$	1	2	1	1	2	2	4
$4/mmm$	1	2	2	2	4	4	8

(iii) Trigonal and hexagonal
(a) Hexagonal cell

Point group	hkl	$hk0$	$hhl, h, 2\bar{h}, l$ $hh0, h, 2\bar{h}, 0$ $h0l, 0kl$			$h00, 0k0$	
			$2h, \bar{h}, l$	$2h, \bar{h}, 0$	$h\bar{h}l$	$h\bar{h}0$	$00l$
3	1	1	1	1	1	1	3
$\bar{3}$	1	1	1	1	1	1	3
312	1	1	1	1	1	2	3
321	1	1	1	2	1	1	3
$31m$	1	1	2	2	1	1	6
$3m1$	1	1	1	1	2	2	6
$\bar{3}1m$	1	1	2	2	1	2	6
$\bar{3}m1$	1	1	1	2	2	2	6
6	1	1	1	1	1	1	6
$\bar{6}$	1	2	1	2	1	2	3
$6/m$	1	2	1	2	1	2	6
622	1	1	1	2	1	2	6
$6mm$	1	1	2	2	2	2	12
$\bar{6}m2$	1	2	1	2	2	4	6
$\bar{6}2m$	1	2	2	4	1	2	6
$6/mmm$	1	2	2	4	2	4	12

Wilson statistics

Table 2.1 (Continued)

(b) Primitive rhombohedral cell

Point group	hkl 0kl, h0l, hk0 $hh\bar{l}$, $hk\bar{h}$, $h\bar{l}l$	hhl, hkh, hll hh0, h0h, 0ll $\bar{h}hh$, $h\bar{h}h$, $hh\bar{h}$ h00, 0k0, 00l	$h\bar{h}0$, $h0\bar{h}$, $0\bar{l}l$	hhh
3	1	1	1	3
$\bar{3}$	1	1	1	3
32	1	1	2	3
3m	1	2	1	6
$\bar{3}m$	1	2	2	6

(iv) Cubic

Point group	hkl	hhl, $hh\bar{l}$, hkh, $hk\bar{h}$, hll, $h\bar{l}l$	hhh $\bar{h}hh$, $h\bar{h}h$, $hh\bar{h}$	0kl h0l hk0	hh0, $h\bar{h}0$ h0h, $h0\bar{h}$ 0ll, $0\bar{l}l$	h00, 0k0, 00l
23	1	1	3	1	1	2
m3	1	1	3	2	2	4
432	1	1	3	1	2	4
$\bar{4}3m$	1	2	6	1	2	4
m3m	1	2	6	2	4	8

† b-axis unique.
‡ c-axis unique.

If $<|F|^2>$ is calculated for distributions (2.10) and (2.11) we obtain the following relation:

$$<|F_\mathbf{h}|^2> = p_\mathbf{h} \sum. \qquad (2.13)$$

If we normalize the variable $|F|^2$ according to (2.13), the normalized structure factor may be defined in the most general way as

$$E_\mathbf{h} = F_\mathbf{h} / \left(p_\mathbf{h} \sum\right)^{1/2}. \qquad (2.14)$$

The definition (2.6), valid for triclinic symmetry, is a special case of (2.14). Accordingly, if F-distributions are transformed into E-distributions according to (2.14), equations (2.10) and (2.11) reduce to the acentric distribution (2.7) and to the centric distribution (2.8), respectively. *It may be concluded that the distributions (2.7) and (2.8) are valid for any space group provided that the normalization is made according to (2.14).*

It is important to advise the reader that normalizing according to (2.14) cannot be performed in practice without having first solved two problems: how to estimate the scale of the experimental intensities (these are usually on an arbitrary scale) and temperature displacements of the atoms (indeed $\sum$ includes the scattering factors modified by thermal movements). Both of these problems will be solved in Section 2.8.

2.4 The $P(z)$ distributions

When y is a function of x, x being a random variable, the equality

$$P(y)\,dy = P(x)\,d\,[y(x)]$$

holds. The above relation may help us to derive the so-called $P(z)$ distributions where $z = R^2$. We have

$$_1P(z) = \exp(-z) \tag{2.15}$$

and

$$_{\bar{1}}P(z) = (2\pi z)^{-1/2} \exp\left(-\frac{z}{2}\right), \tag{2.16}$$

which are illustrated in Fig. 2.2.

2.5 Cumulative distributions

Let us calculate the expected percentage of reflections with R smaller than a given threshold (*cumulative distribution*). This is defined as

$$N(R) = \int_0^R P(t)\,dt$$

and is easily calculated for acentric and centric distributions (2.7) and (2.8):

$$_1N(R) = \int_0^R 2t\exp(-t^2)\,dt = 1 - \exp(-R^2) \tag{2.17}$$

$$_{\bar{1}}N(R) = \sqrt{\frac{2}{\pi}} \int_0^R \exp\left(-\frac{t^2}{2}\right) dt = erf\left(\frac{R}{\sqrt{2}}\right), \tag{2.18}$$

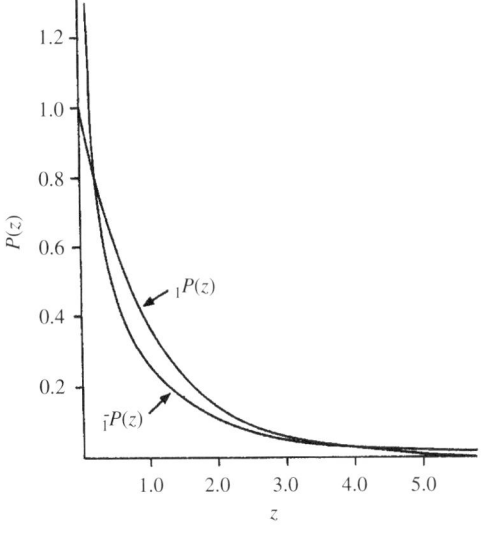

Fig. 2.2
The $P(z)$ distribution for cs. and n.cs. crystals.

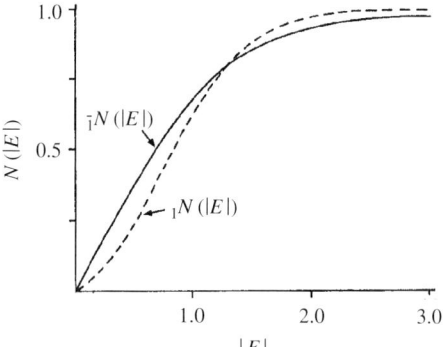

Fig. 2.3
The $N(|E|)$ distribution for cs. and n.cs. crystals.

where $erf(x)$ is the error function, defined by

$$\frac{2}{\sqrt{\pi}} \int_0^x \exp(-t^2)dt. \tag{2.19}$$

erf is extensively tabulated in the literature (see for example, Abramowitz and Stegun 1972, pp. 310–311). $_1N(R)$ and $_{\bar{1}}N(R)$ are illustrated in Fig. 2.3.

The corresponding cumulative distributions for $z = R^2$ are (see equations (2.15) and (2.16)):

$$_1N(z) = \int_0^z \exp(-t)dt = 1 - \exp(-z) \tag{2.20}$$

$$_{\bar{1}}N(z) = \frac{1}{\sqrt{2\pi}} \int_0^z t^{-1/2} \exp\left(-\frac{t}{2}\right) dt$$

$$= \frac{2}{\sqrt{2\pi}} \int_0^{\sqrt{z}} \exp\left(-\frac{u^2}{2}\right) du$$

$$= erf\sqrt{\frac{z}{2}}. \tag{2.21}$$

The respective plots are illustrated in Fig. 2.4.

Cumulative distributions are less sensitive to sampling problems, and may be used to discriminate centric from acentric structures; they are also used in histogram matching procedures (see Section 8.2 and Appendix 8.B).

2.6 Space group identification

Space group identification is one of the first problems encountered during phasing procedures. It is based on the relations which allow the discovery of the symmetry from the diffraction data analysis, and in a more general sense, on the distribution of the normalized structure factors.

Let us suppose that the unit cell has been determined from the diffraction experiment (in practice, from the positions of the reciprocal lattice points). Then the crystal system may tentatively be assigned as the one with the largest

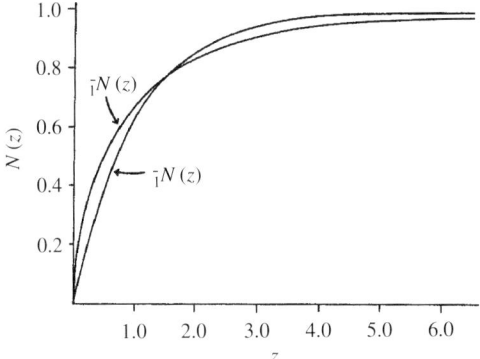

Figure. 2.4
The $N(z)$ distribution for cs. and n.cs. crystals.

symmetry that is compatible with the metric relations between the unit cell parameters. For example, if the following parameters:

$$a = 11.10(2), \quad b = 11.11(3), \quad c = 9.72(2)$$

$$\alpha = 90.25(37), \quad \beta = 89.48(30), \quad \gamma = 90.32(35) \quad (2.22)$$

have been measured, then the conclusion may be: *the unit cell is probably tetragonal*. The unit cell geometry, however, may simulate a lattice symmetry higher than the structure symmetry: e.g. the crystal may be orthorhombic with, occasionally, a close to b, or monoclinic with, occasionally, a close to b and the β angle occasionally close to 90°. At this stage, therefore, it is not possible to make a definitive decision about the crystal system: none of the crystal systems with unit cells compatible with that experimentally estimated (we will call them *feasible systems*) can be excluded. In the case above, besides the orthogonal pseudo-tetragonal cell, all monoclinic settings, $112/m$, $12/m1$, $2/m11$, compatible with the proposed unit cell, should also be considered to be feasible. A triclinic pseudo-tetragonal cell is also feasible!

To confirm the identity of the crystal system it is necessary to identify the Laue group (see Section 1.5): this requires that all the reflections contained in the allowed reciprocal lattice sphere (eventually, Friedel opposites excluded) should be measured. For each feasible crystal system the list of compatible Laue groups (we will denote them as *admitted* Laue groups) should be checked. For example, in the case of (2.22) above, the tetragonal Laue groups $4/m$ and $4/mmm$, the orthorhombic group mmm, and the monoclinic settings $112/m$, $12/m1$, $2/m11$ should be checked. Only if the Laue groups $4/m$ or $4/mmm$ are confirmed by the analysis of data may the tetragonal crystal system be accepted.

How is the Laue group identified? For each ith admitted Laue group, the internal residual factor
$R_{int}(i)$, given by

$$R_{int}(i) = \frac{\sum ||F_{obsh}| - \langle |F_{obsh}| \rangle|}{\sum |F_{obsh}|}, \quad (2.23)$$

is calculated, where

$$\langle |F_{obs}| \rangle = \frac{\sum_{\mathbf{h}} w_{\mathbf{h}} |F_{obs\mathbf{h}}|}{\sum w_{\mathbf{h}}}$$

is the average diffraction modulus. The summation in equation 2.23 goes over all of the measured reflections, and provides an overall estimate of the discrepancy among symmetry equivalent reflections; the second is over the subset of symmetry equivalent reflections (as defined by the current ith Laue group) to the unique reflection $\mathbf{h}$. w is a weight dependent upon measurement accuracy. The most probable Laue group is expected to be that with the smallest value of R_{int}.

For example, for the unit cell (2.22), the symmetry equivalent reflections for the Laue class $4/m$ are

$$(hkl), (\bar{h}\bar{k}l), (\bar{k}hl), (k\bar{h}l), (\bar{h}\bar{k}\bar{l}), (hk\bar{l}), (k\bar{h}\bar{l}), (\bar{k}h\bar{l})$$

They are expected to have the same diffraction intensity, and therefore $R_{int}(4/m)$ is expected to be quite small. If we obtain, experimentally, a small value of R_{int} (e.g. $R_{int}(4/m) = 0.04$), then we can accept the Laue group and, simultaneously, the tetragonal crystal system.

But, if the 16 reflections

$$(hkl), (\bar{h}\bar{k}l), (\bar{k}hl), (k\bar{h}l), (\bar{h}\bar{k}\bar{l}), (hk\bar{l}), (k\bar{h}\bar{l}), (\bar{k}h\bar{l}),$$

$$(\bar{h}k\bar{l}), (h\bar{k}\bar{l}), (kh\bar{l}), (\bar{k}\bar{h}\bar{l}), (h\bar{k}l), (\bar{h}kl), (\bar{k}\bar{h}l), (khl)$$

symmetry equivalent in $4/mmm$, have nearly the same intensity (and we therefore obtain for them $R_{int}(4/mmm) = 0.05$), then the Laue group $4/mmm$ may be preferred and the tetragonal system is accepted.

The above analysis may be performed automatically and the procedure may end with a probability $P_L(i)$, based on the corresponding R_{int} value, for each ith admitted Laue group.

A further step is needed for identification of the space group. Indeed, if the phases could be observed in the diffraction experiment, the space group symmetry information should be totally transferred in the reciprocal space. But, since the phases cannot be observed (at least for X-ray and neutron diffraction), the symmetry of the diffraction moduli does not completely define the space group symmetry of the crystal. *Diffraction symbols* were introduced by Buerger (1942) in order to specify which symmetries were able to be revealed from a diffraction pattern. These were subsequently thoroughly discussed by numerous authors, among whom are West (1954), Nowacki (1955), Donnay and Kennard (1964), and Buerger himself (Buerger, 1969). In the *International Tables for Crystallography*, vol. A (2005), a total of 122 *extinction symbols* are reported (see Table 2.2 for some examples), each formed by a short string indicating, successively, the lattice type, and, in the same order of the symmetry elements defining the Laue group, the symmetry elements. Symmetry directions without reflection conditions are represented by a dash; a symmetry direction with reflection condition is represented by the symbol of the glide plane or of the screw axis. Only 58 space groups and 11 enantiomorph pairs may be determined uniquely by examination of the diffraction pattern and of systematic absences.

Space group identification

Table 2.2 Some examples of extinction symbols and corresponding space groups. It may be observed that the same extinction symbol may be used for different crystal systems

In the orthorhombic system

P - - -	P222,	Pmmm,	Pm2m,	P2mm, Pmm2
P - - a	Pm2a,	P2$_1ma$,	Pmma	

In the tetragonal system

P - - -	P4,	P-4, P4/m,	P422,	P4mm,	P-42m,	P-4m2, P4/mmm

In the hexagonal system

P6$_1$ - -	P6$_1$,	P6$_5$,	P6$_1$22, P6$_5$22

To identify the extinction symbol a probabilistic approach may be followed: for single crystal data the corresponding algorithms are usually implemented within the diffractometer software. We will refer here to the algorithm proposed by Altomare et al. (2004, 2005, 2007) and by Camalli et al. (2012): for each ith admitted Laue group, the z statistics ($z = R^2$) may be used to define the probability $P_{EX}(i,j)$ for each associated jth extinction symbol ($P_{EX}(i,j)$ emphasizes the connection with the Laue group probability). Then,

$$P_{EX}(i,j) = P_L(i)P_{EX}(j) \qquad (2.24)$$

The most probable extinction symbol is expected to be associated with the largest value of $P_{EX}(i,j)$.

A simple example illustrates how $P_{EX}(j)$ may be obtained. Let us consider the orthorhombic system, where any space group may be represented by the general string

$$M\ r_1/s_1\ r_2/s_2\ r_3/s_3. \qquad (2.25)$$

M denotes the unit cell type (primitive or centred), $r_j, j = 1, \ldots, 3$ are the symmetry elements along the three axes and $s_j, j = 1, \ldots, 3$ are the symmetry elements perpendicular to the axes. To identify the correct extinction symbol the z-distributions along reciprocal axes and in reciprocal lattice planes perpendicular to direct axes should be checked. Such distributions provide the occurrence probability for screw axes and glide planes, and, equivalently, complementary probabilities for rotation axes and mirror planes.

Let us consider the extinction symbol Bb-b. Its probability P_{EX} (Bb-b) is defined as (see *International Tables for Crystallography*, 2005, p. 49, Table 3.1.4.1)

$P_{EX}(Bb\text{-}b) = p(B)\ p(2_{1[100]})\ p(b,c\perp\mathbf{a})\ p(2_{1[010]})\ p(n\perp\mathbf{b})\ p(2_{1[001]})\ p(a,b\perp\mathbf{c}).$

$p(B)$, the probability of the B-cell, is obtained by first calculating

$$p'(A) = 1 - \frac{<z_{hkl}>_{k+l=2n+1}}{<z_{hkl}>_{k+l=2n}}, \quad p'(B) = 1 - \frac{<z_{hkl}>_{h+l=2n+1}}{<z_{hkl}>_{h+l=2n}}, \quad p'(C) = 1 - \frac{<z_{hkl}>_{h+k=2n+1}}{<z_{hkl}>_{h+k=2n}},$$

from which the probability

$$p(B) = p'(B)\left[1 - p'(A)\right]\left[1 - p'(C)\right] \qquad (2.26)$$

Wilson statistics

Table 2.3 *PMRU52*. Statistical analysis for deriving the elementary probabilities for symmetry operators with translational component, according to the program *SIR2011* (Burla et al., 2012a). For each class of reflections containing potential systematic absent reflections (*Class*), the table shows: the number of measured reflections (NR_{cl}), the corresponding $<z>_{cl}$ value, and, for each subclass defined by *condition*, the corresponding number of reflections (NR_{scl}) and the average value $<z>_{scl}$ are calculated. In the last column the resulting symmetry operator is given

Class	NR_{cl}	$<z>_{cl}$	condition	NR_{scl}	$<z>_{scl}$	Sym.op.
h 0 0	18	0.366				
			$h \neq 2n$	7	0.01	2₁ _ _
			$h = 2n$	11	0.60	
			$h \neq 4n$	11	0.42	_____
			$h = 4n$	7	0.28	
0 k 0	33	0.506				
			$k \neq 2n$	16	0.01	_ 2₁ _
			$k = 2n$	17	0.98	
			$k \neq 4n$	23	0.16	_____
			$k = 4n$	10	1.30	
0 0 l	26	0.364				
			$l \neq 2n$	13	0.01	_ _ 2₁
			$l = 2n$	13	0.72	
			$l \neq 4n$	19	0.39	_____
			$l = 4n$	7	0.28	
0 k l	869	0.494				
			$k \neq 2n$	385	0.01	b _ _
			$k = 2n$	484	0.88	
			$l \neq 2n$	433	0.51	c _ _
			$l = 2n$	436	0.47	
			$k \neq 2n, l \neq 2n$	630	0.36	_____
			$k = 2n, l = 2n$	239	0.86	
			$k + l \neq 2n$	442	0.50	n _ _
			$k + l = 2n$	427	0.48	
			$k \neq 2n, l \neq 2n, k + l \neq 4n$	750	0.45	d _ _
			$k = 2n, l = 2n, k + l = 4n$	119	0.80	
h 0 l	494	0.530				
			$h \neq 2n$	239	0.56	_ a _
			$h = 2n$	255	0.50	
			$l \neq 2n$	218	0.01	_ c _
			$l = 2n$	276	0.94	
			$h \neq 2n, l \neq 2n$	349	0.39	_____
			$h = 2n, l = 2n$	145	0.87	
			$h + l \neq 2n$	241	0.56	_ n _
			$h + l = 2n$	253	0.50	
			$h \neq 2n, l \neq 2n, h + l \neq 4n$	418	0.47	_ d _
			$h = 2n, l = 2n, h + l = 4n$	76	0.88	
h k 0	625	0.526				
			$h \neq 2n$	273	0.01	_ _ a
			$h = 2n$	352	0.93	
			$k \neq 2n$	313	0.51	_ _ b
			$k = 2n$	312	0.54	
			$h \neq 2n, k \neq 2n$	447	0.36	_____
			$h = 2n, k = 2n$	178	0.94	
			$h + k \neq 2n$	308	0.52	_ _ n
			$h + k = 2n$	317	0.53	
			$h \neq 2n, k \neq 2n, h + k \neq 4n$	533	0.49	_ _ d
			$h = 2n, k = 2n, h + k = 4n$	92	0.71	

Space group identification

Table 2.3 (Continued)

Class	NR_{cl}	$<z>_{cl}$	condition	NR_{scl}	$<z>_{scl}$	Sym.op.
h k l	15395	0.969				
			$h+k \neq 2n$	7627	1.00	C lattice
			$h+k = 2n$	7768	0.94	
			$k+l \neq 2n$	7710	0.98	A lattice
			$k+l = 2n$	7685	0.96	
			$h+l \neq 2n$	7639	0.95	B lattice
			$h+l = 2n$	7756	0.98	
			$h+k+l \neq 2n$	7692	0.99	I lattice
			$h+k+l = 2n$	7703	0.95	
			$(h+k, k+l, h+l) \neq 2n$	11 488	0.98	F lattice
			$(h+k, k+l, h+l) = 2n$	3907	0.95	

is obtained. The multiplication on the right-hand side of equation (2.26) arises from the fact that in the orthorhombic system an F-cell is also allowed. Furthermore,

$$p(2_{1_{[100]}}) = 1 - \frac{<z_{h00}>_{h=2n+1}}{<z_{h00}>_{h=2n}}, \quad p(2_{1_{[010]}}) = 1 - \frac{<z_{0k0}>_{k=2n+1}}{<z_{0k0}>_{k=2n}}, \quad p(2_{1_{[001]}}) = 1 - \frac{<z_{00l}>_{l=2n+1}}{<z_{00l}>_{l=2n}}$$

and equivalently,

$$p(2_{[100]}) = 1 - p(2_{1_{[100]}}), \; p(2_{[010]}) = 1 - p(2_{1_{[010]}}), \; p(2_{[001]}) = 1 - p(2_{1_{[001]}}),$$

$$p(n \perp b) = 1 - \frac{<z_{h0l}>_{h+l=2n+1}}{<z_{h0l}>_{h+l=2n}}, \text{ etc.}$$

The composition of the above elementary probabilities provides the value of P_{EX}, and therefore, according to (2.24), of $P_{EX}(i,j)$.

To provide experimental evidence of the algorithm described above, we apply it to the diffraction data for *PMRU52*, C_{26} H_{48} Cl_4 N_8 O_5 Ru_2 S_8, space group *Pbca*. In Table 2.3 we show the statistical results obtained for each class of reflections with potential systematic absent reflections. The elementary probabilities described above may easily be calculated from the entries in the table.

For example, the elementary probabilities

$$p(2_1[001]) = 1 - \frac{<z_{00l}>_{l=2n+1}}{<z_{00l}>_{l=2n}},$$

$$p(2_{1_{[100]}}) = 1 - \frac{<z_{h00}>_{h=2n+1}}{<z_{h00}>_{h=2n}}, \quad p(2_{1_{[010]}}) = 1 - \frac{<z_{0k0}>_{k=2n+1}}{<z_{0k0}>_{k=2n}}, \quad p(2_{1_{[001]}}) = 1 - \frac{<z_{00l}>_{l=2n+1}}{<z_{00l}>_{l=2n}}$$

are very close to 1 for the three screw axes: indeed the $<z>_{scl}$ are very close to zero for the reflections conditioned by the presence of the screw axes.

As a result, *SIR2011* provides a list of the most probable extinction symbols: the first ten in the list, together with their probabilities, are shown in Table 2.4.

Table 2.4 *PMRU52*. $P_{EX}(i,j)$ for the 10 most probable extinction symbols, as estimated by *SIR2011*

Extinction symbol	$P_{EX}(i,j)$
P b c a	0.639
P b n a	0.065
P n c a	0.064
P b c n	0.064
P b _ a	0.007
P b a a	0.007
P _ c a	0.007
P c c a	0.007
P b c b	0.007
P b c _	0.007

2.7 The centric or acentric nature of crystals: Wilson statistical analysis

As stated in Section 2.6, a limited number of space groups may be uniquely determined by examination of the diffraction pattern and of systematic absences; it is not possible (on the basis of the above theory) to select a space group among those belonging to the same extinction symbol. In earlier times, when crystal structure solution for small molecules was time consuming, knowledge about which was the correct space group was a great help towards avoiding long phasing attempts in the wrong space groups. Thus, physical experiments were frequently used to reveal the acentric nature of the crystals, such as piro- and piezoelectric effects, optical activity, second harmonic effects, etc.

Nowadays, phasing is much more straightforward, particularly for small molecules and a few attempts in the wrong space groups that are compatible with the diffraction symbol are not generally considered to be a great waste of time and ancillary physical experiments are unusual. All permitted space groups are attempted in some preferred sequence. Still frequently used today, Wilson statistics (originally proposed by Foster and Hargreaves (1963a,b) and by Srinivasan and Subramanian (1964)) define the centric or acentric nature of the crystal. For example, the test allows us to distinguish between P1 and P-1, or between the centric P2/m and the pair of acentric space groups P2 and Pm. The simplest procedure involves calculation of the first moments of the experimental distribution of the normalized data, followed by comparison with theoretical moments calculated for the ideal centric and acentric distributions. In Table 2.5 the most useful moments of $P(R)$ are quoted. The reader may either derive these from appropriate integrations or by following the techniques defined in Appendix M.B.

Table 2.5 Statistical criteria (in terms of $|E|$) based on low-order moments to discriminate cs. from n.cs. distributions.: $R(|E|) = 1 - N(|E|)$ is the percentage of normalized structure factors with amplitude greater than the threshold $|E|$

Criterion	cs. distribution	n.cs. distribution
$\langle |E| \rangle$	$0.798 = (2/\pi)^{1/2}$	$0.886 = \pi^{1/2}/2$
$\langle |E|^2 \rangle$	1.000	1.000
$\langle |E|^3 \rangle$	$1.596 = 2^{3/2}/\pi^{1/2}$	$1.329 = 3\pi^{1/2}/4$
$\langle |E|^4 \rangle$	3.000	2.000
$\langle |E|^5 \rangle$	6.383	3.323
$\langle |E|^6 \rangle$	15.000	6.000
$\langle |E^2 - 1| \rangle$	0.968	0.736
$\langle (E^2 - 1)^2 \rangle$	2.000	1.000
$\langle (E^2 - 1)^3 \rangle$	8.000	2.000
$\langle |E^2 - 1|^3 \rangle$	8.691	2.415
$R(1)$	0.320	0.368
$R(2)$	0.050	0.018
$R(3)$	0.003	0.0001

It is apparent from Table 2.5 that for higher-order moments, differences between the corresponding values of the two distributions show a very marked increase. At first sight one might infer that, in order to identify the type of experimental distribution, the indications from the low-order moments are of little or no use. In practice, the effects of errors in experimental measurements, presence of pseudosymmetry, errors in the Wilson analysis, etc., become more critical with increasing order of moments, and therefore it is not safe to consider moments of order greater than four or five.

It is worth mentioning that all of the reflections should be used in the above statistical calculations; indeed systematic errors arise if (in order to spare data collection time) the weakest reflections remain unmeasured, and therefore out of the calculations (Rogers et al., 1955; Vicković and Viterbo, 1979; Cascarano et al., 1991).

Another practical problem concerns negative intensity measurements: what to do with them? They cannot be ignored or omitted from the dataset: usually they are set to zero (Hirshfeld and Rabinovich, 1973; Wilson, 1978). French and Wilson (1978) provided, by means of Bayesian statistics, a technique for obtaining an a posteriori positive estimation of the intensity from the negative intensity and its associated standard deviation.

2.8 Absolute scaling of intensities: the Wilson plot

It is well known that experimental intensities are on a relative scale. Their values depend on several factors: the direct beam intensity, radiation wavelength, film or counter efficiency, exposure time, etc. In this section we show how it is possible, from the experimental intensities I_h (already corrected for Lorentz polarization, absorption effect, etc.), to obtain $|F_\mathbf{h}|^2$ on an absolute scale:

$$I_\mathbf{h} = K|F_\mathbf{h}|^2, \quad (2.27)$$

where K is the required scale factor.

Following Wilson (1942) we will show that it is *also* possible to infer an approximate value for the average thermal factor. In this context it is necessary to assume that the thermal parameter is isotropic and constant for each atom, i.e. with obvious notation,

$$f_j = {}^0f_j \exp\left(-\frac{B \sin^2 \theta}{\lambda^2}\right). \quad (2.28)$$

In reality, this is not strictly true, since atoms belonging to the same atomic species may have quite different thermal displacements. Furthermore there are usually different atomic species present in the unit cell, chemical bonds are directional, ellipsoids rather than spheres would be more representative of thermal displacements.

Let us divide the reciprocal space into shells (spherical shells for three-dimensional data; see Fig. 2.5 for a two-dimensional lattice), each shell being sufficiently thin such that the variation of f_j in each is negligible, and let us take the average of $I_\mathbf{h}$ over the vth shell. If we assume, for the moment, that all

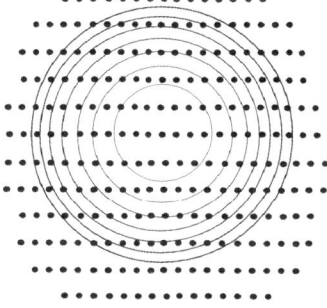

Fig. 2.5
Partition of a two-dimensional lattice into resolution shells, to approximately satisfy the following conditions: (a) the variation of f in each shell is negligible; (b) the number of lattice points in each shell is approximately constant.

of the reflections to be averaged are statistically homogeneous, with statistical weight p = 1, we obtain

$$<I> = K \sum_v.$$

The subscript v denotes that the summation

$$\sum_{j=1}^{N} f_j^2$$

is calculated by assigning to each f_j the value corresponding to the particular $\{\sin^2\theta/\lambda^2\}_v$ representative of the vth shell. According to (2.13) we can write

$$<I>_v = K \sum_{j=1}^{N} \left({}^0 f_j^2\right)_v \exp\left(-2B \left\{\frac{\sin^2\theta}{\lambda^2}\right\}_v\right),$$

which may be rewritten in logarithmic form as

$$\ln\left(\frac{<I>_v}{{}^0\sum_v}\right) = \ln K - 2B \left\{\frac{\sin^2\theta}{\lambda^2}\right\}_v,$$

where

$${}^0\sum_v = \sum_{j=1}^{N} \left({}^0 f_j^2\right)_v.$$

If we plot the values $\ln\left(<I>_v / {}^0\sum_v\right)$ versus the values $\{\sin^2\theta/\lambda^2\}_v$ we should obtain a straight line with equation

$$\ln\left(\frac{<I>_v}{{}^0\sum_v}\right) = \ln K - 2B \left\{\frac{\sin^2\theta}{\lambda^2}\right\}. \quad (2.29)$$

The intercept of (2.29) on the vertical axis yields the value of the scale factor K:

$$K = \lim_{\sin^2\theta/\lambda^2 \to 0} \left(\frac{<I>_v}{{}^0\sum_v}\right);$$

$2B$ is the angular coefficient of the straight line (2.29).

In practice, for various reasons (some of which will be examined in the next section), the experimental values are not rigorously aligned: the best evaluation of K and B is therefore obtained by deriving the least squares line.

Derivation of (2.29) is not so simple when, as is usually the case, we have coexistence of groups of reflections which are not statistically homogeneous, that is, groups of reflections having different statistical weights. It is possible to overcome this difficulty by introducing the *reduced intensity* $I_{r\mathbf{h}} = I_\mathbf{h}/p_\mathbf{h}$; in this way we may obtain the same mean value for each group. Equation 2.29 may then be rewritten as:

$$\ln\left(\frac{<I_r>_v}{{}^0\sum_v}\right) = \ln K - 2B \left\{\frac{\sin^2\theta}{\lambda^2}\right\}. \quad (2.30)$$

Absolute scaling of intensities: the Wilson plot

We will now consider in more detail the practical procedure usually employed by the Wilson method. This may be described schematically as follows.

Firstly, the reciprocal space is divided into n shells of approximately equal volume and including, on average, equal numbers of reflections. The number of reflections must be sufficient to be statistically representative but, at the same time, the variation of scattering factor with $\sin\theta/\lambda$ must be negligible. The limiting radii of the spherical shells usually satisfy the formula $(m/n)(\sin\theta_l/\lambda)^3$, where θ_l is the largest θ angle for the measured reflections, n is the number of intervals into which the observed reciprocal space is subdivided, and $m = 1, \ldots, n$.

Secondly, the $\{\sin\theta/\lambda\}_v$ value, representative of the vth shell, may be approximated in different ways (Rogers, 1965; Parthasarathi, 1975): i.e.

$$\frac{1}{2}(s_1 + s_2), \left[\frac{1}{2}(s_1^2 + s_2^2)\right]^{1/2},$$

where s_1 and s_2 are the extreme values of $\sin\theta/\lambda$ for the reflections within the shell.

Thirdly, within each spherical shell (particularly those characterized by high $\sin\theta/\lambda$ values), the number of observed reflections cannot be too small compared to the number expected for that shell. Shells with too few observed reflections must be left out of the statistical calculations, therefore θ_l has to be redefined. Control of the number of observed reflections is particularly important for organic compounds, which show a rapid decrease in intensities with $\sin\theta/\lambda$, owing to the decrease in atomic scattering factors and to the relatively high thermal vibration.

Fourthly, for each shell the average of the reduced intensities is calculated and the values of

$$\ln\left(\frac{<I_r>_v}{^0\Sigma_v}\right)$$

versus $\sin^2\theta/\lambda^2$ should be displayed graphically. The least-squares line will then give the best K and B values. Deviation of the experimental points from the straight line (those due to sampling effects) may be reduced by combining the quantities

$$\ln\left(\frac{<I_r>_v}{^0\Sigma_v}\right) \text{ and } \sin^2\theta/\lambda^2$$

for contiguous shells.

Finally, reflections with very low values of $\sin^2\theta/\lambda^2$ must be left out of the statistics because the of the anomalous distribution of the variable $\psi = f\cos(2\pi\mathbf{h}\cdot\mathbf{r})$. In fact, when the modulus of $\mathbf{h}$ is small, the cosine will take up essentially positive values, so that $<\psi>$ is no longer equal to zero even if $<\eta> = 0$.

As a numerical example, in Table 2.6, we describe the calculations necessary to obtain the Wilson plot of *LOGANIN* [$P2_12_12_1$, $C_{17}H_{26}O_{10}$, $Z = 4$], shown in Fig. 2.6. We observe the following:

1. Sixteen ranges of $\sin\theta/\lambda$ are defined: each range overlaps with its neighbours (columns 1 and 2 of the table) to make the plot less sensitive to sampling effects.

Table 2.6 *LOGANIN*: numerical values on which Fig. 2.6 is based. See the text for explanation of the symbols

Range	$(\sin\theta/\lambda)^2_v - (\sin\theta/\lambda)^2_{v+1}$	(NREFL)$_v$	$\{(\sin\theta/\lambda)^2\}_v$	$\langle I_r\rangle_v$	$^0\Sigma_v$	$\ln\left(\frac{\langle I_r\rangle_v}{^0\Sigma_v}\right)$
1	0.0000–0.0563	397	0.0342	2027.0	2595.4	−0.2472
2	0.0281–0.0844	605	0.0589	1109.1	1652.8	−0.3989
3	0.0563–0.1125	742	0.0858	634.0	1228.1	−0.6612
4	0.0844–0.1406	865	0.1140	336.2	896.2	−0.9805
5	0.1125–0.1688	977	0.1414	286.2	703.2	−0.8990
6	0.1406–0.1969	1068	0.1698	276.2	588.8	−0.7569
7	0.1688–0.2250	1150	0.1978	232.7	514.4	−0.7932
8	0.1969–0.2532	1224	0.2260	159.9	458.5	−1.0534
9	0.2250–0.2813	1301	0.2540	99.0	411.6	−1.4250
10	0.2532–0.3094	1368	0.2821	68.3	377.2	−1.7089
11	0.2813–0.3375	1395	0.3098	53.6	350.3	−1.8771
12	0.3094–0.3657	1443	0.3382	43.6	328.7	−2.0201
13	0.3375–0.3938	1499	0.3663	37.6	309.1	−2.1066
14	0.3657–0.4219	1434	0.3940	34.3	294.8	−2.1511
15	0.3938–0.4500	1085	0.4176	31.8	282.6	−2.1847
16	0.4219–0.4500	402	0.4334	33.0	276.1	−2.1243

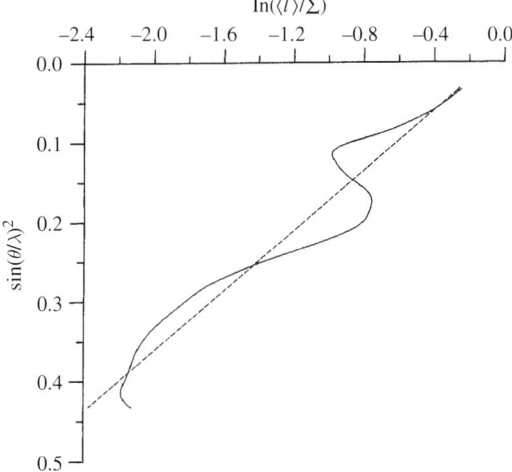

Fig. 2.6
Wilson plot for *LOGANIN*.

2. The number of reflections *NREFL* (symmetry dependent included) and the $(\sin\theta/\lambda)^2$ value representative of each v shell are shown in columns 3 and 4.

3. Both the mean intensity $<I_r>_v$ (column 5) and $^0\Sigma_v$ (column 6) decrease with $\sin\theta/\lambda$. $<I_r>_v$ decreases because of the combined effect of the intrinsic atomic scattering decay and the thermal motion, $^0\Sigma_v$ decreases solely as an effect of the atomic scattering decay. The difference between columns 5 and 6 provides information necessary to establish the average thermal motion.

4. The set of points $\{\ln(<I_r>_v/^0\Sigma_v), (\sin^2\theta/\lambda^2)_v\}$ are plotted in Fig. 2.6, together with the leastsquares straight line (2.30).

2.9 Shape of the Wilson plot

The Wilson approach is based on the assumption that the atomic positions are uniformly distributed over the unit cell: in this case, and in the absence of other sources of error, the Wilson plot should be close to a straight line. In real cases, Wilson plots are observed to deviate from a straight line. When some interatomic distances are very frequent because of structural regularities, or when a few heavy atoms dominate the diffraction pattern, the deviations will be stronger. In Figs. 2.7 and 2.8 we show the Wilson plots of two small molecules, in code *POCRO* [$B112/m$, $K_2 Se_{16} Cr_{10}$, $Z = 1$] and *HOV1* [$C2/m$, $Pr_{14} Ni_6 Si_{11}$, $Z = 4$], which contain heavy atoms and/or suffer through pseudo-translational symmetry: their plots show large deviations from linearity, much larger than with *LOGANIN*.

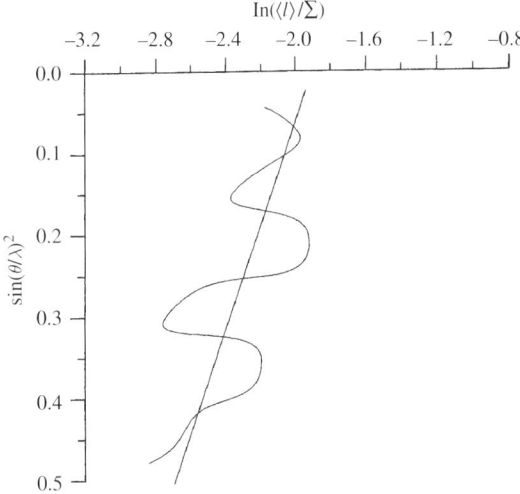

Fig. 2.7
Wilson plot for *POCRO*.

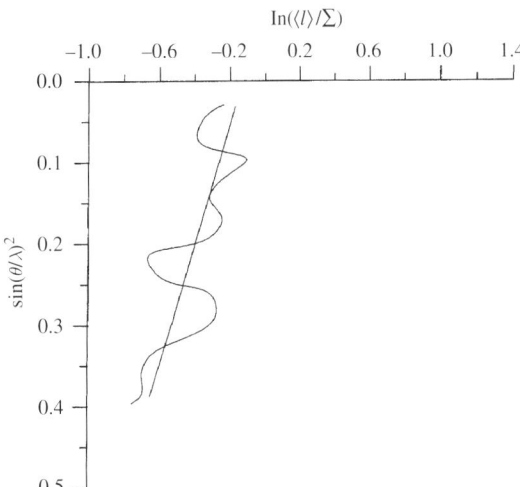

Fig. 2.8
Wilson plot of *HOV1*.

Cascarano et al. (1992b) observed that the mathematical basis for explaining deviations of the Wilson plot from a straight line, is the Debye formula, a tool widely used in the diffraction theory of amorphous solids and liquids (James, 1962; Guinier, 1963; Klug and Alexander, 1974; Magini et al., 1988). According to Debye the following relation holds (see Appendix 2.C):

$$<|F_\mathbf{h}|^2> = \sum_{j=1}^{N} f_j^2 + \sum_{j_1 \neq j_2 = 1}^{N} f_{j_1} f_{j_2} \frac{\sin(2\pi q |\mathbf{r}_{j_1} - \mathbf{r}_{j_2}|)}{2\pi q |\mathbf{r}_{j_1} - \mathbf{r}_{j_2}|}, \qquad (2.31)$$

where $q = |\mathbf{h}|$. The average is calculated by assuming $\mathbf{h}$ to be fixed, while the positional vectors are random variables under the condition that the moduli $|\mathbf{r}_{j_1} - \mathbf{r}_{j_2}|$ are fixed (see (3) in Section 2.2). Equation (2.31) clearly shows that $<|F_\mathbf{h}|^2>$ is an oscillating function whose maxima are in correspondence with the strongest interatomic distances. Conversely, the most frequent interatomic distances may be calculated by inverse Fourier transform of the Wilson plot.

A special study of protein Wilson plots has been made by Morris et al. (2004). They examined plots of 700 high-resolution proteins by taking data from the *Protein Data Bank*, using the *PDB* isotropic displacement factors, and by taking occupancy into account. Wilson plots are very similar to each other, in spite of the wide spread of secondary-structure characteristics: a typical curve is shown in Fig. 2.9 (solid line). All proteins show a local maximum at ∼1.1 Å, a minimum at ∼1.67 Å, a small local maximum at ∼2.1 Å, a strong local maximum at ∼4.5 Å, and a local minimum at 6.25 Å. The same authors performed a similar analysis of 200 *DNA/RNA*-only structures sampled from *PDB*. The curves are again similar to each other and a typical curve is shown in Fig. 2.9 (dashed line).

Hall and Subramanian (1982a,b) showed that the Wilson procedure is rather inefficient when data are truncated at values of $\sin\theta/\lambda$ where the Debye effects are large; in this case an error in the estimated overall temperature factor, and consequently in the scaling factor K, is obtained. In Fig. 2.10, two Wilson plots are shown for the small protein APP: the first (plot a) relates to data up to derivative resolution (∼2 Å) and the second (plot b) is calculated by using

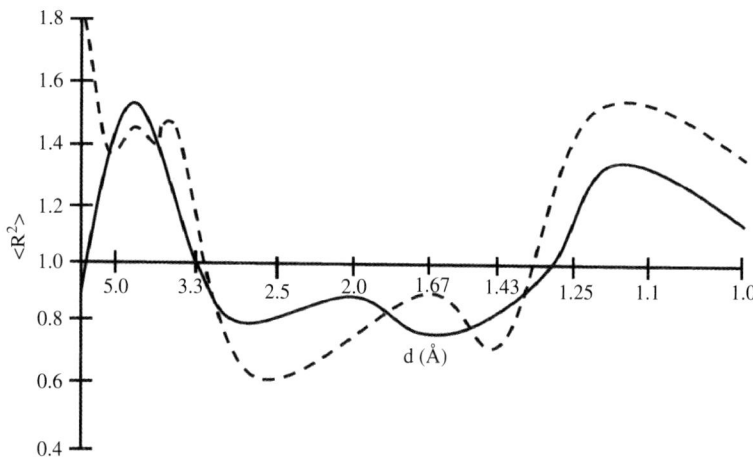

Fig. 2.9
Typical $<R^2>$ curve versus d(Å) for proteins (solid line). Typical $<R^2>$ curve versus d(Å) for nucleic acid structures (dashed line).

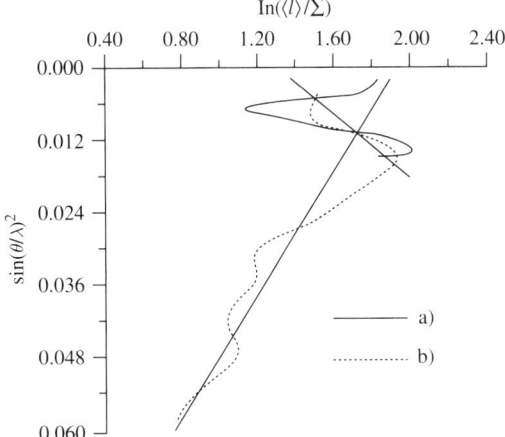

Fig. 2.10
APP: (a) Wilson plot for native diffraction data up to derivative resolution (∼2 Å); (b) Wilson plot for native diffraction data truncated at 4 Å resolution.

the same data, but truncated at 4 Å resolution. The quite different slopes of the least squares straight lines indicate strong differences between the estimated overall temperature factors B, and consequently, between the corresponding scale factors K. The reason for the misfit must be found in the Debye effects: as previously stated, the radial distribution of the diffracted intensities of proteins always shows a strong local maximum at 4.5 Å and a local minimum at 6.25 Å. These Debye effects are responsible for the wrong slope of the least squares straight line calculated at 4 Å resolution.

If we generalize the above example, it may be concluded that shortage of data (a frequent occurrence in macromolecular and occasionally in small molecule crystallography) may strongly influence the accuracy of a Wilson procedure. Since probabilistic prediction of structure factor moduli beyond the measurement limit is still an inaccurate process in the absence of a high-quality model, the most effective way to overcome the resolution problem is to optimize experimental conditions (e.g. by using low temperature apparatus, synchrotron radiation, etc.).

2.10 Unit cell content

The reader will certainly have noticed that it is necessary to know the *chemical content of the unit cell* in order to calculate the Wilson plot; indeed such content defines the parameters $^0\sum_v$ and $\sum_v$ at each Bragg angle. This chemical content is also useful in the next steps of the phasing process. Indeed it allows: (a) better estimation of the reliability of the structure invariants when direct methods are used for crystal structure determination; (b) labelling (in a chemical sense) of the electron density peaks and therefore easier interpretation of electron density maps; (c) convenience in performing crystallographic least squares, where a structural model is refined against the experimental diffraction data.

The chemical composition of a molecule is usually known before the corresponding crystal is submitted to a diffraction experiment. For *LOGANIN*, the example given in Section 2.8, the molecular chemical formula, $C_{17} H_{26} O_{10}$, was probably known before structure solution was undertaken: but the value of Z (i.e. the number of molecules in the unit cell) cannot be known before the unit cell is determined using a diffraction experiment. The Z value depends on how many molecules are in the asymmetric unit, and how many symmetry equivalent copies are in the unit cell, governed by the space group symmetry. Since we have learnt how to determine the space group symmetry, we now have to learn how to define how many molecules are to be expected in the asymmetric unit (sometimes there may be more than ten molecules present).

There is a simple algebraic tool for defining how many molecules are in the unit cell. It has been shown that, for small- to medium-sized molecules, the *volume per non-hydrogen atom* V_{at} should lie in the range 15–19 Å^3; for very compact structures, a value of 14 Å^3 may be reached. Let us apply this criterion to *LOGANIN*, space group $P2_1 2_1 2_1$. Its unit cell is defined by

$$a = 8.187 \text{ Å}, b = 14.277 \text{ Å}, c = 15.693 \text{ Å}, \text{ and volume } V = 1834.29 \text{ Å}^3.$$

There are 27 non-hydrogen atoms in the molecular formula. If we choose $Z = 4$ (i.e. one molecule in the asymmetric unit), then $V_{at} = 1834.29/(27 \times 4) = 16.98$ Å^3, in full agreement with expectations. Obviously, if we had chosen $Z = 8$ (two molecules in the asymmetric unit) then, $V_{at} = 1834.29/(27 \times 8) = 8.49$ Å^3, in strong disagreement with expectations.

Let us consider the *AZET* structure, with molecular chemical formula $C_{21} H_{16} N_1 O_1 Cl_1$, space group $Pca2_1$, $a = 36.042$ Å, $b = 8.730$ Å, $c = 11.084$ Å, $V = 3487.54$ Å^3. There are 24 non-hydrogen atoms in the molecular formula: if we choose $Z = 4$ (that is one molecule per asymmetric unit), then $V_{at} = 36.3$ Å^3, far from expectations. If we choose $Z = 8$ (two molecules in the asymmetric unit), then $V_{at} = 18.16$ Å^3, in full agreement with expectations.

A different method should be applied to proteins, because their unit cell contains a large solvent volume: the reader is referred to Appendix 8.A.

APPENDIX 2.A STATISTICAL CALCULATIONS IN P1 AND P$\bar{1}$

2.A.1 Structure factor statistics in P1

According to equation (1.19), the structure factor in P1 may be partitioned into a real and an imaginary component. Let us define

$$F = \sum_{j=1}^{N} f_j \exp(2\pi i \mathbf{h} \cdot \mathbf{r}_j) = A + iB,$$

and

$$f_j \exp(2\pi i \mathbf{h} \cdot \mathbf{r}_j) = \psi_j + i\eta_j,$$

with

$$\psi_j = f_j \cos\left(2\pi \mathbf{h} \cdot \mathbf{r}_j\right), \quad \eta_j = f_j \sin\left(2\pi \mathbf{h} \cdot \mathbf{r}_j\right).$$

Let us evaluate the means $\langle \psi_j \rangle$ and $\langle \eta_j \rangle$ when $\boldsymbol{h}$ is fixed and $\boldsymbol{r}_j$ uniformly varies in the unit cell. Since $2\pi \boldsymbol{h} \cdot \boldsymbol{r}_j$ is uniformly distributed over the trigonometric circle, ψ_j and η_j have equal chances of assuming negative or positive values, and therefore

$$\langle \psi_j \rangle = 0,$$
$$\langle \eta_j \rangle = 0.$$

The variances of ψ_j and η_j are, respectively,

$$\alpha_j^2 = \langle (\psi_j - \langle \psi_j \rangle)^2 \rangle = \langle \psi_j^2 \rangle = f_j^2 \langle \cos^2(2\pi \boldsymbol{h} \cdot \boldsymbol{r}_j) \rangle = \frac{1}{2} f_j^2,$$

$$\beta_j^2 = \langle (\eta_j - \langle \eta_j \rangle)^2 \rangle = \frac{1}{2} f_j^2.$$

Provided that N is large enough, from the *central limit* theorem (see Appendix M.A.6) the sum variables,

$$A = \sum_{j=1}^{N} \psi_j, \quad B = \sum_{j=1}^{N} \eta_j,$$

turn out have a normal distribution around the mean values,

$$\sum_{j=1}^{N} \langle \psi_j \rangle = 0, \quad \sum_{j=1}^{N} \langle \eta_j \rangle = 0,$$

and the α^2 and β^2 variances are

$$\alpha^2 = \sum_{j=1}^{N} \alpha_j^2 = \frac{1}{2} \sum_{j=1}^{N} f_j^2, \quad \beta^2 = \sum_{j=1}^{N} \beta_j^2 = \frac{1}{2} \sum_{j=1}^{N} f_j^2.$$

Introducing the symbol

$$\Sigma = \sum_{j=1}^{N} f_j^2,$$

we may write

$$\alpha^2 = \beta^2 = \frac{1}{2} \Sigma.$$

Therefore (see equation (M.A.9)), the probability of the real part of a structure factor ranging between A and $A + dA$ is

$$_1P(A)dA = \frac{1}{\sqrt{\pi \Sigma}} \exp\left(-\frac{A^2}{\Sigma}\right) dA,$$

where the prefixed subscript 1 indicates that the space group is P1. The probability that the imaginary part of F_h lies between B and $B + dB$ is

$$_1P(B)dB = \frac{1}{\sqrt{\pi \Sigma}} \exp\left(-\frac{B^2}{\Sigma}\right) dB,$$

Wilson statistics

To a first approximation, A and B may be assumed to be independent random variables; the joint probability that the real part of F_h is restricted between A and $A + dA$ and the complex part between B and $B + dB$ is then

$$_1P(A,B)\mathrm{d}A\mathrm{d}B = {}_1P(A){}_1P(B)\mathrm{d}A\mathrm{d}B = \frac{1}{\pi \Sigma} \exp\left(-\frac{A^2 + B^2}{\Sigma}\right) \mathrm{d}A\mathrm{d}B. \tag{2.A.1}$$

If the structure factor is expressed in polar coordinates according to

$$A = |F|\cos\phi, \quad B = |F|\sin\phi,$$

then the relation (M.A.22) should be applied. Since the Jacobian of the transformation is $J = |F|$, (2.A.1) becomes

$$_1P(|F|,\phi)\mathrm{d}|F|\mathrm{d}\phi = \frac{|F|}{\pi \Sigma} \exp\left(-\frac{|F|^2}{\Sigma}\right) \mathrm{d}|F|\mathrm{d}\phi. \tag{2.A.2}$$

which coincides with equation (2.2).

2.A.2 Structure factor statistics in $P\bar{1}$

In $P\bar{1}$, for each atom in $\mathbf{r}_j$ there is a symmetry equivalent atom in $-\mathbf{r}_j$. Therefore $F_\mathbf{h}$ may be considered as the sum of $N/2$ random variables,

$$F_\mathbf{h} = \sum_{j=1}^{N/2} 2f_j \cos(2\pi \mathbf{h} \cdot \mathbf{r}_j).$$

By analogy with the previously adopted notations, we define

$$\psi_j = 2f_j \cos(2\pi \mathbf{h} \cdot \mathbf{r}_j).$$

It is easily seen that

$$<\psi_j> = 0, \quad \alpha_j^2 = <(\psi_j - <\psi_j>)^2> = 2f_j^2$$

and hence,

$$\alpha^2 = \sum_{j=1}^{N/2} \alpha_j^2 = \sum_{j=1}^{N} f_j^2 = \Sigma.$$

On applying the central limit theorem once again, we deduce the relationship

$$_{\bar{1}}P(F) = \left(2\pi \Sigma\right)^{-1/2} \exp\left(-\frac{|F|^2}{2\Sigma}\right) \tag{2.A.3}$$

where the prefixed subscript -1 indicates that we are referring to the space group $P-1$. F has the same probability of being either positive or negative, therefore the probability for its modulus is

$$_{\bar{1}}P(|F|) = 2 \cdot {}_{\bar{1}}P(F) = \left(\frac{2}{\pi \Sigma}\right)^{1/2} \exp\left(-\frac{|F|^2}{2\Sigma}\right). \tag{2.A.4}$$

Equation (2.A.4) is known as the *centric distribution* of the structure factors, and coincides with distribution (2.5).

APPENDIX 2.B STATISTICAL CALCULATIONS IN ANY SPACE GROUP

2.B.1 The algebraic form of the structure factor

In this section, the algebraic form of the structure factor will be described in order to obtain, in Section 2.B.2, statistical formulas valid in all space groups. Let us see how the space group symmetry specifies the algebraic form of $F_\mathbf{h}$.

The symmetry elements in a space group can be represented (Zachariasen, 1945; MacGillavry, 1950) by means of suitable operators C_s. These operators will in general contain a (proper or improper) rotational component $\mathbf{R}_s$ and a translational component $\mathbf{T}_s$. It is possible to represent all positions $\mathbf{r}_{js}$ symmetrically equivalent to any initial position $\mathbf{r}_j$ by means of the operators C_s. Let m be the number of these operators; then

$$\mathbf{r}_{js} = \mathbf{R}_s \mathbf{r}_j + \mathbf{T}_s, \quad s = 1, 2, \ldots, m. \tag{2.B.1}$$

More explicitly,

$$\mathbf{r}_{js} = \begin{bmatrix} x_{js} \\ y_{js} \\ z_{js} \end{bmatrix} = \begin{bmatrix} R^s_{11} & R^s_{12} & R^s_{13} \\ R^s_{21} & R^s_{22} & R^s_{23} \\ R^s_{31} & R^s_{32} & R^s_{33} \end{bmatrix} \begin{bmatrix} x_j \\ y_j \\ z_j \end{bmatrix} + \begin{bmatrix} T^s_x \\ T^s_y \\ T^s_z \end{bmatrix}, \quad s = 1, 2, \ldots, m,$$

or, in a shortened form emphasizing coordinates,

$$\mathbf{X}_{js} = \mathbf{R}_s \mathbf{X}_j + \mathbf{T}_s.$$

m can assume values from 1 to 192. From now on the symbol $\mathbf{R}_1$ will indicate the identity matrix $\mathbf{I}$.

Thus, in space group P3$_1$, shown in Fig. 2.B.1, when the origin is chosen to lie on a threefold axis, the equivalent positions are given by

$$(x, y, z), \quad \left(-y, x-y, \frac{1}{3}+z\right), \quad \left(-x+y, -x, \frac{2}{3}+z\right)$$

and the corresponding symmetry operators are

$$\mathbf{R}_1 = \begin{bmatrix} 1 & 0 & 0 \\ 0 & 1 & 0 \\ 0 & 0 & 1 \end{bmatrix}, \quad \mathbf{T}_1 = \begin{bmatrix} 0 \\ 0 \\ 0 \end{bmatrix},$$

$$\mathbf{R}_2 = \begin{bmatrix} 0 & \bar{1} & 0 \\ 1 & \bar{1} & 0 \\ 0 & 0 & 1 \end{bmatrix}, \quad \mathbf{T}_2 = \begin{bmatrix} 0 \\ 0 \\ \frac{1}{3} \end{bmatrix},$$

$$\mathbf{R}_3 = \begin{bmatrix} \bar{1} & 1 & 0 \\ \bar{1} & 0 & 0 \\ 0 & 0 & 1 \end{bmatrix}, \quad \mathbf{T}_3 = \begin{bmatrix} 0 \\ 0 \\ \frac{2}{3} \end{bmatrix},$$

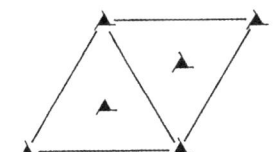

Fig. 2.B.1 P3$_1$ diagram.

From (1.19), the contribution to the structure factor of the jth atom and all its equivalents can be expressed as $\xi_j(\mathbf{h})$, where

$$\xi_j(\mathbf{h}) = f_j \sum_{s=1}^{m_j} \exp(2\pi i \bar{\mathbf{h}} \mathbf{C}_s \mathbf{r}_j) = f_j \sum_{s=1}^{m_j} \cos\left(2\pi \bar{\mathbf{h}} \mathbf{C}_s \mathbf{r}_j\right)$$
$$+ i f_j \sum_{s=1}^{m_j} \sin\left(2\pi \bar{\mathbf{h}} \mathbf{C}_s \mathbf{r}_j\right) = \psi_j(\mathbf{h}) + i \eta_j(\mathbf{h}). \quad (2.B.2)$$

The concise expression $\bar{\mathbf{h}} \mathbf{C} \mathbf{r}$ stands for

$$(h\ k\ l) \begin{bmatrix} R_{11} & R_{12} & R_{13} \\ R_{21} & R_{22} & R_{23} \\ R_{31} & R_{32} & R_{33} \end{bmatrix} \begin{bmatrix} x \\ y \\ z \end{bmatrix} + (h\ k\ l) \begin{bmatrix} T_x \\ T_y \\ T_z \end{bmatrix} = \bar{\mathbf{H}} \mathbf{R} \mathbf{X} + \bar{\mathbf{H}} \mathbf{T}.$$

The equivalent notation $\bar{\mathbf{H}} \mathbf{R} \mathbf{X} + \bar{\mathbf{H}} \mathbf{T}$ emphasizes the components of the vector $\mathbf{h}$ and the coordinates of $\mathbf{r}$.

The bar on top of a matrix symbol indicates the transpose, and m_j is the number of distinct equivalent positions of the jth atom. In a form more suitable for automatic computing (2.B.2) can be rewritten as

$$\xi_j(\mathbf{h}) = f_j \frac{m_j}{m} \sum_{s=1}^{m} \exp(2\pi i \bar{\mathbf{h}} \mathbf{C}_s \mathbf{r}_j), \quad (2.B.3)$$

where m_j/m is the *crystallographic occupancy factor* of the jth atom. Thus, with equation 2.B.3, it is possible to let s range from 1 to m, even when the jth atom is in a special position. When $m_j < m$ the jth atom is said to be in a *special position*.

$\xi_j(\mathbf{h})$ is the Fourier transform of the set of symmetry equivalent atoms located at $\mathbf{C}_s \mathbf{r}_j$, $s = 1, \ldots, m$, and we shall refer to it as the *trigonometric structure factor*.

When all the atoms are in a general position then $t = N/m$ is the number of symmetry independent atoms, and (1.19) becomes

$$F_\mathbf{h} = \sum_{j=1}^{t} \xi_j(\mathbf{h}) = \sum_{j=1}^{t} f_j(\mathbf{h}) \sum_{s=1}^{m} \exp\left[2\pi i \bar{\mathbf{h}} (\mathbf{R}_s \mathbf{r}_j + \mathbf{T}_s)\right]. \quad (2.B.4)$$

In the case of a centrosymmetric (cs.) space group, $\xi_j(\mathbf{h})$ can be expressed in the simplified form,

$$\xi_j(\mathbf{h}) = 2 f_j \sum_{s=1}^{m/2} \cos\left(2\pi \bar{\mathbf{h}} \mathbf{C}_s \mathbf{r}_j\right)$$

and the algebraic form of $F_\mathbf{h}$ will be

$$F_\mathbf{h} = 2 \sum_{j=1}^{t} f_j(\mathbf{h}) \sum_{s=1}^{m/2} \cos\left(2\pi \bar{\mathbf{h}} \mathbf{C}_s \mathbf{r}_j\right). \quad (2.B.5)$$

Statistical calculations in any space group

Let us consider a space group of order m with a primitive unit cell. If a reflection $\boldsymbol{h}$ has general indices (h, k, l), m distinct equivalent reflections $\bar{\boldsymbol{h}}\boldsymbol{R}$ can be found. For special values of (h, k, l) the number of distinct equivalent reflections may be m/ε_h, where ε_h is the order of the point group which leaves $\boldsymbol{h}$ unchanged. In this case,

$$\xi_j(\boldsymbol{h}) = \varepsilon_h f_j \sum_{s=1}^{m/\varepsilon_h} \exp(2\pi i \bar{\boldsymbol{h}} \boldsymbol{C}_s \boldsymbol{r}_j). \qquad (2.B.6)$$

For example, for all the space groups with point symmetry 222, $\varepsilon = 2$ for the reflections $(h00)$, $(0k0)$, and $(00l)$; for all the space groups with point symmetry mmm, $\varepsilon = 2$ for the reflections $(hk0)$, $(h0l)$, and $(0kl)$, and $\varepsilon = 4$ for the reflections $(h00)$, $(0k0)$, and $(00l)$ (see Table 2.1).

For space groups with a centred unit cell we denote by τ the centring order of the cell ($\tau = 2$ for a face-centred or body-centred cell; $\tau = 4$ for all-face-centred cells; $\tau = 3$ for a rhombohedral cell in hexagonal reference). Then, the m rotation matrices will coincide in groups of τ and therefore only m/τ matrices will be distinct. Accordingly,

$$F_{\boldsymbol{h}} = \sum_{j=1}^{t} \xi_j(\boldsymbol{h}) = \sum_{j=1}^{t} \psi_j(\boldsymbol{h}) + i \sum_{j=1}^{t} \eta_j(\boldsymbol{h}), \qquad (2.B.7)$$

where

$$\xi_j(\boldsymbol{h}) = f_j p_{\boldsymbol{h}} \sum_{s=1}^{m/p_{\boldsymbol{h}}} \exp(2\pi i \bar{\boldsymbol{h}} \boldsymbol{C}_s \boldsymbol{r}_j)$$

$$\psi_j(\boldsymbol{h}) = f_j p_{\boldsymbol{h}} \sum_{s=1}^{m/p_{\boldsymbol{h}}} \cos(2\pi \bar{\boldsymbol{h}} \boldsymbol{C}_s \boldsymbol{r}_j)$$

$$\eta_j(\boldsymbol{h}) = f_j p_{\boldsymbol{h}} \sum_{s=1}^{m/p_{\boldsymbol{h}}} \sin(2\pi \bar{\boldsymbol{h}} \boldsymbol{C}_s \boldsymbol{r}_j).$$

$p_{\boldsymbol{h}} = \tau \varepsilon_h$ has an algebraic origin; however, since it plays an important role in the statistics of the structure factor, it will be called the *statistical weight* of the reflection $\boldsymbol{h}$ or the *Wilson coefficient*. Equation 2.B.7 is the most general expression we will use for the trigonometric structure factor.

2.B.2 Structure factor statistics for centric and acentric space groups

In the presence of symmetry elements, the N atomic positions can no longer be considered as randomly distributed within the unit cell; indeed an m-fold symmetry element relates the coordinates of m atoms in the unit cell through precise operations. However, if expression (2.B.7) is used, we may consider as random variables only the positions of the t atoms in the asymmetric unit. If t is large enough to warrant a uniform distribution of the arguments of ξ over the trigonometric circle, the basic assumptions valid for space group P1 are also valid for all of the other space groups.

We first apply the central limit theorem to the acentric space groups with primitive unit cell.

(a) *Structure factor statistics for space groups with a primitive unit cell.*

We will show that different distributions $P(|F|)$ exist depending on the type of reflection. Let us first consider a reflection of the general type. According to equation (2.B.7),

$$\psi_j = f_j \sum_{s=1}^{m} \cos\left(2\pi \bar{\mathbf{h}} \mathbf{C}_s \mathbf{r}_j\right), \quad \eta_j = f_j \sum_{s=1}^{m} \sin\left(2\pi \bar{\mathbf{h}} \mathbf{C}_s \mathbf{r}_j\right).$$

If the $\mathbf{r}_j$s are uniformly distributed in the unit cell,

$$\langle \psi_j \rangle = \langle \eta_j \rangle = 0$$

$$\alpha_j^2 \equiv \langle \psi_j^2 \rangle = f_j^2 \sum_{s_1,s_2=1}^{m} \langle \cos\left(2\pi \bar{\mathbf{h}} \mathbf{C}_{s_1} \mathbf{r}_j\right) \cos\left(2\pi \bar{\mathbf{h}} \mathbf{C}_{s_2} \mathbf{r}_j\right) \rangle$$

$$= \frac{1}{2} f_j^2 \sum_{s_1,s_2=1}^{m} \left\{ \langle \cos\left[2\pi \bar{\mathbf{h}}(\mathbf{C}_{s_1} - \mathbf{C}_{s_2})\mathbf{r}_j\right] \rangle + \langle \cos\left[2\pi \bar{\mathbf{h}}(\mathbf{C}_{s_1} - \mathbf{C}_{s_2})\mathbf{r}_j\right] \rangle \right\}.$$

The terms for which $s_1 \neq s_2$ do not contribute to the average, while for each of the m terms for which $s_1 = s_2$, $\cos\left[2\pi \bar{\mathbf{h}}(\mathbf{C}_{s_1} - \mathbf{C}_{s_2})\mathbf{r}_j\right]$ is always equal to unity, no matter what is the value of $\mathbf{r}_j$. Thus,

$$\alpha_j^2 = m f_j^2 / 2.$$

Similarly,

$$\beta_j^2 = \langle \eta_j^2 \rangle = f_j^2 \sum_{s_1,s_2=1}^{m} \langle \sin\left(2\pi \bar{\mathbf{h}} \mathbf{C}_{s_1} \mathbf{r}_j\right) \sin\left(2\pi \bar{\mathbf{h}} \mathbf{C}_{s_2} \mathbf{r}_j\right) \rangle = m f_j^2 / 2.$$

Accordingly,

$$\alpha^2 = \sum_{j=1}^{t} \alpha_j^2 = \tfrac{1}{2} N f_j^2 = \tfrac{1}{2} \Sigma$$

$$\beta^2 = \tfrac{1}{2} \Sigma,$$

as in $P1$. In conclusion, for a reflection of the general type, the joint probability densities $P(A, B)$ and $P(|F|)$ are identical to (2.A.1) and (2.4) respectively.

If $F_\mathbf{h}$ is a systematically absent reflection, then $\xi_j(\mathbf{h}) = 0$ and therefore

$$P(F_\mathbf{h}) = \delta(F_\mathbf{h}),$$

where δ is the Dirac delta function.

Statistical calculations in any space group

Let us now consider a systematically non-absent reflection with statistical weight equal to ε (i.e. m/ε is the number of distinct symmetry equivalent reflections). Then, according to (2.B.6),

$$\xi_j(\boldsymbol{h}) = \varepsilon f_j \sum_{s=1}^{m/\varepsilon} \exp(2\pi i \boldsymbol{\bar{h}} \boldsymbol{C}_s \boldsymbol{r}_j). \tag{2.B.8}$$

By applying the central limit theorem, as in the preceding cases, we obtain

$$\psi_j = f_j \varepsilon \sum_{s=1}^{m/\varepsilon} \cos\left(2\pi \boldsymbol{\bar{h}} \boldsymbol{C}_s \boldsymbol{r}_j\right),$$

$$\eta_j = f_j \varepsilon \sum_{s=1}^{m/\varepsilon} \sin\left(2\pi \boldsymbol{\bar{h}} \boldsymbol{C}_s \boldsymbol{r}_j\right),$$

$$\langle \psi_j \rangle = \langle \eta_j \rangle = 0,$$

$$\alpha_j^2 = \langle \psi_j^2 \rangle = \varepsilon m f_j^2 / 2, \quad \beta_j^2 = \langle \eta_j^2 \rangle = \varepsilon m f_j^2 / 2,$$

$$\alpha^2 = \varepsilon \sum /2, \quad w\beta^2 = \varepsilon \sum /2,$$

$$_1P(A)\mathrm{d}A = \frac{1}{\sqrt{\pi \varepsilon \sum}} \exp - \left(\frac{A^2}{\varepsilon \sum}\right) \mathrm{d}A,$$

$$_1P(B)\mathrm{d}B = \frac{1}{\sqrt{\pi \varepsilon \sum}} \exp - \left(\frac{B^2}{\varepsilon \sum}\right) \mathrm{d}B,$$

$$_1P(|F|) \simeq \frac{2|F|}{\varepsilon \sum} \exp\left(-\frac{|F|^2}{\varepsilon \sum}\right). \tag{2.B.9}$$

For reflections with $\varepsilon > 1$, (2.B.9) is seen to be different from (2.5). In particular (2.5) may be considered to be a special case of (2.B.9), obtainable when $\varepsilon = 1$.

The above approach can be also applied to cs, space groups, for reflections with $\varepsilon > 1$. One obtains

$$_{\bar{1}}P(|F|) = \left(\frac{2}{\pi \varepsilon \sum}\right)^{1/2} \exp\left(-\frac{|F|^2}{2\varepsilon \sum}\right). \tag{2.B.10}$$

(b) *Structure factor statistics for space groups with a centred unit cell*

According to (2.B.6) in an n.cs. space group with a centred cell,

$$\xi_j(\boldsymbol{h}) = p f_j \sum_{s=1}^{m/p} \exp(2\pi i \boldsymbol{\bar{h}} \boldsymbol{C}_s \boldsymbol{r}_j),$$

Wilson statistics

where $p = \tau\varepsilon$ is the statistical weight or Wilson coefficient. If we denote

$$\psi_j = f_j p \sum_{s=1}^{m/p} \cos\left(2\pi \bar{\mathbf{h}} \mathbf{C}_s \mathbf{r}_j\right), \quad \eta_j = f_j p \sum_{s=1}^{m/p} \sin\left(2\pi \bar{\mathbf{h}} \mathbf{C}_s \mathbf{r}_j\right),$$

then

$$\langle \psi_j \rangle = \langle \eta_j \rangle = 0, \quad \alpha^2 = \beta^2 = p\Sigma/2,$$

$$_1P(A)dA = \frac{1}{\sqrt{\pi p \Sigma}} \exp-\left(\frac{A^2}{p\Sigma}\right),$$

$$_1P(B)dB = \frac{1}{\sqrt{\pi p \Sigma}} \exp-\left(\frac{B^2}{p\Sigma}\right),$$

$$_1P(|F|) \simeq \int_0^{2\pi} P(|F|, \phi) d\phi = \frac{2|F|}{p\Sigma} \exp\left(-\frac{|F|^2}{p\Sigma}\right). \qquad (2.B.11)$$

Equation (2.B.11) is the most general distribution for an n.cs. space group. For $\tau = 1$, it reduces to (2.B.9) and for $t = 1$ and $\varepsilon = 1$, it reduces to (2.4).

Similar considerations for a cs. space group will lead to the distribution

$$_{\bar{1}}P(|F|) = \left(\frac{2}{\pi p \Sigma}\right)^{1/2} \exp\left(-\frac{|F|^2}{2p\Sigma}\right). \qquad (2.B.12)$$

This reduces to (2.5) for $\tau = 1$ and to (1.25) for $t = 1$ and $\varepsilon = 1$.

APPENDIX 2.C   THE DEBYE FORMULA

We want to calculate the expected value

$$<|F_\mathbf{h}|^2> = <\sum_{j_1,j_2=1}^N f_{j_1} f_{j_2} \exp\left[2\pi i \mathbf{h} \cdot (\mathbf{r}_{j_1} - \mathbf{r}_{j_2})\right]>$$

$$= \sum_{j=1}^N f_j^2 + <\sum_{j_1 \neq j_2=1}^N f_{j_1} f_{j_2} \exp\left[2\pi i \mathbf{h} \cdot (\mathbf{r}_{j_1} - \mathbf{r}_{j_2})\right]>,$$

when $\mathbf{h}$ is fixed and the atomic positions are random variables under the condition that the moduli $r_{j_1 j_2} = |\mathbf{r}_{j_1} - \mathbf{r}_{j_2}|$ are fixed. To this purpose we introduce the polar coordinates (see Fig. 2.C.1) defined by

$$x_{j_1 j_2} = r_{j_1 j_2} \sin\phi \cos\theta$$
$$y_{j_1 j_2} = r_{j_1 j_2} \sin\phi \sin\theta$$
$$z_{j_1 j_2} = r_{j_1 j_2} \cos\phi$$

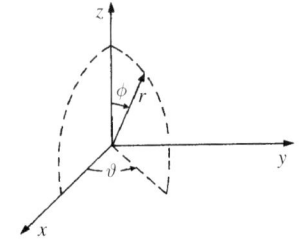

Fig. 2.C.1
Polar and Cartesian coordinates.

If we choose the z-axis along the $\mathbf{h}$ direction, then $\mathbf{h} \cdot (\mathbf{r}_{j_1} - \mathbf{r}_{j_2})$ is equal to $qr_{j_1 j_2} \cos\phi$, where $q = |\mathbf{h}|$ and

$$<\exp[2\pi i \mathbf{h} \cdot (\mathbf{r}_{j_1} - \mathbf{r}_{j_2})]> = \frac{1}{A} \int_0^\pi \int_0^{2\pi} \exp(2\pi i q r_{j_1 j_2} \cos\phi) \, r_{j_1 j_2}^2 \sin\phi \, d\phi \, d\theta.$$

$r_{j_1j_2}^2 \sin\phi$ is the Jacobian of the transformation from Cartesian to polar axes, and $A = 4\pi r_{j_1j_2}^2$ is the area of the sphere of radius $r_{j_1j_2}$. Simple calculations bring us to

$$<\exp\left[2\pi i\mathbf{h}\cdot(\mathbf{r}_{j_1}-\mathbf{r}_{j_2})\right]> \;=\; \frac{1}{2}\int_0^\pi \exp(2\pi iqr_{j_1j_2}\cos\phi)\sin\phi\,d\phi$$

$$= \frac{1}{4\pi qir_{j_1j_2}}\exp(2\pi iqr_{j_1j_2}\cos\phi)\Big]_\pi^0 = \frac{\sin(2\pi qr_{j_1j_2})}{2\pi qr_{j_1j_2}},$$

from which,

$$<|F_\mathbf{h}|^2> = \sum_{j=1}^N f_j^2 + \sum_{j_1\neq j_2=1}^N f_{j_1}f_{j_2}\frac{\sin(2\pi qr_{j_1j_2})}{2\pi qr_{j_1j_2}} \qquad (2.C.1)$$

The second term in (2.C.1) modulates the average value of $|F_\mathbf{h}|^2$ and is responsible for the fluctuations (called Debye effects) in the Wilson plot.

3 The origin problem, invariants, and seminvariants

3.1 Introduction

In Section 1.6, the *basic postulate of structural crystallography* was formulated, according to which there is a biunique correspondence between the crystal structure and the set of experimental diffraction data:

$$\text{crystal structure} \Leftrightarrow \{|F_\mathbf{h}|\}.$$

Can this property be extended to phases? In more simple terms, is the following logical relation valid?

$$\text{crystal structure} \Leftrightarrow \{\phi_\mathbf{h}\}\ ? \tag{3.1}$$

If relation (3.1) is valid, only one set of phases should be compatible with a given crystal structure. We will show in Section 3.2 that relation (3.1) is false; there are more sets of phases that are compatible with the same crystal structure. It will also be shown that such behaviour arises due to *the origin problem*. The above conclusion suggests that there is no sense in declaring: *the phase of the reflection $\mathbf{h}$ is 35°*, without first specifying the origin with respect to which the phase is estimated.

This conclusion presents a new basic question: since the origin is arbitrarily chosen by the crystallographer, how can we hope to directly derive phases from experimental amplitudes? This seems to be a contradiction in logic; indeed, the experimental amplitudes, which are invariant with respect to origin translations, cannot define the phases, since their values depend on the arbitrary origin chosen by the crystallographer. This contradiction may be overcome since combinations of phases exist whose values do not depend on the origin, but only on the structure. They may therefore be estimated from the diffraction amplitudes and are called *structure invariants* (shortened to *s.i.*; see Section 3.3).

The reader should not be surprised that invariants are more than just necessary to our treatment of the phase problem. Indeed they will guide our approach, providing a general and simplified description of the properties of phase space. This situation is similar to that met in the study of mathematics, where we learn that the geometrical properties of curves may be more easily studied via invariant properties. The only difference here is that we are dealing with phases and with the invariant properties of phase space.

The presence of symmetry in the crystal space allows us to describe another concept related to invariance, the *seminvariance* concept. The algebraic form of the symmetry operators depends on the class of origin chosen to describe the structure (e.g. in $P\bar{1}$, the translational components of the symmetry operators vanish if an inversion centre is chosen as origin). Usually one accepts the class of origin suggested by the *International Tables for Crystallography*, thus, simultaneously fixing the form of the symmetry operators; however, different choices can be made. Once the symmetry operators have been fixed, a class of *allowed origins* (see Section 3.4) is simultaneously selected. For example, if in $P\bar{1}$ we choose for operator $\bar{1}$ the relation $(x, y, z) \rightarrow (\bar{x}, \bar{y}, \bar{z})$, then the fixed class of origins coincides with the eight inversion centres. Analogously, if in P2 we choose the symmetry relation $(x, y, z) \rightarrow (\bar{x}, y, \bar{z})$, then the origin is chosen on the binary axes. We will see that phases or combinations of phases exist, the values of which do not depend on the specific origin selected inside the permissible class (in the above examples, phases do not change if the origin is moved from one inversion centre to another or from one binary axis to the other); they are called *structure seminvariants* (shortened to *s.s.*; they are characterized in Sections 3.5 and 3.6) and may be estimated from the diffraction moduli provided the algebraic form of the symmetry operators has been fixed.

A third question will also be answered in this chapter: is it possible to fix the origin by fixing some subset of phases? Assigning coordinates to atoms is only possible if the origin has been specified. Conversely, the phases of the structure factors define, through the Fourier transform, the atomic positions with respect to a given origin. Thus a mechanism must exist by which phasing the reflections simultaneously fixes the origin in direct space. This intriguing correspondence will be elucidated in Section 3.7 by making use of the concepts of *s.i.* and *s.s.*

3.2 Origin, phases, and symmetry operators

Let O be the origin of our reference system, and $\mathbf{r}_j$ the atomic positional vector of the jth atom; in Fig. 3.1, P_j marks its location. Then,

$$F_\mathbf{h} = \sum_{j=1}^{N} f_j \exp(2\pi i \mathbf{h} \cdot \mathbf{r}_j)$$

is the structure factor. Let us now translate the origin by the vector $\mathbf{x}_0$ (the new reference axes remain parallel to the original ones). Then the new origin moves to O', the atomic positional vector of the jth atom is $\mathbf{r}'_j = \mathbf{r}_j - \mathbf{x}_0$, and the structure factor becomes

$$F'_\mathbf{h} = \sum_{j=1}^{N} f_j \exp(2\pi i \mathbf{h} \cdot \mathbf{r}'_j) = \sum_{j=1}^{N} f_j \exp[2\pi i \mathbf{h} \cdot (\mathbf{r}_j - \mathbf{x}_0)] = F_\mathbf{h} \exp(-2\pi i \mathbf{h} \cdot \mathbf{x}_0). \tag{3.2}$$

From (3.2) we deduce that: *the modulus of any structure factor does not change with a shift in origin, while the phase value changes according to*

$$\phi'_\mathbf{h} = \phi_\mathbf{h} - 2\pi \mathbf{h} \cdot \mathbf{x}_0. \tag{3.3}$$

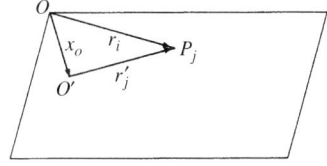

Fig. 3.1
$\mathbf{r}_j$ transforms into $\mathbf{r}'_j$ after a change of origin from O to O'.

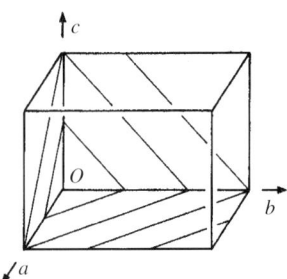

Fig. 3.2
Planes (232) constitute an equiphase surface for F_{232}.

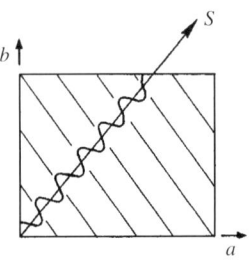

Fig. 3.3
Equiphase surface for F_{530}.

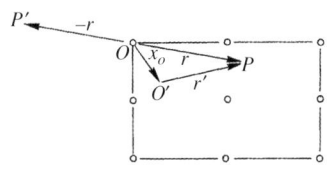

Fig. 3.4
Change of origin in $P\bar{1}$.

Relation (3.3) suggests that the origin shift produces a phase shift that is proportional (modulo 2π) to the scalar product $\mathbf{h} \cdot \mathbf{x}_0$. The phase shift vanishes if

$$\mathbf{h} \cdot \mathbf{x}_0 = n, \qquad (3.4)$$

with n an integer value. Since (3.4) is the classical equation defining the set of lattice planes with Miller indices $\mathbf{h} \equiv (h, k, l)$, the conclusion is: if the origin is moved from O to any point O' lying on the lattice planes $\mathbf{h}$, then $F_\mathbf{h}$ does not change its phase value. In a concise way, we say that *the lattice planes $\mathbf{h}$ are an equiphasic surface for the reflection $\mathbf{h}$* (see Fig. 3.2).

Figure 3.3 shows the equiphasic surface for the reciprocal vector $\mathbf{h} = (530)$. This also illustrates that the phase variation (for shifts $\mathbf{x}_0$ normal to the lattice planes) must be faster for higher reflection indices; this result can also be derived from relation (3.3). If the origin is moved along any vector S not lying on the lattice planes (h, k, l), from one equiphasic plane to those adjacent, then the phase of $F_\mathbf{h}$ will assume all possible values in the range $(-\pi, \pi)$.

Let us now investigate how a change in origin modifies the matrix representation of the symmetry operators for a space group with point group order equal to m. In their daily work, crystallographers use the origins tabulated in the *International Tables for Crystallography*; e.g. in $P\bar{1}$ the origin coincides with an inversion centre, in $P2$ with a binary axis, etc. If, for some reason, they are obliged to move the origin, they should know how the symmetry operators change due to the origin shift. For a primitive unit cell with origin at O, the symmetry operators $\mathbf{C}_s$ are defined by the relationship

$$\mathbf{r}_{js} = \mathbf{C}_s \mathbf{r}_j = \mathbf{R}_s \mathbf{r}_j + \mathbf{T}_s, s = 1, \ldots, m. \qquad (3.5)$$

If $\mathbf{x}_0$ is the origin translation vector, in the new reference system, symmetry equivalent points will be related by the relationship

$$\mathbf{r}'_{js} = \mathbf{C}'_s \mathbf{r}'_j = \mathbf{R}'_s \mathbf{r}'_j + \mathbf{T}'_s, s = 1, \ldots, m. \qquad (3.6)$$

In order to derive the relationship between each $\mathbf{C}_s$ and its corresponding $\mathbf{C}'_s$ we simply substitute into (3.6) the values

$$\mathbf{r}'_{js} = \mathbf{r}_{js} - \mathbf{x}_0 \quad \text{and} \quad \mathbf{r}'_j = \mathbf{r}_j - \mathbf{x}_0,$$

obtaining

$$\mathbf{r}_{js} - \mathbf{x}_0 = \mathbf{R}'_s \mathbf{r}_j - \mathbf{R}'_s \mathbf{x}_0 + \mathbf{T}'_s,$$

or

$$\mathbf{r}_{js} = \mathbf{R}'_s \mathbf{r}_j - (\mathbf{R}'_s - \mathbf{I})\mathbf{x}_0 + \mathbf{T}'_s, s = 1, \ldots, m. \qquad (3.7)$$

Since (3.7) and (3.5) must be identical whatever the value of $\mathbf{r}_j$, it follows that $\mathbf{R}'_s = \mathbf{R}_s$. Thus, *a change of origin does not affect the rotation matrices but only the translational components of the symmetry operators, and the translation matrix changes according to*

$$\mathbf{T}'_s = \mathbf{T}_s + (\mathbf{R}_s - \mathbf{I})\mathbf{x}_0, s = 1, \ldots, m. \qquad (3.8)$$

For instance, in the space group $P\bar{1}$, if we choose a new origin at a distance $\mathbf{x}_0$ from a centre of symmetry (see Fig. 3.4) then a point P, defined by the positional vector $\mathbf{r}' = \mathbf{r} - \mathbf{x}_0$, will correspond to an equivalent point P' at $-\mathbf{r} - \mathbf{x}_0 \equiv -(\mathbf{r}' + 2\mathbf{x}_0)$. Since $\mathbf{R}'_s = \mathbf{R}_s$, from (3.8) the new symmetry operators arise:

$$\mathbf{R}'_1 = \begin{bmatrix} 1 & 0 & 0 \\ 0 & 1 & 0 \\ 0 & 0 & 1 \end{bmatrix}, \quad \mathbf{T}'_1 = \begin{bmatrix} 0 \\ 0 \\ 0 \end{bmatrix}$$

$$\mathbf{R}'_2 = \begin{bmatrix} \bar{1} & 0 & 0 \\ 0 & \bar{1} & 0 \\ 0 & 0 & \bar{1} \end{bmatrix}, \quad \mathbf{T}'_2 = \begin{bmatrix} -2x_0 \\ -2y_0 \\ -2z_0 \end{bmatrix},$$

provided that $\mathbf{x}_0 \equiv (x_o, y_o, z_o)$.

3.3 The concept of structure invariant

In Section 3.2 we showed that the phase of the reflection $\mathbf{h}$ changes if the origin is shifted; consequently, $\phi_\mathbf{h}$ cannot be directly determined from the experimental data. The opposite statement (i.e. $\phi_\mathbf{h}$ may be determined from the data) should be illogical; indeed the amplitudes are fixed by the structure, the phases by our arbitrary choice of the origin.

How can we determine phases from experimental data? The only way is to check if some products of structure factors can be identified which remain invariant whatever the origin translation. In this case the values of such combination are origin independent and therefore depend on the structure. Obviously we have to consider products of structure factors which contain phase information. Let us consider the product

$$F_{\mathbf{h}_1} F_{\mathbf{h}_2} \cdot \ldots \cdot F_{\mathbf{h}_n} = |F_{\mathbf{h}_1} F_{\mathbf{h}_2} \cdot \ldots \cdot F_{\mathbf{h}_n}| \exp[i(\phi_{\mathbf{h}_1} + \phi_{\mathbf{h}_2} + \cdots + \phi_{\mathbf{h}_n})]. \tag{3.9}$$

According to (3.2) an origin translation will modify (3.9) into

$$F'_{\mathbf{h}_1} F'_{\mathbf{h}_2} \cdot \ldots \cdot F'_{\mathbf{h}_n} = F_{\mathbf{h}_1} F_{\mathbf{h}_2} \cdot \ldots \cdot F_{\mathbf{h}_n} \exp[-2\pi i(\mathbf{h}_1 + \mathbf{h}_2 + \cdots + \mathbf{h}_n) \cdot \mathbf{x}_0]. \tag{3.10}$$

Relation (3.10) suggests that *the product of structure factors (3.9) is invariant under origin translation if*

$$\mathbf{h}_1 + \mathbf{h}_2 + \cdots + \mathbf{h}_n = \mathbf{0}. \tag{3.11}$$

Products of structure factors which satisfy (3.11) are called structure invariants (s.i.), since their values do not depend on the origin, and therefore depend only on the structure (Hauptman and Karle, 1953).

The simplest examples of s.i. are:

1. For $n = 1$, relation (3.11) confirms F_{000} as the simplest structure invariant (it is equal to the number of electrons in the unit cell).
2. For $n = 2$, relation (3.11) reduces to $\mathbf{h}_1 + \mathbf{h}_2 = \mathbf{0}$ or, in other notation, $\mathbf{h}_2 = -\mathbf{h}_1$. Accordingly, the product $F_\mathbf{h} F_{-\mathbf{h}} = |F_\mathbf{h}|^2$ is a structure invariant (which agrees well with the obvious expectation that an observation does not depend on the origin we choose).

3. For $n = 3$, relation (3.11) reduces to $\mathbf{h}_1 + \mathbf{h}_2 + \mathbf{h}_3 = \mathbf{0}$. Accordingly,

$$F_{\mathbf{h}_1} F_{\mathbf{h}_2} F_{-(\mathbf{h}_1+\mathbf{h}_2)} = |F_{\mathbf{h}_1} F_{\mathbf{h}_2} F_{-(\mathbf{h}_1+\mathbf{h}_2)}| \exp\left[i(\phi_{\mathbf{h}_1} + \phi_{\mathbf{h}_2} - \phi_{\mathbf{h}_1+\mathbf{h}_2})\right] \quad (3.12)$$

is a *s.i.*, specifically called *triplet invariant*.

4. For $n = 4$, relation (3.11) defines the *quartet invariant*,

$$F_{\mathbf{h}_1} F_{\mathbf{h}_2} F_{\mathbf{h}_3} F_{-(\mathbf{h}_1+\mathbf{h}_2+\mathbf{h}_3)} = |F_{\mathbf{h}_1} F_{\mathbf{h}_2} F_{\mathbf{h}_3} F_{-(\mathbf{h}_1+\mathbf{h}_2+\mathbf{h}_3)}|$$
$$\times \exp\left[i(\phi_{\mathbf{h}_1} + \phi_{\mathbf{h}_2} + \phi_{\mathbf{h}_3} - \phi_{\mathbf{h}_1+\mathbf{h}_2+\mathbf{h}_3})\right].$$

Quintet, sextet, etc. s.i.s are defined by analogy.

Frequently the terms triplet, quartet, quintet invariant are referred to as:

(a) a product of normalized structure factors like $E_{\mathbf{h}_1} E_{\mathbf{h}_2} E_{-(\mathbf{h}_1+\mathbf{h}_2)}$, $E_{\mathbf{h}_1} E_{\mathbf{h}_2} E_{\mathbf{h}_3} E_{-(\mathbf{h}_1+\mathbf{h}_2+\mathbf{h}_3)}$, etc.;

(b) the sum of phases rather than to the product of structure factors. For example, we will refer to $(\phi_{\mathbf{h}_1} + \phi_{\mathbf{h}_2} - \phi_{\mathbf{h}_1+\mathbf{h}_2})$ as a triplet invariant, to $(\phi_{\mathbf{h}_1} + \phi_{\mathbf{h}_2} + \phi_{\mathbf{h}_3} - \phi_{\mathbf{h}_1+\mathbf{h}_2+\mathbf{h}_3})$ as a quartet invariant, and so on. In equivalent notation, we can also write triplet invariants as $(\phi_\mathbf{h} + \phi_\mathbf{k} - \phi_{\mathbf{h}+\mathbf{k}})$ or $(\phi_\mathbf{h} - \phi_\mathbf{k} - \phi_{\mathbf{h}-\mathbf{k}})$, and quartet invariants as $(\phi_\mathbf{h} + \phi_\mathbf{k} + \phi_\mathbf{l} - \phi_{\mathbf{h}+\mathbf{k}+\mathbf{l}})$ or $(\phi_\mathbf{h} - \phi_\mathbf{k} - \phi_\mathbf{l} - \phi_{\mathbf{h}-\mathbf{k}-\mathbf{l}})$.

Let us now suppose that, at a certain step of the phasing process, a model structure is available and that F_p is the corresponding structure factor. Then a new type of s.i. may be devised which simultaneously contains F and F_p structure factors (see Sections 7.2 and 7.5); we will see that such invariants are very useful for facilitating the passage from the model to the target structure. Examples of this second type of invariant (the reader will easily see below that origin translations do not modify the value of the invariants) are:

$n = 2$: $F_\mathbf{h} F_{-p\mathbf{h}}$, or in terms of phase cosine $\cos(\phi_\mathbf{h} - \phi_{p\mathbf{h}})$;

$n = 3$: $F_\mathbf{h} F_\mathbf{k} F_{-\mathbf{h}-\mathbf{k}}$ or $(\phi_\mathbf{h} + \phi_\mathbf{k} + \phi_{-\mathbf{h}-\mathbf{k}})$,

$F_\mathbf{h} F_\mathbf{k} F_{p-\mathbf{h}-\mathbf{k}}$ or $(\phi_\mathbf{h} + \phi_\mathbf{k} + \phi_{p-\mathbf{h}-\mathbf{k}})$,

$F_{p\mathbf{h}} F_\mathbf{k} F_{-\mathbf{h}-\mathbf{k}}$ or $(\phi_{p\mathbf{h}} + \phi_\mathbf{k} + \phi_{-\mathbf{h}-\mathbf{k}})$,

$F_\mathbf{h} F_{p\mathbf{k}} F_{-\mathbf{h}-\mathbf{k}}$ or $(\phi_\mathbf{h} + \phi_{p\mathbf{k}} + \phi_{-\mathbf{h}-\mathbf{k}})$,

$F_\mathbf{h} F_{p\mathbf{k}} F_{p-\mathbf{h}-\mathbf{k}}$ or $(\phi_\mathbf{h} + \phi_{p\mathbf{k}} + \phi_{p-\mathbf{h}-\mathbf{k}})$,

$F_{p\mathbf{h}} F_\mathbf{k} F_{p-\mathbf{h}-\mathbf{k}}$ or $(\phi_{p\mathbf{h}} + \phi_\mathbf{k} + \phi_{p-\mathbf{h}-\mathbf{k}})$,

$F_{p\mathbf{h}} F_{p\mathbf{k}} F_{-\mathbf{h}-\mathbf{k}}$,   or $(\phi_{p\mathbf{h}} + \phi_{p\mathbf{k}} + \phi_{-\mathbf{h}-\mathbf{k}})$,

$F_{p\mathbf{h}} F_{p\mathbf{k}} F_{p-\mathbf{h}-\mathbf{k}}$ or $(\phi_{p\mathbf{h}} + \phi_{p\mathbf{k}} + \phi_{p-\mathbf{h}-\mathbf{k}})$.

Any of the above invariants may be estimated from the amplitudes of the corresponding observed and calculated structure factors.

Similar expressions may be obtained for quartets, quintets, etc.

3.4 Allowed or permissible origins in primitive space groups

In Section 3.2 it has been shown that fixing the symmetry operators $\mathbf{C}_s$ (and through this the algebraic form of the structure factor) is equivalent to selecting the class of allowed origin. In order to simplify the calculations during structural analysis and in order to handle the symmetry more easily it is convenient, in practice, to choose the origin on one or more of the symmetry elements. Thus, it is usual to choose the origin on high-order symmetry elements when they are present: in cs. (centrosymmetric) space groups it may be convenient to locate the origin on an inversion centre. This frequently corresponds with the choices given in the *International Tables for Crystallography*.

Moving the origin from one site to another usually modifies the algebraic representation of the symmetry operators. We define an *allowed* or *permissible origin* as all those points in direct space which, when taken as the origin, maintain the same symmetry operators $\mathbf{C}_s$. The allowed origins will therefore correspond to points having the same 'symmetry environment', in the sense that they are related to the symmetry elements in the same way. For instance, if the origin is located on an inversion centre, all the inversion centres in $P\text{-}1$ that are compatible with symmetry operators $\mathbf{C}_s$, given by

$$\mathbf{R}_1 = \begin{bmatrix} 1 & 0 & 0 \\ 0 & 1 & 0 \\ 0 & 0 & 1 \end{bmatrix}, \quad \mathbf{T}_1 = \begin{bmatrix} 0 \\ 0 \\ 0 \end{bmatrix}, \quad \mathbf{R}_2 = \begin{bmatrix} \bar{1} & 0 & 0 \\ 0 & \bar{1} & 0 \\ 0 & 0 & \bar{1} \end{bmatrix}, \quad \mathbf{T}_2 = \begin{bmatrix} 0 \\ 0 \\ 0 \end{bmatrix},$$

will be permissible origins. To each functional form of the structure factor there will be a class of permissible origins which, since they are all related to the symmetry elements in the same way, will be said to be *equivalent*. These constitute a class of *equivalent origins* or *equivalence class*.

Recognizing permissible origins is in general quite simple, through visual inspection of the space group diagram in the *International Tables for Crystallography*. We shall now see how to define permissible origins using an algebraic procedure.

Let O be an origin compatible with a fixed algebraic form of the structure factor; all other origins belonging to the same equivalence class can be defined in a very simple way using relation (3.7). Since a shift of origin must leave $\mathbf{R}'_s = \mathbf{R}_s$, it will be sufficient, in order to keep the symmetry operators $\mathbf{C}_s$ and thus the algebraic form of the structure factor unchanged, to have $\mathbf{T}'_s = \mathbf{T}_s$ for all values of s. More generally, because of the periodicity of crystal lattices, it will be sufficient to have $\mathbf{T}'_s - \mathbf{T}_s = \mathbf{V}$, where V is a vector with zero or integer components. All origins allowed by a fixed functional form of the structure factor will be connected by translational vectors x_0 such that

$$(\mathbf{R}_s - \mathbf{I})x_0 = \mathbf{V}, \qquad s = 1, 2, \ldots, m. \qquad (3.13)$$

A translation between permissible origins will be called a *permissible or allowed translation*. Trivial allowed translations correspond to the lattice periods or to their multiples.

Let us now consider some examples of the above concepts. In the space group P2/m (compare Fig. 3.5) the origin is chosen on an inversion centre with b as a twofold axis; the general equivalent positions in the unit cell are

$$(x, y, z), \quad (\bar{x}, \bar{y}, \bar{z}), \quad (\bar{x}, y, \bar{z}), \quad (x, \bar{y}, z).$$

The symmetry operators are then

$$\mathbf{R}_1 = \mathbf{I} = \begin{bmatrix} 1 & 0 & 0 \\ 0 & 1 & 0 \\ 0 & 0 & 1 \end{bmatrix}, \quad \mathbf{R}_2 = \begin{bmatrix} \bar{1} & 0 & 0 \\ 0 & \bar{1} & 0 \\ 0 & 0 & \bar{1} \end{bmatrix},$$

$$\mathbf{R}_3 = \begin{bmatrix} \bar{1} & 0 & 0 \\ 0 & 1 & 0 \\ 0 & 0 & \bar{1} \end{bmatrix}, \quad \mathbf{R}_4 = \begin{bmatrix} 1 & 0 & 0 \\ 0 & \bar{1} & 0 \\ 0 & 0 & 1 \end{bmatrix},$$

$$\mathbf{T}_1 = \mathbf{T}_2 = \mathbf{T}_3 = \mathbf{T}_4 = 0.$$

The $\mathbf{R}_s - \mathbf{I}$ matrices are

$$\mathbf{R}_1 - \mathbf{I} = \begin{bmatrix} 0 & 0 & 0 \\ 0 & 0 & 0 \\ 0 & 0 & 0 \end{bmatrix}, \quad \mathbf{R}_2 - \mathbf{I} = \begin{bmatrix} \bar{2} & 0 & 0 \\ 0 & \bar{2} & 0 \\ 0 & 0 & \bar{2} \end{bmatrix},$$

$$\mathbf{R}_3 - \mathbf{I} = \begin{bmatrix} \bar{2} & 0 & 0 \\ 0 & 0 & 0 \\ 0 & 0 & \bar{2} \end{bmatrix}, \quad \mathbf{R}_4 - \mathbf{I} = \begin{bmatrix} 0 & 0 & 0 \\ 0 & \bar{2} & 0 \\ 0 & 0 & 0 \end{bmatrix}.$$

It is easy to verify that the translations

$$\begin{bmatrix} n_1/2 \\ 0 \\ 0 \end{bmatrix} \quad \begin{bmatrix} 0 \\ n_2/2 \\ 0 \end{bmatrix} \quad \begin{bmatrix} 0 \\ 0 \\ n_3/2 \end{bmatrix},$$

with n_1, n_2, n_3 integer numbers, satisfy (3.13) and therefore connect origins belonging to the same equivalence class. Any combination of the above three

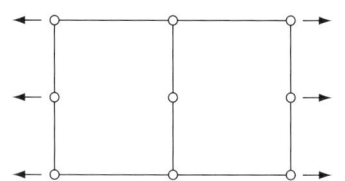

Fig. 3.5
Space group P2/m.

Allowed or permissible origins in primitive space groups

translations will also connect origins allowed by the given functional form of the structure factor. From the three basic translations

$$\begin{bmatrix} \frac{1}{2} \\ 0 \\ 0 \end{bmatrix}, \quad \begin{bmatrix} 0 \\ \frac{1}{2} \\ 0 \end{bmatrix}, \quad \begin{bmatrix} 0 \\ 0 \\ \frac{1}{2} \end{bmatrix},$$

we can derive a sort of lattice of permissible translations. Within a single unit cell the permissible origins will be defined by the translation vectors

$$\begin{bmatrix} 0 \\ 0 \\ 0 \end{bmatrix}, \begin{bmatrix} \frac{1}{2} \\ 0 \\ 0 \end{bmatrix}, \begin{bmatrix} 0 \\ \frac{1}{2} \\ 0 \end{bmatrix}, \begin{bmatrix} 0 \\ 0 \\ \frac{1}{2} \end{bmatrix}, \begin{bmatrix} 0 \\ \frac{1}{2} \\ \frac{1}{2} \end{bmatrix}, \begin{bmatrix} \frac{1}{2} \\ 0 \\ \frac{1}{2} \end{bmatrix}, \begin{bmatrix} \frac{1}{2} \\ \frac{1}{2} \\ 0 \end{bmatrix}, \begin{bmatrix} \frac{1}{2} \\ \frac{1}{2} \\ \frac{1}{2} \end{bmatrix}. \tag{3.14}$$

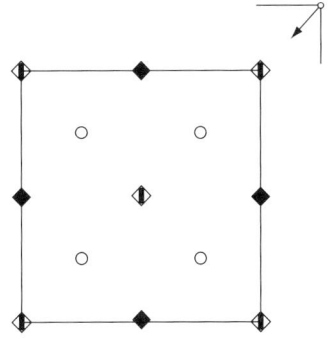

Fig. 3.6
Space group P4/n.

In the space group P4/n, when the origin is chosen on $\bar{4}$ (compare Fig. 3.6) at $\left(\frac{\bar{1}}{4}, \frac{1}{4}, 0\right)$ from an inversion centre, the general equivalent positions are

$(x, y, z), \quad \left(\frac{1}{2} - x, \frac{1}{2} - y, \bar{z}\right), \quad (\bar{x}, \bar{y}, z), \quad \left(\frac{1}{2} + x, \frac{1}{2} + y, \bar{z}\right), \quad (\bar{y}, x, \bar{z}),$

$(y, \bar{x}, \bar{z}), \quad \left(\frac{1}{2} - y, \frac{1}{2} + x, z\right), \quad \left(\frac{1}{2} + y, \frac{1}{2} - x, z\right),$

The $\mathbf{R}_s - \mathbf{I}$ matrices are

$$\mathbf{R}_1 - \mathbf{I} = \begin{bmatrix} 0 & 0 & 0 \\ 0 & 0 & 0 \\ 0 & 0 & 0 \end{bmatrix}, \quad \mathbf{R}_2 - \mathbf{I} = \begin{bmatrix} \bar{2} & 0 & 0 \\ 0 & \bar{2} & 0 \\ 0 & 0 & \bar{2} \end{bmatrix},$$

$$\mathbf{R}_3 - \mathbf{I} = \begin{bmatrix} \bar{2} & 0 & 0 \\ 0 & \bar{2} & 0 \\ 0 & 0 & 0 \end{bmatrix}, \quad \mathbf{R}_4 - \mathbf{I} = \begin{bmatrix} 0 & 0 & 0 \\ 0 & 0 & 0 \\ 0 & 0 & \bar{2} \end{bmatrix}.$$

$$\mathbf{R}_5 - \mathbf{I} = \begin{bmatrix} \bar{1} & \bar{1} & 0 \\ 1 & \bar{1} & 0 \\ 0 & 0 & \bar{2} \end{bmatrix}, \quad \mathbf{R}_6 - \mathbf{I} = \begin{bmatrix} \bar{1} & 1 & 0 \\ \bar{1} & \bar{1} & 0 \\ 0 & 0 & \bar{2} \end{bmatrix},$$

$$\mathbf{R}_7 - \mathbf{I} = \begin{bmatrix} \bar{1} & \bar{1} & 0 \\ 1 & \bar{1} & 0 \\ 0 & 0 & 0 \end{bmatrix}, \quad \mathbf{R}_8 - \mathbf{I} = \begin{bmatrix} \bar{1} & 1 & 0 \\ \bar{1} & \bar{1} & 0 \\ 0 & 0 & 0 \end{bmatrix}.$$

A lattice of permissible origins with basic translations,

$$\begin{bmatrix} \frac{1}{2} \\ 0 \\ 0 \end{bmatrix}, \begin{bmatrix} 0 \\ \frac{1}{2} \\ 0 \end{bmatrix}, \begin{bmatrix} 0 \\ 0 \\ \frac{1}{2} \end{bmatrix},$$

satisfies relations (3.13) for $s = 1, 2, 3, 4$; but in order to also satisfy (3.13) for $s \geq 5$, the sum and the difference of the components of the translation vector x_0 in the (a, b) plane must be integer numbers. The allowed translations defining the permissible origins will be

$$\begin{bmatrix} 0 \\ 0 \\ 0 \end{bmatrix}, \begin{bmatrix} 0 \\ 0 \\ \frac{1}{2} \end{bmatrix}, \begin{bmatrix} \frac{1}{2} \\ \frac{1}{2} \\ 0 \end{bmatrix}, \begin{bmatrix} \frac{1}{2} \\ \frac{1}{2} \\ \frac{1}{2} \end{bmatrix}. \tag{3.15}$$

Let us now apply equation (3.13) to some n.cs. (non-centrosymmetric) space groups. In the space group P3, with the origin on a threefold axis (compare Fig. 3.7), the general equivalent positions are

$$(x, y, z), \quad (\bar{y}, x - y, z), \quad (y - x, \bar{x}, z).$$

The $\mathbf{R}_s - \mathbf{I}$ matrices are

$$\mathbf{R}_1 - \mathbf{I} = \begin{bmatrix} 0 & 0 & 0 \\ 0 & 0 & 0 \\ 0 & 0 & 0 \end{bmatrix}, \quad \mathbf{R}_2 - \mathbf{I} = \begin{bmatrix} \bar{1} & \bar{1} & 0 \\ 1 & \bar{2} & 0 \\ 0 & 0 & 0 \end{bmatrix}, \quad \mathbf{R}_3 - \mathbf{I} = \begin{bmatrix} \bar{2} & 1 & 0 \\ \bar{1} & \bar{1} & 0 \\ 0 & 0 & 0 \end{bmatrix}.$$

We can easily verify that origin translations with components $\left(\frac{1}{2}, \frac{1}{3}\right)$ in the (a, b) plane satisfy (3.13) for all values of s. Note, however, that relations (3.13) do not imply any restriction to shifts in the z direction. The allowed translations in the unit cell are then given by

$$\begin{bmatrix} 0 \\ 0 \\ z \end{bmatrix}, \begin{bmatrix} \frac{2}{3} \\ \frac{1}{3} \\ z \end{bmatrix}, \begin{bmatrix} \frac{1}{3} \\ \frac{2}{3} \\ z \end{bmatrix}. \tag{3.16}$$

Let us now consider, as a further example, space group R3 (compare Fig. 3.8). With rhombohedral axes and the origin on a threefold axis, the general equivalent positions are

$$(x, y, z), \quad (z, x, y), \quad (y, z, x).$$

The $\mathbf{R}_s - \mathbf{I}$ matrices are

$$\mathbf{R}_1 - \mathbf{I} = \begin{bmatrix} 0 & 0 & 0 \\ 0 & 0 & 0 \\ 0 & 0 & 0 \end{bmatrix}, \quad \mathbf{R}_2 - \mathbf{I} = \begin{bmatrix} \bar{1} & 0 & 1 \\ 1 & \bar{1} & 0 \\ 0 & 1 & \bar{1} \end{bmatrix}, \quad \mathbf{R}_3 - \mathbf{I} = \begin{bmatrix} \bar{1} & 1 & 0 \\ 0 & \bar{1} & 1 \\ 1 & 0 & \bar{1} \end{bmatrix}.$$

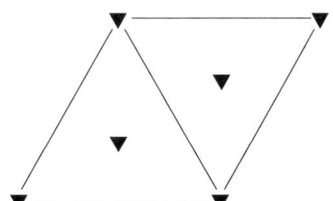

Fig. 3.7
Space group P3.

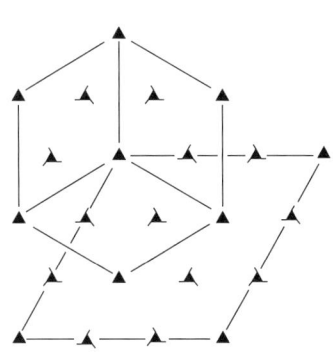

Fig. 3.8
Space group R3.

In order to satisfy relationships (3.13), one has to choose translation vectors with equal components, i.e.

$$x_0 = \begin{bmatrix} x \\ x \\ x \end{bmatrix}. \qquad (3.17)$$

The reader will easily be able to find the permissible translations for any other primitive space group using the algebraic procedure described in this section.

There are two important points that should also be noted. The first is related to the absence in equation 3.13 of the translational components of the symmetry operators. The distribution in direct space of the allowed origins is only dependent on the rotation matrices; for a given space group the allowed translations are therefore independent of the chosen form of the structure factor. More explicitly, let us consider the space group *P4/n*, mentioned above. When the origin is chosen at $\bar{1}$ at $(\frac{1}{4}, \frac{1}{4}, 0)$ from $\bar{4}$, the general equivalent positions become

$$(x, y, z), \ (\bar{x}, \bar{y}, \bar{z}), \ \left(\frac{1}{2} - x, \frac{1}{2} - y, z\right), \left(\frac{1}{2} + x, \frac{1}{2} + y, \bar{z}\right), \ \left(\bar{y}, \frac{1}{2} + x, \bar{z}\right),$$
$$\left(\frac{1}{2} + y, \bar{x}, \bar{z}\right), \ \left(\frac{1}{2} - y, x, z\right), \ \left(y, \frac{1}{2} - x, z\right);$$

The reader can immediately verify that the rotation matrices are unchanged in the new reference system, even if the algebraic form of the structure factor is modified; (3.15) are therefore the only allowed translations in this case, as well. Note that the four inversion centres in Fig. 3.6 do not all belong to the same equivalence class, because of the presence of the **n** glide.

The second point to stress is that the permissible translations, depending on the rotation matrices only, will remain the same in those space groups in which simple axes or planes are substituted by screw axes or glide planes (for instance, P2/m, P2$_1$/m, P2/c, and P2$_1$/c).

We will call any set of cs. or n.cs. space groups having the same allowed origin translation a *Hauptman–Karle* (*H–K*) *family*. The 94 n.cs. primitive space groups can be collected in 13 H–K families and the 62 primitive cs. groups in 4 H–K families. The second and the third rows of Tables 3.1 and 3.2 show these results explicitly.

It is worth emphasizing, for the benefit of the reader who wishes to refer to the literature for further details, that this concept of permissible origins is formally different from that given in the classical papers of Hauptman and Karle (1953b, 1956). We have followed the treatment described by Giacovazzo (1974a) because this provides a more general treatment.

3.5 The concept of structure seminvariant

In Section 3.3 we showed that products of structure factors (or the sum of phases) exist, the values of which do not depend on the origin, but only on the structure; we called them *structure invariant* (*s.i.*). Their importance is related to the following specific property: their phase values may be estimated from measured amplitudes.

Table 3.1 Allowed origin translations, seminvariant moduli, and phases for centrosymmetric primitive space groups

	H–K family			
	$(h,k,l)\underline{P}(2,2,2)$	$(h+k,l)\underline{P}(2,2)$	$(l)\underline{P}(2)$	$(h+k+l)\underline{P}(2)$
Space group	$P\bar{1}$	$Pmna$	$P\frac{4}{m}$ $P\frac{4}{n}mm$	$P\bar{3}$	$R\bar{3}$
	$P\frac{2}{m}$	$Pcca$	$P\frac{4_2}{m}$ $P\frac{4}{n}cc$	$P\bar{3}1m$	$R\bar{3}m$
	$P\frac{2_1}{m}$	$Pbam$	$P\frac{4}{n}$ $P\frac{4_2}{n}mc$	$P\bar{3}1c$	$R\bar{3}c$
	$P\frac{2}{c}$	$Pccn$	$P\frac{4_2}{n}$ $P\frac{4_2}{n}cm$	$P\bar{3}m1$	$Pm\bar{3}$
	$P\frac{2_1}{c}$	$Pbcm$	$P\frac{4}{m}mm$ $P\frac{4_2}{n}bc$	$P\bar{3}c1$	$Pn\bar{3}$
	$Pmmm$	$Pnnm$	$P\frac{4}{m}cc$ $P\frac{4_2}{n}nm$	$P\frac{6}{m}$	$Pa\bar{3}$
	$Pnnn$	$Pmmn$	$P\frac{4}{m}bm$ $P\frac{4_2}{n}bc$	$P\frac{6_3}{m}$	$Pm\bar{3}m$
	$Pccm$	$Pbcn$	$P\frac{4}{n}nc$ $P\frac{4_2}{m}nm$	$P\frac{6}{m}mm$	$Pn\bar{3}n$
	$Pban$	$Pbca$	$P\frac{4}{n}bm$ $P\frac{4_2}{m}mc$	$P\frac{6}{m}cc$	$Pm\bar{3}n$
	$Pmma$	$Pnma$	$P\frac{4}{m}nc$ $P\frac{4_2}{m}cm$	$P\frac{6_3}{m}cm$	$Pn\bar{3}m$
	$Pnna$			$P\frac{6_3}{m}mc$	
Allowed origin translations	$(0,0,0);$	$(0,\frac{1}{2},\frac{1}{2})$	$(0,0,0)$	$(0,0,0)$	$(0,0,0)$
	$(\frac{1}{2},0,0);$	$(\frac{1}{2},0,\frac{1}{2})$	$(0,0,\frac{1}{2})$	$(0,0,\frac{1}{2})$	$(\frac{1}{2},\frac{1}{2},\frac{1}{2})$
	$(0,\frac{1}{2},0);$	$(\frac{1}{2},\frac{1}{2},0)$	$(\frac{1}{2},\frac{1}{2},0)$		
	$(0,0,\frac{1}{2});$	$(\frac{1}{2},\frac{1}{2},\frac{1}{2})$	$(\frac{1}{2},\frac{1}{2},\frac{1}{2})$		
Vector $\boldsymbol{h}_s$ seminvariantly associated with $\boldsymbol{h}=(h,k,l)$	(h,k,l)	$(h+k,l)$	(l)	$(h+k+l)$
Seminvariant modulus $\boldsymbol{\omega}_s$	$(2,2,2)$	$(2,2)$	(2)	(2)
Seminvariant	ϕ_{eee}	$\phi_{eee}; \phi_{ooe}$	$\phi_{eee}; \phi_{eoe}$ $\phi_{oee}; \phi_{ooe}$	$\phi_{eee}; \phi_{ooe}$ $\phi_{oeo}; \phi_{eoo}$
Number of semi-independent phases to be specified	3	2	1	1

We now introduce a related concept, *structure seminvariant* (*s.s.*). Let us suppose that, for a given space group, the symmetry operators $\boldsymbol{C}_s$, and therefore the allowed origins, have been fixed. If $\boldsymbol{x}_0$ is a generic translation, equation (3.3) suggests that the phase of a structure factor will change, after an origin shift, by a finite quantity. The central question for this section is: *do particular phases or combinations of phases exist, the values of which do not change when the origin is moved within the set of allowed origins? If yes, we will say that the phases or the combination of phases are s.s.* (Hauptman and Karle, 1956).

The s.s. are important because they can be estimated from measurements, provided that the symmetry operators have been fixed (and they usually remain fixed during the full phasing process). Indeed, the fixed nature of the symmetry operators restricts the allowed origins to a specific subset of points, within

Table 3.2 Allowed origin translations, seminvariant moduli, and phases for non-centrosymmetric primitive space groups

	H–K family										
	(h, k, l) $P(0, 0, 0)$	(h, k, l) $P(2, 0, 2)$	(h, k, l) $P(0, 2, 0)$	(h, k, l) $P(2, 2, 2)$	(h, k, l) $P(2, 2, 0)$	$(h + k, l)$ $P(2, 0)$	$(h + k, l)$ $P(2, 2)$	$(h - k, l)$ $P(3, 0)$	$(2h + 4k + 3l)$ $P(6)$	$(l)P(0)$	$(l)P(2)$	$(h + k + l)$ $P(0)$	$(h + k + l)$ $P(2)$

Space group	P1	P2	Pm	P222	Pmm2	P4	P$\bar{4}$	P3	P312	P31m	P321	R3	R32
		P2$_1$	Pc	P222$_1$	Pmc2$_1$	P4$_1$	P422	P3$_1$	P3$_1$12	P31c	P3$_1$21	R3m	P23
				P2$_1$2$_1$2	Pcc2	P4$_2$	P42$_1$2	P3$_2$	P3$_2$12	P6	P3$_2$21	R3c	P2$_1$3
				P2$_1$2$_1$2$_1$	Pma2	P4$_3$	P4$_1$22	P3m1	P6	P6$_1$	P622		P432
					Pca2$_1$	P4mm	P4$_1$2$_1$2	P3c1	P$\bar{6}$m2	P6$_5$	P6$_1$22		P4$_2$32
					Pnc2	P4bm	P4$_2$22		P$\bar{6}$c2	P6$_4$	P6$_5$22		P4$_3$32
					Pmn2$_1$	P4$_2$cm	P4$_2$2$_1$2			P6$_3$	P6$_2$22		P4$_1$32
					Pba2	P4$_2$nm	P4$_3$22			P6$_2$	P6$_4$22		P$\bar{4}$3m
					Pna2$_1$	P4cc	P4$_3$2$_1$2			P6mm	P6$_3$22		P$\bar{4}$3n
					Pnn2	P4nc	P$\bar{4}$2m			P6cc	P$\bar{6}$2m		
						P4$_2$mc	P$\bar{4}$2c			P6$_3$cm	P$\bar{6}$2c		
						P4$_2$bc	P$\bar{4}$2$_1$m			P6$_3$mc			
							P$\bar{4}$2$_1$c						
							P$\bar{4}$m2						
							P$\bar{4}$c2						
							P$\bar{4}$b2						
							P$\bar{4}$n2						

| Allowed origin translations | (x, y, z) | $(0, y, 0)$ $(0, y, \frac{1}{2})$ $(\frac{1}{2}, y, 0)$ $(\frac{1}{2}, y, \frac{1}{2})$ | $(x, 0, z)$ $(x, \frac{1}{2}, z)$ | $(0, 0, z)$ $(\frac{1}{2}, 0, 0)$ $(0, \frac{1}{2}, 0)$ $(0, 0, \frac{1}{2})$ $(0, \frac{1}{2}, \frac{1}{2})$ $(\frac{1}{2}, 0, \frac{1}{2})$ $(\frac{1}{2}, \frac{1}{2}, 0)$ $(\frac{1}{2}, \frac{1}{2}, \frac{1}{2})$ | $(0, 0, z)$ $(0, \frac{1}{2}, z)$ $(\frac{1}{2}, 0, z)$ $(\frac{1}{2}, \frac{1}{2}, z)$ | $(0, 0, z)$ $(\frac{1}{2}, \frac{1}{2}, z)$ | $(0, 0, 0)$ $(0, 0, \frac{1}{2})$ $(\frac{1}{2}, \frac{1}{2}, 0)$ $(\frac{1}{2}, \frac{1}{2}, \frac{1}{2})$ | $(0, 0, z)$ $(\frac{1}{3}, \frac{2}{3}, z)$ $(\frac{2}{3}, \frac{1}{3}, z)$ | $(0, 0, 0)$ $(0, 0, \frac{1}{2})$ $(\frac{1}{3}, \frac{2}{3}, 0)$ $(\frac{1}{3}, \frac{2}{3}, \frac{1}{2})$ $(\frac{2}{3}, \frac{1}{3}, 0)$ $(\frac{2}{3}, \frac{1}{3}, \frac{1}{2})$ | $(0, 0, z)$ | $(0, 0, 0)$ $(0, 0, \frac{1}{2})$ | (x, x, x) | $(0, 0, 0)$ $(\frac{1}{2}, \frac{1}{2}, \frac{1}{2})$ |

Table 3.2 (Continued)

	H–K family													
	(h, k, l) P(0, 0, 0)	(h, k, l) P(2, 0, 2)	(h, k, l) P(0, 2, 2)	(h, k, l) P(0, 2, 0)	(h, k, l) P(2, 2, 2)	(h, k, l) P(2, 2, 0)	$(h + k, l)$ P(2, 0)	$(h + k, l)$ P(2, 2)	$(h − k, l)$ P(3, 0)	$(2h + 4k + 3l)$ P(6)	(l)P(0)	(l)P(2)	$(h + k + l)$ P(0)	$(h + k + l)$ P(2)
Vector $\mathbf{h}_s$ seminvariantly associated with $\mathbf{h} = (h, k, l)$	(h, k, l)	(h, k, l)	(h, k, l)	(h, k, l)	(h, k, l)	(h, k, l)	$(h + k, l)$	$(h + k, l)$	$(h − k, l)$	$(2h + 4k + 3l)$	(l)	(l)	$(h + k + l)$	$(h + k + l)$
Seminvariant modulus ω_s	(0, 0, 0)	(2, 0, 2)	(0, 2, 2)	(0, 2, 0)	(2, 2, 2)	(2, 2, 0)	(2, 0)	(2, 2)	(3, 0)	(6)	(0)	(2)	(0)	(2)
Seminvariant phases	ϕ_{000}	ϕ_{e0e}	ϕ_{0e0}	ϕ_{eee}	ϕ_{ee0}	ϕ_{ee0} ϕ_{oo0}	ϕ_{ee0} ϕ_{oo0}	ϕ_{eee} ϕ_{ooe}	ϕ_{hk0} if $h − k \equiv 0$ (mod 3)	ϕ_{hkl} if $2h + 4k + 3l \equiv 0$ (mod 6)	ϕ_{hk0}	ϕ_{hke}	$\phi_{h,k,\bar{h}+\bar{k}}$	$\phi_{eee}; \phi_{ooe}$ $\phi_{oeo}; \phi_{ooe}$
Allowed variations for the semi-independent phases	$\|\infty\|$	$\|\infty\|,$ $\|2\|$ if $k = 0$	$\|\infty\|,$ $\|2\|$ if $h = l = 0$	$\|2\|$	$\|\infty\|,$ $\|2\|$ if $l = 0$	$\|\infty\|,$ $\|2\|$ if $l = 0$	$\|2\|$	$\|\infty\|,$ $\|3\|$ if $l = 0$	$\|2\|$ if $h \equiv k$ (mod 3) $\|3\|$ if $l \equiv 0$ (mod 2)	$\|\infty\|$	$\|2\|$	$\|\infty\|$	$\|2\|$
Number of semi-independent phases to be specified	3	3	3	3	3	2	2	2	1	1	1	1	1

which the seminvariants do not change. We can then state that *the s.s. values only depend on the structure, provided that the class of allowed origins has been fixed.*

We now show how to recognize s.s.s via Tables 3.1–3.4 (we will see in Chapter 4 that other appropriate definitions also exist).

Because of (3.3), the condition for ϕ_h to be a s.s. is that, for all permissible translations x_p (Hall, 1970a),

$$\boldsymbol{h} \cdot \boldsymbol{x}_p = r, \quad p = 1, 2, \ldots, \tag{3.18}$$

where r is a positive, null, or negative integer. Equation (3.18) shows that the s.s.s are determined by the permissible translations only. Therefore, if a phase is an s.s. for a given space group, it will also be an s.s. for all other space groups belonging to the same H–K family.

Let us now apply the above considerations to some H–K families. For the families indicated in Table 3.1 by the symbol $(h, k, l)P(2, 2, 2)$, equation (3.18) suggests immediately that a phase ϕ_h with $\boldsymbol{h} = (h, k, l)$ is an s.s. when

$$h = 2n_1, \quad k = 2n_2, \quad l = 2n_3,$$

with n_1, n_2, n_3 integers. Introducing the symbolism of modular algebra, these conditions can be expressed in the form,

$$\boldsymbol{h} \equiv 0 \ (\text{mod } (2, 2, 2)). \tag{3.19}$$

In fact, the congruence of $\boldsymbol{h}$ to zero modulo $(2, 2, 2)$ is equivalent to all components of $\boldsymbol{h}$ being even.

For the H–K family indicated in Table 3.1 by the symbol $(h + k, l)P(2, 2)$, we derive from equation (3.18) that to a seminvariant phase ϕ_h indices h, k, l must correspond such that

$$h + k = 2n_1, \quad l = 2n_2. \tag{3.20}$$

In this family, the seminvariance conditions do not apply to the individual components h, k, l of the reciprocal vector $\boldsymbol{h}$, but to the two-dimensional vector $(h + k, l)$. Using the relations of modular congruence we can express condition (3.20) in the form,

$$\boldsymbol{h}_s \equiv 0 \quad (\text{mod } (2, 2)),$$

where $\boldsymbol{h}_s = (h + k, l)$. The vector $\boldsymbol{h}_s$, so defined, is said to be *seminvariantly associated* with $\boldsymbol{h}$ and the vector $\boldsymbol{\omega}_s = (2, 2)$ will be the *seminvariant modulus of the family* $(h + k, l)P(2, 2)$. Condition (3.19) can also be rewritten in terms of $\boldsymbol{h}_s$ and $\boldsymbol{\omega}_s$ as,

$$\boldsymbol{h}_s \equiv 0 \quad (\text{mod } \boldsymbol{\omega}_s),$$

with $\boldsymbol{h}_s = (h, k, l)$ and $\boldsymbol{\omega}_s = (2, 2, 2)$.

We can therefore associate with each H–K family, the corresponding vectors $\boldsymbol{\omega}_s$ and $\boldsymbol{h}_s$. In general, a phase ϕ_h is an s.s. for a given H–K family when the seminvariantly associated vector $\boldsymbol{h}_s$ is congruent to zero modulo $\boldsymbol{\omega}_s$, where $\boldsymbol{\omega}_s$ is the seminvariant modulus of the group. We write

$$\boldsymbol{h}_s \equiv 0 \quad (\text{mod } \boldsymbol{\omega}_s). \tag{3.21}$$

Thus, in the group $(h + k + l)$P2, the seminvariance condition for ϕ_h is $h + k + l = 2n$, and we will say that ϕ_h is an s.s. if (3.21) holds, where $\boldsymbol{h}_s$ and $\boldsymbol{\omega}_s$ are the one-dimensional vectors with components $(h + k + l)$ and (2) respectively.

So far we have only used cs. groups as examples, but the results obtained are entirely general. In fact the reader can easily verify that, for the family $(h + k, l)$P$(2, 0)$, the seminvariance conditions for the phase ϕ_h are

$$h + k = 2n \text{ and } l = 0. \tag{3.22}$$

Conditions (3.22) can be expressed in a more concise form by equation (3.21), assuming

$$\boldsymbol{h}_s = (h + k, l), \qquad \boldsymbol{\omega}_s = (2, 0).$$

In Tables 3.1 and 3.2, for each H–K family, the seminvariant modulus $\boldsymbol{\omega}_s$ and the vector $\boldsymbol{h}_s$ seminvariantly associated with the vector $\boldsymbol{h}$ are given; it is then very simple to select, for each primitive space group, those phases which are s.s.s. The notation used to indicate H–K families is that proposed by Giacovazzo (1974a), giving, in turn, the vector seminvariant associated with each phase ϕ_h, the lattice type, and the seminvariant modulus. We shall underline the lattice symbol to indicate an H–K family that includes cs. space groups.

Recalling the idea of an equiphase surface, mentioned in Section 3.2, we can derive an immediate geometrical interpretation of the seminvariance condition for a given phase ϕ_h. *The necessary and sufficient condition for a phase ϕ_h to be unchanged for any possible origin translation is that the lattice planes (h, k, l) shall contain all the origins allowed by the functional form of the s.f.* Thus, for instance, in the H–K family (h, k, l)P$(2, 2, 2)$, lattice planes (h, k, l), having

$$h = 2n_1, \quad k = 2n_2, \quad l = 2n_3,$$

contain all permissible origins; lattice planes (h, k, l) with

$$h - k = 3n, \quad l = 0$$

contain all permissible origins of the group $(h-k, l)$P$(3, 0)$, and similarly for all other groups.

It is now clear that when a phase is not an s.s. its value will change, passing from one origin to another within the equivalence class. We will define as *allowed or permissible phase changes* for linearly semi-independent phases, the phase shifts corresponding to allowed origin translations. Following Karle (1970a,b), in Table 3.2 (in Table 3.1 it would be trivial), we indicate by the symbol $\|\infty\|$ the permissible phase changes assuming any value between $-\pi$ and π, and by the symbols $\|2\|, \|3\|, \|4\|$ the permissible phase changes of $2n\pi/2, 2n\pi/3, 2n\pi/4$ respectively. Thus in the group $(h-k, l)$P$(3, 0)$, if F_h is not an s.s. the phase ϕ_h can take any value between $-\pi$ and π for any allowed origin translation; but, when $l = 0$, ϕ_h can only have three discrete values,

$$\phi, \qquad \phi + \frac{2\pi}{3}, \qquad \phi + \frac{4\pi}{3}.$$

To show this we will, for simplicity, assume the origin on a threefold axis. In this case, phase $\phi_{h, k, 0}$ (not an s.s.) has the same value for all permissible translations along the *c*-axis; in other words, all points on a threefold axis, when chosen as origins, will attribute the same value to $\phi_{h,k,0}$. As the origin translates from one threefold axis to another (there are three such axes in the cell), this implies a phase shift of $2n\pi/3$.

We shall now see how we can extend this idea for the definition of two-phase, three-phase, and *n*-phase seminvariants.

Let us consider the product of s.f.s,

$$F_{h_1} F_{h_2} \cdots F_{h_n} = |F_{h_1} F_{h_2} \cdots F_{h_n}| \exp\left[i(\phi_{h_1} + \phi_{h_2} + \cdots + \phi_{h_n})\right]. \quad (3.23)$$

The general product of s.f.s (3.23), or, more specifically, its phase

$$\Phi = \phi_{h_1} + \phi_{h_2} + \cdots + \phi_{h_n},$$

is an s.s. for a given space group (or for a given H–K family) if its value does not change when the origin is moved within the same equivalence class. By analogy with (3.18), the necessary and sufficient condition for Φ to be an s.s. is that, for all allowed translations, the following condition is satisfied:

$$\left(\sum_{j=1}^{n} \mathbf{h}_j\right) \mathbf{x}_p = r, \quad p = 1, 2, \ldots, \quad (3.24)$$

where *r* is a positive, null, or negative integer. By analogy with (3.21), definition (3.24) is equivalent to the following: the sum of phases

$$\sum_{j=1}^{n} \phi_{h_j}$$

is an s.s. if condition

$$\sum_{j=1}^{n} \mathbf{h}_{js} \equiv 0 \quad (\text{mol } \boldsymbol{\omega}_s) \quad (3.25)$$

is satisfied. Depending on the value of *n*, we speak of one-phase seminvariant, two-phase seminvariant, etc. . . . . We will now give a few examples.

1. In $P\bar{1}$ (as for all space groups belonging to the H–K family $(h, k, l)P(2, 2, 2)$), $\phi_{h,k,l}$ is a structure seminvariant if h, k, l are all even numbers. Two-phase seminvariants will be the phases

    $\phi_{h_1, k_1, l_1} + \phi_{h_2, k_2, l_2}$ provided

    $h_1 + h_2 \equiv 0 \,(\text{mod } 2), \quad k_1 + k_2 \equiv 0 \,(\text{mod } 2), \quad l_1 + l_2 \equiv 0 \,(\text{mod } 2).$

2. In P2 (and also in P2$_1$), ϕ_{e0e} is an s.s. (*e* stands for even). Also,

    $\phi_{h_1, k_1, l_1} + \phi_{h_2, \bar{k}_1, l_2}$ is an s.s if $h_1 + h_2 \equiv 0 \,(\text{mod } 2)$ and $l_1 + l_2 \equiv 0 \,(\text{mod } 2)$.

Similarly,

$\phi_{h_1 k_1 l_1} + \phi_{h_2 k_2 l_2} + \phi_{h_3 \overline{k_1 + k_2} l_3}$ is a three-phase s.s. if

$h_1 + h_2 + h_3 \equiv 0 \,(\text{mod } 2)$ and $l_1 + l_2 + l_3 \equiv 0 \,(\text{mod } 2).$

We may now note an interesting observation. We have seen that the seminvariance condition does not change when we choose a different functional form of the s.f. (that is, when we fix another equivalence class for the origins). Thus a phase or a linear combination of phases which is a seminvariant with respect to a given equivalent class of origins will remain a semi-invariant for any other equivalence class.

3.6 Allowed or permissible origins in centred cells

The theory developed over the preceding paragraphs may be adapted, with minor changes, to space groups with a non-primitive unit cell. In fact it is well known that it is always possible to reduce a centred cell to a primitive one using a suitable transformation. Once the transformation has been performed, there is no further difficulty in applying the above theory to determine the conditions for origin definition. The results obtained in this way must be transformed again in terms valid for a centred cell if one wishes to operate in this reference system (Hauptman and Karle (1956, 1959); Karle and Hauptman (1961); see also Lessinger and Wondratschek (1975)).

An alternative and simpler way was suggested by Giacovazzo (1974a), which uses directly algebraic relations specific for non-primitive cells. We will follow this approach.

Let us suppose that, for a given non-primitive cell with centring order τ, we have chosen the algebraic form of the structure factor; the equivalent positions not related by the centring of the cell can be expressed by means of

$$\mathbf{r}_{js} = \mathbf{C}_s \mathbf{r}_j = \mathbf{R}_s \mathbf{r}_j + \mathbf{T}_s, s = 1, 2, \ldots, \frac{m}{\tau}.$$

The complete set of general equivalent positions can be obtained by applying the centring vectors $\mathbf{B}_v$, since for each atom $\mathbf{r}_{js}$ there will be $\tau-1$ equivalent atoms at $\mathbf{r}_{js} + \mathbf{B}_v$ with $v = 2, 3, \ldots, \tau$.

Since an origin translation corresponding to a centring vector $\mathbf{B}_v$ does not change the functional form of the s.f., all vectors $\mathbf{B}_v$ represent permissible translations. Therefore, $\mathbf{x}_p$ will be an allowed translation, not only when, as stated in Section 3.4, the difference $\mathbf{T}'_s - \mathbf{T}_s$ is equal to one or more lattice units, but also when, for any s, the condition

$$\mathbf{T}'_s - \mathbf{T}_s = \mathbf{V} + \alpha \mathbf{B}_v, \quad s = 1, 2, \ldots, \quad \alpha = 0, 1$$

or

$$(\mathbf{R}_s - \mathbf{I})\mathbf{X}_p = \mathbf{V} + \alpha \mathbf{B}_v, \quad s = 1, 2, \ldots, \quad \alpha = 0, 1 \quad (3.26)$$

is satisfied.

Let us consider the example of the space group C2/c. From Fig. 3.9, we easily see that the only allowed origin translations are

$$(0,0,0), \quad \left(0, \frac{1}{2}, 0\right) \quad \left(\frac{1}{2}, \frac{1}{2}, 0\right) \quad \left(\frac{1}{2}, 0, 0\right),$$

$$\left(0, 0, \frac{1}{2}\right), \quad \left(0, \frac{1}{2}, \frac{1}{2}\right), \quad \left(\frac{1}{2}, \frac{1}{2}, \frac{1}{2}\right), \quad \left(\frac{1}{2}, 0, \frac{1}{2}\right).$$

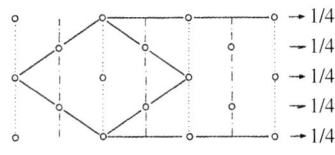

Fig. 3.9
Space group C2/c and the corresponding standard primitive cell.

Allowed or permissible origins in centred cells

In the case of primitive cells we neglect all allowed translations corresponding to lattice translations; similarly, here we will neglect the trivial translations coinciding with the centring vectors $\boldsymbol{B}_v$. The only permissible origins are therefore assumed to be

$$(0,0,0), \quad \left(\frac{1}{2},0,0\right), \quad \left(0,0,\frac{1}{2}\right), \quad \left(\frac{1}{2},0,\frac{1}{2}\right).$$

We can thus derive the results reported in Table 3.3.

Note that the centring of the cell in space group C2/c does not add any new allowed translation with respect to those found for the group P2/c. This is not always the case; the presence of the translation vectors $\boldsymbol{B}$ with non-integer components (or, in the reciprocal lattice, the absence of certain classes of reflections) can sometimes generate new allowed translations which cannot be foreseen by the use of rotation matrices alone. We can, in general, state that *in centred cells the point group symmetry no longer defines all permissible origins*. To show this, let us consider the space group F222, illustrated in Fig. 3.10. When the origin is chosen at 222, the general equivalent positions are

$$\{(x,y,z),(\bar{x},\bar{y},z),(x,\bar{y},\bar{z}),(\bar{x},y,\bar{z})\} + \left[(0,0,0), \left(0,\frac{1}{2},\frac{1}{2}\right), \left(\frac{1}{2},0,\frac{1}{2}\right), \left(\frac{1}{2},\frac{1}{2},0\right)\right].$$

Table 3.3 Allowed origin translations seminvariant moduli, and phases for centrosymmetric non-primitive space groups

	H–K family				
	$(h, l)\underline{C}(2, 2)$	$(k, l)\underline{I}(2, 2)$	$(h + k + l)\underline{F}(2)$	$(l)\underline{I}(2)$	$\underline{I}$
Space groups	$C\frac{2}{m}$	$Immm$	$Fmmm$	$I\frac{4}{m}$	$Im\bar{3}$
	$C\frac{2}{c}$	$Ibam$	$Fddd$	$I\frac{4_1}{a}$	$Ia\bar{3}$
	$Cmcm$	$Ibca$	$Fm\bar{3}$	$I\frac{4}{m}mm$	$Im\bar{3}m$
	$Cmca$	$Imma$	$Fd\bar{3}$	$I\frac{4}{m}cm$	$Ia\bar{3}d$
	$Cmmm$		$Fm\bar{3}m$	$I\frac{4_1}{a}md$	
	$Cccm$		$Fm\bar{3}c$	$I\frac{4_1}{a}cd$	
	$Cmma$		$Fd\bar{3}m$		
	$Ccca$		$Fd\bar{3}c$		
Allowed origin translations	$(0, 0, 0)$	$(0, 0, 0)$	$(0, 0, 0)$	$(0, 0, 0)$	$(0, 0, 0)$
	$(0, 0, \frac{1}{2})$	$(0, 0, \frac{1}{2})$	$(\frac{1}{2}, \frac{1}{2}, \frac{1}{2})$	$(0, 0, \frac{1}{2})$	
	$(\frac{1}{2}, 0, 0)$	$(0, \frac{1}{2}, 0)$			
	$(\frac{1}{2}, 0, \frac{1}{2})$	$(\frac{1}{2}, 0, 0)$			
Vector $\boldsymbol{h}_s$ seminvariantly associated with $\boldsymbol{h} = (h, k, l)$	(h, l)	(k, l)	$(h + k + l)$	(l)	(h, k, l)
Seminvariant modulus $\boldsymbol{\omega}_s$	$(2, 2)$	$(2, 2)$	(2)	(2)	$(1, 1, 1)$
Seminvariant phases	ϕ_{eee}	ϕ_{eee}	ϕ_{eee}	$\phi_{eoe}; \phi_{eee}$	All
				$\phi_{ooe}; \phi_{oee}$	
Number of semi-independent phases to be specified	2	2-	1	1	0

Fig. 3.10
Space group F222.

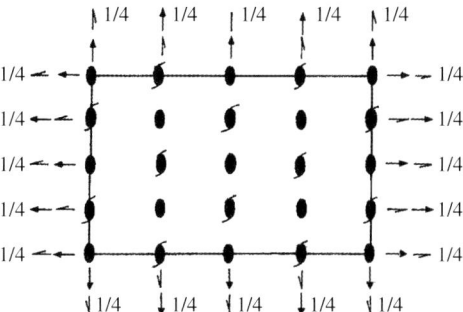

The $\mathbf{R}_s - \mathbf{I}$ matrices are therefore

$$\mathbf{R}_1 - \mathbf{I} = \begin{bmatrix} 0 & 0 & 0 \\ 0 & 0 & 0 \\ 0 & 0 & 0 \end{bmatrix}, \quad \mathbf{R}_2 - \mathbf{I} = \begin{bmatrix} \bar{2} & 0 & 0 \\ 0 & \bar{2} & 0 \\ 0 & 0 & 0 \end{bmatrix},$$

$$\mathbf{R}_3 - \mathbf{I} = \begin{bmatrix} 0 & 0 & 0 \\ 0 & \bar{2} & 0 \\ 0 & 0 & \bar{2} \end{bmatrix}, \quad \mathbf{R}_4 - \mathbf{I} = \begin{bmatrix} \bar{2} & 0 & 0 \\ 0 & 0 & 0 \\ 0 & 0 & \bar{2} \end{bmatrix},$$

Substitution in (3.25) gives the following allowed translations:

$$(0,0,0), \quad \left(\tfrac{1}{2},\tfrac{1}{2},0\right), \quad \left(0,\tfrac{1}{2},\tfrac{1}{2}\right), \quad \left(\tfrac{1}{2},0,\tfrac{1}{2}\right),$$

$$\left(\tfrac{1}{4},\tfrac{1}{4},\tfrac{1}{4}\right), \quad \left(\tfrac{3}{4},\tfrac{3}{4},\tfrac{1}{4}\right), \quad \left(\tfrac{1}{4},\tfrac{3}{4},\tfrac{3}{4}\right), \quad \left(\tfrac{3}{4},\tfrac{1}{4},\tfrac{3}{4}\right),$$

$$\left(\tfrac{3}{4},\tfrac{3}{4},\tfrac{3}{4}\right), \quad \left(\tfrac{1}{4},\tfrac{1}{4},\tfrac{3}{4}\right), \quad \left(\tfrac{3}{4},\tfrac{1}{4},\tfrac{1}{4}\right), \quad \left(\tfrac{1}{4},\tfrac{3}{4},\tfrac{1}{4}\right),$$

$$\left(\tfrac{1}{2},0,0\right), \quad \left(\tfrac{1}{2},\tfrac{1}{2},\tfrac{1}{2}\right), \quad \left(0,\tfrac{1}{2},0\right), \quad \left(0,0,\tfrac{1}{2}\right).$$

We can see that the centring condition of the cell adds, in this case, further allowed origins with respect to P222. Neglecting those corresponding to the vectors $\mathbf{B}_v$, the permissible translations become

$$(0,0,0), \quad \left(\tfrac{1}{4},\tfrac{1}{4},\tfrac{1}{4}\right), \quad \left(\tfrac{1}{2},\tfrac{1}{2},\tfrac{1}{2}\right), \quad \left(\tfrac{3}{4},\tfrac{3}{4},\tfrac{3}{4}\right),$$

in accordance with Table 3.4.

In non-primitive space groups, the systematic absences due to the centring of the cell can conveniently be associated with the conditions defining the seminvariant moduli and vectors. For instance, in the space group I222 equations (3.18) or (3.21) alone would define a seminvariant vector $\mathbf{h}'_s = (h, k, l)$ and a modulus $\boldsymbol{\omega}'_s = (2, 2, 2)$. Considering also that the reflections with

Table 3.4 Allowed origin translations, seminvariant moduli, and phases for non-centrosymmetric, non-primitive space groups

	H–K family												
	(k, l) $C(0, 2)$	(h, l) $C(0, 0)$	(h, l) $C(2, 0)$	(h, l) $C(2, 2)$	(h, l) $A(2, 0)$	(h, l) $I(2, 0)$	(h, l) $I(2, 2)$	$(h + k + l)$ $F(2)$	$(h + k + l)$ $F(4)$	$(l)I(0)$	$(l)I(2)$	$(2k - l)$ $I(4)$	$(l)F(0)$	I
Space group	$C2$	Cm Cc	$Cmm2$ $Cmc2_1$ $Ccc2$	$C222$ $C222_1$	$Amm2$ $Abm2$ $Ama2$ $Aba2$	$Imm2$ $Iba2$ $Ima2$	$I222$ $I2_12_12_1$	$F432$ $F4_132$	$F222$ $F23$ $F\bar{4}3m$ $F\bar{4}3c$	$I4$ $I4_1$ $I4mm$ $I4cm$ $I4_1md$ $I4_1cd$	$I422$ $I4_122$ $I\bar{4}2m$ $I\bar{4}2d$	$I\bar{4}$ $I\bar{4}m2$ $I\bar{4}c2$	$Fmm2$ $Fdd2$	$I23$ $I2_13$ $I432$ $I4_132$ $I\bar{4}3m$ $I\bar{4}3d$
Allowed origin translations	$(0, y, 0)$ $(0, y, \frac{1}{2})$	$(x, 0, z)$	$(0, 0, z)$ $(\frac{1}{2}, 0, z)$	$(0, 0, 0)$ $(0, 0, \frac{1}{2})$ $(\frac{1}{2}, 0, 0)$ $(\frac{1}{2}, 0, \frac{1}{2})$	$(0, 0, z)$ $(\frac{1}{2}, 0, z)$	$(0, 0, z)$ $(\frac{1}{2}, 0, z)$	$(0, 0, 0)$ $(0, 0, \frac{1}{2})$ $(0, \frac{1}{2}, 0)$ $(\frac{1}{2}, 0, 0)$	$(0, 0, 0)$ $(\frac{1}{4}, \frac{1}{4}, \frac{1}{4})$ $(\frac{1}{2}, \frac{1}{2}, \frac{1}{2})$	$(0, 0, 0)$ $(\frac{1}{4}, \frac{1}{4}, \frac{1}{4})$ $(\frac{1}{2}, \frac{1}{2}, \frac{1}{2})$ $(\frac{3}{4}, \frac{3}{4}, \frac{3}{4})$	$(0, 0, z)$	$(0, 0, 0)$ $(0, 0, \frac{1}{2})$	$(0, 0, 0)$ $(0, 0, \frac{1}{2})$ $(\frac{1}{2}, 0, \frac{1}{4})$ $(\frac{1}{2}, 0, \frac{3}{4})$	$(0, 0, z)$	$(0, 0, 0)$
Vector $\mathbf{h}_s$, seminvariantly associated with $\mathbf{h} = (h, k, l)$	(k, l)	(h, l)	(h, l)	(h, l)	(h, l)	(h, l)	(h, l)	$(h + k + l)$	$(h + k + l)$	(l)	(l)	$(2k - l)$	(l)	(h, k, l)
Seminvariant modulus $\boldsymbol{\omega}_s$	$(0, 2)$	$(0, 0)$	$(2, 0)$	$(2, 2)$	$(2, 0)$	$(2, 0)$	$(2, 2)$	(2)	(4)	(0)	(2)	(4)	(0)	$(1, 1, 1)$

Table 3.4 (Continued)

	H–K family										
	(k, l) C(0, 2)	(h, l) C(0, 0)	(h, l) C(2, 0)	(h, l) C(2, 2)	(h, l) A(2, 0)	(h, l) I(2, 0)	(h, l) I(2, 2)	$(h + k + l)$ F(2)	$(h + k + l)$ F(4)	(l)I(0)	(l)I(2)	$(2k − l)$ I(4)	(l)F(0)	I
Seminvariant phases	ϕ_{eVe}	ϕ_{0e0}	ϕ_{ee0}	ϕ_{eee}	ϕ_{ee0}	ϕ_{ee0}	eee	ϕ_{eee}	ϕ_{hkl} with $h+k+l \equiv 0$ (mod 4)	ϕ_{hk0}	ϕ_{hke}	ϕ_{hkl} with $(2k − l) \equiv 0$ (mod 4)	ϕ_{hk0}	All
Allowed variations for the semi-independent phases	$\|\infty\|$, $\|2\|$ if $k = 0$	$\|\infty\|$	$\|\infty\|$, $\|2\|$ if $l = 0$	$\|2\|$	$\|\infty\|$, $\|2\|$ if $l = 0$	$\|\infty\|$, $\|2\|$ if $l = 0$	$\|2\|$	$\|2\|$	$\|2\|$ if $h + k + l \equiv 0$ (mod 2) $\|4\|$ if $h + k + l \equiv 1$ (mod 2)	$\|\infty\|$	$\|2\|$	$\|2\|$ if $h + k + l \equiv 0$ (mod 2) $\|4\|$ if $2k − l \equiv 1$ (mod 2)	$\|\infty\|$	All
Number of seminvariant phases to be specified as family	2	2	2	2	2	2	2	1	1	1	1	1	1	0

$h + k + l \not\equiv 0$ (mod 2) are systematic absences, it will be sufficient to choose $\boldsymbol{h}_s = (h, k)$ and $\boldsymbol{\omega}_s = (2, 2)$. In this way we find that the dimensions of vectors $\boldsymbol{h}_s$ and ω_s correspond with the number of phases to be specified in order to fix the origin.

3.7 Origin definition by phase assignment

Direct methods (see later in this book) are a typical phasing approach working in reciprocal space; they associate phase values with structure factors, and a subsequent Fourier synthesis may reveal the crystal structure. Since no structure is geometrically defined without having fixed in advance the origin of the reference system, one must deduce that phase assignment in some way implicitly includes the origin definition. When and how is this done during the phasing process? In order to clarify this point, we first discuss the following question: *is it possible to assign arbitrary phases to one or more structure factors* (Zachariasen (1952); Hauptman and Karle (1953))?

We should specify what we mean by arbitrarily assigning one or more phases. Suppose that we have chosen a class of origins; at the same time, the symmetry operators and the functional form of the structure factors are fixed. As we have seen in the preceding sections, the choice of symmetry operators leads to some restrictions on the phase values of certain classes of reflections. To give some trivial examples, we recall that in space group $P\bar{1}$, when the origin is chosen on an inversion centre, the phases can only assume values 0 or π; similarly, in space group P2, if the origin is chosen on a twofold axis (coinciding with **b**) the $(h0l)$ reflections will have phases 0 or π. Therefore, when we talk of arbitrary assignment of the phase to one or more reflections, we will always refer to the phase values allowed by the chosen functional form of the structure factor.

From the theory so far developed, it is obvious that a phase value can be arbitrarily assigned to one or more structure factors if there is at least one allowed origin which, fixed as the origin of the unit cell, will give those phase values to the chosen reflections. Let us suppose that, for a given H–K family, $\phi_{\boldsymbol{h}_1}$ is not a s.s. Its value can be fixed arbitrarily (within the limitations given above), because there exists at least one origin compatible with the given value. Once $\phi_{\boldsymbol{h}_1}$ is assigned, a second phase $\phi_{\boldsymbol{h}_2}$ cannot be fixed arbitrarily if $\phi_{\boldsymbol{h}_2}$ is a s.s., or if $\phi_{\boldsymbol{h}1} + \phi_{\boldsymbol{h}2}$ is a s.s.

Suppose now that neither $\phi_{\boldsymbol{h}_1}$, nor $\phi_{\boldsymbol{h}_2}$, nor $\phi_{\boldsymbol{h}_1} + \phi_{\boldsymbol{h}_2}$ are s.s.s. The necessary condition to be able to arbitrarily assign a third phase $\phi_{\boldsymbol{h}3}$ is that none of $\phi_{\boldsymbol{h}_1}$, $\phi_{\boldsymbol{h}_2}$, $\phi_{\boldsymbol{h}_3}$, $\phi_{\boldsymbol{h}_1} + \phi_{\boldsymbol{h}_2}$, $\phi_{\boldsymbol{h}_1} + \phi_{\boldsymbol{h}_3}$, or $\phi_{\boldsymbol{h}_2} + \phi_{\boldsymbol{h}_3}$, $\phi_{\boldsymbol{h}_1} + \phi_{\boldsymbol{h}_2} + \phi_{\boldsymbol{h}_3}$ is a s.s. As a rule for fixing the origin, three phases can be assigned in those H–K families for which $\boldsymbol{\omega}_s$ is of type (l_1, l_2, l_3), two phases when $\boldsymbol{\omega}_s = (l_1, l_2)$, one phase when $\boldsymbol{\omega}_s = (l)$, and no phase when all ϕ are s.s.s.

As an example, in $P\bar{1}$ the origin may be defined by assigning three phases, provided that none of the following indices,

$$\boldsymbol{h}_1, \boldsymbol{h}_2, \boldsymbol{h}_3, \boldsymbol{h}_1 + \boldsymbol{h}_2, \boldsymbol{h}_1 + \boldsymbol{h}_3, \boldsymbol{h}_2 + \boldsymbol{h}_3, \boldsymbol{h}_1 + \boldsymbol{h}_2 + \boldsymbol{h}_3$$

is of type (even, even, even).

But, how can we arbitrarily fix the enantiomorph? We observe that, if a s.s. (single phase or a combination of phases) has a value Φ for a given structure, it will have the value $-\Phi$ for the enantiomorph structure. However, if Φ is equal to zero, the s.s. has the same value for both enantiomorphs. Therefore, once the origin has been defined, the sign of a s.s., different from zero or π, can be arbitrarily assigned.

So far, we have seen that up to three phases may be freely assigned for defining the origin, and one phase should be fixed for defining the enantiomorph. After that, the phasing process may progressively determine the other phases, which, if the phasing procedure succeeds, will be consistent with a unique origin. It is, however, useful to stress that the procedure quite often starts with an origin definition and stops with phases consistent with another origin. It is nowadays impossible to foresee which origin the phasing process will choose.

The problem of fixing the origin in direct space by assigning phases in reciprocal space impassioned many scientists in the second half of the past century; indeed, the most popular phasing procedures started with a small set of phases, including those for origin definition, which, suitably extended, could lead to the crystal structure solution. Today, this logical but primitive approach is no longer followed; thus the origin definition problem is a curiosity rather than an essential for phasers. The reader will find a thorough treatment of this subject in *Direct Phasing in Crystallography*, Chapter 2.

The method of joint probability distribution functions, neighbourhoods, and representations

4.1 Introduction

Wilson statistics, described in Chapter 2, aims at calculating the distribution of the structure factor $P(F) \equiv P(|F|, \phi)$ when nothing is known about the structure; the positivity and atomicity of the electron density (both promoted by the positive nature of the atomic scattering factors f_j) are the only necessary assumptions. Wilson results may be synthesized as follows:

the modulus $R = |E|$ is distributed according to equations (2.7) or (2.8), while no prevision is possible about ϕ, which is distributed with constant probability $1/(2\pi)$.

In other words, knowledge of the R moduli does not provide information about a phase; this agrees well with Section 3.3, according to which experimental data only allow an estimate of s.i. (and also s.s. if the algebraic form of the symmetry operators has been fixed).

Let us now consider $P(F_{\mathbf{h}_1}, F_{\mathbf{h}_2}) \equiv P(|F_{\mathbf{h}_1}|, |F_{\mathbf{h}_2}|, \phi_{\mathbf{h}_1}, \phi_{\mathbf{h}_2})$, the joint probability distribution function of two structure factors. If the two structure factors are uncorrelated (i.e. no relation is expected between their moduli and between their phases), P will coincide with the product of two Wilson distributions (2.7) or (2.8), say,

$$P(F_{\mathbf{h}_1}, F_{\mathbf{h}_2}) \equiv P(|F_{\mathbf{h}_1}|, \phi_{\mathbf{h}_1}) \cdot P(|F_{\mathbf{h}_2}|, \phi_{\mathbf{h}_2}) = \frac{1}{4\pi^2} P(|F_{\mathbf{h}_1}|) P(|F_{\mathbf{h}_2}|),$$

which is useless (because the two Wilson distributions are useless) for solving the phase problem; indeed, the relation does not provide any phase information.

The question is now: if two structure factors are correlated, may their joint probability distribution function be used for solving the phase problem? Let us first use a simple example to show how much additional information (i.e. that is not present in the two elementary distributions) may be stored in a joint probability distribution function; then we will answer the question.

Let us suppose that the human population of a village has been submitted to statistical analysis to define how weight and height are distributed. From the study of the frequencies, two distributions have been obtained, called P_w and P_h, the first for weight and the second for height. Let us also suppose that the following numerical values have been found:

$$P_w(30)\, dw = P_w(70)\, dw = 0.15$$

and that

$$P_h(100)\, dh = P_h(170)\, dh = 0.15.$$

$P_w(X)\, dw$ is the probability of finding a person with weight X (in kilos), in the interval dw around X, and $P_h(Y)\, dh$ is the probability of finding a person with height Y (in cm) in the interval dh around Y.

If height and weight were statistically independent, their joint probability should be

$$P(X, Y)\, dX\, dY = P_w(X)\, P_h(Y)\, dw\, dh,$$

and therefore the following equalities should arise:

$$P(30, 100)\, dh\, dw = P(30, 170)\, dh\, dw = P(70, 100)\, dh\, dw = P(70, 170)\, dh\, dw = 0.0225.$$

We know, however, that for humans, height and weight are correlated and therefore the above equalities cannot be true. Indeed, it may be expected that the probability values

$$P(30, 100)\, dh\, dw \text{ and } P(70, 170)\, dh\, dw$$

should be larger than the values

$$P(30, 170)\, dh\, dw \text{ and } P(70, 100)\, dh\, dw.$$

Accordingly, the correct joint probability function cannot be the product of the two elementary distributions, but should take into account the covariance of the variables weight and height in order to be a realistic statistical description of the population.

Let us now transfer the above example to the joint distribution of structure factors. A correlation among structure factors implies that some combination of phases should have a non-vanishing probability of assuming a given value; this is just the basis for the solution of the phase problem. For example, let us suppose that the following relations hold:

(a) $\ <F_h F_k> \equiv <|F_h F_k|\exp i(\phi_h + \phi_k)> \approx sF_l \equiv s|F_l|\exp(i\phi_l)$
(b) $\ <F_h F_k F_l> \equiv <|F_h F_k F_l|\exp i(\phi_h + \phi_k + \phi_l)> \approx tF_m \equiv t|F_m|\exp(i\phi_m),$

where s and t are suitable scaling factors. In the first case we may conclude that F_h and F_k are not statistically independent and that the phase relationship

$$(\phi_h + \phi_k) \approx \phi_l$$

may be expected. In case (b) it may be concluded that F_h, F_k, and F_l are not statistically independent, and that the phase relationship

$$(\phi_h + \phi_k + \phi_l) > \approx \phi_m$$

may be expected. The above results would suggest using the joint probability distributions of structure factors to discover the correlations between structure factors.

Introduction

The introduction into crystallography of the joint probability distribution of a set of n normalized structure factors was the great goal of Hauptman and Karle (1953); they showed that moduli and phases belonging to different structure factors are correlated when their indices are chosen in such a way as to constitute s.i.s or s.s.s. After this, other important contributions in the area of joint probability distributions followed, e.g. Bertaut (1955a,b, 1960a,b), Klug (1958), Naya et al. (1964, 1965), and by Giacovazzo and co-workers in a wide series of papers that will be referred to in various chapters of this book.

So, how do we choose structure factors for the study of the statistical properties of s.i.s (and s.s.s)? The simplest approach is as follows: if one wants to study the s.i.,

$$\Phi = \sum_{i=1}^{n} \phi_{\mathbf{h}_i}, \text{ with } \mathbf{h}_1 + \mathbf{h}_2 + \cdots + \mathbf{h}_n = 0 \quad (4.1)$$

the set of normalized structure factors $\{E\} = \{E_{\mathbf{h}_1}, E_{\mathbf{h}_2}, \ldots, E_{\mathbf{h}_n}\}$ should be considered. In terms of joint probability distribution functions, the probability

$$P(E_{\mathbf{h}_1}, E_{\mathbf{h}_2}, \ldots, E_{\mathbf{h}_n}) \equiv P(R_{\mathbf{h}_1}, R_{\mathbf{h}_2}, \ldots, R_{\mathbf{h}_n}, \phi_{\mathbf{h}_1}, \phi_{\mathbf{h}_2}, \ldots, \phi_{\mathbf{h}_n}) \quad (4.2)$$

should be studied. If we apply equation (4.2) to the triplet invariant

$$\Phi = (\phi_{\mathbf{h}} + \phi_{\mathbf{k}} + \phi_{-\mathbf{h}-\mathbf{k}}),$$

the joint probability

$$P(E_{\mathbf{h}}, E_{\mathbf{k}}, E_{\mathbf{h}+\mathbf{k}}) \equiv P(R_{\mathbf{h}}, R_{\mathbf{k}}, R_{\mathbf{h}+\mathbf{k}}, \phi_{\mathbf{h}}, \phi_{\mathbf{k}}, \phi_{\mathbf{h}+\mathbf{k}}) \quad (4.3)$$

should be considered (see Chapter 5).

This simple approach can be generalized as follows. It may be supposed that the n-phase s.i. (4.1) can be better estimated by exploiting more experimental information, e.g. via a number of diffraction amplitudes $p>n$. The general mathematical machinery could be:

1. The set of reflections

$$\{E\} = \{E_{\mathbf{h}_1}, E_{\mathbf{h}_2}, \ldots, E_{\mathbf{h}_n}, \ldots, E_{\mathbf{h}_p}\},$$

   which is considered to be useful for the estimation of the s.i. or s.s. (4.1) is defined. Appropriate sets can be chosen according to the methods described in Section 4.2.

2. The joint probability

$$\begin{aligned}&P(E_{\mathbf{h}_1}, E_{\mathbf{h}_2}, \ldots, E_{\mathbf{h}_n}, \ldots, E_{\mathbf{h}_p}) \\ &\equiv P(R_{\mathbf{h}_1}, R_{\mathbf{h}_2}, \ldots, R_{\mathbf{h}_n}, \ldots, R_{\mathbf{h}_p}, \phi_{\mathbf{h}_1}, \phi_{\mathbf{h}_2}, \ldots, \phi_{\mathbf{h}_n}, \ldots, \phi_{\mathbf{h}_p})\end{aligned} \quad (4.4)$$

   is calculated.

3. The conditional distribution (i.e. probability of the phases, given the moduli)

$$P(\phi_{\mathbf{h}_1}, \phi_{\mathbf{h}_2}, \ldots, \phi_{\mathbf{h}_n}, \ldots, \phi_{\mathbf{h}_p} | R_{\mathbf{h}_1}, R_{\mathbf{h}_2}, \ldots, R_{\mathbf{h}_n}, \ldots, R_{\mathbf{h}_p}) \quad (4.5)$$

   is derived (indeed, we are interested to guess about the phases while $R_{\mathbf{h}_1}, R_{\mathbf{h}_2}, \ldots, R_{\mathbf{h}_n}, \ldots, R_{\mathbf{h}_p}$ are known from experiments, therefore constituting the primary information to exploit).

4. Since we want to estimate $\Phi = \sum_{i=1}^{n} \phi_{\mathbf{h}_i}$ and we have no interest in the unknown phases $\phi_{\mathbf{h}_{n+1}}, \ldots, \phi_{\mathbf{h}_p}$, we can integrate the conditional distribution (4.5) over the phases $\phi_{\mathbf{h}_{n+1}}, \ldots, \phi_{\mathbf{h}_p}$ to obtain the marginal distribution,

$$P(\phi_{\mathbf{h}_1}, \phi_{\mathbf{h}_2}, \ldots, \phi_{\mathbf{h}_n} | R_{\mathbf{h}_1}, R_{\mathbf{h}_2}, \ldots, R_{\mathbf{h}_n}, \ldots, R_{\mathbf{h}_p}). \qquad (4.6)$$

Expression (4.6) provides the distribution of the n phases $\phi_{\mathbf{h}_1}, \phi_{\mathbf{h}_2}, \ldots, \phi_{\mathbf{h}_n}$ no matter what are the values of $\phi_{\mathbf{h}_{n+1}}, \ldots, \phi_{\mathbf{h}_p}$. From (4.6), integration over $\phi_{\mathbf{h}_1}, \phi_{\mathbf{h}_2}, \ldots, \phi_{\mathbf{h}_n}$ under the condition that $\phi_{\mathbf{h}_1} + \phi_{\mathbf{h}_2} + \cdots + \phi_{\mathbf{h}_n} = \Phi$ provides the required s.i. phase distribution

$$P(\Phi | R_{\mathbf{h}_1}, R_{\mathbf{h}_2}, \ldots, R_{\mathbf{h}_n}, \ldots, R_{\mathbf{h}_p}) \qquad (4.7)$$

given the chosen set of moduli. In a more general way we can represent (4.7) by

$$P(\Phi | \{R\}), \qquad (4.8)$$

where $\{R\}$ represents the appropriate set of diffraction amplitudes.

In order to illustrate how to move from (4.6) to (4.7) we consider a two-dimensional example. Let us suppose that we have derived the distribution $P(\phi_{\mathbf{h}_1}, \phi_{\mathbf{h}_2} | R_{\mathbf{h}_1}, R_{\mathbf{h}_2})$ and that we want to derive $P(\Phi | R_{\mathbf{h}_1}, R_{\mathbf{h}_2})$, where $\Phi = \phi_{\mathbf{h}_1} + \phi_{\mathbf{h}_2}$. We then set $\phi_{\mathbf{h}_2} = \Phi - \phi_{\mathbf{h}_1}$ and we integrate $P(\phi_{\mathbf{h}_1}, \phi_{\mathbf{h}_2} | R_{\mathbf{h}_1}, R_{\mathbf{h}_2})$ over $\phi_{\mathbf{h}_1}$, according to

$$P(\Phi | R_{\mathbf{h}_1}, R_{\mathbf{h}_2}) = \int_0^{2\pi} P(\phi_{\mathbf{h}_1}, \Phi - \phi_{\mathbf{h}_1} | R_{\mathbf{h}_1}, R_{\mathbf{h}_2}) d\phi_{\mathbf{h}_1}.$$

The result is the probability distribution of the phase $\Phi = \phi_{\mathbf{h}_1} + \phi_{\mathbf{h}_2}$, given the two magnitudes $R_{\mathbf{h}_1}, R_{\mathbf{h}_2}$. To provide further insight into its meaning in practice, let us suppose that, according to $P(\Phi | R_{\mathbf{h}_1}, R_{\mathbf{h}_2})$, the most probable value of Φ is zero, and that the reliability of the relation $\Phi \approx 0$ is proportional to the product $R_{\mathbf{h}_1} R_{\mathbf{h}_2}$. In this case, if we want to select the cases in which relation $\Phi \approx 0$ is more reliable, we need only consider those in which both $R_{\mathbf{h}_1}$ and $R_{\mathbf{h}_2}$ are sufficiently large.

The mathematical approach described so far is very flexible. For example, if we suppose that of the n phases, $\phi_{\mathbf{h}_1}, \phi_{\mathbf{h}_2}, \ldots, \phi_{\mathbf{h}_n}$, $n-1$ are known (i.e. $\phi_{\mathbf{h}_2}, \ldots, \phi_{\mathbf{h}_n}$), then, from (4.6) the conditional distribution

$$P(\phi_{\mathbf{h}_1} | \phi_{\mathbf{h}_2}, \ldots, \phi_{\mathbf{h}_n}, R_{\mathbf{h}_1}, R_{\mathbf{h}_2}, \ldots, R_{\mathbf{h}_n}, \ldots, R_{\mathbf{h}_p}) \qquad (4.9)$$

may be derived. Prior information now consists of $n-1$ phases and p magnitudes, all used to estimate the single phase $\phi_{\mathbf{h}_1}$. We will see in the next part of the book that distributions (4.7) and (4.9) are equally useful.

Which mathematical techniques should be used in deriving the joint probability distributions of structure factors? The method should not be a surprise to crystallographers; they well know that, if the electron density $\rho(\mathbf{r})$ is inaccessible, while access is allowed to its Fourier transform $F(\mathbf{r}^*)$, then the necessary operation for recovery $\rho(\mathbf{r})$ is to calculate the inverse Fourier transform of $F(\mathbf{r}^*)$. By analogy, for each joint distribution (4.4), its Fourier transform C is first calculated (called the *characteristic function* in statistics), then (4.4) is obtained via an inverse Fourier transform of C. The mathematical techniques used to calculate the two Fourier transforms are described in Appendix 4.A.

A further point to define is how to choose the primitive variables in such probabilistic approaches. In accordance with Section 2.2, the crystallographic problem is quite clear: the joint distribution $P(E_{\mathbf{h}_1}, E_{\mathbf{h}_2}, \ldots, E_{\mathbf{h}_n})$ involves a number of normalized structure factors each of which is a linear sum of random variables; i.e. the atomic contributions to the structure factors. So, for a statistical interpretation of the phase problem, the atomic positions may be considered as primitive random variables.

An alternative basis (see Section 2.2) is that the crystal structure is assumed to be fixed while the reciprocal vectors are assumed to be the random variables. Although the two types of distribution are distinct, we can expect almost equivalent results in many cases, because of the symmetrical role played by $\mathbf{h}$ and $\mathbf{r}$ in the algebraic form of the structure factor. Thus, for derivation of the phase probability for various s.i.s often only one kind of distribution is needed. *As a general rule, in this book the atomic position vectors will be assumed to be primitive random variables*. This is due to the fact that when prior structural information is available (e.g. a group of atoms has been well oriented and positioned), it is more easily treated if the atomic positional vectors are the primitive random variables.

The reader should not think that this method only concerns the structure factors of one crystal structure. Indeed, if a model structure and/or a derivative structure are correlated with the structure under study, then the corresponding structure factors are expected to be correlated, thus suggesting the study of joint probability distributions involving the *E*s of the structure under study and the *E*s of the model and/or the derivative structure. This type of distribution is very useful in macromolecular crystallography, and in general, for modern phasing methods, and is dealt with in some detail in this book.

In this chapter, for simplicity, we will not describe the *saddle point method* (Riemann, 1892; Debye, 1909; de Bruijn, 1970; Kyrala, 1972; Bricogne, 1984), an alternative way to calculate joint probability distribution functions. The interested reader is referred to Appendix 5A of *Direct Phasing in Crystallography* or to the original papers.

4.2 Neighbourhoods and representations

As anticipated in Section 4.1, for any s.i. or s.s., Φ we have to solve the problem of fixing the set of magnitudes which provides the most reliable estimates of Φ. In principle, the set may not be unique; several sets could be found, each giving rise to a different dependence of Φ on the selected magnitudes. Formulation of the nested neighbourhood principle (Hauptman, 1975b) first fixed the idea of defining a sequence of sets of reflections (a sequence of nested neighbourhoods) each contained within the succeeding one and with the property that any s.i. or s.s. may be estimated via the magnitudes constituting any neighbourhood. A practical application of the idea had already been performed, independently, by Giacovazzo (1975b), who calculated in $P\bar{1}$ the one-phase s.s.s via the magnitudes of the second neighbourhood. Subsequently, a more general theory was formulated (the *representation theory*: Giacovazzo, 1977a, 1980), which is able, for any Φ, to arrange the set of magnitudes, provided by the diffraction experiment, into a sequence of subsets, called *phasing shells*, in order of their expected effectiveness (in the statistical sense) for the estimation of Φ.

Here, we will provide a simplified description of the representation theory in the space group $P1$ and $P\bar{1}$. The reader interested in more details is referred to the original papers or to Chapter 6 of *Direct Phasing in Crystallography*.

First representation of a structure invariant. Let

$$\Phi = \phi_{\mathbf{h}_1} + \phi_{\mathbf{h}_2} + \ldots + \phi_{\mathbf{h}_n} \quad (\mathbf{h}_1 + \mathbf{h}_2 + \ldots\ldots + \mathbf{h}_n = 0) \quad (4.10)$$

be an s.i. Its first representation coincides with Φ itself, and its *first phasing shell* is given by the collection of magnitudes $\{B_1\}$, which are *basis magnitudes* (i.e. $\{R_{\mathbf{h}_1}, R_{\mathbf{h}_2}, \ldots, R_{\mathbf{h}_n}\}$) or *cross-magnitudes* of the s.i. The cross-magnitudes satisfy the relation

$$m_1\mathbf{h}_1 + m_2\mathbf{h}_2 + \ldots\ldots + m_n\mathbf{h}_n = 0, \quad \text{with} \quad m_p = 0, 1$$

All the cross-vectors are linear combinations of the basis vectors. We give below three examples:

1. $\Phi = \phi_{\mathbf{h}_1} + \phi_{\mathbf{h}_2} + \phi_{\mathbf{h}_3} + \phi_{\mathbf{h}_4}$ with $\mathbf{h}_1 + \mathbf{h}_2 + \mathbf{h}_3 + \mathbf{h}_4 = 0$ is a quartet invariant. Its basis magnitudes are $\{R_{\mathbf{h}_1}, R_{\mathbf{h}_2}, R_{\mathbf{h}_3}, R_{\mathbf{h}_4}\}$, the cross-magnitudes are $\{R_{\mathbf{h}_1+\mathbf{h}_2}, R_{\mathbf{h}_1+\mathbf{h}_3}, R_{\mathbf{h}_2+\mathbf{h}_3}\}$. Then the first phasing shell consists of seven magnitudes:

$$\{B_1\} = \{R_{\mathbf{h}_1}, R_{\mathbf{h}_2}, R_{\mathbf{h}_3}, R_{\mathbf{h}_4}, R_{\mathbf{h}_1+\mathbf{h}_2}, R_{\mathbf{h}_1+\mathbf{h}_3}, R_{\mathbf{h}_2+\mathbf{h}_3}\} \quad (4.11)$$

2. $\Phi = \phi_{\mathbf{h}_1} + \phi_{\mathbf{h}_2} + \phi_{\mathbf{h}_3}$ is a triplet invariant. The basis magnitudes of the first representation of Φ are $\{R_{\mathbf{h}_1}, R_{\mathbf{h}_2}, R_{\mathbf{h}_3}\}$, the cross terms coincide with them.
3. $\Phi = \phi_{\mathbf{h}_1} + \phi_{\mathbf{h}_2} + \phi_{\mathbf{h}_3} + \phi_{\mathbf{h}_4} + \phi_{\mathbf{h}_5}$ is a quintet invariant. Its basis magnitudes are $\{R_{\mathbf{h}_1}, R_{\mathbf{h}_2}, R_{\mathbf{h}_3}, R_{\mathbf{h}_4}, R_{\mathbf{h}_5}\}$, the cross-magnitudes are $\{R_{\mathbf{h}_1+\mathbf{h}_2}, R_{\mathbf{h}_1+\mathbf{h}_3}, R_{\mathbf{h}_1+\mathbf{h}_4}, R_{\mathbf{h}_1+\mathbf{h}_5}, R_{\mathbf{h}_2+\mathbf{h}_3}, R_{\mathbf{h}_2+\mathbf{h}_4}, R_{\mathbf{h}_2+\mathbf{h}_5}, R_{\mathbf{h}_3+\mathbf{h}_4}, R_{\mathbf{h}_3+\mathbf{h}_5}, R_{\mathbf{h}_4+\mathbf{h}_5}\}$.

The first phasing shell $\{B_1\}$ then consists of 15 magnitudes:

$$\{B_1\} = \{R_{\mathbf{h}_1}, R_{\mathbf{h}_2}, R_{\mathbf{h}_3}, R_{\mathbf{h}_4}, R_{\mathbf{h}_5}, R_{\mathbf{h}_1+\mathbf{h}_2}, R_{\mathbf{h}_1+\mathbf{h}_3}, R_{\mathbf{h}_1+\mathbf{h}_4}, R_{\mathbf{h}_1+\mathbf{h}_5}, \\ R_{\mathbf{h}_2+\mathbf{h}_3}, R_{\mathbf{h}_2+\mathbf{h}_4}, R_{\mathbf{h}_2+\mathbf{h}_5}, R_{\mathbf{h}_3+\mathbf{h}_4}, R_{\mathbf{h}_3+\mathbf{h}_5}, R_{\mathbf{h}_4+\mathbf{h}_5}\} \quad (4.12)$$

As in music, *n-tet* invariants may also be considered; each can be estimated from its first phasing shell.

Second representation of a structure invariant. If Φ, as defined by (4.1), is our structure invariant, its second representation is the collection of $(n + 2)$–*tet* invariants

$$\Psi = \{\phi_{\mathbf{h}_1} + \phi_{\mathbf{h}_2} + \ldots + \phi_{\mathbf{h}_n} + \phi_{\mathbf{k}} - \phi_{\mathbf{k}}\}, \quad (4.13)$$

obtained when $\mathbf{k}$ freely varies over reciprocal space. The *second phasing shell* is the set of magnitudes consisting of the basis and cross-magnitudes of the s.i.s (4.13). We give two examples:

1. The second phasing shell of the triplet invariant $\Phi = \phi_{\mathbf{h}_1} + \phi_{\mathbf{h}_2} + \phi_{\mathbf{h}_3}$ is

$$\{B_2\} = \{R_{\mathbf{h}_1}, R_{\mathbf{h}_2}, R_{\mathbf{h}_3}, R_{\mathbf{k}}, R_{\mathbf{h}_1+\mathbf{k}}, R_{\mathbf{h}_1-\mathbf{k}}, R_{\mathbf{h}_2+\mathbf{k}}, R_{\mathbf{h}_2-\mathbf{k}}, R_{\mathbf{h}_3+\mathbf{k}}, R_{\mathbf{h}_3-\mathbf{k}}\},$$

where $\mathbf{k}$ is a free vector.

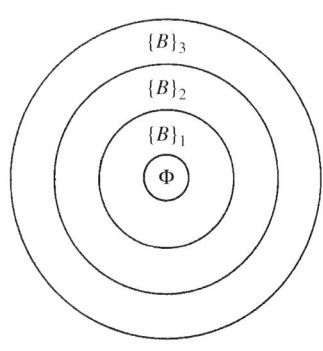

Fig. 4.1
A general scheme of the phasing shell for any s.i., Φ.

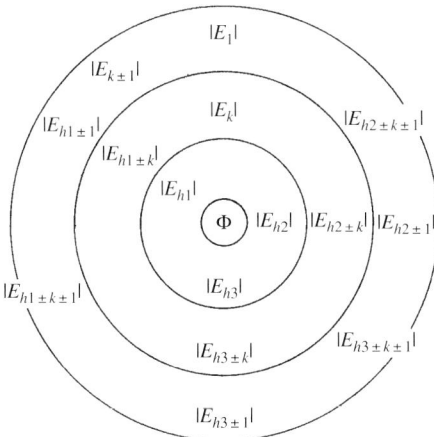

Fig. 4.2
The sequence of the first three phasing shells in P1 and P$\bar{1}$ for the s.i., $\Phi = \phi_{\mathbf{h}_1} + \phi_{\mathbf{h}_2} + \phi_{\mathbf{h}_3}$.

2. The second phasing shell of the quartet invariant $\Phi = \phi_{\mathbf{h}_1} + \phi_{\mathbf{h}_2} + \phi_{\mathbf{h}_3} + \phi_{\mathbf{h}_4}$ is

$\{B_2\} = \{R_{\mathbf{h}_1}, R_{\mathbf{h}_2}, R_{\mathbf{h}_3}, R_{\mathbf{h}_4}, R_{\mathbf{h}_1+\mathbf{h}_2}, R_{\mathbf{h}_1+\mathbf{h}_3}, R_{\mathbf{h}_2+\mathbf{h}_3}, R_{\mathbf{k}}, R_{\mathbf{h}_1\pm\mathbf{k}}, R_{\mathbf{h}_2\pm\mathbf{k}}, R_{\mathbf{h}_3\pm\mathbf{k}},$
$R_{\mathbf{h}_4\pm\mathbf{k}}, R_{\mathbf{h}_1+\mathbf{h}_2\pm\mathbf{k}}, R_{\mathbf{h}_1+\mathbf{h}_3\pm\mathbf{k}}, R_{\mathbf{h}_2+\mathbf{h}_3\pm\mathbf{k}}\}$.

Third representation of a structure invariant. If Φ, as defined by (4.1), is our structure invariant, its third representation is the collection of $(n+4)$–tet invariants,

$$\Psi = \{\phi_{\mathbf{h}_1} + \phi_{\mathbf{h}_2} + \cdots + \phi_{\mathbf{h}_n} + \phi_{\mathbf{k}} - \phi_{\mathbf{k}} + \phi_{\mathbf{l}} - \phi_{\mathbf{l}}\}, \quad (4.14)$$

where $\mathbf{k}$ and $\mathbf{l}$ are free vectors in the reciprocal space. The *third phasing shell* may be defined similarly to above.

These definitions suggest that an s.i. of order n is represented in its second representation by special invariants of order $n + 2$, in its third representation by special invariants of order $n + 4$, etc., and that the reciprocal space may be arranged into subsets of measured reflections, the *phasing shells*, each contained within the succeeding one and having the property that any s.i. or s.s. may be estimated via the magnitudes of the shell.

A symbolic way of representing such an arrangement of the reciprocal space in phasing shells is shown in Figs. 4.1–4.3. Figure 4.1 is a symbolic way of partitioning the reciprocal space according to the subsets $\{B_j\}$; Figs. 4.2 and 4.3 show the first phasing shells for a triplet and for a quartet invariant respectively.

Extension of the concept of s.i. representation is more complicated when symmetry is taken into account (but also more rewarding, because symmetry allows us to exploit a larger number of cross-magnitudes). We refer the interested reader to Chapter 6 of *Direct Phasing in Crystallography*, or to the original papers.

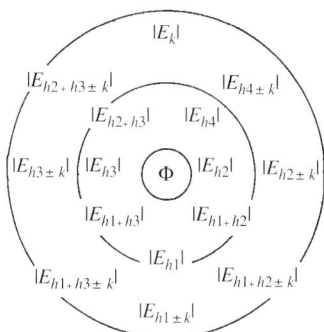

Fig. 4.3
The sequence of the first two phasing shells in P1 and P$\bar{1}$ for the s.i., $\Phi = \phi_{\mathbf{h}_1} + \phi_{\mathbf{h}_2} + \phi_{\mathbf{h}_3} + \phi_{\mathbf{h}_4}$.

4.3 Representations of structure seminvariants

It was shown in Section 3.5 that $\Phi = \phi_{\mathbf{u}_1} + \phi_{\mathbf{u}_2} + \cdots + \phi_{\mathbf{u}_n}$, where $\mathbf{u}_j$ is the generic reciprocal vector, is an s.s. if its value does not change when the origin moves from one to another *allowed* or *permissible origin*. In other words, Φ is an s.s. if,

$$\left(\sum_{j=1}^{n} \mathbf{u_j}\right)\mathbf{x}_p = r, \quad p = 1, 2, \ldots,$$

where r is a positive, null, or negative integer, and $\{\mathbf{x}_p, p = 1, 2, \ldots\}$ are the *allowed translations* between permissible origins.

The theory of representations provides more useful (for phasing purposes) definitions of the s.s.s. In particular, the theory identifies two types of s.s.: first and second rank seminvariants. To define the s.s.s of the first rank:

$\Phi = \phi_{\mathbf{u}_1} + \phi_{\mathbf{u}_2} + \cdots + \phi_{\mathbf{u}_n}$ is an s.s. of the first rank if at least one phase, $\phi_{\mathbf{h}_1}$, and a symmetry operator, $\mathbf{C}_p \equiv (\mathbf{R}_p, \mathbf{T}_p)$, exist such that

$$\Psi_1 = \phi_{\mathbf{u}_1} + \phi_{\mathbf{u}_2} + \cdots + \phi_{\mathbf{u}_n} + \phi_{\mathbf{h}_1} - \phi_{\mathbf{h}_1 \mathbf{R}_p} \quad (4.15)$$

is an s.i. Once Ψ_1 is estimated from the observed amplitudes, Φ is estimated simultaneously. Indeed, because of equation (1.25), we may rewrite (4.15) in the form,

$$\Psi_1 = \Phi + \Delta_p,$$

where $\Delta_p = 2\pi \mathbf{h}_1 \mathbf{T}_p$ is the phase shift arising because of translational symmetry. For the same Φ, further $\mathbf{h}_1$ reflections and therefore more invariants Ψ_1 may be found. The set $\{\Psi_1\}$ is called the first representation of Φ.

Let us give some examples.

1. $\Phi = \phi_{\mathbf{u}_1} = \phi_{406}$ is an s.s. of first rank in P2; in fact (4.15) is verified by $\mathbf{R}_s = \mathbf{I}$, $\mathbf{h}_1 = (\overline{2k3})$, $\mathbf{R}_p = \mathbf{R}_2$. Accordingly, the first representation of Φ is the collection of the special triplet invariants

$$\Psi_1 = \phi_{406} - \phi_{2k3} + \phi_{\overline{2}k\overline{3}},$$

with k a free index

2. The first representation in P$\overline{1}$ of $\Phi = \phi_{2h}$ coincides with one special triplet only:

$$\Psi_1 = \phi_{2h} - \phi_h - \phi_h.$$

The estimate of ϕ_{2h} will depend on R_{2h} and R_h only.

3. $\Phi = \phi_{123} + \phi_{\overline{7}2\overline{5}}$ is an s.s. of the first rank in P222; in fact (4.15) is verified when

$$\mathbf{R}_s = \mathbf{R}_t = \mathbf{I} \quad \mathbf{R}_p = \begin{vmatrix} \overline{1} & 0 & 0 \\ 0 & 1 & 0 \\ 0 & 0 & \overline{1} \end{vmatrix}, \quad \mathbf{h}_1 = (3, k, 1),$$

or

$$\mathbf{R}_s = \mathbf{I} \quad \mathbf{R}_t = \mathbf{R}_p = \begin{vmatrix} \overline{1} & 0 & 0 \\ 0 & 1 & 0 \\ 0 & 0 & \overline{1} \end{vmatrix}, \quad \mathbf{h}_1 = (\overline{4}, k, \overline{4}).$$

Accordingly, the first representation of Φ is the collection of special quartets,

$$\Psi_1 = \phi_{123} + \phi_{\overline{7}2\overline{5}} + \phi_{3k1} - \phi_{\overline{3}k\overline{1}} \quad (4.16a)$$

and
$$\Psi'_1 = \phi_{123} + \phi_{7\bar{2}5} + \phi_{\bar{4}k\bar{4}} - \phi_{4k4}. \quad (4.16b)$$

Again, k is a free index.

4. The first representation in $P\bar{1}$ of the two-phase s.s. of first rank, $\Phi = \phi_{u_1} + \phi_{u_2} = \phi_{h+k} + \phi_{h-k}$, is the collection of two special quartets:
$$\begin{aligned}\Psi'_1 &= \phi_{h+k} + \phi_{h-k} - \phi_h - \phi_h, \\ \Psi''_1 &= \phi_{h+k} - \phi_{h-k} - \phi_k - \phi_k.\end{aligned} \quad (4.17)$$

Suppose now that Φ is an s.s. for which (4.15) cannot be satisfied, but two phases, ϕ_{h_1} and ϕ_{h_2}, and four symmetry operators C_p, C_q, C_i, C_j can be found such that
$$\Psi_1 = \phi_{u_1 R_s} + \phi_{u_2 R_t} + \cdots + \phi_{u_n R_v} + \phi_{h_1 R_p} - \phi_{h_1 R_q} + \phi_{h_2 R_i} - \phi_{h_2 R_j} \quad (4.18)$$

is an s.i. Then Φ is an s.s. of second rank. In this case, the first representation of Φ is the collection of s.i.s, $\{\Psi\}_1$, defined by (4.18).

As an example, for space groups with point group 222, the s.s.s $\Phi = \phi_{u_1}$ for which
$$u_1 \equiv 0 \bmod (0, 2, 2) \text{ or } (2, 0, 2) \text{ or } (2, 2, 0)$$
are of first rank, whereas the s.s.s for which this condition is not fulfilled, but the condition
$$u_1 \equiv 0 \bmod (2, 2, 2)$$
is satisfied, are of the second rank. Thus, $\Phi = \phi_{246}$ and $\Phi = \phi_{146} + \phi_{3\bar{6}2}$ are both s.s.s of the second rank.

The reader can easily verify the following rule: *an n-phase s.s. of the first rank is represented in its first representation by a collection of* n + 2-*phase invariants if it is of first rank, by a collection of* n + 4-*phase invariants if it is of second rank.*

For both first and second rank s.s.s the first phasing shell $\{B\}_1$ is the collection of the R magnitudes which are basis or cross-magnitudes of at least one $\Psi_1 \in \{\Psi\}_1$. For example:

1. The first phasing shell $\{B\}_1$ of $\Phi = \phi_{123} + \phi_{7\bar{2}5}$ in P222 is (see the relations (4.16))
$$\{B\}_1 = \{R_{123}, R_{725}, R_{3k1}, R_{4k4}, R_{\bar{6}0\bar{2}}, R_{808}\},$$
with k the free index.

2. The first phasing shell corresponding to the first representation (4.17) of $\Phi = \phi_{h+k} + \phi_{h-k}$ is
$$\{B\}_1 = \{R_{h+k}, R_{h-k}, R_h, R_k, R_{2h}, R_{2k}\}.$$

4.4 Representation theory for structure invariants extended to isomorphous data

It will be shown in various chapters of this book that isomorphous data (say data corresponding to crystal structures with correlated electron density) play a central role both in small molecules and in macromolecular crystallography. A typical pair of isomorphous structures consists of the *target structure* (the structure of which we want to know the electron density) and a *model structure*, eventually available at a given step of the phasing process. Other types of isomorphous data are those available in anomalous dispersion or in isomorphous derivative techniques, as well as in molecular replacement applications (see Chapters 13 to 15).

In general, for a given index, **h**, a collection of moduli $R_{j,h}$, say

$$R_{j,h} \quad j = 1, \ldots, b$$

might be available. For example:

1. If a model structure is available, both the measured modulus R_h (of the target structure) and the calculated (from the model) R_{ph} values are disposable. Then $b = 2$ and

$$R_{1,h} = R_h, \quad R_{2,h} = R_{ph}.$$

2. If diffraction data from two heavy atom derivatives are available (besides native protein data), then $b = 3$; $R_{1,h}$ may correspond to the native protein while $R_{2,h}$ and $R_{3,h}$ refer to isomorphous derivatives.

3. If measurements are made at a specific wavelength generating anomalous dispersion effects, then $b = 2$. The isomorphous pairs are then the Friedel related reflections

$$R_{1,h} = R_h, \quad R_{2,h} = R_{-h}.$$

If two wavelengths λ_1 and λ_2 are used, then $b = 4$ and

$$R_{1,\mathbf{h}} = R_{\lambda_1,\mathbf{h}}, \quad R_{2,\mathbf{h}} = R_{\lambda_1,-\mathbf{h}},$$

$$R_{3,\mathbf{h}} = R_{\lambda_2,\mathbf{h}}, \quad R_{4,\mathbf{h}} = R_{\lambda_2,-\mathbf{h}}.$$

We now state here, for $P1$ and $P\bar{1}$, the necessary definitions for s.i.s and s.s.s and their main properties (for simplicity, we do not use the symmetry, even if useful).

First representation of a structure invariant. For a given s.i.,

$$\Phi = \phi_{i,\mathbf{h}_1} + \phi_{i,\mathbf{h}_2} + \ldots + \phi_{i,\mathbf{h}_n},$$

the collection of distinct s.i.s

$$\Psi = \left\{ \phi_{i,\mathbf{h}_1} + \phi_{j,\mathbf{h}_2} + \ldots + \phi_{r,\mathbf{h}_n} \right\},$$

obtained when $i, j, \ldots, r$ vary over the different isomorphous data set is called the first representation of Φ.

We will now give some practical examples.

1. The phase

$$\Phi = \phi_{i,\mathbf{h}} + \phi_{j,-\mathbf{h}} = \phi_{i,\mathbf{h}} - \phi_{j,\mathbf{h}}$$

is an s.i. i and j correspond with the pair (model structure, target structure), (protein, isomorphous derivative), or (first isomorphous derivative, second isomorphous derivative), etc.

2.
$$\Phi = \phi_{i,\mathbf{h}} + \phi_{i,-\mathbf{h}}$$

corresponds with data collected at the same wavelength (anomalous dispersion exciting) for two Friedel related reflections.

3.
$$\Phi = \phi_{i,\mathbf{h}} + \phi_{j,-\mathbf{h}}$$

or

$$\Phi = \phi_{i,\mathbf{h}} - \phi_{j,\mathbf{h}}$$

corresponds to data collected at two different wavelengths, both exciting anomalous dispersion.

4. In the case of a native protein and one heavy atom derivative, the first representation of the triplet invariant

$$\Phi = (\phi_{1,\mathbf{h}_1} + \phi_{1,\mathbf{h}_2} + \phi_{1,\mathbf{h}_3})$$

is the collection of eight triplets,

$$(\phi_{1,\mathbf{h}_1} + \phi_{1,\mathbf{h}_2} + \phi_{1,\mathbf{h}_3}), (\phi_{1,\mathbf{h}_1} + \phi_{1,\mathbf{h}_2} + \phi_{2,\mathbf{h}_3})$$

$$(\phi_{1,\mathbf{h}_1} + \phi_{2,\mathbf{h}_2} + \phi_{1,\mathbf{h}_3}), (\phi_{2,\mathbf{h}_1} + \phi_{1,\mathbf{h}_2} + \phi_{1,\mathbf{h}_3})$$

$$(\phi_{1,\mathbf{h}_1} + \phi_{2,\mathbf{h}_2} + \phi_{2,\mathbf{h}_3}), (\phi_{2,\mathbf{h}_1} + \phi_{1,\mathbf{h}_2} + \phi_{2,\mathbf{h}_3})$$

$$(\phi_{2,\mathbf{h}_1} + \phi_{2,\mathbf{h}_2} + \phi_{1,\mathbf{h}_3}), (\phi_{2,\mathbf{h}_1} + \phi_{2,\mathbf{h}_2} + \phi_{2,\mathbf{h}_3}),$$

The same eight triplets are the first representation of the target triplet invariant,

$$\Phi = (\phi_{\mathbf{h}_1} + \phi_{\mathbf{h}_2} + \phi_{-\mathbf{h}_1-\mathbf{h}_2}),$$

when a model structure exists (the subscript 1 then corresponds to the target and the subscript 2 to the model structure).

For simplicity, we do not involve examples of quartet invariants, we will not introduce the concept of upper representation (which may easily be deduced from the definitions given in Section 4.2), and we also avoid introducing the symmetry. Interested readers are referred to Chaper 6 of *Direct Phasing in Crystallography*.

APPENDIX 4.A THE METHOD OF STRUCTURE FACTOR JOINT PROBABILITY DISTRIBUTION FUNCTIONS

4.A.1 Introduction

The method of joint probability distributions plays a special role in phasing approaches. As stated in the main text, it was introduced into crystallography by Hauptman and Karle (1953) and has since been successfully applied to an extraordinary variety of cases. Even if most of the phase relationships we will treat in this book may be obtained via algebraic considerations, the probabilistic techniques are the only methods capable of also associating reliability criteria. The general reader, if only involved in practical phasing applications, may not be interested in going deeper into this topic. However, a brief description of its bases is mandatory, in order to inform other readers more interested in aspects of theoretical phasing.

4.A.2 Multivariate distributions in centrosymmetric structures: the case of independent random variables

Let us consider a cs. space group of order m. We wish to develop an expression for the joint probability distribution,

$$P(E_{h_1}, E_{h_2}, \ldots, E_{h_n}), \qquad (4.A.1)$$

which does not depend on the choice of primitive random variables. Then, it could be applied to the wide variety of situations frequently encountered in the course of direct phasing procedures. We will use the notation

$$F_h = \sum_{j=1}^{t} \xi_j(h),$$

where

$$\xi_j(h) = f_j(h) \sum_{s=1}^{m} \exp(2\pi i \bar{h} C_s r_j),$$

and t is the number of atoms in the asymmetric unit. For the sake of simplicity, we will assume that no atom is in a special position. Then, according to (2.6),

$$E_h = \frac{F_h}{\langle |F_h|^2 \rangle^{1/2}} = \sum_{j=1}^{t} \xi'_j(h),$$

where

$$\xi'_j(h) = \frac{\xi_j(h)}{\langle |F_h|^2 \rangle^{1/2}}.$$

$\langle |F_h|^2 \rangle$ is calculated according to available specific prior information. We make the explicit assumption, valid for Sections 4.A.2 to 4.A.4, that the primitive random variables are statistically independent of each other.

The distribution (4.A.1) may be found (refer to Appendix M.A) by taking the Fourier transform of the characteristic function

$$C(u_1, u_2, \ldots, u_n),$$

where u_r is the carrying variable associated with E_{h_r}. In accordance with Klug (1958) and Karle and Hauptman (1958), this is given by

$$C(u_1, \ldots, u_n) = \langle \exp[i(u_1 E_{h_1} + u_2 E_{h_2} + \cdots + u_n E_{h_n})] \rangle$$
$$= \left\langle \exp\left[i \left(u_1 \sum_{j=1}^{t} \xi'_j(h_1) + u_2 \sum_{j=1}^{t} \xi'_j(h_2) + \cdots + u_n \sum_{j=1}^{t} \xi'_j(h_n) \right) \right] \right\rangle. \qquad (4.A.2)$$

The method of structure factor joint probability distribution functions

Since each random variable is supposed to be statistically independent, equation (4.A.2) may be written as

$$\begin{aligned}
C(u_1,\ldots,u_n) &= \langle \exp\{i[u_1\xi_1'(\boldsymbol{h}_1) + u_2\xi_1'(\boldsymbol{h}_2) + \cdots + u_n\xi_1'(\boldsymbol{h}_n)]\}\rangle \\
&\quad \cdot \langle \exp\{i[u_1\xi_2'(\boldsymbol{h}_1) + u_2\xi_2'(\boldsymbol{h}_2) + \cdots + u_n\xi_2'(\boldsymbol{h}_n)]\}\rangle \\
&\quad \vdots \quad \vdots \quad \vdots \quad \vdots \quad \vdots \quad \vdots \quad \vdots \\
&\quad \cdot \langle \exp\{i[u_1\xi_t'(\boldsymbol{h}_1) + u_2\xi_t'(\boldsymbol{h}_2) + \cdots + u_n\xi_t'(\boldsymbol{h}_n)]\}\rangle \\
&= \prod_{j=1}^{t} {}^{j}C(u_1, u_2, \ldots, u_n).
\end{aligned} \quad (4.A.3)$$

In (4.A.3), ${}^{j}C$ is the characteristic function of the joint probability distribution of the contributions of the jth atom to $E_{\boldsymbol{h}_1}, \ldots, E_{\boldsymbol{h}_n}$. Equation (4.A.3) is the required expression of the characteristic function. The averages in (4.A.3) should be taken over $\boldsymbol{h}$ or over $\boldsymbol{r}$, depending on whether $\boldsymbol{h}$ or $\boldsymbol{r}$ is the primitive random variable. In this last case, each average in (4.A.3) may be accomplished by integrating over the space spanned by $\boldsymbol{r}$, provided the prior probability $q_j(\boldsymbol{r})$ is introduced ($q_j(\boldsymbol{r})$ is the probability that the jth atom is in $\boldsymbol{r}$):

$${}^{j}C(u_1, u_2, \ldots, u_n) = \int q_j(\boldsymbol{r}) \exp\left\{i\left[u_1\xi_j'(\boldsymbol{h}_1) + \cdots + u_n\xi_j'(\boldsymbol{h}_n)\right]\right\} d\boldsymbol{r}. \quad (4.A.4)$$

In this last case, ξ_j' should be expressed as a function of $\boldsymbol{r}$. In general (but not always) we will assume that $q_j(\boldsymbol{r})$ is a uniform distribution, that $q_i(\boldsymbol{r}) = q_j(\boldsymbol{r}) \equiv q(\boldsymbol{r})$ for any pair i, j, and that each atom is statistically independent of all others.

For each ${}^{j}C$, the multivariate *cumulant generating function* (see Sections M.A.4 and M.A.7),

$${}^{j}K(iu_1, iu_2, \ldots, iu_n) = \log {}^{j}C(u_1, u_2, \ldots, u_n),$$

may be introduced, which, when expanded in the series of cumulants, gives

$$\begin{aligned}
C(u_1, u_2, \ldots, u_n) &= \exp\left(\sum_{j=1}^{t} {}^{j}K(iu_1, iu_2, \ldots, iu_n)\right) \\
&= \exp\left(\sum_{r+s+\cdots+w=2}^{\infty} \sum_{j=1}^{t} \frac{{}^{j}k_{rs\cdots w}}{r!s!\cdots w!}(iu_1)^r (iu_2)^s \cdots (iu_n)^w\right).
\end{aligned} \quad (4.A.5)$$

The cumulants ${}^{j}k_{rs\cdots w}$ can be expressed in terms of joint moments $m_{pq\cdots t}$ via the relations quoted in Sections M.A.4 and M.A.7. In their turn the joint moments may be evaluated according to the technique described in Section M.A.7.

If we emphasize the standardized cumulants of second order in the right-hand side of (4.A.5) and we denote by S_v the contributions of all the terms with $r + s + \cdots + w = v$, we obtain

$$\begin{aligned}
C(u_1, u_2, \ldots, u_n) &= \exp\left\{-\frac{1}{2}\left[\left(\sum_{j=1}^{t} {}^{j}k_{200\cdots}\right)u_1^2 \right.\right. \\
&\quad \left. + \left(\sum_{j=1}^{t} {}^{j}k_{020\cdots}\right)u_2^2 + \cdots + \left(\sum_{j=1}^{t} {}^{j}k_{110\cdots}\right)u_1 u_2 \right. \\
&\quad \left.\left. + \left(\sum_{j=1}^{t} {}^{j}k_{1010\cdots}\right)u_1 u_3 + \cdots\right]\right\} \cdot \exp\left(\sum_{v=3}^{\infty} S_v\right),
\end{aligned} \quad (4.A.6)$$

where

$$S_v = \sum_{r+s+\cdots+w=v} \left(\sum_{j=1}^{t} \frac{{}^j k_{rs\cdots w}}{r!s!\cdots w!} (iu_1)^r (iu_2)^s \cdots (iu_n)^w \right). \quad (4.A.7)$$

The joint probability density (4.A.1) is given by the Fourier transform of (4.A.6). If $\bar{E}$ and $\bar{u}$ are the n-dimensional column vectors, that is,

$$\bar{E} = (E_{h_1}, E_{h_2}, \ldots, E_{h_n}), \quad \bar{u} = (u_1, u_2, \ldots, u_n),$$

we have

$$P(E) = \frac{1}{(2\pi)^n} \int C(u) \exp(-i\bar{u}E) \, du$$
$$\simeq \frac{1}{(2\pi)^n} \int_{-\infty}^{+\infty} \cdots \int_{-\infty}^{+\infty} \exp\left(-i\bar{u}E - \frac{1}{2}\bar{u}\lambda u\right) \exp\left(\sum_{v=3}^{\infty} S_v\right) d\bar{u}, \quad (4.A.8)$$

where

$$\lambda = \begin{bmatrix} \left(\sum_{j=1}^{t} {}^j k_{200\cdots}\right) & \left(\sum_{j=1}^{t} {}^j k_{110\cdots}\right) & \left(\sum_{j=1}^{t} {}^j k_{1010\cdots}\right) & \cdots & \left(\sum_{j=1}^{t} {}^j k_{10\cdots 1}\right) \\ \left(\sum_{j=1}^{t} {}^j k_{110\cdots}\right) & \left(\sum_{j=1}^{t} {}^j k_{020\cdots}\right) & \left(\sum_{j=1}^{t} {}^j k_{0110\cdots}\right) & \cdots & \left(\sum_{j=1}^{t} {}^j k_{01\cdots 1}\right) \\ \vdots & \vdots & \vdots & & \vdots \\ \left(\sum_{j=1}^{t} {}^j k_{10\cdots 1}\right) & \left(\sum_{j=1}^{t} {}^j k_{010\cdots 1}\right) & \left(\sum_{j=1}^{t} {}^j k_{0010\cdots 1}\right) & \cdots & \left(\sum_{j=1}^{t} {}^j k_{0\cdots 2}\right) \end{bmatrix}. \quad (4.A.9)$$

If the values of S_v are estimated, (4.A.1) may, in principle, be calculated by means of (4.A.8).

We should note explicitly that, in most practical applications (e.g. when no *prior* information is available about the positional atomic vectors), (4.A.9) may be considered to be the covariance matrix of the distribution expressed in terms of second-order standardized cumulants. In fact, its diagonal elements equal

$$\langle E_{h_p} E_{h_p} \rangle = \left\langle \sum_{i,j=1}^{t} \xi'_i(\boldsymbol{h}_p)\xi'_j(\boldsymbol{h}_p) \right\rangle = \left\langle \sum_{j=1}^{t} \xi'^2_j(\boldsymbol{h}_p) \right\rangle = 1$$

and its off-diagonal elements equal

$$\langle E_{h_p} E_{h_q} \rangle = \left\langle \sum_{i,j=1}^{t} \xi'_i(\boldsymbol{h}_p)\xi'_j(\boldsymbol{h}_q) \right\rangle = \left\langle \sum_{j=1}^{t} \xi'_j(\boldsymbol{h}_p)\xi'_j(\boldsymbol{h}_q) \right\rangle = \sum_{j=1}^{t} {}^j k_{0\cdots 10\cdots 10\cdots}$$

4.A.3 Multivariate distributions in non-centrosymmetric structures: the case of independent random variables

In accordance with Section 4.A.2,

$$F_h = \sum_{j=1}^{t} \xi_j(h) = \sum_{j=1}^{t} \psi_j(h) + i \sum_{j=1}^{t} \eta_j(h),$$

where

$$\psi_j(h) = f_j(h) \sum_{s=1}^{m} \cos\left(2\pi \bar{h} C_s r_j\right)$$

$$\eta_j(h) = f_j(h) \sum_{s=1}^{m} \sin\left(2\pi \bar{h} C_s r_j\right).$$

$E_h = A_h + iB_h$ is the normalized structure factor whose algebraic expression may be written as

$$E_h = \sum_{j=1}^{t} \xi'_j(h) = \sum_{j=1}^{t} \psi'_j(h) + i\eta'_j(h),$$

where

$$\xi'_j(h) = \frac{\xi_j(h)}{\langle |F_h|^2 \rangle^{1/2}}.$$

Accordingly,

$$A_h = \sum_{j=1}^{t} \psi'_j(h)$$

$$B_h = \sum_{j=1}^{t} \eta'_j(h),$$

are the real and imaginary parts, respectively, of E_h.

We wish to develop, for a n.cs. space group of order m, an expression for the joint probability distribution

$$P(E_{h_1}, E_{h_2}, \ldots, E_{h_n}) = P(A_{h_1}, \ldots, A_{h_n}, B_{h_1}, \ldots, B_{h_n}) \quad (4.A.10)$$

which does not formally depend on the choice of the primitive random variables.

On the assumption that A_j and B_j are independent variables, let us first calculate the characteristic function

$$C(u_1, u_2, \ldots, u_n, v_1, \ldots, v_n),$$

where u_r and v_r are carrying variables associated with A_{h_r} and B_{h_r} respectively. In accordance with Appendix M.A and Section 4.A.2 we find that

$$\begin{aligned}
C(u_1, u_2, &\ldots, u_n, v_1, \ldots, v_n) \\
&= \langle \exp[i(u_1 A_1 + u_2 A_2 + \cdots + v_n B_n)] \rangle \\
&= \left\langle \exp\left[i\left(u_1 \sum_{j=1}^{t} \psi_j'(\boldsymbol{h}_1) + u_2 \sum_{j=1}^{t} \psi_j'(\boldsymbol{h}_2) + \cdots + v_n \sum_{j=1}^{t} \eta_j'(\boldsymbol{h}_n)\right)\right] \right\rangle \\
&= \langle \exp\{i[u_1 \psi_1'(\boldsymbol{h}_1) + u_2 \psi_1'(\boldsymbol{h}_2) + \cdots + v_n \eta_1'(\boldsymbol{h}_n)]\} \rangle \\
&\quad \times \langle \exp\{i[u_1 \psi_2'(\boldsymbol{h}_1) + u_2 \psi_2'(\boldsymbol{h}_2) + \cdots + v_n \eta_2'(\boldsymbol{h}_n)]\} \rangle \\
&\quad \cdots \\
&\quad \times \langle \exp\{i[u_1 \psi_t'(\boldsymbol{h}_1) + u_2 \psi_t'(\boldsymbol{h}_2) + \cdots + v_n \eta_t'(\boldsymbol{h}_n)]\} \rangle \\
&= \prod_{j=1}^{t} {}^j C(u_1, u_2, \ldots, v_n).
\end{aligned}$$
(4.A.11)

As in (4.A.3), ${}^j C$ is the characteristic function of the joint probability distribution of the contributions of the jth atom to $E_{h_1}, E_{h_2}, \ldots, E_{h_n}$. Again, $\boldsymbol{h}$ or $\boldsymbol{r}$ may be the primitive random variables. In this last case each average in (4.A.11) may be accomplished by integrating over the space spanned by $\boldsymbol{r}$ provided the prior probability $q_j(\boldsymbol{r})$ is introduced ($q_j(\boldsymbol{r})$ is the probability that the jth atom is in $\boldsymbol{r}$):

$${}^j C(u_1, \ldots, v_n) = \int q_j(\boldsymbol{r}) \exp\{i[u_1 \psi_j'(\boldsymbol{h}_1) + \cdots + v_n \eta_j'(\boldsymbol{h}_n)]\} \, d\boldsymbol{r}. \quad (4.A.12)$$

In this last case, ψ_j' and η_j' should be expressed as functions of $\boldsymbol{r}$. In their turn the moments of the distribution should be calculated via the formula

$$\begin{aligned}
{}^j m'_{rs\cdots w} &= \langle [\psi_j'(\boldsymbol{h}_1)]^r [\psi_j'(\boldsymbol{h}_2)]^s \cdots [\eta_j'(\boldsymbol{h}_n)]^w \rangle \\
&= \int q_j(\boldsymbol{r}) [\psi_j'(\boldsymbol{h}_1)]^r [\psi_j'(\boldsymbol{h}_2)]^s \cdots [\eta_j'(\boldsymbol{h}_n)]^w \, d\boldsymbol{r}.
\end{aligned} \quad (4.A.13)$$

Details about the use of the symmetry in (4.A.13) are given in Section M.A.8.

In general (but not always), we will assume that $q_j(\boldsymbol{r})$ is a uniform distribution, that $q_i(\boldsymbol{r}) \equiv q_j(\boldsymbol{r}) = q(\boldsymbol{r})$ for any pair i,j, and that each atom is statistically independent of the others. As in Section 4.A.2, for each ${}^j C$ the multivariate cumulant generating function is introduced, which, when expanded in a series of cumulants, leads to

$$C(u_1, u_2, \ldots, u_n, v_1, \ldots, v_n) = \exp\left(\sum_{v=2}^{\infty} S_v\right), \quad (4.A.14)$$

where

$$S_v = \sum_{r+s+\cdots+w=v} \left(\sum_{j=1}^{t} \frac{{}^j k_{rs\cdots w}}{r! s! \cdots w!} (iu_1)^r (iu_2)^s \cdots (iv_n)^w \right). \quad (4.A.15)$$

The method of structure factor joint probability distribution functions

To assist the reader we will express the lower-order cumulants in terms of moments (see Sections M.A.4 and M.A.7):

$$^j k_{20\cdots 0} = \langle [\psi'_j(\mathbf{h}_1)]^2 \rangle, ^j k_{020\cdots 0} = \langle [\psi'_j(\mathbf{h}_2)]^2 \rangle, \ldots,$$

$$^j k_{110\cdots 0} = \langle \psi'_j(\mathbf{h}_1)\psi'_j(\mathbf{h}_2) \rangle, \ldots, ^j k_{0\cdots 02} = \langle [\eta'_j(\mathbf{h}_n)]^2 \rangle,$$

$$^j k_{1110\cdots 0} = \langle \psi'_j(\mathbf{h}_1)\psi'_j(\mathbf{h}_2)\psi'_j(\mathbf{h}_3) \rangle, \text{ etc.}$$

The distribution (4.A.10) is given by the Fourier transform of (4.A.14). If $\mathbf{E}$ and $\mathbf{u}$ are n-dimensional vectors, that is,

$$\bar{\mathbf{E}} = (A_{\mathbf{h}_1}, \ldots, A_{\mathbf{h}_n}, B_{\mathbf{h}_1}, \ldots, B_{\mathbf{h}_n}),$$

$$\bar{\mathbf{u}} = (u_{\mathbf{h}_1}, \ldots, u_{\mathbf{h}_n}, v_{\mathbf{h}_1}, \ldots, v_{\mathbf{h}_n}),$$

we have,

$$P(\mathbf{E}) = \frac{1}{(2\pi)^{2n}} \int C(\mathbf{u}) \exp(-i\bar{\mathbf{u}}\bar{\mathbf{E}}) \, d\bar{\mathbf{u}}$$

$$\simeq \frac{1}{(2\pi)^{2n}} \int_{-\infty}^{+\infty} \cdots \int_{-\infty}^{+\infty} \exp\left(-i\bar{\mathbf{u}}\bar{\mathbf{E}} - \frac{1}{2}\bar{\mathbf{u}}\lambda\mathbf{u}\right) \exp\left(\sum_{\nu=3}^{\infty} S_\nu\right) d\bar{\mathbf{u}}. \qquad (4.A.16)$$

4.A.4 Simplified joint probability density functions in the absence of prior information

Let the positional atomic vectors be the primitive random variables. If nothing is known about the distribution of atoms in the asymmetric unit of the cell, we can assume that all the non-symmetry related atoms are independently distributed through the asymmetric unit.

Let us first consider the cs. case. Then (see Section M.A.8),

$$\sum_{j=1}^{t} {}^j k_{200\cdots} = \sum_{j=1}^{t} |\xi'_j(\mathbf{h}_1)|^2 = 1 = \sum_{j=1}^{t} {}^j k_{020\cdots} = \sum_{j=1}^{t} {}^j k_{00\cdots 2},$$

while all the mixed cumulants of order two vanish:

$$\sum_{j=1}^{t} {}^j k_{110\cdots} = \sum_{j=1}^{t} {}^j k_{0110\cdots} = \cdots = 0.$$

Then in (4.A.8), λ reduces to a diagonal matrix with diagonal elements equal to unity, so that (4.A.8) may be written as

$$P(\mathbf{E}) \simeq \frac{1}{(2\pi)^n} \int_{-\infty}^{+\infty} \cdots \int_{-\infty}^{+\infty}$$
$$\times \exp\left[-i(u_1 E_{\mathbf{h}_1} + \cdots + u_n E_{\mathbf{h}_n}) - \frac{1}{2}(u_1^2 + \cdots u_n^2)\right] \qquad (4.A.17)$$
$$\times \exp\left(\sum_{\nu=3}^{\infty} S_\nu\right) du_1 \cdots du_n.$$

Under the same conditions, in a n.cs. case,

$$\sum_{j=1}^{t} {}^{j}k_{200\cdots} = \sum_{j=1}^{t} {}^{j}k_{020\cdots} = \frac{1}{2},$$

while the mixed moments of order two all vanish. Then (4.A.16) becomes

$$P(E_{h_1},\ldots,E_{h_n}) \simeq \frac{1}{(2\pi)^{2n}} \int_{-\infty}^{+\infty} \cdots \int_{-\infty}^{+\infty}$$
$$\times \exp\left[-i(u_1 A_1 + u_2 A_2 + \cdots + v_n B_n) - \frac{1}{4}(u_1^2 + \cdots + v_n^2)\right]$$
$$\times \exp\left(\sum_{v=3}^{\infty} S_v\right) du_1 \cdots dv_n.$$
(4.A.18)

If, in (4.A.18), we make the variable changes

$$\begin{cases} u_j = \sqrt{2}\rho_j \cos\theta_j, & v_j = \sqrt{2}\rho_j \sin\theta_j, \quad j=1,\ldots,n \\ A_{h_j} = R_{h_j} \cos\phi_{h_j}, & B_{h_j} = R_{h_j} \sin\phi_{h_j} \quad j=1,\ldots,n, \end{cases}$$
(4.A.19)

we can express (see equation (M.A.22)) the joint probability density function in terms of the magnitude R and the phase ϕ of the normalized s.f.s:

$$P(R_{h_1},\ldots,R_{h_n},\phi_{h_1},\ldots,\phi_{h_n})$$
$$\simeq \frac{1}{(2\pi)^{2n}} 2^n R_{h_1} \cdots R_{h_n} \int_0^{\infty} \cdots \int_0^{\infty} \int_0^{2\pi} \cdots \int_0^{2\pi} \rho_1 \rho_2 \cdots \rho_n$$
$$\times \exp\left\{-i\left[\sqrt{2}\rho_1 R_{h_1} \cos(\phi_{h_1} - \theta_1)\right.\right.$$
$$\left.\left. + \cdots + \sqrt{2}\rho_n R_{h_n} \cos(\phi_{h_n} - \theta_n)\right] - \frac{1}{2}(\rho_1^2 + \cdots + \rho_n^2)\right\}$$
(4.A.20)
$$\times \exp\left(\sum_{v=3}^{\infty} S_v\right) d\rho_1 \cdots d\rho_n d\theta_1 \cdots d\theta_n,$$

where

$$S_v = \sum_{r+s+\cdots+w=v} 2^{v/2} \left(\sum_{j=1}^{t} \frac{{}^{j}k_{rs\cdots w}}{r!s!\cdots w!} i^v \cdot (\rho_1 \cos\theta_1)^r (\rho_2 \cos\theta_2)^s \cdots (\rho_n \sin\theta_n)^w\right).$$
(4.A.21)

The distributions (4.A.17) and (4.A.20) are basic for the following chapters. In practice, their calculation may often be arduous, particularly when the contributions of terms in S_v with $v \geq 5$ have to be estimated. In this case it seems likely that, even if it were possible to perform the integrations exactly in the right-hand sides of (4.A.17) and (4.A.20), the results would be too intractable for actual use. To simplify the calculations, the following approximation technique may be used. The second exponential functions in the right-hand sides of (4.A.17) or (4.A.20) are expanded in power series.

The problem now is, how should the terms of the expansion be arranged? Let us consider for the cs. case the generic term S_v, and inside it, the contribution from the typical joint moment

$${}^{j}m'_{rs\cdots w} = \langle [\xi'_j(\boldsymbol{h}_1)]^r [\xi'_j(\boldsymbol{h}_2)]^s \cdots [\xi'_j(\boldsymbol{h}_n)]^w \rangle.$$

The contribution to S_v from this moment is

$$\frac{\sum_{j=1}^{t} {}^{j}m'_{rs\cdots w}}{r!s!\cdots w!} (iu_1)^r (iu_2)^s \cdots (iu_n)^w,$$

which may be written as

$$\frac{1}{r!s!\cdots w!} \frac{\sum_{j=1}^{t} {}^j m_{rs\cdots w}}{\left(\sum_{j=1}^{t} {}^j\xi_j^2(\mathbf{h}_1)\right)^{r/2} \left(\sum_{j=1}^{t} {}^j\xi_j^2(\mathbf{h}_2)\right)^{s/2} \cdots \left(\sum_{j=1}^{t} {}^j\xi_j^2(\mathbf{h}_n)\right)^{w/2}} (iu_1)^r (iu_2)^s \cdots (iu_n)^w,$$

(4.A.22)

where ${}^j m_{rs\cdots w}$ is the joint moment for the ξ function.

If ${}^j m_{rs\cdots w} \neq 0$ for any j, then there are t contributions at the numerator of (4.A.22), while at the denominator there are $t^{(r+s+\cdots w)/2} = t^{\nu/2}$ terms. In conclusion (4.A.22), as well as S_ν, is a term of order $t^{1-\nu/2}$. This suggests that the terms of the expansion (Cramér, 1951; Klug, 1958) should be regrouped against the order of t (Edgeworth series). We then write,

$$\exp\left(\sum_{\nu=3}^{\infty} S_\nu\right) \simeq 1 + S_3 + \left(S_4 + \frac{1}{2}S_3^2\right) + \left(S_5 + S_3 S_4 + \frac{1}{6}S_3^3\right)$$

$$+ \left(S_6 + \frac{1}{2}S_4^2 + S_3 S_5 + \frac{1}{2}S_3^2 S_4 + \frac{1}{24}S_3^4\right)$$

$$+ \left(S_7 + S_3 S_6 + S_4 S_5 + \frac{1}{2}S_3^2 S_5 + \frac{1}{2}S_3 S_4^2 + \frac{1}{6}S_3^3 S_4 + \frac{1}{120}S_3^5\right)$$

$$+ \left(S_8 + \frac{1}{2}S_5^2 + S_3 S_7 + S_4 S_6 + \frac{1}{6}S_4^3 + \frac{1}{2}S_3^2 S_6 + S_3 S_4 S_5\right.$$

$$\left. + \frac{1}{6}S_3^3 S_5 + \frac{1}{4}S_3^2 S_4^2 + \frac{1}{24}S_3^4 S_4 + \frac{1}{720}S_3^6\right) + \cdots.$$

(4.A.23)

Relation (4.A.23) is an expansion in terms of increasing t order; for example, S_3 is of order $t^{-1/2}$, $(S_4 + \frac{1}{2}S_3^2)$ is of order t^{-1}, $(S_5 + S_3 S_4 + \frac{1}{6}S_3^3)$ is of order $t^{-3/2}$, etc. If t is large enough, each term is negligible compared with the preceding ones, but if large normalized s.f.s are involved, this may not always be true. Therefore, only a limited number of terms in the series can be taken in such a way that the estimate of only a small number of moments is necessary.

The above considerations suggest the following machinery.

1. The characteristic function is calculated in the exponential form or as an Edgeworth series expansion (4.A.23). The function is truncated at a convenient power of t, say at order $t^{-\beta}$.
2. Only the terms S_ν must be calculated which contribute to the truncated series. For example, if the Edgeworth series is chosen and $\beta = 2$, then we only need to calculate S_2, S_3, S_4, S_5, S_6.
3. Each term S_i is estimated from (4.A.7) or (4.A.15), according to whether the space group is c.s. or n.c.s. respectively.
4. In their turn, the cumulants ${}^j k_{rs\cdots w}$, on which each S depends, can be calculated from the moments through the relation described in Sections M.A.4 and M.A.7.
5. The integrations in the right-hand side of (4.A.17) or (4.A.18) are made. If the space group is n.c.s., (4.A.18) should be transformed into (4.A.20).

An important limitation of the method has to be outlined. Since the cumulant generating function is expanded in a Taylor series in the vicinity of

$$u_1 = u_2 = \cdots = u_n = 0,$$

the corresponding probability distribution will be accurate only in the vicinity of

$$E_1 = E_2 = \cdots = E_n = 0.$$

4.A.5 The joint probability density function when some prior information is available

We are interested in the calculation of

$$P(E_1, E_2, \ldots, E_n)$$

when some prior information is available, including that involving correlation among the primitive random variables. If such a correlation is present, (4.A.3) will not hold and a more general procedure for the calculation of P will be necessary.

We first deal with the cs. case. The characteristic function can again be defined as

$$C(u_1, u_2, \ldots, u_n) = \langle \exp[i(u_1 E_1 + u_2 E_2 + \cdots + u_n E_n)] \rangle,$$

which in its turn, may be expanded in series:

$$C(u_1, u_2, \ldots, u_n) = 1 + i\langle X \rangle - \frac{\langle X^2 \rangle}{2!} - i\frac{\langle X^3 \rangle}{3!} + \frac{\langle X^4 \rangle}{4!} + \cdots, \quad (4.A.24)$$

where

$$X = u_1 E_2 + u_2 E_2 + \cdots + u_n E_n.$$

As in (4.A.23), the terms in the series expansion of $C(u_1, \ldots, u_n)$ can be arranged in groups, so as to improve the convergence of the series. Then,

$$P(E_1, E_2, \ldots, E_n) \simeq (2\pi)^{-n} \int_{-\infty}^{+\infty} \cdots \int_{-\infty}^{+\infty} \exp[-i(u_1 E_1 + \cdots + u_n E_n)]$$
$$\cdot C(u_1, u_2, \ldots, u_n) \, du_1 du_2 \cdots du_n.$$

In order to emphasize the difference between this procedure and that described in Sections 4.A.2–4.A.4, we will explain the terms $\langle X \rangle$ and $\langle X^2 \rangle$. We have,

1. $\langle X \rangle = u_1 \langle E_1 \rangle + u_2 \langle E_2 \rangle + \cdots + u_n \langle E_n \rangle$. While $\langle X \rangle = 0$ in the absence of prior information on the phase values, $\langle E_i \rangle \neq 0$ if this information is available.
2. $\langle X^2 \rangle = u_1^2 \langle E_1^2 \rangle + u_2^2 \langle E_2^2 \rangle + \cdots + u_n^2 \langle E_n^2 \rangle + 2u_1 u_2 \langle E_1 E_2 \rangle + \cdots + 2u_{n-1} u_n \langle E_{n-1} E_n \rangle$. Since $E_i = F_i/\langle |F_i|^2 \rangle^{1/2}$, by definition $\langle E_i^2 \rangle = 1$, but $\langle E_r E_s \rangle$ may be different from zero when supported by the prior information. Accordingly,

$$\langle X^2 \rangle = u_1^2 + u_2^2 + \cdots + u_n^2 + 2u_1 u_2 \langle E_1 E_2 \rangle + \cdots + 2u_{n-1} u_n \langle E_{n-1} E_n \rangle.$$

The calculation of $P(E_1, E_2, \ldots, E_n)$ in an n.cs. space group involves similar steps. First,

$$C(u_1, u_2, \ldots, v_n) = \langle \exp\{i[(u_1 A_1 + u_2 A_2 + \cdots + v_n B_n)]\} \rangle$$

The method of structure factor joint probability distribution functions

is derived through the series expansion

$$C(u_1, u_2, \ldots, v_n) = 1 + i\langle X \rangle - \frac{\langle X^2 \rangle}{2!} - i\frac{\langle X^3 \rangle}{3!} + \frac{\langle X^4 \rangle}{4!} + \cdots, \quad (4.A.25)$$

where

$$X = u_1 A_1 + u_2 A_2 + \cdots + v_n B_n.$$

The variable changes (4.A.19) are then made, and the terms in the series expression are arranged in groups so as to improve the convergence of the series. Then,

$$P(R_{h_1}, \ldots, R_{h_n}, \phi_{h_1}, \ldots, \phi_{h_n})$$

$$\simeq \frac{1}{(2\pi)^{2n}} 2^n R_{h_1} \cdots R_{h_n} \int_0^\infty \cdots \int_0^\infty \int_0^{2\pi} \cdots \int_0^{2\pi}$$

$$\cdot \rho_1 \rho_2 \cdots \rho_n \exp\left\{-i\left[\sqrt{2}\,\rho_1 R_{h_1} \cos(\phi_{h_1} - \theta_1)\right.\right. \quad (4.A.26)$$

$$\left.\left. + \cdots + \sqrt{2}\rho_n R_{h_n} \cos(\phi_{h_n} - \theta_n)\right]\right\}$$

$$\cdot C(\rho_1, \ldots, \rho_n, \theta_1, \ldots, \theta_n)\, d\rho_1 \cdots d\rho_n d\theta_1 \cdots d\theta_n.$$

4.A.6 The calculation of P(E) in the absence of prior information

Even if the method of joint probability distribution function displays its full power when more structure factors are considered, it may also be applied for the calculation of $P(E)$ (i.e. the Wilson type distribution). For the centric case, the following distribution is obtained:

$$_{\bar{1}}P(R) = \sqrt{\frac{2}{\pi}} \exp\left(-\frac{R^2}{2}\right)\left[1 - \frac{1}{8N_{eff}} H_4(R)\right], \quad (4.A.27)$$

where

$$H_4(R) = R^4 - 6R^2 + 3$$

is the Hermite polynomial of order four (see Appendix M.D). Compared with (2.8) equation (4.A.27) contains a correction term of order $1/N$. Supplementary terms of higher order could be obtained if the Edgeworth series (4.A.23) were truncated at higher powers of t.

For the acentric case, the following distribution is obtained:

$$_1 P(R) = 2R \exp(-R^2)\left[1 - \frac{1}{4N_{eff}} L_4(R)\right], \quad (4.A.28)$$

where

$$L_4(R) = R^4 - 4R^2 + 2$$

is the Laguerre polynomial of order 4 (see Appendix M.D). Compared with (2.7), equation (4.A.28) contains a correction term of order $1/N$. Supplementary terms could be obtained if the Edgeworth expansion (4.A.23) were truncated at higher powers of t.

5 The probabilistic estimation of triplet and quartet invariants

5.1 Introduction

This chapter describes how to estimate, by probabilistic approaches, triplet and quartet invariants from diffraction magnitudes. We will skip quintet (Fortier and Hauptman, 1977a,b,c; Hauptman and Fortier, 1977a,b; Van der Putten and Schenk, 1977; Giacovazzo, 1977b, 1980; Burla et al., 1977) and higher-order s.i.s, because their usefulness in modern phasing procedures is entirely marginal. For simplicity, we will also skip the mathematics necessary to obtain conclusive formulas (the general approach is described in Appendix 4.A), except for the triplet invariants, first representation, because of their prominent role. Triplet and quartet estimates will be discussed, particularly in relation to their impact on phasing procedures.

For simplicity, some other specialized topics will also be skipped, even if theoretically relevant. For example: results obtained by Shmueli and Weiss (1986, 1992), who used Fourier series representations of joint probability density functions to estimate triplets; the effect of pseudotranslational symmetry on the triplet phase estimates, as described by Cascarano et al. (1985a,b, 1987b, 1988a,b); algebraic formulas obtained by Karle and Hauptman (1957), Vaughan (1958), Hauptman et al. (1969), Hauptman (1970), Fischer et al. (1970a,b), all related to (and encompassed by) the estimation of triplet phases via their second representation. Interested readers are referred to the original papers.

5.2 Estimation of the triplet structure invariant via its first representation: the P1 and the P$\bar{1}$ case

Let us first consider the space group P1. According to Chapter 4, the simplest way to estimate the triplet s.i.

$$\Phi = \phi_{\mathbf{h}_1} + \phi_{\mathbf{h}_2} + \phi_{\mathbf{h}_3} \text{ with } \mathbf{h}_1 + \mathbf{h}_2 + \mathbf{h}_3 = 0 \tag{5.1}$$

is to study the joint probability distribution

$$P(E_{\mathbf{h}_1}, E_{\mathbf{h}_2}, E_{\mathbf{h}_3}) \equiv P(R_{\mathbf{h}_1}, R_{\mathbf{h}_2}, R_{\mathbf{h}_3}, \phi_{\mathbf{h}_1}, \phi_{\mathbf{h}_2}, \phi_{\mathbf{h}_3}). \quad (5.2)$$

According to Section 4.1 we must first calculate the *characteristic function* C and then, by Fourier inversion, recover the distribution (5.2). Because of the importance of the triplet invariant, we report the necessary calculations in Appendix 5.A. The resulting distribution is

$$P(R_1, R_2, R_3, \phi_1, \phi_2, \phi_3) = \frac{R_1 R_2 R_3}{\pi^3} \exp\left\{-R_1^2 - R_2^2 - R_3^2 + C\,\cos(\phi_1 + \phi_2 + \phi_3)\right\}, \quad (5.3)$$

where $R_1, R_2, R_3, \phi_1, \phi_2, \phi_3$ stand for $R_{\mathbf{h}_1}, R_{\mathbf{h}_2}, R_{\mathbf{h}_3}, \phi_{\mathbf{h}_1}, \phi_{\mathbf{h}_2}, \phi_{\mathbf{h}_3}$, respectively,

$$C = \frac{2 R_1 R_2 R_3}{\sqrt{N_{eq}}} \quad (5.4)$$

$$\sqrt{N_{eq}} = \sigma_2^{3/2}/\sigma_3, \quad (5.5)$$

and

$$\sigma_n = \sum_{j=1}^{N} Z_j^n.$$

N is the number of atoms in the unit cell and Z_j is the atomic number of the jth atom. If all of the atoms are of the same species (and have similar thermal displacement), then $N_{eq} \equiv N$ and

$$C = \frac{2 R_1 R_2 R_3}{\sqrt{N}}.$$

The simultaneous presence of heavy and light atoms in the unit cell makes $N_{eq} < N$ (see Section 5.3).

From (5.3) the conditional distribution, $P(\Phi|R_{\mathbf{h}_1}, R_{\mathbf{h}_2}, R_{\mathbf{h}_3})$, may be obtained (abbreviated to $P(\Phi)$; Cochran, 1955):

$$P(\Phi) = [2\pi I_0(C)]^{-1} \exp(C \cos \Phi), \quad (5.6)$$

which may also be written as

$$P(\Phi) = M(\Phi; 0, C),$$

where

$$M(\Phi; \theta, C) = [2\pi I_0(C)]^{-1} \exp[C \cos(\Phi - \theta)]$$

is the von Mises distribution for the variable Φ, centred at θ, with *concentration parameter* equal to C.

Equation (5.6) is plotted in Fig. 5.1, from which we observe:

(i) I_0 is the modified Bessel function of order 0 (see Appendix M.E). We have to think of $[2\pi I_0(C)]^{-1}$ as a scaling factor, allowing $\int_0^{2\pi} P(\Phi) d\Phi = 1$.
(ii) Equation (5.6) has its maximum at $\Phi = 0$ (where $\cos \Phi = 1$). It may be concluded that the expected value of Φ is always zero.
(iii) The sharpest curves are obtained in correspondence with the largest values of C. Thus the statistical indication $\Phi \approx 0$ is reliable only if C is sufficiently large. This condition is satisfied if all three Rs are sufficiently large and N is sufficiently small.

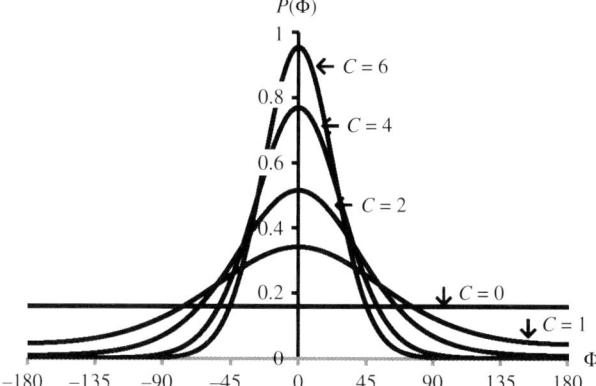

Fig. 5.1
The Cochran distribution $P(\Phi)$ for a triplet phase invariant, for different values of parameter C.

(iv) If at least one of the Rs is zero, then $P(\Phi) = (2\pi)^{-1}$: no phase indication is obtained.

The statement $\Phi \approx 0$ is a statistical expectation; it does not mean that Φ must be zero. To better understand this point, let us calculate the following mean values:

$$<\cos(\Phi)> = \int_0^{2\pi} \cos(\Phi)P(\Phi)d\Phi = D_1(C) \tag{5.7}$$

$$<\sin(\Phi)> = \int_0^{2\pi} \sin(\Phi)P(\Phi)d\Phi = 0, \tag{5.8}$$

where $D_1(C) = I_1(C)/I_0(C)$ is the ratio of the two modified Bessel functions of order 1 and zero, respectively (see Fig. 5.2). According to (5.7), the average value of $<\cos(\Phi)>$ is smaller than 1, and is sufficiently close to 1 if C is large. However, as for any statistical indication, it may also be that $\cos(\Phi)$ is actually negative, even if C is positive and large.

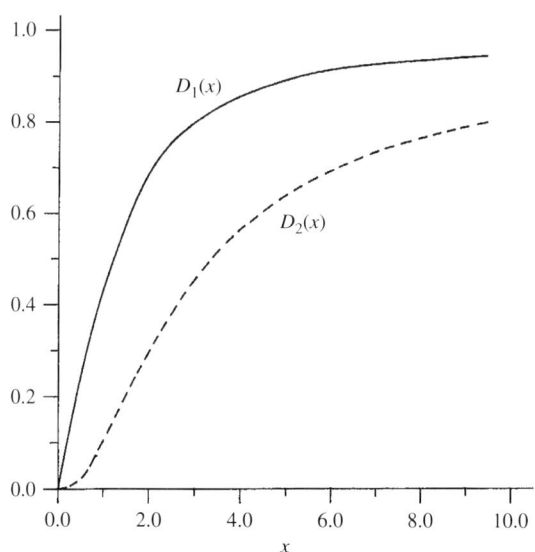

Fig. 5.2
The functions $D_1(x)$ and $D_2(x)$.

According to (5.8), the Cochran relationship is unable to fix the enantiomorph. Thus, if $<\cos(\Phi)> = \cos q$, equation (5.8) says that $+q$ and $-q$ have equal probability, and coherently gives $<\Phi> = 0$ and $<\sin\Phi> = 0$.

A last remark may be useful for readers not familiar with the von Mises distribution. For circular variables, it plays a role similar to that played by the Gaussian function for linear variables. In particular, the von Mises distribution is marked by maximum likelihood and by maximum entropy characterization (Mardia, 1972). While the normal distribution along a line has useful mathematical and statistical properties, this is not true for a normal distribution along a circle (i.e. in the case of directional data). Indeed, in the theory of circular variables, a normal distribution (as well as the most significant distributions on a line, e.g. Cauchy, Poisson, etc.) is wrapped around the circumference of a circle of unit radius, thus producing the so-called wrapped distribution.

Let us now consider the $P\bar{1}$ case (Cochran and Woolfson, 1955): E_1, E_2, E_3 are now real numbers, and according to Appendix 5.A the following joint probability distribution is obtained:

$$P(E_1, E_2, E_3) = \frac{1}{(2\pi)^{3/2}} \exp\left\{-\frac{1}{2}(E_1^2 + E_2^2 + E_3^2) + \frac{E_1 E_2 E_3}{\sqrt{N_{eq}}}\right\}.$$

In this case the phase problem reduces to a sign problem. The probability that the sign of $E_1 E_2 E_3$ is plus, is given (but for a scaling term) by

$$P_+ \approx \exp\left(+\frac{R_1 R_2 R_3}{\sqrt{N_{eq}}}\right),$$

and the probability that it is minus is given (but for a scaling term) by

$$P_- \approx \exp\left(-\frac{R_1 R_2 R_3}{\sqrt{N_{eq}}}\right).$$

Since it must be that $P_+ + P_- = 1$, the rescaled value of the positive sign probability is

$$P^+ = (1 + P_-/P_+)^{-1} = \left[1 + \exp\left(-\frac{2R_1 R_2 R_3}{\sqrt{N_{eq}}}\right)\right]^{-1}. \qquad (5.9)$$

Since

$$(1 + e^{-2x})^{-1} = e^x/(e^x + e^{-x}) = \frac{1}{2} + \frac{1}{2}\tanh x,$$

from (5.9),

$$P^+ = \frac{1}{2} + \frac{1}{2}\tanh\frac{R_1 R_2 R_3}{\sqrt{N_{eq}}} \qquad (5.10)$$

is obtained (see Fig. 5.3). As for the acentric case we notice:

(i) P^+ is always larger than $\frac{1}{2}$, unless some of R_1, R_2, R_3 are vanishing.
(ii) The reliability of sign indication is large only for large values of $\frac{R_1 R_2 R_3}{\sqrt{N_{eq}}}$.
(iii) The efficiency of (5.10) decays with the size of the structure.

Triplet estimation in space groups with symmetry higher than triclinic is described briefly in Appendix 5.B.

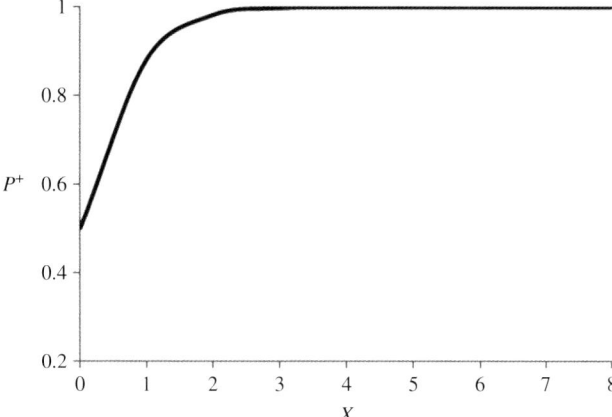

Fig. 5.3
Centrosymmetric space groups. $P^+(X)$ is the probability that the triplet sign is positive, according to equation (5.10), and $X = \frac{R_1 R_2 R_3}{\sqrt{N_{eq}}}$. P^+ is always equal to or larger than $1/2$.

5.3 About triplet invariant reliability

The relationships (5.6) and (5.10) have been obtained by making use of two basic assumptions: the structure is composed of discrete atoms (*atomicity postulate*) and the electron density is everywhere real and positive (*positivity postulate*). For X-rays, positivity and atomicity are implicit in the positivity of the atomic scattering factor f. It is, however, worthwhile noticing that when triplets are to be estimated for neutron diffraction data (see Chapter 11), the positivity postulate may be violated and relations (5.6) and (5.10) are no longer valid. In an analogous way dispersion effects could introduce complex scattering factors, $(f_j = f'_j + if''_j)$: in this case also, the probabilistic theory for triplet estimation should be reformulated (Hauptman, 1982a,b; Giacovazzo, 1983b; see Chapter 15).

In this section we focus our attention only on X-ray data: we wish to enquire about the range of structural complexity inside which equations (5.6) and (5.10) may be usefully applied. Since $<R^2> = 1$ by definition, the R values do not change their order of magnitude, no matter how complex is the structure. Therefore, the only parameter in C which changes size with structural complexity is $2/\sqrt{N_{eq}}$: this parameter influences the average efficiency of the triplet relationships. In more detail:

1. For crystal structures where non-hydrogen atoms are nearly equal, N_{eq} is almost equal to the number of non-hydrogen atoms in the unit cell (this is only valid for X-ray data). Therefore, hydrogen atoms could even be omitted from the calculation of N_{eq}.
2. $N > N_{eq}$ when heavy and light atoms coexist in the unit cell. The difference becomes large with increasing values of the ratio
    atomic number of heavy atom:atomic number of light atoms.
    For example, *JAMILAS* [$K_4C_{64}H_{68}N_8O_{20}S_4$, space group P1] is a small structure with $N = 100$ non-hydrogen atoms in the unit cell; the corresponding value of N_{eq} is 55. The above result indicates that crystal structures with a large number of light atoms and a few heavy atoms are more easily

Table 5.1 *Schwarz* [$C_{46}H_{70}O_{27}$, P1]: statistical results on triplet estimates (Cochran formula). nr is the number of triplets with Cochran parameter $C > THR$, $<|\Phi|>$ is the corresponding average value of $|\Phi|$, and % is the percentage of triplets with positive value of $\cos \Phi$

THR	nr	$<|\Phi|>^o$	%
0.4	5117	41	90
1.2	4572	40	91
2.0	1552	30	96
2.6	570	27	98
3.8	81	19	100

solvable by direct methods than structures of the same size but without heavy atoms.

3. For unit cells with a large number of atoms, C is small for most of the triplets; correspondingly, extremely broad probability distributions (5.6) are expected. The consequence is that few triplet phases Φ are really close to zero, the majority are dispersed in the interval $(0, 2\pi)$. If the structure size is small, a high percentage of triplet phases will be close to zero.

Table 5.1 shows some statistical calculations for the *Schwarz* [$C_{46} H_{70} O_{27}$, space group P1] structure, showing how Φ is distributed versus C. The table entries may be interpreted as follows:

(i) There are 81 triplets for which $C > 3.8$; for these, the average value of $|\Phi|$ is 19° (in this case the condition $C > 3.8$ selects triplets with phase Φ really close to zero), and $\cos \Phi$ is always positive.

(ii) There are 570 triplets with $C > 2.6$; for these, the average value of $|\Phi|$ is 27°.

Data in Table 5.1 may be usefully compared with data in Table 5.2, where we show similar statistics for a small protein (*1e8a*; space group R3, 182 residues, corresponding to 1472 non-hydrogen atoms in the asymmetric unit. Data resolution: 1.95 Å). Only 92 triplets reach a C value larger than 0.5, the percentage of triplets which deviate from the Cochran expectation $\Phi \approx 0$ is very high.

Table 5.2 *1e8a*. Statistical results on triplet estimates (Cochran formula). nr is the number of triplets with Cochran parameter $C > THR$, $<|\Phi|>$ is the corresponding average value of $|\Phi|$, and % is the percentage of triplets with positive value of $\cos \Phi$

THR	nr	$<|\Phi|>^o$	%
0.1	300000	86	54
0.2	79494	84	55
0.3	7355	83	56
0.4	759	78	59
0.5	92	78	59

Apparently, the structural complexity does not allow selection of reliable triplet invariants, with obvious consequences in the phasing steps.

5.4 The estimation of triplet phases via their second representation

The Cochran formula (5.6) estimates triplet phases (5.1) by exploiting only the information contained in three diffraction moduli; any Φ is expected to be close to 2π, and there is no chance of recognizing *bad triplets* (i.e. triplet phases close to $\pm\pi/2$ or with negative cosine values). This is of paramount importance to the efficiency of the phasing process. We will see in the Chapter 6 that the occurrence of a relatively large number of bad triplets in the phasing process can lead to its failure. Alternatively, the probability of finding the correct set of phases is enhanced if bad triplets are recognized; they should be excluded from the structure solving process or actively used in a correct manner.

The representation theory, described in Chapter 4, indicates how information contained in all of the reciprocal space may be used to improve the Cochran estimates. In accordance with Section 4.2, the second representation of Φ is a collection of special quintets,

$$\{\Psi\}_2 = \{\Phi + \phi_\mathbf{k} - \phi_\mathbf{k}\}, \qquad (5.11)$$

where $\mathbf{k}$ is a free vector in reciprocal space. The basis magnitudes of any Ψ_2 are

$$R_{\mathbf{h}_1}, R_{\mathbf{h}_2}, R_{\mathbf{h}_3}, R_\mathbf{k}$$

and the cross-magnitudes are

$$R_{\mathbf{h}_1 \pm \mathbf{k}}, R_{\mathbf{h}_2 \pm \mathbf{k}}, R_{\mathbf{h}_3 \pm \mathbf{k}}.$$

The collection of the basis and cross-magnitudes of the various quintets Ψ_2 is $\{B\}_2$, and is called the *second phasing shell* of Φ:

$$\{B\}_2 = \{R_{\mathbf{h}_1}, R_{\mathbf{h}_2}, R_{\mathbf{h}_3}, R_\mathbf{k}, R_{\mathbf{h}_1 \pm \mathbf{k}}, R_{\mathbf{h}_2 \pm \mathbf{k}}, R_{\mathbf{h}_3 \pm \mathbf{k}}\}.$$

These results suggest, for P1 and P$\bar{1}$, a study of the ten-variate probability distribution

$$P(E_{\mathbf{h}_1}, E_{\mathbf{h}_2}, E_{\mathbf{h}_3}, E_\mathbf{k}, E_{\mathbf{h}_1+\mathbf{k}}, E_{\mathbf{h}_2+\mathbf{k}}, E_{\mathbf{h}_3+\mathbf{k}}, E_{\mathbf{h}_1-\mathbf{k}}, E_{\mathbf{h}_2-\mathbf{k}}, E_{\mathbf{h}_3-\mathbf{k}}), \qquad (5.12)$$

from which the conclusive conditional distribution,

$$P(\Phi|10 \text{ moduli}), \qquad (5.13)$$

is obtained. Equations (5.12) and (5.13) may be calculated by means of the techniques described in Chapter 4. Since $\mathbf{k}$ is a free vector, a formula can be found which provides the conditional probability distribution of Φ given the basis and cross-moduli of any quintet Ψ_2. We will denote such a probability $P_{10}(\Phi)$, in order to emphasize the fact that the formula explores the reciprocal space by means of a ten-node figure. Three nodes (i.e. $\mathbf{h}_1, \mathbf{h}_2, \mathbf{h}_3$) are fixed while $\mathbf{k}$ varies; the remaining seven nodes sweep out reciprocal space.

The estimation of triplet phases via their second representation

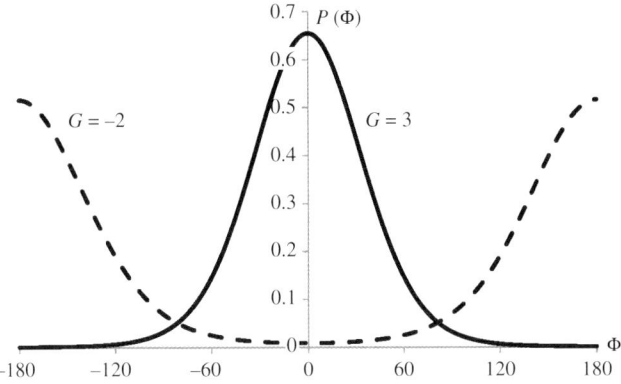

Fig. 5.4
$P_{10}(\Phi)$ according to equation (5.14). We choose $G = 3$ (continuous line) and $G = -2$ (dashed line).

The final probabilistic formula (Cascarano et al., 1984; Burla et al., 1989a) is of a von Mises type, and may be written as

$$P_{10}(\Phi) = [2\pi I_0(G)]^{-1} \exp(G \cos \Phi), \quad (5.14)$$

where G is a concentration parameter which depends on hundreds or thousands of magnitudes, and may be positive or negative. If $G > 0$, the expected value of Φ is zero, if negative, the expected value of Φ is π; unlike the Cochran relationship, $P_{10}(\Phi)$ is able to identify negative triplet cosines. Two distributions (5.14), one corresponding to a positive and the other to a negative value of G are shown in Fig. 5.4: it is evident that, when $G < 0$, the value of Φ is probably closer to π than to 0.

For cs. space groups the triplet sign may be estimated by equation (5.15),

$$P^+ = \frac{1}{2} + \frac{1}{2} \tanh\left(\frac{G}{2}\right) \quad (5.15)$$

as a substitute for equation (5.10). Since G may also be negative, positive and negative triplets may be identified. Correspondingly, Fig. 5.3 may be generalized into Fig. 5.5, allowing values of P^+ smaller than $\frac{1}{2}$.

For the interested reader, a formal expression of G, including symmetry effects, is given in Appendix 5.C, where we also compare the efficiencies of the Cochran and the P_{10} formulas. Because of its superiority, the P_{10} formula

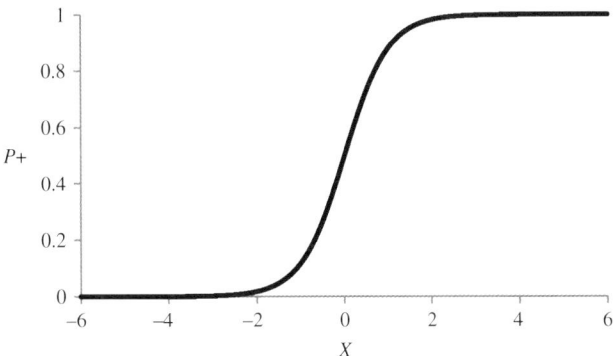

Fig. 5.5
P^+ in accordance with equation (5.15). P^+ is larger or smaller than $\frac{1}{2}$, according to whether G is positive or negative.

has been fully integrated in the *SIR* suite of phasing programs starting from SIR88 (Burla et al., 1989a).

5.5 Introduction to quartets

Four phases are said to form the quartet invariant,

$$\Phi = \phi_{\mathbf{h}_1} + \phi_{\mathbf{h}_2} + \phi_{\mathbf{h}_3} + \phi_{\mathbf{h}_4},$$

if

$$\mathbf{h}_1 + \mathbf{h}_2 + \mathbf{h}_3 + \mathbf{h}_4 = 0.$$

Hauptman and Karle (1953) and Simerska (1956), independently, suggested that Φ would be approximately zero for large values of $R_{\mathbf{h}_1}R_{\mathbf{h}_2}R_{\mathbf{h}_3}R_{\mathbf{h}_4}$. The use of quartets in direct procedures for phase solution was first introduced by Schenk (1973a,b, 1974), who, from semi-empirical observations on the moduli $R_{\mathbf{h}_1+\mathbf{h}_2}, R_{\mathbf{h}_1+\mathbf{h}_3}, R_{\mathbf{h}_2+\mathbf{h}_3}$, derived useful conditions for improving estimation of the relation $\Phi \approx 0$. Probabilistic theories for quartet estimation from the first phasing shell were, independently, described for P1 by Hauptman (1975a,b) and by Giacovazzo (1976b,c). Theories for $P\bar{1}$ were given by Giacovazzo (1975a, 1976a), Green and Hauptman (1976), and Hauptman and Green (1976). A general probabilistic theory of quartets valid in all space groups was given by Giacovazzo (1976d).

Both Hauptman's and Giacovazzo's approaches use the first phasing shell, $\{R_{\mathbf{h}_1}, R_{\mathbf{h}_2}, R_{\mathbf{h}_3}, R_{\mathbf{h}_4}, R_{\mathbf{h}_1+\mathbf{h}_2}, R_{\mathbf{h}_1+\mathbf{h}_3}, R_{\mathbf{h}_2+\mathbf{h}_3}\}$, to estimate quartets; mainly, they differ because the second author has used the Gram–Charlier expansion of the characteristic function (see Appendix 4.A). For brevity we will use the following notation:

$$R_i = R_{\mathbf{h}_i}, \phi_i = \phi_{\mathbf{h}_i} \text{ for } i = 1, \ldots, 4,$$

$$R_5 = R_{\mathbf{h}_1+\mathbf{h}_2}, R_6 = R_{\mathbf{h}_1+\mathbf{h}_3}, R_7 = R_{\mathbf{h}_2+\mathbf{h}_3},$$

$$\phi_5 = \phi_{\mathbf{h}_1+\mathbf{h}_2}, \phi_6 = \phi_{\mathbf{h}_1+\mathbf{h}_3}, \phi_7 = \phi_{\mathbf{h}_2+\mathbf{h}_3}.$$

5.6 The estimation of quartet invariants in P1 and P$\bar{1}$ via their first representation: Hauptman approach

Hauptman derived in P1 the following conditional distribution:

$$P(\Phi|R_1, \ldots, R_7) \simeq \frac{1}{L} \exp(-4C \cos \Phi) I_0(R_5 Z_5 \cdot) I_0(R_6 Z_6) I_0(R_7 Z_7), \quad (5.16)$$

where $I_0(x)$ is the modified Bessel function of order zero,

$$C = R_1 R_2 R_3 R_4 / N, \quad (5.17)$$

$$Z_5 = \frac{2}{\sqrt{N}} (R_1^2 R_2^2 + R_3^2 R_4^2 + 2NC \cos \Phi)^{1/2}, \quad (5.18a)$$

$$Z_6 = \frac{2}{\sqrt{N}}(R_1^2 R_3^2 + R_2^2 R_4^2 + 2NC\cos\Phi)^{1/2}, \quad (5.18b)$$

$$Z_7 = \frac{2}{\sqrt{N}}(R_2^2 R_3^2 + R_1^2 R_4^2 + 2NC\cos\Phi)^{1/2}. \quad (5.18c)$$

As for the triplet invariants, distribution (5.16) depends on $\cos\Phi$; therefore only $\cos\Phi$ may be estimated, it being impossible to distinguish between $+\Phi$ an $-\Phi$ (or, in other words, to distinguish between the two enantiomorphs).

Since L, the scaling factor, has a rather complicated expression, one might use numerical methods for calculating:

1. the scaling factor L, via the condition

$$\int_0^\pi P(\Phi)d\Phi = 1;$$

2. the mode Φ_m of $P(\Phi)$;
3. the mean value, given by

$$\langle\Phi\rangle = \int_0^\pi \Phi P(\Phi)d\Phi;$$

4. the variance, V, as given by

$$V = \int_0^\pi (\Phi - \langle\Phi\rangle)^2 P(\Phi)d\Phi.$$

5. $\sigma_\Phi = \sqrt{V}$.

Estimation of $|\Phi|$, via (5.16), depends on an intricate interrelationship among all the seven magnitudes. However, some working rules can be stated:

1. $P(\Phi)$ is unimodal between 0 and π, and Φ_m can, in principle, lie anywhere between 0 and π;
2. if the cross-magnitudes are large, Φ is expected to be close to zero;
3. if the cross-magnitudes are small, Φ is expected to be close to π;
4. if the cross-magnitudes are of medium size and N is sufficiently small, then Φ is expected to be close to $\pm\pi/2$;
5. the larger N, the larger the overall variance associated with quartet phase estimation.

Figures 5.6 and 5.7 show (broken curves) the distribution (5.16) for some values of the seven magnitudes when $N = 47$. In Fig. 5.6, where all the cross-magnitudes are large, $\Phi_m = 0.0, \langle\Phi\rangle \simeq 29°, \sigma_\Phi = V^{1/2} = 21.9°$. In Fig. 5.7 where all the cross-magnitudes are small, $\Phi_m = 180°, \langle\Phi\rangle \simeq 142°, \sigma_\Phi = 32.7°$.

It is clear from the figures that cosines estimated near π will (on average) be in poorer agreement with the true values than the cosines estimated near 0, because of the relatively larger value of the variance. Even poorer will be the estimates of the cosines located in the middle range (usually called *enantiomorph sensitive quartets*); no useful application has been found for them.

The three cross-magnitudes are not always in the set of measured reflections. Then, some marginal joint probability distributions must be considered in order

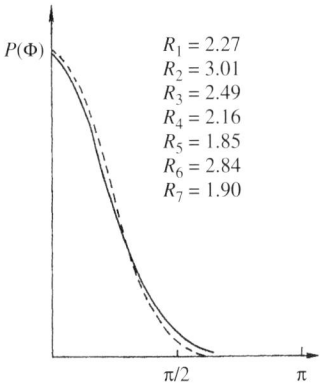

Fig. 5.6
Distribution (5.16) (broken curve) and (5.22) (continuous curve) for the indicated |E| values in a structure with $N = 47$ atoms in the unit cell.

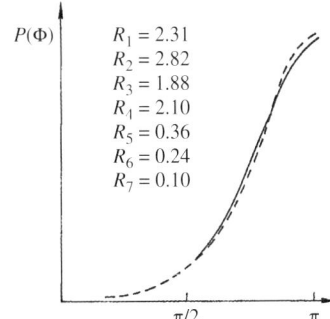

Fig. 5.7
Distribution (5.16) (broken curve) and (5.22) (continuous curve) for the indicated |E| values in a structure with $N = 47$ atoms in the unit cell.

to derive useful information in these less favourable cases. In accordance with equation (M.A.16), marginal distributions may be obtained by performing the integration of (5.16) with respect to $R5$, $R6$, $R7$, depending on which cross-magnitudes are not measured. The resulting general formula is

$$P(\Phi|\ldots) \approx \frac{1}{L}\exp[-2C(n-1)\cos\Phi]I_0(w_5R_5Z_5)I_0(w_6R_6Z_6)I_0(w_7R_7Z_7), \quad (5.19)$$

where n is the number of known cross-magnitudes and $I_0(x)$ is the modified Bessel function of order zero. w_i is equal to zero if the cross term R_i is unknown, otherwise $w_i = 1$.

A last observation concerns the generalization of (5.19) to unequal atom structures. This can be made by replacing: in equation (5.17), N by σ_2^2/σ_4; in equations (5.18), N by N_{eq}, as given by equation (5.5). It has been verified that, in practice, one can deal with unequal atom structures by replacing N by N_{eq} in the full expression (5.19).

For centric space groups, Hauptman and Green (1976) obtained the sign probability

$$P^{\pm} \approx \frac{1}{L}\exp(\mp 2C)\cosh(R_5Z_5^{\pm})\cosh(R_6Z_6^{\pm})\cosh(R_7Z_7^{\pm}), \quad (5.20)$$

where

$$L = \exp(-2C)\cosh(R_5Z_5^{+})\cosh(R_6Z_6^{+})\cosh(R_7Z_7^{+})$$

$$\exp(+2C)\cosh(R_5Z_5^{-})\cosh(R_6Z_6^{-})\cosh(R_7Z_7^{-})$$

and

$$Z_5^{\pm} = \frac{1}{\sqrt{N}}(R_1R_2 \pm R_3R_4), \quad Z_6^{\pm} = \frac{1}{\sqrt{N}}(R_1R_3 \pm R_2R_4)$$

$$Z_7^{\pm} = \frac{1}{\sqrt{N}}(R_1R_4 \pm R_2R_3).$$

P^+ and P^- are the probabilities that $E_1E_2E_3E_4$ is positive or negative; they may lie anywhere between 0 and 1. P^- is close to 1 or close to 0 if R_5, R_6, and R_7 are either all relatively small or all relatively large, respectively. In order to derive useful information, even in the less favourable cases in which some of the cross-magnitudes are not among the measurements, conditional probability values may be derived, leading to the general formula,

$$P^{\pm} \approx \frac{1}{L}\exp[\mp C(n-1)]\cosh(w_5R_5Z_5^{\pm})\cosh(w_6R_6Z_6^{\pm})\cosh(w_7R_7Z_7^{\pm}), \quad (5.21)$$

where n is the number of known cross-magnitudes, $w_i = 0$ if the cross term R_i is unknown, otherwise $w_i = 1$. As in the acentric case, we can generalize (5.21) to unequal atom structures by replacing N by N_{eq}, as given by equation (5.5).

5.7 The estimation of quartet invariants in P1 and P$\bar{1}$ via their first representation: Giacovazzo approach

Giacovazzo expressions for estimating quartet invariants via their first phasing shell are simpler than Hauptman formulas, but equally efficient. His final expression for an acentric space group is a von Mises formula,

$$P(\Phi|R_1, R_2, \ldots, R_7) = (2\pi I_0(G))^{-1} \exp(G \cos \Phi), \quad (5.22)$$

where

$$G = \frac{2C(1 + \varepsilon_5 + \varepsilon_6 + \varepsilon_7)}{1 + Q}, \quad (5.23)$$

$$Q = [(\varepsilon_1\varepsilon_2 + \varepsilon_3\varepsilon_4)\varepsilon_5 + (\varepsilon_1\varepsilon_3 + \varepsilon_2\varepsilon_4)\varepsilon_6 + (\varepsilon_1\varepsilon_4 + \varepsilon_2\varepsilon_3)\varepsilon_7]/2N,$$

and $\varepsilon_i = R_i^2 - 1$.

It should be noted that:

1. It is convenient to set $Q = 0$ when $Q \leq 0$.
2. G is positive if $\varepsilon_5 + \varepsilon_6 + \varepsilon_7 > 1$, and negative in the opposite case. If G is negative, the most probable value of Φ is π, and the quartet cosine is estimated to be negative.
3. Large values of the cross-magnitudes (no matter whether the basis magnitudes are large or small) will correspond to positive estimated quartets; small values of the cross-magnitudes will mark negative estimated quartets.
4. The marginal distributions of Φ, corresponding to cases in which some of the cross-magnitudes are unknown, can be obtained by setting the corresponding terms ε_i to zero (this corresponds to making E^2 equal to 1, its expected value). In mathematical notation, a general expression may be used for G, covering all cases,

$$G = \frac{2C(1 + w_5\varepsilon_5 + w_6\varepsilon_6 + w_7\varepsilon_7)}{1 + Q}, \quad (5.24)$$

where

$$Q = [w_5(\varepsilon_1\varepsilon_2 + \varepsilon_3\varepsilon_4)\varepsilon_5 + w_6(\varepsilon_1\varepsilon_3 + \varepsilon_2\varepsilon_4)\varepsilon_6 + w_7(\varepsilon_1\varepsilon_4 + \varepsilon_2\varepsilon_3)\varepsilon_7]/2N.$$

w_i is equal to one, but for the case in which the ith cross-reflection is not measured. In this last case, $w_i = 0$.
5. The generalization of (5.23) to unequal atom structures can be done by replacing N with N_{eq} (as given by (5.5)).

Distribution (5.22) is drawn (full curve) in Figs. 5.6 and 5.7, for the same values for which distribution (5.16) is calculated.

If an analogous approach is applied in P$\bar{1}$, we have

$$P^+ \simeq 0.5 + 0.5 \tanh(G/2), \quad (5.25)$$

where G is given by (5.23) (or (5.24)) and $P+$ is the probability that the sign of $E_{\mathbf{h}_1} E_{\mathbf{h}_2} E_{\mathbf{h}_3} E_{\mathbf{h}_4}$ is positive. If $G > 0$, then $P+ > 1/2$.

The accuracy of the Hauptman and Giacovazzo formulations is discussed in Section 5.8.

Quartet estimates may improve if full use is made of the symmetry. Readers interested in this topic will find preliminary information in Appendix 5.D and a more general description in Giacovazzo (1976d).

5.8 About quartet reliability

There are three basic questions concerning the use of quartet invariants in phasing procedures. Let us consider these in the following order:

(i) How many quartets can be found among a selected (and sufficiently large) number of reflections? Is this number larger than the number of triplet invariants? The answer is yes; the number of quartets is usually much larger than the number of reflections. The practical aspects concerning triplet and quartet identification are discussed in Appendix 6.A, but, even intuitively, the reader can easily understand why the quartet number is much larger than the triplet number (i.e. quartets have a one degree of freedom more).

(ii) In its first representation a triplet invariant depends on three diffraction amplitudes only, while a quartet invariant depends on (at least) seven magnitudes. Does this mean that quartets should be preferable to triplets in the phasing process? The answer is no; indeed, quartets are phase relationships of order N, while triplets are phase relationships of order $\sqrt{N}$. Thus, for medium-sized or large crystal structures it may be expected that the number of reliable quartets may be a small percentage of the large total number of estimated quartets. The number of reliable quartets decreases with structural complexity much more rapidly than the number of reliable triplets. Thus, in spite of the large number of quartets, the number of reliable ones is usually smaller than the corresponding number of triplet invariants.

(iii) Can quartets and triplets be used together in phasing procedures? To answer this question, some preliminary considerations should be made. As soon as a phasing procedure progresses, the number of estimated phases increases and at a certain step, the seven phases of the reflections belonging to the first phasing shell of the quartet Φ are estimated. The following *tripoles* may then be established:

$$\begin{cases} \Phi_1 = \Phi \\ t_1 = -\phi_{h_1} - \phi_{h_2} + \phi_{h_1+h_2} \\ t_2 = -\phi_{h_3} + \phi_{h_1+h_2+h_3} - \phi_{h_1+h_2} \end{cases}$$

$$\begin{cases} \Phi_1 = \Phi \\ t_3 = -\phi_{h_1} - \phi_{h_3} + \phi_{h_1+h_3} \\ t_4 = -\phi_{h_2} + \phi_{h_1+h_2+h_3} - \phi_{h_1+h_3} \end{cases}$$

$$\begin{cases} \Phi_1 = \Phi \\ t_5 = -\phi_{h_2} - \phi_{h_3} + \phi_{h_2+h_3} \\ t_2 = -\phi_{h_1} + \phi_{h_1+h_2+h_3} - \phi_{h_2+h_3}. \end{cases}$$

Then,

$$-\Phi = t_1 + t_2 = t_3 + t_4 = t_5 + t_6.$$

Thus, the expected value of a quartet estimates three sums of two triplets. Let us now describe two typical cases:

(a) *Strong positive quartets*. We suppose that all the reflections involved in the tripoles have magnitudes larger than E_t, where E_t is the minimum value of $|E|$ chosen when carrying out phase determination by means of triplets. If E_t is large enough, then

$$\Phi \simeq t_1 \simeq t_2 \simeq \cdots \simeq 0.$$

In this case, the correlation between triplet and quartet information is very high. Therefore, a phase refinement which uses triplet and quartet relationships as though they were independent could emphasize the inadequacies of the standard tangent formula, instead of improving it. From the point of view of direct space, a 'cubing effect' should be added to the 'squaring effect' of the triplet relationships, so that the procedure should tend to strengthen the dominant features of the structure.

(b) *Negative quartets*. If all the basis magnitudes of the quartet are larger than E_t, but the cross-magnitudes are smaller, no triplet appearing in the tripoles is estimated in the direct procedure. In this case, $\Phi = \pi$ is a phase indication uncorrelated with triplet phase assignment.

The above considerations suggest that the combined use of triplets and of negative quartet relationships is more useful than the combined use of positive triplets and quartets. Several phasing procedures have benefited from the supplementary information provided by quartets (see Chapter 6); in particular, we refer to the computer programs *SAYTAN* (Debaerdemaeker et al., 1985, 1988), *SHELX* (Sheldrick, 1990), *SIR92* (Giacovazzo et al., 1992a; Burla et al., (1992)) and its heirs.

APPENDIX 5.A   THE PROBABILISTIC ESTIMATION OF THE TRIPLET INVARIANTS IN P1

We will use the notation previously employed in Appendix 4.A, and we will assume that no prior information is available, but for the chemical content of the unit cell and for the space group. Then, equation (4.A.18), limited up to terms of order $N^{-1/2}$, may be rewritten in the form,

$$P(A_1, A_2, A_3, B_1, B_2, B_3) \simeq \frac{1}{(2\pi)^6} \times \int_{-\infty}^{+\infty} \cdots \int_{-\infty}^{+\infty} \exp\left[-i(u_1 A_1 + u_2 A_2 + \cdots + v_3 B_3) - \frac{1}{4}(u_1^2 + \cdots + v_3^2) + S_3 + \cdots\right] du_1\, du_2 \ldots dv_3,$$

(5.A.1)

where $A_1, A_2, A_3, B_1, B_2,$ and B_3 stand for $A_{h_1}, A_{h_2}, A_{h_3}, B_{h_1}, B_{h_2},$ and B_{h_3}, respectively, and

$$S_3 = -\mathrm{i}\left(\sum_{j=1}^{N} {}^j k_{111000}\, u_1 u_2 u_3 + \sum_{j=1}^{N} {}^j k_{001110}\, u_3 v_1 v_2 \right.$$
$$\left. + \sum_{j=1}^{N} {}^j k_{010101}\, u_2 v_1 v_3 + \sum_{j=1}^{N} {}^j k_{100011}\, u_1 v_2 v_3 \right).$$

The explicit expression for the first cumulants is

$$\sum_{j=1}^{N} {}^j k_{111000} = \sum_{j=1}^{N} \langle \psi_j'(\mathbf{h}_1)\, \psi_j'(\mathbf{h}_2)\, \psi_j'(\mathbf{h}_3) \rangle = \sum_{j=1}^{N} v_j(\mathbf{h}_1)\, v_j(\mathbf{h}_2)\, v_j(\mathbf{h}_3)$$
$$\times \langle \cos(2\pi \mathbf{h}_1 \cdot \mathbf{r}_j)\cos(2\pi \mathbf{h}_2 \cdot \mathbf{r}_j)\cos(2\pi \mathbf{h}_3 \cdot \mathbf{r}_j) \rangle = \frac{1}{4}\sum_{j=1}^{N}\frac{1}{\sqrt{N_{\mathrm{eq}}}},$$

where (see equation (2.9)),

$$v_j(\mathbf{h}) = f_j(\mathbf{h})\left[\sum_{j=1}^{N} f_j^2(\mathbf{h})\right]^{-1/2},$$

and (see equation (5.5)), $N_{\mathrm{eq}} = \sigma_2^3/\sigma_3^2$.

Similar expressions hold for the other cumulants. In particular,

$${}^j k_{001110} = {}^j k_{010101} = {}^j k_{100011} = \frac{1}{4}\frac{1}{\sqrt{N_{\mathrm{eq}}}}.$$

Then,

$$S_3 = \frac{-\mathrm{i}}{4\sqrt{N_{\mathrm{eq}}}}\left(u_1 u_2 u_3 - u_3 v_1 v_2 - u_2 v_1 v_3 - u_1 v_2 v_3\right).$$

If we neglect in our calculations terms of order higher than $N^{-1/2}$, and we apply the variable change (4.A.18), then (5.A.1) reduces to

$$P \equiv P(\phi_1, \phi_2, \phi_3, R_1, R_2, R_3)$$
$$\simeq \frac{1}{(2\pi)^6}2^3 R_1 R_2 R_3 \int_0^\infty \int_0^\infty \int_0^\infty \int_0^{2\pi}\int_0^{2\pi}\int_0^{2\pi} \rho_1 \rho_2 \rho_3$$
$$\times \exp\left(-\mathrm{i}\sqrt{2}\sum_{j=1}^{3}\rho_j R_j \cos(\phi_j - \theta_j) - \frac{1}{2}\sum_{j=1}^{3}\rho_j^2 - \mathrm{i}\frac{\rho_1 \rho_2 \rho_3}{\sqrt{2 N_{\mathrm{eq}}}}\right. \quad (5.A.2)$$
$$\left. \times \cos(\theta_1 + \theta_2 + \theta_3)\right) d\rho_1\, d\rho_2\, d\rho_3\, d\theta_1\, d\theta_2\, d\theta_3,$$

where $\phi_1, \phi_2,$ and ϕ_3 stand for $\phi_{h_1}, \phi_{h_2},$ and ϕ_{h_3} respectively. The terms in the exponent involving ρ_1 and θ_1 are first combined by means of equation (M.F.8):

$$-\tfrac{1}{2}\rho_1^2 - \mathrm{i}\rho_1\left(\sqrt{2}R_1 \cos(\theta_1 - \phi_1) + \frac{\rho_3 \rho_3}{\sqrt{2 N_{\mathrm{eq}}}}\cos(\theta_1 + \theta_2 + \theta_3)\right)$$
$$= -\tfrac{1}{2}\rho_1^2 - \mathrm{i}X\rho_1 \cos(\theta_1 + y),$$

where,

$$X^2 \simeq 2R_1^2 + \frac{2}{\sqrt{N_{\mathrm{eq}}}}R_1 \rho_2 \rho_3 \cos(\phi_1 + \theta_2 + \theta_3).$$

X and y do not depend on θ_1. Then, using (M.E.16),

$$P \simeq \frac{1}{(2\pi)^5} 2^3 R_1 R_2 R_3 \exp(-R_1^2) \int_0^\infty \ldots \int_0^{2\pi} \times \exp\left\{-i\sqrt{2}\sum_{j=2}^3 \rho_j R_j \cos(\phi_j - \theta_j)\right.$$
$$\left.-\frac{1}{2}\sum_{j=2}^3 \rho_j^2 - \frac{R_1 \rho_2 \rho_3}{\sqrt{N_{\text{eq}}}} \cos(\phi_1 + \theta_2 + \theta_3)\right\} \rho_2 \rho_3 \, d\rho_2 \, d\rho_3 \, d\theta_2 \, d\theta_3.$$

The same procedure is followed for ρ_2, θ_2 and ρ_3, θ_3, with the final result,

$$P \equiv P(\phi_1, \phi_2, \phi_3, R_1, R_2, R_3)$$
$$\simeq \frac{R_1 R_2 R_3}{\pi^3} \exp\left(-R_1^2 - R_2^2 - R_3^2 + \frac{2R_1 R_2 R_3}{\sqrt{N_{\text{eq}}}} \cos(\phi_1 + \phi_2 + \phi_3)\right). \quad (5.\text{A}.3)$$

Since $\Phi = \phi_1 + \phi_2 + \phi_3$, elementary statistics gives,

$$P(\Phi, R_1, R_2, R_3) \simeq \int_0^{2\pi} \int_0^{2\pi} P(\phi_1, \phi_2, \Phi - \phi_1 - \phi_2, R_1, R_2, R_3) \, d\phi_1 \, d\phi_2$$
$$\simeq \tfrac{4}{\pi} R_1 R_2 R_3 \exp(-R_1^2 - R_2^2 - R_3^2 + G \cos \Phi),$$

where,

$$G = 2R_1 R_2 R_3 / \sqrt{N_{\text{eq}}}.$$

The conditional distribution (Cochran, 1955)

$$P(\Phi | R_1, R_2, R_3)$$

is obtained from (M.A.17) with the use of (M.E.25),

$$P(\Phi | R_1, R_2, R_3) \simeq (2\pi I_0(G))^{-1} \exp(G \cos \Phi). \quad (5.\text{A}.4)$$

The conditional expectations and the conditional variance of Φ, $\sin(n\Phi)$ and $\cos(n\Phi)$, where n is an arbitrary integral parameter, may readily be derived by applying (M.E.26) and (M.E.27):

$$\langle \Phi | G \rangle = [2\pi I_0(G)]^{-1} \int_{-\pi}^{\pi} \Phi \exp(G \cos \Phi) d\Phi = 0$$
$$\langle \sin(n\Phi) | G \rangle = 0; \quad \langle \cos(n\Phi) | G \rangle = D_n(G)$$
$$\text{var}[\sin(n\Phi)|G] = \langle \sin^2(n\Phi)|G \rangle - \langle \sin(n\Phi)|G \rangle^2 = \frac{1}{2} - \frac{1}{2} D_{2n}(G) \quad (5.\text{A}.5)$$
$$\text{var}[\cos(n\Phi)|G] = \frac{1}{2} + \frac{1}{2} D_{2n}(G) - D_n^2(G),$$

where, $D_i(x) = I_i(x)/I_0(x)$. When $n = 1$ we have, from (M.E.22),

$$\text{var}[\sin \Phi | G] = \frac{D_1(G)}{G},$$
$$\text{var}[\cos \Phi | G] = 1 - \frac{D_1(G)}{G} - D_1^2(G).$$

The conditional distribution, $P(\phi_1 | \phi_2, \phi_3, R_1, R_2, R_3)$ may readily be obtained by application of (M.A.17) to (5.A.3).

APPENDIX 5.B SYMMETRY INCONSISTENT TRIPLETS

Let us consider the triplet invariant

$$\Phi = \phi_{\mathbf{h}_1} + \phi_{\mathbf{h}_2} + \phi_{\mathbf{h}_3} \quad \text{with} \quad \mathbf{h}_1 + \mathbf{h}_2 + \mathbf{h}_3 = 0$$

and let us suppose that one or more additional triplet invariants can be found of type

$$\Psi = \phi_{\mathbf{h}_1} + \phi_{\mathbf{h}_2 \mathbf{R}_s} + \phi_{\mathbf{h}_3 \mathbf{R}_r},$$

with

$$\mathbf{h}_1 + \mathbf{h}_2 \mathbf{R}_s + \mathbf{h}_3 \mathbf{R}_r = 0, \mathbf{h}_2 \mathbf{R}_s \neq \mathbf{h}_2, \text{ and } \mathbf{h}_3 \mathbf{R}_r \neq \mathbf{h}_3. \tag{5.B.1}$$

We define the first representation of Φ as the collection of triplet invariants (5.B.1). Because of relationship (1.25),

$$\Psi = \Phi - \Delta_{s,r},$$

where,

$$\Delta_{s,r} = 2\pi(\mathbf{h}_2 \mathbf{T}_s + \mathbf{h}_3 \mathbf{T}_r).$$

Ψ and Φ are *symmetry consistent* triplets if $\Delta_{s,r} = 2n\pi$, otherwise, they are said to be *symmetry inconsistent*.

The existence of inconsistent triplets in $P2_12_12_1$ was pointed out by Hauptman and Karle (1956). A probabilistic approach for the estimation of triplet invariants, which takes into account the space group symmetry, was described by Giacovazzo (1974a,b). His results were confirmed and extended by Pontenagel and Krabbendam (1983); they found that in the eleven pairs of enantiomorphically related space groups there are triplet phases which are not expected to be close to 2π. Han and Langs (1988) examined the 230 space groups in order to identify conditions which permit symmetry inconsistent triplets. A table describing such conditions for the various space groups was presented. The matter was re-examined by Giacovazzo (1989) in order to:

1. complete the results by Han and Langs, who missed conditions for cubic space groups and neglected triplets with symmetry restricted phase values;
2. provide an algorithm for the discovery of symmetry inconsistent triplets in any space group, in order to avoid the use of the large Han and Langs table.

The estimation of triplet invariants by making full use of the symmetry poses the following questions:

(i) is the distribution of Φ a von Mises distribution?
(ii) is the standard Cochran parameter C, as defined by (5.3), the correct one or should it be modified? Indeed, if we apply the Cochran distribution first to Φ and then to any Ψ, both are expected to be close to zero, but this is contradictory when $\Delta_{s,r} \neq 2n\pi$.

The theory of representations is able to show that Φ is still distributed according to a von Mises function, but proper weights (smaller or larger than unity) should be associated with the parameter C: in particular, weights equal to zero should be introduced when $\Delta_{s,r} = \pi$.

We show below some examples:

1. In P3$_1$,
$$\Phi = \phi_{30\bar{3}} + \phi_{\bar{3}31} + \phi_{0\bar{3}2}$$
and the symmetry equivalent triplet,
$$\Psi = \phi_{30\bar{3}} + \phi_{0\bar{3}1} + \phi_{\bar{3}32} = \phi_{30\bar{3}} + \phi_{(\bar{3}31)\mathbf{R}_3} + \phi_{(0\bar{3}2)\mathbf{R}_2} = \Phi - 2\pi/3$$
can be found. The most probable value of Φ is $\pi/3$ and $G' = G$.

2. In P4$_1$
$$\left[(x,y,z), \left(\bar{x},\bar{y},z+\frac{1}{2}\right), \left(\bar{y},x,z+\frac{1}{4}\right) \left(y,\bar{x},z-\frac{1}{4}\right)\right],$$
the equivalent pairs of triplets may be found:
$$\Phi = \phi_{221} + \phi_{\bar{4}01} + \phi_{2\bar{2}\bar{2}},$$
$$\Psi = \phi_{221} + \phi_{0\bar{4}1} + \phi_{\bar{2}2\bar{2}} = \phi_{221} + \phi_{(\bar{4}01)\mathbf{R}_4} + \phi_{(2\bar{2}\bar{2})\mathbf{R}_2} = \Phi + \pi/2.$$
The most probable value of Φ is $-\pi/4$ and $G' = G\sqrt{2}$.

3. Let us consider in space group P4$_1$2$_1$2, the triplet
$$\Phi = \phi_{1,\bar{1},\overline{19}} + \phi_{1,1,12} + \phi_{\bar{2}07}.$$
If the Cochran distribution is taken into account, Φ is expected to be close to 2π. However, the values of $\phi_{1,\bar{1},\overline{19}}, \phi_{1,1,12}$, and $\phi_{\bar{2}07}$ are symmetry restricted to $(0, \pi), (0, \pi), (-\pi/4, 3\pi/4)$, respectively, so that the triplet Φ is itself restricted to $(-\pi/4, 3\pi/4)$. We see that the Cochran distribution is inadequate to describe this situation.

If representation theory is adopted, a triplet symmetry equivalent to Φ may be taken into consideration:
$$\Psi = \phi_{1,\bar{1},\overline{19}} + \phi_{\bar{1},\bar{1},12} + \phi_{207} = \phi_{1,\bar{1},\overline{19}} + \phi_{(1,1,12)\mathbf{R}_2} + \phi_{(\bar{2}07)\mathbf{R}_8} = \Phi + \pi/2.$$
The most probable value of Φ is $-\pi/4$ and $G' = G\sqrt{2}$.

4. Let us consider in P2$_1$2$_1$2$_1$ all the triplets constituted by reflections with restricted phase values, which sum to $\pm \pi/2$. For example, let $\mathbf{h}_1 = (0, e, 0), \mathbf{h}_2 = (e, e, 0)$, and $\mathbf{h}_3 = (e, 0, 0)$. The crystallographic symmetry restricts $\phi_{\mathbf{h}_1}$ and $\phi_{\mathbf{h}_2}$ to $n\pi$, and $\phi_{\mathbf{h}_3}$ to $\pm \pi/2$, so that $\Phi = \pm \pi/2$. We cannot know from the moduli alone if Φ is $+\pi/2$ or $-\pi/2$ (one of them defines the enantiomorph, and this choice is arbitrary). Accordingly, $G' = 0$.

Inconsistent triplets are routinely searched in some direct methods programs (e.g. *SIR2011*) and eliminated from the set of triplets actively used in the phasing process.

APPENDIX 5.C THE P_{10} FORMULA

In accordance with Section 5.4, the distribution (5.12), with **k** a free vector, may be used to calculate the conditional distribution (5.14). If the space group symmetry is taken into account, definition (5.11) may be generalized as follows: the first representation of the triplet invariant is the collection of the special quintets,

$$\Psi_2 = \Phi + \phi_{k\mathbf{R}_i} - \phi_{k\mathbf{R}_i}, i = 1, 2, \ldots m,$$

where m is the number of symmetry operators, $C_j \equiv (\mathbf{R}_j, \mathbf{T}_j)$, not related by a centre of symmetry ($\mathbf{R}_j$ is the rotational part and $\mathbf{T}_j$ the translational part of the symmetry operator). In conclusion, any quintet Ψ_2 depends on, in addition to the basis magnitudes $R_{h_1}, R_{h_2}, R_{h_3}, R_k$, the $6m$ cross-magnitudes:

$$\begin{array}{l} R_{h_1+k\mathbf{R}_1}, R_{h_1-k\mathbf{R}_1}, R_{h_2+k\mathbf{R}_2}, R_{h_2-k\mathbf{R}_1}, R_{h_3+k\mathbf{R}_1}, R_{h_3-k\mathbf{R}_1} \\ R_{h_1+k\mathbf{R}_2}, R_{h_1-k\mathbf{R}_2}, R_{h_2+k\mathbf{R}_2}, R_{h_2-k\mathbf{R}_2}, R_{h_3+k\mathbf{R}_2}, R_{h_3-k\mathbf{R}_2} \\ \vdots \hspace{4cm} \vdots \\ R_{h_1+k\mathbf{R}_m}, R_{h_1-k\mathbf{R}_m}, R_{h_2+k\mathbf{R}_m}, R_{h_2-k\mathbf{R}_m}, R_{h_3+k\mathbf{R}_m}, R_{h_3-k\mathbf{R}_m}. \end{array} \quad (5.\text{C}.1)$$

In this case we write,

$$\{B\}_2 = \{R_{h_1}, R_{h_2}, R_{h_3}, R_k, R_{h_1 \pm k\mathbf{R}_i}, R_{h_2 \pm k\mathbf{R}_i}, R_{h_3 \pm k\mathbf{R}_i} \ \ i = 1, 2, \ldots m\}.$$

The conditional probability, $P(\Phi|\{B\}_2)$, now provides an estimate for Φ in any space group. Since its exact expression is not easy to obtain, we prefer to introduce a simple approximation of $P(\Phi|\{B\}_2)$, which may be derived as a proper combination of the various ten-variate distributions,

$$P(\Phi|R_{h_1}, R_{h_2}, R_{h_3}, R_k, R_{h_1 \pm k\mathbf{R}_i}, R_{h_2 \pm k\mathbf{R}_i}, R_{h_3 \pm k\mathbf{R}_i}). \quad (5.\text{C}.2)$$

The P_{10} distribution was derived by Cascarano et al. (1984). For n.cs. space groups, P_{10} has a von Mises expression:

$$P_{10}(\Phi) \simeq (2\pi I_0(G))^{-1} \exp(G \cos \Phi), \quad (5.\text{C}.3)$$

where

$$\begin{array}{l} G = C(1+Q) \\ C = 2R_{h_1} R_{h_2} R_{h_3} / \sqrt{N_{\text{eq}}}, \end{array} \quad (5.\text{C}.4)$$

C is the Cochran reliability parameter,

$$Q = \sum_k \left(\frac{\Sigma_{i=1}^{\prime m} A_{k,i}/N}{1 + (\varepsilon_{h_1}\varepsilon_{h_2}\varepsilon_{h_3} + \Sigma_{i=1}^{\prime m} B_{k,i})/2N} \right), \quad (5.\text{C}.5)$$

$$\begin{array}{l} A_{k,i} = \varepsilon_k [\varepsilon_{h_1+k\mathbf{R}_i}(\varepsilon_{h_2-k\mathbf{R}_i} + \varepsilon_{h_3-k\mathbf{R}_i}) \\ \hspace{1.2cm} + \varepsilon_{h_2+k\mathbf{R}_i}(\varepsilon_{h_1-k\mathbf{R}_i} + \varepsilon_{h_3-k\mathbf{R}_i}) + \varepsilon_{h_3+k\mathbf{R}_i}(\varepsilon_{h_1-k\mathbf{R}_i} + \varepsilon_{h_2-k\mathbf{R}_i})], \\ B_{k,i} = \varepsilon_{h_1}[\varepsilon_k(\varepsilon_{h_1+k\mathbf{R}_i} + \varepsilon_{h_1-k\mathbf{R}_i}) + \varepsilon_{h_2+k\mathbf{R}_i}\varepsilon_{h_3-k\mathbf{R}_i} + \varepsilon_{h_2-k\mathbf{R}_i}\varepsilon_{h_3+k\mathbf{R}_i}] \\ \hspace{1cm} + \varepsilon_{h_2}[\varepsilon_k(\varepsilon_{h_2+k\mathbf{R}_i} + \varepsilon_{h_2-k\mathbf{R}_i}) + \varepsilon_{h_1+k\mathbf{R}_i}\varepsilon_{h_3-k\mathbf{R}_i} + \varepsilon_{h_1-k\mathbf{R}_i}\varepsilon_{h_3+k\mathbf{R}_i}] \\ \hspace{1cm} + \varepsilon_{h_3}[\varepsilon_k(\varepsilon_{h_3+k\mathbf{R}_i} + \varepsilon_{h_3-k\mathbf{R}_i}) + \varepsilon_{h_1+k\mathbf{R}_i}\varepsilon_{h_2-k\mathbf{R}_i} + \varepsilon_{h_1-k\mathbf{R}_i}\varepsilon_{h_2+k\mathbf{R}_i}]; \\ \varepsilon = |E|^2 - 1. \end{array}$$

$(\varepsilon_{h_1}\varepsilon_{h_2}\varepsilon_{h_3} + \Sigma_{i=1}^{\prime m} B_{k,i})$ is assumed to be zero, if it is experimentally negative. The prime on the summation warns the reader that precautions have to be taken in order to avoid duplications in the contributions when k sweeps out reciprocal space.

Distribution (5.C.3) is unimodal and takes its maximum at $\Phi = 0$ if $G > 0$, and at $\Phi = \pi$ if $G < 0$. Therefore, it is able in principle to estimate with high reliability only triplets with phase values around 0 or π. In accordance with the theory, enantiomorph-sensitive triplets should present a rather flat distribution ($G \simeq 0$), so that they cannot be reliably fixed. This is the most important limitation of the theory; however, triplets with $G \simeq 0$ can be used in phasing procedures for recognizing the correct solution in multisolution approaches (Cascarano et al., 1987a).

The accuracy with which the value of Φ is estimated by (5.C.5) depends strongly on ε_k. Thus, in practice only a subset of reciprocal space (the reflection k with large values of ε) may be used for estimating Φ.

To provide some insight into the relative efficiency of the Cochran and the P_{10} formulas, we show some simple statistical tests in Tables 5.C.1 and 5.C.2.

Line four of Table 5.C.1 may be interpreted as follows: there are 905 triplet invariants with $C > 1.4$ (and therefore estimated positive by the Cochran formula), among which there are 30 negative triplets (which are therefore wrongly estimated). There are 1460 triplets with $G > 1.4$ (and therefore estimated positive by P_{10}), among which there are only four negative triplets.

Line four of Table 5.C.2 may be interpreted as follows: there are 201 triplets with $|G| > 0.4$, estimated negative by P_{10} (and therefore with a negative value of G), among which there are only 23 wrongly estimated triplets (triplets which are really positive).

Similar advantages are obtained when P_{10} is applied to n.cs. space groups. The ability of the P_{10} formula to identify negative triplets (and to better rank the positive ones) is a great advantage in direct phasing procedures (see Chapter 6).

APPENDIX 5.D THE USE OF SYMMETRY IN QUARTET ESTIMATION

The probabilistic theories described in Sections 5.5–5.8 do not fully exploit the space group symmetry. Since this strongly influences quartet estimates, we will make some introductory remarks in this section (Giacovazzo, 1976d).

According to Section 4.2, the first representation of

$$\Phi = \phi_{h_1} + \phi_{h_2} + \phi_{h_3} + \phi_{\overline{h_1+h_2+h_3}}$$

is the collection of the quartets,

$$\phi_{h_1} + \phi_{h_2 R_i} + \phi_{h_3 R_j} + \phi_{\overline{h_1+h_2 R_i+h_3 R_j}}, \qquad (5.D.1)$$

obtained when $\mathbf{R}_i$ and $\mathbf{R}_j$ vary over the set of rotation matrices of the space group under the condition that $h_1 + h_2 \mathbf{R}_i + h_3 \mathbf{R}_j$ is symmetry equivalent to $h_1 + h_2 + h_3$. The first representation will consist of more than one quartet if at least one of the cross terms has a statistical weight different from unity. In this case, the number of cross-magnitudes will exceed three. For example, let us consider in space group P2/m, the quartet

$$\Phi = \phi_{234} + \phi_{112} + \phi_{2\bar{1}3} - \phi_{539}, \qquad (5.D.2)$$

which depends on the cross-magnitudes

$$R_{346}, R_{427}, R_{305}. \qquad (5.D.3)$$

Table 5.C.1 *Quinol* [$C_{18}H_{18}O_6$, R-3]: cs. case. Statistical results on triplet estimates according to the Cochran parameter C and the P_{10} parameter G. nr is the number of triplets with C or G larger than *THR*, nr_w is the number of negative triplets (i.e. triplets for which $\Phi = \pi$). In the last column, only triplets estimated positive by P_{10} are quoted

THR	C	G
	$nr\ (nr_w)$	$nr\ (nr_w)$
0.4	5974 (972)	4174 (145)
0.8	3150 (322)	2940 (37)
1.4	905 (30)	1460 (4)
2.0	329 (0)	642 (0)
2.8	96 (1)	198 (0)

Table 5.C.2 *Quinol* [$C_{18}H_{18}O_6$, R-3]: cs. case. Statistical results on the triplets estimated negative by the P_{10} formula (i.e. with a negative value of G). nr is the number of triplets with $|G|$ larger than *THR*, nr_w is the number of wrongly estimated triplets (i.e. triplets for which $\Phi = 0$)

THR	$nr\ (nr_w)$
0.0	500 (121)
0.2	412 (81)
0.4	201 (23)
0.8	43 (2)

Besides (5.D.2), the first representation includes the quartet

$$\Phi' = \phi_{234} + \phi_{1\bar{1}2} + \phi_{2\bar{1}3} - \phi_{539} \equiv \Phi, \qquad (5.D.4)$$

which depends on the cross-magnitudes

$$R_{326}, R_{447}, R_{305}. \qquad (5.D.5)$$

If the magnitudes (5.D.3) are all large, Φ is expected to be close to zero; also, Φ' is expected to be close to zero if the magnitudes (5.D.5) are large. The two phase indications are mostly independent of each other (only R_{305} is shared by both sets (5.D.3) and (5.D.5)). Since $\Phi' \equiv \Phi$, the case in which all five cross-magnitudes in $\{B\}_1$ are large is quite favourable for the reliable phase assignment $\Phi \simeq 0$. Conversely, if all five cross-magnitudes are small we can confidently assign $\Phi = \pi$. The worst situation occurs when R_{346} and R_{427} are large while R_{326} and R_{447} are small, or vice versa, because the indication for Φ from (5.D.2) contradicts the indication for Φ coming from (5.D.4).

Let us now suppose that the space group is P2$_1$/c : then, F_{305} is systematically absent and $\Phi' = \Phi + \pi$. If the non-absent reflections in (5.D.3) and (5.D.5) are large, both Φ and Φ' are expected to be close to zero, but this is contradictory because $\Phi' = \Phi + \pi$. An analogous situation occurs when the non-absent reflections in (5.D.3) and (5.D.5) are small; both Φ and Φ' are expected to be close to π, which is a contradictory expectation. The most favourable case for a reliable estimate of Φ occurs when the non-absent reflections in (5.D.3) are large and those in (5.D.5) are small, or vice versa, then the two phase indications comply. The most correct way of dealing with such a problem is to calculate the distribution

$$P(E_{234}, E_{112}, E_{2\bar{1}3}, E_{539}, E_{346}, E_{427}, E_{326}, E_{447}, E_{305}),$$

which is the joint probability distribution of all the reflections in the first phasing shell. The resulting formula for estimating the sign of the cosine of Φ in the Giacovazzo formulation may be expressed, qualitatively, as follows: (5.25) is still valid but G is now approximately given by

$$G = \frac{2C}{1+Q}(\varepsilon_{346} + \varepsilon_{427} + \varepsilon_{326} + \varepsilon_{447} + \varepsilon_{305}) \quad \text{for} \quad \text{P2/}m$$

$$G = \frac{2C}{1+Q}[\varepsilon_{346} + \varepsilon_{427} - (\varepsilon_{326} + \varepsilon_{447})] \quad \text{for} \quad \text{P2}_1/c.$$

Q is a suitable scaling factor which, for simplicity, is not given.

The above examples show that it is statistically more convenient to estimate Φ from its first representation than from a single quartet of the set. A more formal mathematical treatment of the problem may be found in Giacovazzo (1976d).

Traditional direct phasing procedures

6.1 Introduction

Which phasing methods can be included in the category *direct methods*, and which require a different appellation?

Originally, direct phasing was associated with those approaches which were able to derive phases directly from the diffraction moduli, without passing through deconvolution of the Patterson function. Since a Patterson map provides interatomic distances, and therefore lies in 'direct space', direct methods were also referred to as *reciprocal space methods*, and Patterson techniques as *real-space methods*. Historically, direct methods use 3-,4-,...,n-phase invariants and 1-2-,... phase seminvariants via the tangent formula or its modified algorithms. Since the 1950s, about a half a century of scientific effort has fallen under the above definition. Such approaches are classified here as *traditional direct methods*.

Today, the situation is more ambiguous, because:

(i) modern direct methods programs involve steps operating both in reciprocal space and in direct space, the latter mainly devoted to phase extension and refinement (see Chapter 8);
(ii) in the past decade, new phasing methods for crystal structure solution (see Chapter 9) have been developed, based on the properties of Fourier transforms, which again work both in direct and in reciprocal space. Should they therefore be considered to be outside the direct methods category or not?

Our choice is as follows. Direct methods are all of the approaches which allow us to derive phases from diffraction amplitudes, without passing through a Patterson function deconvolution. Thus, we also include in this category, *charge flipping* and *VLD* (*vive la différence*), here classified as non-traditional direct methods; their description is postponed to Chapter 9.

In accordance with the above assumptions, in this chapter we will shortly illustrate traditional direct phasing procedures, with particular reference to those which are current and in regular use today: mainly the *tangent procedures* (see Section 6.2) and the *cosine least squares technique*, which is the basic tool of the *shake and Bake* method (see Section 6.4). For brevity, we will skip the methods listed below, even if they have fundamentally important properties:

(a) *symbolic addition techniques* (Zachariasen, 1952; Karle and Karle, 1966);
(b) variants of the tangent formula, such as the *primary–secondary* method (Declercq et al., 1975; Hull et al., 1981; Zhang and Woolfson, 1982), the *linear equation method* (Debaerdemaeker and Woolfson, 1975; Woolfson, 1977, 1978; Baggio et al., 1978), and the *X–Y method* (Debaerdemaeker and Woolfson, 1983, 1989).
(c) *probabilistic determinantal approaches* (Lajzerowicz and Lajzerowicz, 1966; de Rango et al., 1969; Tsoucaris, 1970; de Rango et al., 1974; Taylor et al., 1979; de Gelder et al., 1990; de Gelder, 1992).

We now return to the core of traditional direct methods. In Chapter 5, we described how s.i. phases may be estimated *via* the prior information on diffraction moduli. In particular, formulas like $P(\Phi)|\{R\})$ were obtained, where

$$\Phi = (\phi_\mathbf{h} + \phi_\mathbf{k} + \phi_\mathbf{-h-k})$$

or

$$\Phi = (\phi_\mathbf{h} + \phi_\mathbf{k} + \phi_\mathbf{l} + \phi_\mathbf{-h-k-l}),$$

and $\{R\}$ may coincide with the first or the second phasing shell, etc.

The goal of any phasing procedure is to associate phases with reflections, that is, to pass from invariant estimates to single-phase estimates. The mathematical technique is very simple; in order to show how it works we consider the triplet

$$\Phi = (\phi_\mathbf{h} + \phi_\mathbf{k} + \phi_\mathbf{-h-k}).$$

According to the Cochran relationship (5.6), Φ *is expected to be close to zero, say*

$$\phi_\mathbf{h} + \phi_\mathbf{k} + \phi_\mathbf{-h-k} \approx 0, \tag{6.1}$$

with probability fixed by the concentration parameter

$$C = 2R_\mathbf{h} R_\mathbf{k} R_\mathbf{h+k} / \sqrt{N_{eq}}.$$

Relationship (6.1) is perfectly equivalent to the following proposition:

$\phi_\mathbf{h}$ *is expected to be close to* $-(\phi_\mathbf{k} + \phi_\mathbf{-h-k})$, *with reliability fixed by C.*

To simplify the notation, we observe that, owing to the Friedel relationship, Φ may also be written as $\Phi = (\phi_\mathbf{h} + \phi_\mathbf{k} - \phi_\mathbf{h+k})$, or also, by replacing $\mathbf{k}$ by $-\mathbf{k}$, as

$$\Phi = (\phi_\mathbf{h} - \phi_\mathbf{k} - \phi_\mathbf{h-k}). \tag{6.2}$$

The notation (6.2) is more useful for our purposes. Indeed, for the s.i. (6.2), the Cochran relationship states that

$$\phi_\mathbf{h} \approx (\phi_\mathbf{k} + \phi_\mathbf{h-k}), \tag{6.3}$$

with probability fixed by

$$C = 2R_\mathbf{h} R_\mathbf{k} R_\mathbf{h-k} / \sqrt{N_{eq}}.$$

Therefore, the corresponding conditional distribution is

$$P(\phi_{\mathbf{h}}|\phi_{\mathbf{k}}, \phi_{\mathbf{h-k}}, R_{\mathbf{h}}, R_{\mathbf{k}}, R_{\mathbf{h-k}}) = [2\pi I_0(C)]^{-1} \exp[C\cos(\phi_{\mathbf{h}} - \theta_{\mathbf{h}})], \quad (6.4)$$

where

$$\theta_{\mathbf{h}} = (\phi_{\mathbf{k}} + \phi_{\mathbf{h-k}}) \quad (6.5)$$

is the most probable value of $\phi_{\mathbf{h}}$. In other words, from (6.4) and (6.5), the phase relationship

$$\phi_{\mathbf{h}} \approx \theta_{\mathbf{h}} \equiv \phi_{\mathbf{k}} + \phi_{\mathbf{h-k}} \quad (6.6a)$$

is obtained.

A graphical representation of the distribution (6.4) is shown in Fig. 6.1. All of the curves have their maximum at $\theta_{\mathbf{h}}$ and show the same features noticed in the discussion on Fig. 5.1 (e.g. the larger that C becomes, the larger is the probability of relation (6.6a)).

Distribution (6.4) may be generalized by admitting, in accordance with the second representation estimate of the triplet invariant, that the concentration parameter G replaces C, where G may be positive or negative; in the first case (6.3) is still valid, in the second, (6.3) is replaced by

$$\phi_{\mathbf{h}} \approx (\phi_{\mathbf{k}} + \phi_{\mathbf{h-k}}) + \pi. \quad (6.6b)$$

Accordingly, (6.4) is replaced by the more general distribution,

$$P(\phi_{\mathbf{h}}) = [2\pi I_0(G)]^{-1} \exp[G\cos(\phi_{\mathbf{h}} - \theta_{\mathbf{h}})] \quad (6.6c)$$

where the short notation $P(\phi_{\mathbf{h}})$ implies that the diffraction amplitudes belonging to the second phasing shell have been exploited to obtain G, and that $(\phi_{\mathbf{k}} + \phi_{\mathbf{h-k}})$ is known. If G is positive, $\phi_{\mathbf{h}} \approx \theta_{\mathbf{h}}$, if G is negative, $\phi_{\mathbf{h}} \approx \theta_{\mathbf{h}} + \pi$ (see Fig. 6.2).

So far we have described the mathematical machinery for passing from phase invariants to single phases. However, relations (6.6) can define the value of $\phi_{\mathbf{h}}$ only if $\phi_{\mathbf{k}}$ and $\phi_{\mathbf{h-k}}$ are known; but how do we obtain this information? In the following sections we will see how this may be achieved.

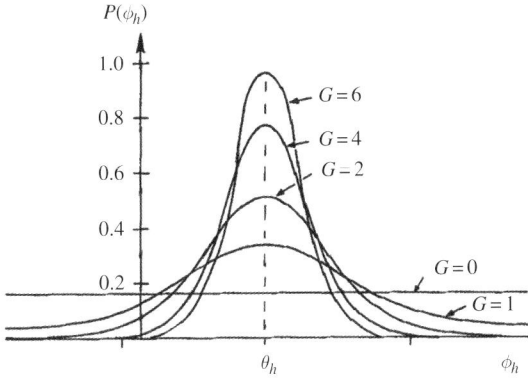

Fig. 6.1
The conditional distribution $P(\phi_{\mathbf{h}}|R_{\mathbf{h}}, E_{\mathbf{k}}, E_{\mathbf{h-k}})$.

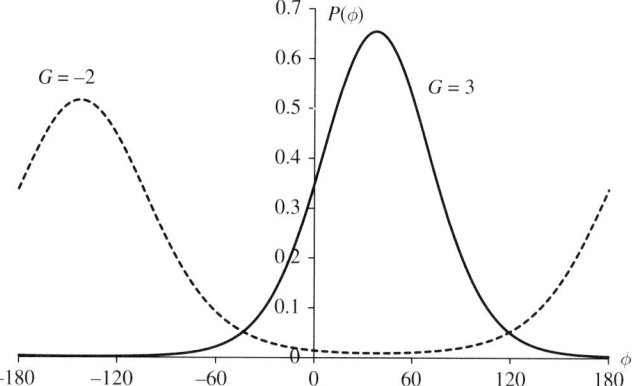

Fig. 6.2
$P(\phi_\mathbf{h})$ in accordance with distribution (6.6c). We choose for the continuous line, $G = 3$ and $\theta_\mathbf{h} = 38°$; the maximum of the curve is at $\theta_\mathbf{h}$. The dashed line represents the case where $G = -2$ and $\theta_\mathbf{h} = 38°$. In this case, the curve attains its maximum at $\phi_\mathbf{h} \approx \theta_\mathbf{h} + \pi$.

6.2 The tangent formula

The main limitation of the Cochran relationship (and therefore of relation (6.4)) is that the C parameter may be small for most triplet relationships, particularly when the number of atoms in the unit cell is large. This may also be the case for concentration parameter G of the P_{10} formula (see equation (5.14)). To make this problem less critical, let us consider the relations (6.6): for a given $\mathbf{h}$, more $\mathbf{k}$ may exist, and, as a consequence, more phase indications for the given $\phi_\mathbf{h}$ may arise, each providing an independent estimate for $\phi_\mathbf{h}$. How to combine these has been the goal of the tangent formula (Hauptman and Karle, 1956; Karle and Karle, 1966).

In more detail, let us suppose that moduli and phases are known for r different products $E_{\mathbf{k}_j} E_{\mathbf{h}-\mathbf{k}_j}$; in particular, we will assume that $\theta_j = (\phi_{\mathbf{k}_j} + \phi_{\mathbf{h}-\mathbf{k}_j})$, with $j = 1, \ldots, r$, are known. Then, to a first approximation, the probability density,

$$P_j(\phi_\mathbf{h}) \approx \left[2\pi I_0\left(G_{\mathbf{h},\mathbf{k}_j}\right)\right]^{-1} \exp\left[G_{\mathbf{h},\mathbf{k}_j} \cos\left(\phi_\mathbf{h} - \theta_j\right)\right] \qquad (6.7)$$

is defined for any jth pair of phases, $(\phi_{\mathbf{k}_j} + \phi_{\mathbf{h}-\mathbf{k}_j})$, where $G_{\mathbf{h},\mathbf{k}_j}$ represents the concentration parameter of the distribution (6.7). This may coincide with the parameter of the Cochran distribution (say C in equation (5.4)), or the parameter of the P_{10} formula (say G in equation (5.14) and in Appendix 5.C). We will use the same letter G as representative of both.

Thus, the total probability, $P(\phi_\mathbf{h})$, that the phase of $E_\mathbf{h}$ is ϕ_h is expressed by the product, suitably normalized, or r specific probabilities:

$$P(\phi_h) \simeq \prod_{j=1}^{r} P_j(\phi_h) \simeq L \exp\left[\sum_{j=1}^{r} G_{h,k_j} \cos(\phi_h - \theta_j)\right], \qquad (6.8)$$

where L is a suitable normalizing factor. Applying (M.F.8) to the right-hand side of (6.8) we get

$$P(\phi_h) \simeq L \exp[\alpha_h \cos(\phi_h - \theta_h)], \qquad (6.9)$$

where

$$\tan\theta_h = \frac{\sum_{j=1}^{r} G_{h,k_j} \sin(\phi_{k_j} + \phi_{h-k_j})}{\sum_{j=1}^{r} G_{h,k_j} \cos(\phi_{k_j} + \phi_{h-k_j})} = \frac{T}{B}, \qquad (6.10)$$

$$\alpha_h^2 = T^2 + B^2 = \left[\sum_{j=1}^{r} G_{h,k_j} \cos(\phi_{k_j} + \phi_{h-k_j})\right]^2 + \left[\sum_{j=1}^{r} G_{h,k_j} \sin(\phi_{k_j} + \phi_{h-k_j})\right]^2. \qquad (6.11)$$

The value of L may be found by requiring that

$$\int_0^{2\pi} P(\phi_h) \, d\phi_h$$

be equal to unity. From (M.E.25), we obtain

$$P(\phi_h) \simeq [2\pi I_0(\alpha_h)]^{-1} \exp[\alpha_h \cos(\phi_h - \theta_h)]. \qquad (6.12)$$

Equation (6.12) represents the probability density function to be found. It is again a von Mises distribution; thus the curves of Fig. 6.1 formally describe (6.12), provided that $G_{h,k}$ is replaced by α_h.

Equation (6.10) is the well-known *tangent formula*; θ_h is the most probable value of ϕ_h, and α_h is the reliability parameter of the phase indication. The role of α_h may be better understood if its r components are plotted on an Argand diagram as vectors of modulus G_j and phase $\phi_{k_j} + \phi_{h-k_j}$. Then, α_h appears to be the resultant of r complex vectors $G_j \exp(i\theta_j)$, and θ_h is its phase angle. In Fig. 6.3, a case is illustrated with $r = 5$. The figure suggests that, if r is sufficiently large, α_h may be large even if the various G_{h,k_j} are relatively small. For example, let us consider the following numerical case:

$$G_1 = 1.6, \quad \theta_1 = 30°; \quad G_2 = 1.1, \quad \theta_2 = 54°;$$
$$G_3 = 1.8, \quad \theta_3 = -30°; \quad G_4 = 0.7, \quad \theta_4 = 99°;$$
$$G_5 = 2.0, \quad \theta_5 = 70°.$$

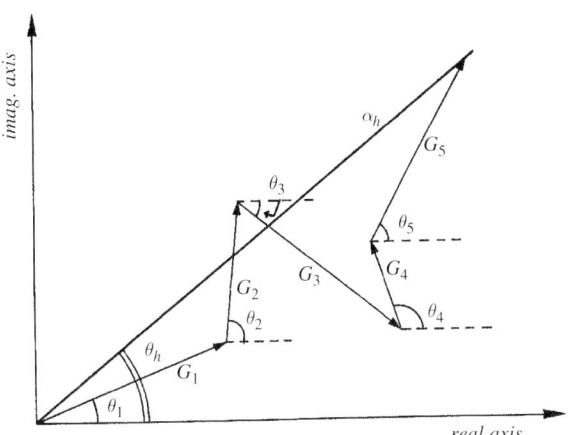

Fig. 6.3
E_h is represented in the Argand plane as the sum of five complex vectors $G_j \exp(i\theta_j)$.

Then, $\theta_h = 39°$ and $\alpha_h = 5.35$. While single-phase relationships are of modest reliability, α_h is large and therefore θ_h is a reliable estimate of ϕ_h.

A similar mathematical procedure for cs. crystals leads to the overall sign probability for the structure factor $E_\mathbf{h}$, when the signs of more products $(E_{\mathbf{k}_j} E_{\mathbf{h}-\mathbf{k}_j})$ are known:

$$P(E_\mathbf{h})_+ = \frac{1}{2} + \frac{1}{2} \tanh \frac{|E_\mathbf{h}| \sum_j E_{\mathbf{k}_j} E_{\mathbf{h}-\mathbf{k}_j}}{\sqrt{N_{eq}}}. \qquad (6.13)$$

According to (6.13), the sign of $E_\mathbf{h}$ will probably be positive if $\sum_j E_{\mathbf{k}_j} E_{\mathbf{h}-\mathbf{k}_j}$ is positive, otherwise it will probably be negative.

The main appeal of the tangent formula (or of its equivalent relationship (6.13) for the cs. case) is as follows. The reliability of the phase indication $\phi_\mathbf{h} \approx \theta_\mathbf{h}$ may be large, even if the single contributors in the tangent formula are small, provided that the number of triplets involved in the formula is large enough. There is, however, a basic assumption which for the moment has not been justified: how do we know the values of $\phi_{\mathbf{k}_j} + \phi_{\mathbf{h}-\mathbf{k}_j}$? We will postpone the answer until Section 6.3, where a typical direct methods procedure will be described.

Here, we consider a problem connected with the potential of the tangent formula and with its efficiency: what is the expected distribution of $\alpha_\mathbf{h}$ when the phases $\phi_{\mathbf{k}_j}$ and $\phi_{\mathbf{h}-\mathbf{k}_j}$, for $j = 1, \ldots, r$ are still unknown? What one actually knows about $\alpha_\mathbf{h}$ is as follows. The parameters $G_j \equiv G_{\mathbf{h},\mathbf{k}_j}$ are given by the experiment, the phases $\phi_{\mathbf{k}_j} + \phi_{\mathbf{h}-\mathbf{k}_j}$ are distributed about $\phi_\mathbf{h}$ according to the von Mises distribution, $M(\phi_{\mathbf{k}_j} + \phi_{\mathbf{h}-\mathbf{k}_j}; \phi_\mathbf{h}, G_j)$. Then, the expected value of $\alpha_\mathbf{h}$ (see Cascarano et al., 1984) is

$$<\alpha_\mathbf{h}> = \sum_{j=1}^{r} G_j D_1(G_j), \qquad (6.14)$$

with variance given by

$$\sigma_\alpha^2 = \frac{1}{2} \sum_{j=1}^{r} G_j^2 [1 + D_2(G_j) - 2D_1^2(G_j)]. \qquad (6.15)$$

$D_i(x) = I_i(x)/I_0(x)$ is the ratio of the modified Bessel function of order i and 0 respectively. If the phases $\phi_{\mathbf{k}_j} + \phi_{\mathbf{h}-\mathbf{k}_j}$, determined by the tangent formula, are distributed about $\phi_\mathbf{h}$ in accordance with the G_j parameters, then it may be expected that, for each $\mathbf{h}$, the experimental value of $\alpha_\mathbf{h}$ (as obtained from equation (6.11)) will be close to the expected (say the value $<\alpha_\mathbf{h}>$ provided by equation (6.14)). Strong deviations between $\alpha_\mathbf{h}$ and $<\alpha_\mathbf{h}>$ usually indicate bad phasing.

6.3 Procedure for phase determination via traditional direct methods

A robust single-solution direct methods procedure which works efficiently in all cases does not exist. The most effective approaches are essentially multisolution techniques whereby several different attempts at phase assignment are performed, among which the correct solution must be found.

A computer is, therefore, a basic requirement for application of the methods. Before describing a typical direct phasing procedure, we must acknowledge Michael Woolfson and his group (among whom we recall Peter Main and Gabriel Germain) for their extraordinary contribution: they led the way from old to modern phasing procedures.

A typical direct procedure for phase assignment may be presented, schematically, as follows:

Step 1: normalization of the structure factors
Step 2: set-up of phase relationships
Step 3: assignment of starting phases
Step 4: phase determination
Step 5: finding out the correct solution
Step 6: E-map interpretation
Step 7: phase extension and refinement.

For Step 1, the reader is referred to Chapter 2. The other steps are treated over the following sections.

6.3.1 Set-up of phase relationships

Here we will mainly focus our attention on triplet invariants; the set-up of quartet invariants is described in Appendix 6.A. The reader should, however, bear in mind that other phase relationships can also be set up: one- and two-phase s.s.s, quintet invariants, etc. For triplet invariants, the basic steps are as follows:

1. First, the *unique list* of reflections is defined (i.e. the list of symmetry-independent reflections). These are ordered according to the normalized amplitude; e.g. the *code* 1 is associated with the reflection with the largest normalized amplitude, etc. The subset of reflections (say the number of NLAR) that one wants to phase is defined; i.e. only reflections with |E| larger than a given threshold E_s, where E_s may vary, according to circumstances, between 1.2 and 1.5. In this way, only those reflections which may be expected to form triplet relationships with a sufficiently high value of G (the concentration parameter associated with the P_{10} formula) or of C (the concentration parameter associated with the Cochran relationship), are selected. It is likely that most of the NLAR reflections will be characterized by high values of $<\alpha>$, and therefore they may be carefully phased.
2. The indices of the reflections symmetry equivalent to the NLARs are calculated, so obtaining the *expanded list*.
3. For a given $\mathbf{h}$, all pairs of $(\mathbf{k}, \mathbf{h} - \mathbf{k})$ with $|E| \geq E_s$ are found; the set of phase relationships so obtained is called the Σ_2 *list*. The search is made by letting $\mathbf{h}$ vary over the set of 'standard' reflections, while $\mathbf{k}$ (and therefore $\mathbf{h} - \mathbf{k}$) can span over the expanded list. In practice, all combinations of $\mathbf{h} \pm \mathbf{k}\mathbf{R}_i$ are checked, where $\mathbf{R}_i$ varies over all rotation matrices of the space group not related by an inversion centre.

4. If $|E_{h \pm kR_i}| > E_s$, the triplet is retained. Any triplet is stored in terms of standard reflections. For example, let us focus our attention on the space group $P2_12_12_1$, where we have to store the following information:

$$\Phi = \phi_{123} + \phi_{\overline{468}} + \phi_{\overline{385}} \simeq 0, \qquad (6.16)$$

with reliability parameter

$$C = 2|E_{123}E_{468}E_{385}|/\sqrt{N_{\text{eq}}}.$$

Because of (1.25),

$$\phi_{\overline{468}} = \phi_{468} - 2\pi(468)\begin{pmatrix}0\\1/2\\1/2\end{pmatrix} = \phi_{468},$$

$$\phi_{\overline{385}} = \phi_{\overline{385}} = -\left[\phi_{385} - 2\pi(385)\begin{pmatrix}0\\1/2\\1/2\end{pmatrix}\right] = -\phi_{385} + \pi.$$

Therefore, (6.16) may be written as

$$\Phi = \phi_{123} + \phi_{468} - \phi_{385} + \pi \simeq 0. \qquad (6.17)$$

Equation (3.48) involves the phases of the standard reflections which may be referred to via their code numbers. Thus, (6.16) may be transformed into the symbolic expression

$$n_1 + n_2 - n_3 + \Delta \simeq 0, \qquad (6.18)$$

where n_1, n_2, n_3 are the code numbers of the standard reflections and Δ is the phase shift due to translational symmetry. In conclusion, complete information on a triplet estimate may be stored by recording, besides n_1, n_2, n_3, Δ, also s_2 and s_3 (the signs of n_2 and n_3), and the value of C. In order to show how a typical Σ_2 list in organized, we show in Table 6.1 the first items of the Σ_2 list of LOGANIN.

5. *NWEAK* reflections, those with the smallest $|E|$ values, are selected for construction of triplets (called *psi-zero triplets*) with two reflections belonging to the *NLAR* subset. Usually *NWEAK* $\sim$ *NLAR*/3. The psi-zero triplets are stored using a technique quite similar to that used for the Σ_2 list and are used as a *figure of merit* for recognizing the correct solution (see Section 6.3.4).

Some considerations now about the NLAR value. A good value should secure a number of triplets per reflection sufficient to generate large α_h parameters for most of the NLAR reflections. If NLAR is small, are the NLAR reflections alone, once phased, able to display a sufficiently accurate electron density (Burgi and Dunitz, 1971; Mö et al., 1973)? To answer this question, let us consider the electron density expression

$$\rho(\mathbf{r}) = \frac{1}{V}\sum_{\mathbf{h}} F_{\mathbf{h}}\exp(-2\pi i\mathbf{h}\cdot\mathbf{r}).$$

Procedure for phase determination via traditional direct methods

Table 6.1 For each reflection h, the code number n_2, n_3, the signs s_2, s_3, the C value, and the Δ shift are given

Reflection (0 11 17)				Reflection (0 13 2)				Reflection (1 14 0)			
s_2n_2	s_3n_3	C	Δ	s_2n_2	s_3n_3	C	Δ	s_2n_2	s_3n_3	C	Δ
3	127	4.61	0	3	242	3.81	180	5	−98	3.73	0
7	−70	4.20	180	−7	−205	3.22	0	11	−56	3.77	180
14	227	3.00	180	−12	248	2.88	180	26	72	3.25	0
−23	93	3.46	0	16	74	3.56	180	−29	−206	2.59	180
33	−93	3.46	180	−18	186	2.90	0	34	−242	2.45	0
−5	84	4.19	180	4	132	3.75	0	6	−130	3.42	180
10	−135	3.54	0	10	−211	3.03	0	−18	−216	2.66	180
−14	158	3.28	180	14	−243	2.84	0	27	66	3.28	0
30	181	2.91	180	17	20	4.28	0	−30	91	3.08	180
39	−87	3.29	180	21	49	3.70	180	35	201	2.53	0
−6	168	3.55	0	7	50	4.20	0	10	106	3.39	180
11	243	3.06	0	−11	57	3.94	0	23	−246	2.54	180
17	−255	2.80	180	−14	15	4.61	180	28	127	2.90	0
31	−36	3.94	0	17	−150	3.08	0	−31	244	2.48	180
40	−174	2.88	180	−21	−210	2.79	180	−39	−50	3.26	180

In principle, the summation contains an infinite number of terms, but in practice the number is finite. The completeness theorem of a Fourier series states that

$$\frac{1}{V}\int_V \rho^2(\mathbf{r})d\mathbf{r} = \sum_{\mathbf{h}} |F_{\mathbf{h}}|^2.$$

If we construct a Fourier series via the NLAR reflections only, we obtain

$$\rho'(\mathbf{r}) = \frac{1}{V}\sum_{\mathbf{h}\in\mathrm{NLAR}} F_{\mathbf{h}}\exp(-2\pi i\mathbf{h}\cdot\mathbf{r}) = \frac{1}{V}\sum_{\mathbf{h}} F'_{\mathbf{h}}\exp(-2\pi i\mathbf{h}\cdot\mathbf{r}),$$

where $F'_{\mathbf{h}} = F_{\mathbf{h}}$ if $\mathbf{h}$ is one of the NLAR reflections, $F'_{\mathbf{h}} = 0$ otherwise. The variance of $\rho'(r)$ with respect to $\rho(r)$ is

$$D^2 = \frac{1}{V}\int [\rho(\mathbf{r}) - \rho'(\mathbf{r})]^2\, d\mathbf{r} = \sum_{\mathbf{h}} |F_{\mathbf{h}} - F'_{\mathbf{h}}|^2 = \sum_{\mathbf{h}\notin\mathrm{NLAR}} |F_{\mathbf{h}}|^2.$$

The variance is minimized just because the last summation involves only weaker reflections (Vainshtein and Kayushina, 1967). This is why we are able to find a sufficiently good model of the electron density from a limited set of high-intensity phased reflections.

In order to obtain interpretable maps, at least six $|E|$ values per atom in the asymmetric unit may be advised. However, a good choice for NLAR should also take into account the space group symmetry (for a fixed value of NLAR the number of triplets which can be constructed among the NLAR reflections is higher for high-symmetry space groups). An empirical formula that works well is

$$\mathrm{NLAR} = [(4t + 100) + x]1.1,$$

where t is the number of atoms in the asymmetric unit, $x = 100$ if the system is triclinic, $x = 50$ if the system is monoclinic.

6.3.2 Assignment of starting phases

The tangent formula is able to fix one phase given the values of other phases, but how can such phases be known? That was the question that was not answered at the end of Section 6.2. There are two alternative ways of proceeding, both belonging to the so-called *multisolution approach*.

The most traditional way is to fix some reflections to define the origin and the emantiomorph (see Section 3.7), and to then derive the other phases from these. To better understand the process we will give a practical example.

Let us suppose that for a crystal structure in the space group P*nma*, the phase values of the reflections with code number $n_1 = 18$, $n_2 = 73$, $n_3 = 96$ have been arbitrarily fixed at 2π to define the origin. The phasing process will aim at determining all other phases, one after the other in a chain process, by cyclic application of the tangent formula. In the example shown in Fig. 6.4, the phase of reflection 24 is estimated because it makes a triplet with 18 and 73. Similarly, 37 is determined since it makes a triplet with 73 and 96. In the second cycle of the tangent process, the phases 65 and 106 are determined, and so on.

A starting set of only three reflections is usually inadequate to establish a good chain; not all of the phase may be accessible through the chain, and weak links in the chain are frequent (i.e. some phases are poorly determined through low-reliability relationships). The reliability of the phases determined in the first steps of the phasing process is critical. Indeed, errors introduced at this point would reflect on the phases determined in the subsequent steps. In the example in Fig. 6.4, all of the phases fixed in the first two cycles are determined by means of single triplets, and therefore phase errors are highly probable.

It is therefore usual to introduce within the starting set some unknown *extra phases*, necessary to provide a more straightforward phasing process. Following the example above, let us suppose that, besides the origin-fixing reflections 18, 73, 96, the extra reflections 27, 38, 63, are included in the starting set and that a procedure is available which is able to provide values for such extra phases. Let us assume that, in this situation, the triplet relationships involving the starting set and the early determined phases are those shown in Fig. 6.5. By comparing Fig. 6.4 with Fig. 6.5 we observe that: (1) in the first cycle of the tangent process 37 is defined from the two pairs (73,96) and (73,27), and 99 is defined from the two pairs (27,63) and (96,38) (on the other hand, in Fig. 6.5, all the phases were determined via single triplets); (2) the

Fig. 6.4
The starting set consists of three reflections: 18, 73, and 96. In the first cycle of the tangent procedure, the phases of reflections 24 and 37 are estimated. These are used in the second cycle to fix phases 65 and 106.

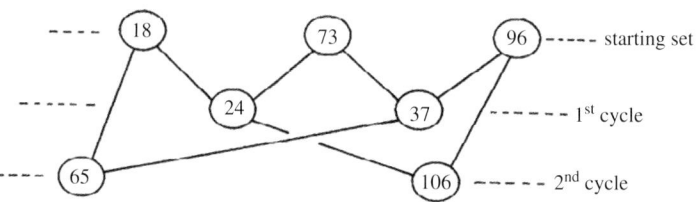

Procedure for phase determination via traditional direct methods

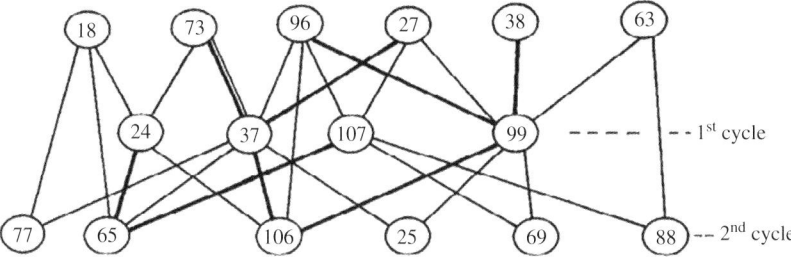

Fig. 6.5
The starting set consist of the origin fixing reflections 18, 73, and 96, plus three extra reflections 27, 38, and 63.

number of phases determined in the first two cycles in Fig. 6.5 is, remarkably, larger than in Fig. 6.4.

By generalizing the above observations, one can conclude that a large starting set favours the use of a large number of triplets right from the beginning. Thus the reliability of the phases determined in the first steps is higher.

Occasionally some reflections in the starting set with high $|E|$ value may enter into very few phase relationships and they are therefore poorly determined in terms of other phases and are of limited usefulness as components of the starting set. A good starting set, however, should guarantee an extended network of triplet relationships; phases should possibly be determined by multiple indications through strong relationships.

An optimal method for securing a good starting set is the *convergence procedure*, proposed by Germain et al. (1970); for further details, the interested reader is referred to the original paper.

An observation is now appropriate. While the three (or less, depending on the space group symmetry) phases defining the origin can be arbitrarily fixed, the three extra phases used in Fig. 6.5, traditionally called *symbols*, cannot. In the case of centric structures, in order to have at least one starting set with correct phases, eight different permutations of the signs of the symbols should be made, and for each permutation, phasing should be attempted (see Table 6.2).

We will obtain eight trial solutions, among which will be the correct one. If one includes n symbols into the starting set, 2^n trials would be performed, among which the correct solution should be found. Obviously it is not advisable to use too large values of n, otherwise the computing time would be too long and the identification of the correct solution difficult.

Let us now move to the acentric case, where phases may continuously vary from 0 to 2π. A simple approach would be the *quadrant permutation technique*, whereby, to unknown symbolic phases four values could be assigned in turn. For example, $\pi/4$, $3\pi/4$, $5\pi/4$, $7\pi/4$, could be the assigned values, each symbolic phase having unit probability of being, at least for one trial, within 45° of its true value. If p symbols are used in the phasing process 4^p trials should be performed, among which the correct solution should be found.

In the technique of quadrant permutation, the introduction of every new extra phase represents a fourfold increase in computing time, which soon leads to a computing time limitation, even for fast computers. An optimal strategy for reducing the number of trials is the *magic integer method* (White

Table 6.2 The starting set for a centric structure for the example depicted in Fig. 6.5. The first three reflections are used to fix the origin, the last three reflections are symbols and their signs are varied. For each of the eight combinations, a solution is attempted

18	73	96	27	38	63
+	+	+	+	+	+
			+	+	−
			+	−	+
			−	+	+
			+	−	−
			−	+	−
			−	−	+
			−	−	−

and Woolfson, 1975; Declercq et al., 1975, 1979; Taylor and Woolfson, 1975; Main, 1977). This technique strongly increased the efficiency of direct methods procedures, but has been replaced these days by the *random phase approach* (see below). For brevity, the magic integer method is not described, but the interested reader is referred to the original papers or to Chapter 3 of *Direct Phasing in Crystallography*.

The most popular approach today is the *random phase approach*. Random phase values are allocated for all of the NLAR reflections, which then constitute at the same time the starting set and the set of reflections to phase (Baggio et al., 1978; Furàsaki, 1979; Yao, 1981). As single triplet phase relationships, the tangent formula also exploits the positivity and the atomicity properties of the electron density, essentially through the positivity of the atomic scattering factors. However, the passage from random phases to the correct ones by cyclic application of the tangent formula is not guaranteed; again, various phasing attempts are necessary, each starting from one set of random phases, among which the correct solution has to be found.

6.3.3 Phase determination

The fundamental tool for phase assignment is the *weighted tangent formula*. Germain and Main (1971) proposed that every time a phase is determined, a weighting factor w_h is simultaneously calculated, which should be proportional to the reliability of the $\phi_\mathbf{h}$ estimate. For example, if $\phi_{\mathbf{k}_j}$ and $\phi_{\mathbf{h}-\mathbf{k}_j}$ have been determined with weights $w_{\mathbf{k}_j}$ and $w_{\mathbf{h}-\mathbf{k}_j}$, then equation (6.10) should be replaced by the weighted tangent formula,

$$\tan \phi_\mathbf{h} = \frac{\sum_j w_{\mathbf{k}_j} w_{\mathbf{h}-\mathbf{k}_j} |E_{\mathbf{k}_j} E_{\mathbf{h}-\mathbf{k}_j}| \sin(\phi_{\mathbf{k}_j} + \phi_{\mathbf{h}-\mathbf{k}_j})}{\sum_j w_{\mathbf{k}_j} w_{\mathbf{h}-\mathbf{k}_j} |E_{\mathbf{k}_j} E_{\mathbf{h}-\mathbf{k}_j}| \cos(\phi_{\mathbf{k}_j} + \phi_{\mathbf{h}-\mathbf{k}_j})} = \frac{T_\mathbf{h}}{B_\mathbf{h}}, \quad (6.19)$$

with a *reliability parameter* given by

$$\alpha_\mathbf{h} = 2 N_{eff}^{-1/2} |E_\mathbf{h}| (T_\mathbf{h}^2 + B_\mathbf{h}^2)^{1/2}. \quad (6.20)$$

The use of weighting factors yields a process of phase assignment which develops rapidly at a good level of reliability, since no triplet $E_\mathbf{h} E_{-\mathbf{k}_j} E_{-(\mathbf{h}-\mathbf{k}_j)}$ is left out. In addition, phases $\phi_{\mathbf{k}_j}$ and $\phi_{\mathbf{h}-\mathbf{k}_j}$, determined at a low level of reliability do not affect the value of $\phi_\mathbf{h}$ very much, provided that a good weighting scheme is adopted. Today, various weighting schemes are adopted in programs that use the tangent formula (Schenk, 1972a,b; Giacovazzo, 1979).

We schematize, in Table 6.3, how the tangent formula drives a set of random phases to the correct values. We suppose that NLAR has been fixed at 482; then, in column 1, the corresponding starting set of random phases is represented (by say $\{\phi\}$). The tangent formula is first applied to estimate a new value for the phase of reflection 1, by using for the phases $\phi_{\mathbf{k}_j}$ and $\phi_{\mathbf{h}-\mathbf{k}_j}$ making a triplet with reflection 1, the randomly assigned values. The new phase, ϕ'_1, is obtained. The same operations are then repeated for reflection 2 and continued up to reflection 482. The resulting phases are in column 2 in the table (say,

Table 6.3 Schematic of the classical tangent procedure. The number of reflections to phase has been fixed to NLAR = 482

$\{\phi\}$	$\{\phi'\}$	$\{\phi''\}$	--	$\{\phi^c\}$
ϕ_1	ϕ'_1	ϕ''_1	--	ϕ^c_1
ϕ_2	ϕ'_2	ϕ''_2	--	ϕ^c_2
ϕ_3	ϕ'_2	ϕ''_2	--	ϕ^c_3
–	–	–	– –	–
ϕ_{482}	ϕ'_{482}	ϕ''_{482}	--	ϕ^c_{482}

the set $\{\phi'\}$). The tangent formula is again applied to column 2, exactly as for column 1; the result is column 3 (say, the set $\{\phi''\}$). These cycles are repeated up to when the phases no longer change (from one column to the next); then we say that *convergence* has been attained. The last column represents the convergence phases (say, the set $\{\phi^c\}$) and since the first phasing attempt is now accomplished, the convergence phases are called *first trial* phases.

Since we cannot be sure that the tangent formula has driven the random phases to the correct values, we can attempt a second trial by assigning to the NLAR reflections a new set of random phases. These are submitted to the tangent process and will end in a convergent set of phases; the *second trial* has then been accomplished. A number of trials, usually between 30 and 500, are attempted, according to the size of the structure.

6.3.4 Finding the correct solution

As stated in the previous section, multisolution procedures provide several trial solutions; it would be too time consuming to only recognize the correct solution after having calculated and interpreted all of the various electron density maps. It is more efficient to compute, as soon as a trial solution is available, some appropriate functions, called figures of merit (*FOMs*), which are expected to be extreme for the correct solution. Since the FOMs increase the efficiency of direct methods, much effort has been dedicated to them: among others, we would mention Karle and Karle (1966), Schenk (1971, 1972a,b), Riche (1973), Germain et al. (1974), Hašek (1974), Kennard et al. (1971), Roberts et al. (1973), Cascarano et al. (1987a, 1992a). These are the most popular (and simplest) examples.

In accordance with the $\alpha_\mathbf{h}$ definition (see equation (6.11)), a trial solution is expected to be reliable if, for all the NLAR phased reflections, the sum

$$Z = \sum_\mathbf{h} \alpha_\mathbf{h}$$

is very large (or extreme with respect to the other trial solutions). Equivalent FOMs are

$$t_\mathbf{h} = \frac{\sum_\mathbf{h} \alpha_\mathbf{h}}{\sum_j G_{\mathbf{h},\mathbf{k}_j}}, \quad \text{with } 0 \leq t_\mathbf{h} \leq 1,$$

or

$$MABS = \frac{\sum_\mathbf{h} \alpha_\mathbf{h}}{\sum_\mathbf{h} <\alpha_\mathbf{h}>},$$

where $<\alpha_\mathbf{h}> = \sum_{j=1}^{r} G_j D_1(G_j)$ is the expected value of $\alpha_\mathbf{h}$ (see equation (6.14)), r is the number of triplets contributing to $\alpha_\mathbf{h}$, G_j is the concentration parameter for the jth triplet, $D_1(x) = I_1(x)/I_0(x)$ is the ratio of the modified Bessel functions I_1 and I_0. Schenk (1972a) observed that in the 73 symmorphic space groups (i.e. those which do not contain any glide or screw axis), a symmetry effect occurs which disturbs the FOM efficiency; the solutions with the highest FOM values may correspond to the so-called *uranium solutions*, that is, a solution for which the phases are maximally consistent but are meaningless from a structural point of view. Two FOMs are very useful for fighting against this undesired tendency: the first, the *psi-0* FOM, was first proposed by Cochran and Douglas (1957) and further elaborated on by Giacovazzo (1993) and by Cascarano and Giacovazzo (1995); the second, *NQ*, based on the negative quartet relationships. These are briefly commented on below.

(i) *The psi-0 FOM*

Let us consider a *psi-0 triplet*, that is a triplet for which $|E_\mathbf{h}| \approx 0$ while $|E_\mathbf{k}|$, $|E_\mathbf{h-k}|$ are large. Since $|E_\mathbf{h}| \approx 0$, such triplets are not actively used within the phasing process, even if $E_\mathbf{k}$ and $E_\mathbf{h-k}$ have been phased via other triplets. It may be expected that the phases of the complex vectors

$$\mathbf{A}_j = \frac{2|E_{\mathbf{k}_j} E_{\mathbf{h-k}_j}|}{\sqrt{N_{eff}}} \exp i(\phi_{\mathbf{k}_j} + \phi_{\mathbf{h-k}_j}), j = 1, \ldots, r$$

are randomly distributed between 0 and 2π. The problem is clearly connected with the classical problem of random walk (Pearson, 1905); according to the theory developed in Chapter 2, the resultant of the *r* vectors is expected to have minimal magnitude for the correct solution.

(ii) *NQ, a FOM based on negative quartet invariants*

Schenk (1974) suggested the use of negative quartets (not usually involved in the phasing process) as a figure of merit for identifying the correct solution. The simplest criterion is (Giacovazzo, 1976b; see also de Titta et al., 1975)

$$NQ = \sum_{\substack{negative \\ quartets}} G_j \cos(\Phi_j) = \max,$$

where Φ_j is the *j*th negative estimated quartet invariant and G_j is its concentration parameter as given by equation (5.23). A large positive value of *NQ* implies that quartet cosines, estimated to be negative (because they have a negative value of *G*) really are negative.

If negative triplet cosine estimates are available (e.g. by application of the P_{10} formula, see Section 5.5) then *NQ* may be combined with a negative triplet FOM:

$$NT + NQ = \sum_{\substack{negative \\ quartets}} G_j \cos(\Phi_j) + \sum_{\substack{negative \\ quartets}} G_j \cos(\Phi_j) = \max$$

NT + NQ is expected to be a maximum for the correct solution.

Once the various FOMs have been calculated, it is possible to combine them in a single figure, usually called *CFOM, combined figure of merit*, which is used to rank the trial solutions.

6.3.5 E-map interpretation

We have seen that direct methods procedures usually end with several sets of approximated phases (trials) by which *E*-maps should be computed. The FOMs in the preceding section are a guide for arranging the solutions according to their probability of being correct, and therefore for setting a priority list for the examination of the Fourier syntheses. However, the solution with the best figure of merit does not always correspond to the true structure. Sometimes, for particularly difficult structures, the correct solution can be found only after some other more plausible solutions have been examined and rejected. In these circumstances, supplementary work is needed and solution of the structure is delayed.

At this stage of the procedure much assistance can be provided to the crystallographer by a program with provision for performing several Fourier syntheses and, for each synthesis, selecting Fourier peaks higher than a specific threshold, examining distance and angles, and performing a graphical representation of consistent groups of Fourier peaks which could be chemically bonded (Declercq et al., 1973; Koch, 1974; Bart and Busetti, 1976; Main and Hull, 1978). A typical procedure for interpretation of the electron density map should involve five steps:

1. Peak search.
2. Separation of peaks into clusters (each peak in a cluster is within chemical bonding distance of at least one other peak in the same cluster). The maximum bond distance depends on the chemical composition of the crystal structure and is fixed as the sum of the largest covalent radii plus a tolerance value (say $\simeq 0.34$ Å) to allow for some distortion in molecular geometry, which is unavoidable in E-maps and in early stages of refinement.
3. Application of stereochemical criteria to produce molecular fragments. These should comply with the chemical composition of the unit cell. For example, if all the atoms in the unit cell are C, N, or O then limit distances between 1.95 Å and 1.19 Å and bond angles between 85° and 145° should be fixed. If heavy atoms are present, then stereochemical criteria should be coherently modified.
4. Labelling of atomic peaks in terms of atomic species. This may be done on the basis of peak intensity, or, even better, of integrated peak intensity, bond distance, and bond angles.
5. (Eventual) comparison of the fragments with the expected molecular structure.

Unfortunately it is not always possible to extract a satisfactory structural model from the analysis of an E-map. Indeed the quality of a map depends on several factors, among which the following three play a prominent role:

(a) *Phase errors*. In most cases these are unavoidable; considerable random errors can be tolerated without too great a loss in structural information in the E-map, but systematic errors have greater destructive effects (Silva and Viterbo, 1980).
(b) *Amplitude truncation effects* in the series representation of electron density. Traditional direct methods do not phase reflections under the minimum threshold value $E_{Tr} \simeq 1.2$. In most practical applications, E_{Tr} lies in the range 1.30–1.50. If phases are determined with sufficient accuracy, amplitude truncation effects are not harmful (this is a necessary condition for the general success of direct methods). However, if this effect is associated with severe phase errors, the final result is often destructive. A classically difficult example is presented by structures suffering pseudotranslation symmetry; if no special action is taken, the reflections

actively used in the phasing process coincide with substructure reflections. Even when these reflections are accurately phased, and this is not generally the rule, the information on superstructure is completely lost in the E-map.

(c) *Fourier coefficients* used for calculating the map. It is traditional practice to use E coefficients at the conclusion of a phasing process; they produce a peaked effect in the map which makes its interpretation easier in terms of atoms. On the other hand, false details, ripples of heavy atoms, etc., are also produced which in some cases make the correct interpretation difficult. Such a behaviour is generated by the intrinsic nature of the $|E|$ coefficients (i.e. real atoms are replaced by unreal point atoms; $<|E|^2>$ is never vanishing for any $\sin\theta/\lambda$ value), and by their imperfect estimation from experimental $|F|$ values.

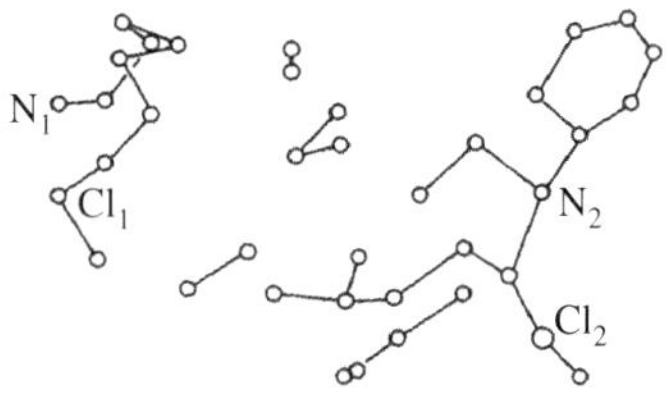

If the trial solution with the highest value of CFOM is unsuccessful, subsequent trials are explored. We show two applications of the procedure in Figs. 6.6 and 6.7:

(a) for *AZET* [$Pca2_1$; C_{21} H_{16} Cl N O, $Z = 8$] NLAR = 342, reflections are phased with an average phase error of 28°: the best E-map is depicted in Fig. 6.6.
(b) for *APAPA* [$P4_12_12$; C_{30} H_{37} N_{15} O_{16} $P_2 \cdot 6H_2O$, $Z = 8$] NLAR = 426; the best E-map is shown in Fig. 6.7.

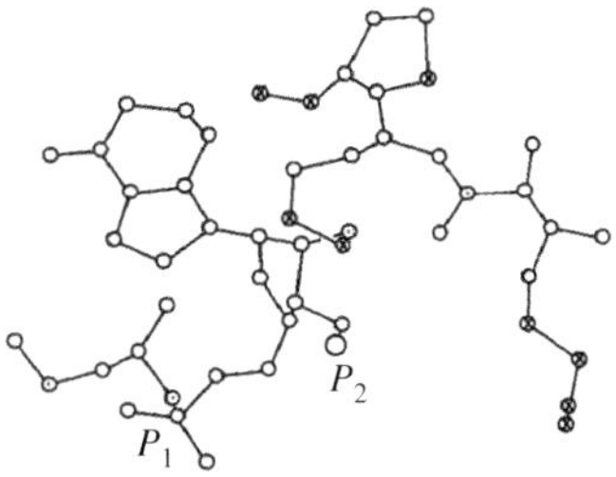

In both cases, distorted molecular geometry, false peaks, and missed atoms can be seen, due to the combination of Fourier series truncation and phase errors. We will see in Section 6.3.6 how such effects may be reduced.

6.3.6 Phase extension and refinement: reciprocal space techniques

A simple way to improve the information contained in a traditional E-map is to refine and extend phases to an $|E|$ smaller than the threshold value normally used in standard direct procedures. Phase extension may imply that there will be some penalty to pay in terms of phase accuracy, but this is largely compensated by smaller amplitude truncation effects; it is really a necessary ingredient for any modern phasing program, because it can change a not interpretable into a perfectly interpretable electron density map. Phase extension may be performed in two different ways, however, the most efficient way is via direct space techniques (so-called *EDM* (*electron density modification*) procedures); these are described in Section 8.2.

Phase extension may also be performed in reciprocal space. A simple and efficient approach is as follows (Sheldrick, 1982; Altomare et al., 1991). A number (say *NEXP*) of reflections immediately following the NLAR reflections in a list sorted in decreasing order of $|E|$, is selected and triplets relating them with the phased NLAR reflections are calculated. Then phase extension is performed via the tangent formula.

The effects of such a procedure may be deduced from Figs. 6.8 and 6.9, which are the molecular models obtained after phase extension for *AZET* and

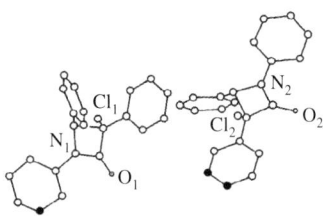

Fig. 6.6
AZET: molecular fragment provided by an E-map obtained from NLAR = 342 reflections.

Fig. 6.7
APAPA: molecular fragment provided by an E-map obtained from NLAR reflections; crossed circles correspond with ghosts.

Fig. 6.8
AZET: molecular fragment provided by an E-map obtained from NLAR+NEXP reflections. Full circles denote atoms not located after application of the procedure.

APAPA respectively. These models are now less distorted and more complete than those shown in Figs. 6.6 and 6.7.

The historical development of phase extension procedures in reciprocal space deserves a short mention, because it is responsible for many phasing advances. Let F_p be the structure factor calculated on the basis of the fragment provided by the first E-map and let F be the structure factor of the target structure. According to Karle (1970b, 1976), a phase is regarded as known if $|F_p| > \eta |F|$ and $|E| > |E_{T1}|$, where $|E_{T1}|$ is a suitable threshold value (usually $\sim$1.5), and η is the fraction of the total scattering power relative to the fragment. These known phases are used as input to the tangent formula and new phases are determined up to a threshold value $|E_{T2}| < |E_{T1}|$. A Fourier synthesis calculated with these phases can yield a more complete model of the structure. The process is iterative.

Beurskens and Noordik (1971), Gould et al. (1975), Beurskens et al. (1976), and van der Hark et al. (1976) all proposed the use of difference structure factors,

$$\Delta F = (|F| - |F_p|)\exp(i\phi_p).$$

After suitable normalization, the ΔF values are submitted to a weighted tangent formula refinement procedure, which usually leads to a significantly improved set of difference structure factors. The procedure is iterative and is the basis of the *DIRDIF* (Beurskens et al., 2008) program.

Similar reciprocal space techniques have been applied to macromolecules (Weinzierl et al. (1969), Coulter and Dewar (1971), Destro (1972), Hendrickson et al. (1973), Hendrickson (1975)). The results are all very similar to each other; the tangent formula (or its variants) performs useful phase extension and refinement when the starting phase error is small and RES is sufficiently high.

The application of the Sayre equation via a least squares procedure (Sayre, 1972, 1973, 1974a,b) shows a similar trend; Cutfield et al. (1975) first extended phases from 1.9 Å to 1.5 Å, and then combined the Sayre equation with the least squares method. Phases were well refined, but the method is not competitive with modern direct space techniques, particularly for the larger proteins.

These last results pose questions about the limits of the tangent formula (see Section 6.3.7).

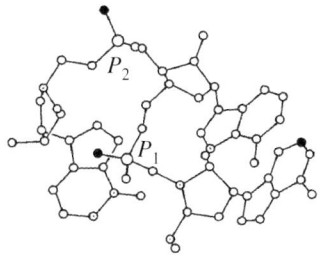

Fig. 6.9
APAPA: molecular fragment provided by an E-map obtained from NLAR+NEXP reflections. Full circles denote atoms not located after application of the procedure.

6.3.7 The limits of the tangent formula

It has been suggested by Sheldrick (1990) that direct methods are not expected to succeed in the solution of small molecule structures if fewer than half of the reflections in the range 1.1–1.2 Å are observed with $F > 4\sigma(|F|)$. This empirical rule is realistic, and, for example, indicates that phasing may also be very difficult for very small molecules when RES >1.2 Å (see also Section 1.6 for a theoretical discussion of this problem).

The state of the art has been described in a paper by Caliandro et al. (2009), where, for a large set of small test structures, data were cut at different resolution limits to check the efficiency of tangent approaches. The overall result was as follows: the success obtained with ab initio protein phasing at non-atomic resolution (see Section 10.4) cannot be extended to small and medium-sized molecules. Unexpectedly, RES is a parameter more critical for such molecules than for proteins. The rationale is that the high percentage of solvent (i.e. from 30 to about 80%) in a protein unit cell is a big source of information. The higher the percentage, the more effective are solvent-flattening techniques, so basic to phase extension and refinement. In small and medium-sized molecules, this source of information is absent and solvent-flattening techniques cannot be used. Thus, solving ab initio these structures at RES ~1.8 Å is outside the limits of the present state of the art. Similar difficulties were described by Palatinus and Chapuis (2007), where a charge-flipping method was used.

A theoretical criterion for guessing whether a structure may be solvable by direct methods has been suggested by Giacovazzo et al. (1994a). They showed that the distribution of the parameter α, as defined by equation (6.11), is the normal distribution,

$$P(\alpha_\mathbf{h}) \approx N\left(\alpha_\mathbf{h}, <\alpha_\mathbf{h}>, \sigma_{\alpha_\mathbf{h}}\right),$$

where $<\alpha_\mathbf{h}>$ is given by equation (6.14) and $\sigma_{\alpha_\mathbf{h}}^2$ by equation (6.15). Both parameters are defined in terms of the experimental data; e.g. as a function of the structural complexity (since G depends on N_{eq}) and as a function of RES (since r increases when RES improves). The ratio

$$z_\mathbf{h} = <\alpha_\mathbf{h}>/\sigma_{\alpha_\mathbf{h}}$$

is very useful for guessing if a given structure is or is not solvable via tangent procedures. z may be considered as a 'signal-to-noise' ratio; large values of $z_\mathbf{h}$ correspond to large values of $<\alpha_\mathbf{h}>$ in front of a small value of $\sigma_{\alpha_\mathbf{h}}$. Diffraction data for which high values of z may be calculated for most of the strong reflections constitute a set suitable for the successful application of direct methods. Indeed, owing to the large values of α (implicit in large z values), corresponding phases can be accurately defined. If small values of z are calculated for most of the strong reflections, then the tangent formula will hardly work at all.

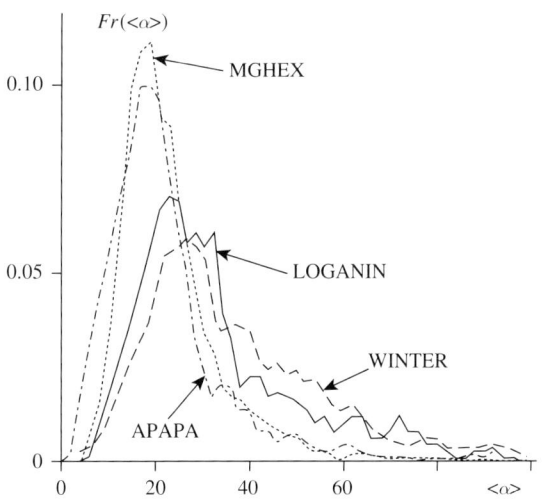

Fig. 6.10
$Fr(<\alpha>)$ is the frequency distribution of $<\alpha>$, calculated for some small-molecule data.

Procedure for phase determination via traditional direct methods

Accordingly, when the size of the structure increases and/or the data resolution becomes poorer, the z values are expected to be small (on average).

In Figs. 6.10 and 6.11, we show the experimental frequency distribution of $<\alpha_h>$ ([say $Fr(<\alpha>)$]) for some small molecule structures and for some small proteins; in Figs. 6.12 and 6.13 are shown the corresponding $Fr(z)$ distributions. The percentage of reflections with large values of $<\alpha_h>$ is sufficiently high for all the small structures in Fig. 6.10 and for most of the proteins in Fig. 6.11. All of the small structures in Fig. 6.12 show a favourable value of z for most of the reflections (they are therefore expected to be easily solved by application of the tangent formula), while the same behaviour cannot be seen in Fig. 6.13 for most of the proteins. *APP* and probably *BPTI* are the only two proteins which seem to be solvable by application of the sole tangent formula. Proteins are therefore outside the potential of the tangent formula, unless some supplementary information is introduced into the phasing process (see Section 6.4 for the combination of tangent techniques with direct space approaches).

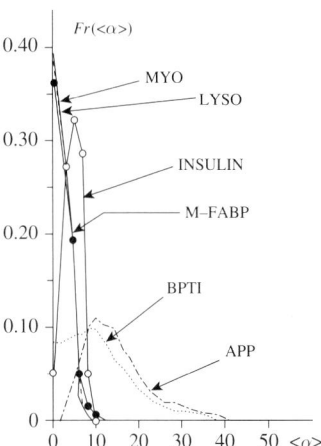

Fig. 6.11
$Fr(<\alpha>)$ is the frequency distribution of $<\alpha>$, calculated for some protein data.

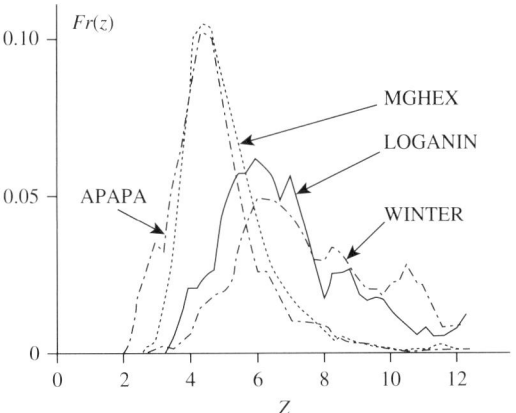

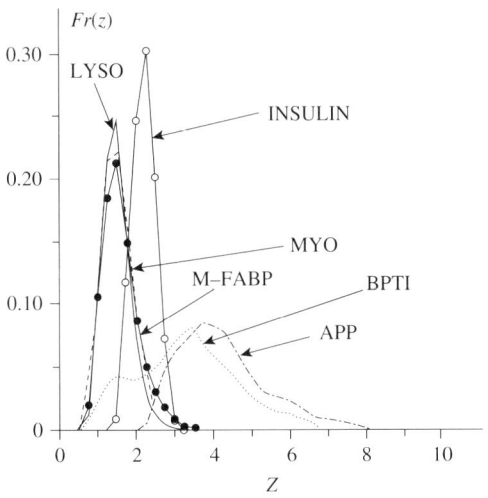

Fig. 6.12
The $P(z)$ distribution for some small-molecule data.

Fig. 6.13
The $P(z)$ distribution for some protein data.

6.4 Third generation direct methods programs

Phase relationships work in reciprocal space; indeed, they establish a probabilistic relation between one phase and another known set of phases. Their validity depends on the positivity and atomicity of the electron density; why then do we not exploit such properties in direct space? This approach has been adopted by the so-called third generation direct methods programs, all combining direct and reciprocal space techniques. We will make reference to *shake and bake*, to *half-bake*, and to the *SIR suite* programs as being representative of the new paradigm.

6.4.1 The shake and bake approach

The first attempt to theorize the combined use of direct and reciprocal space approaches was *Shake and Bake* (de Titta et al., 1994; Weeks et al., 1994; Hauptman, 1995; Miller et al., 1994). This was the first program capable of solving small protein structures ab initio. It is based on the cosine least squares method, mentioned in Section 6.1 and on repeated cycles of electron density modification (EDM). The reader is referred to Section 8.2 for preliminary information on this technique.

The triplet invariants

$$t_{ijl} = \phi_i + \phi_j + \phi_l$$

and the negative quartet invariants

$$q_{ijlm} = \phi_i + \phi_j + \phi_l + \phi_m$$

are generated from a specified basis set of reflections which have the largest R values. The Cochran concentration parameters C (as given by (5.4)) and the Giacovazzo parameter G (as defined by (5.23)) are here renamed T_{ijl} and Q_{ijlm}, respectively; they are associated with triplets and quartets respectively. The phases are required to minimize the function

$$R(\phi) = \frac{\Sigma_{i,j,l} T_{ijl} \left(\cos \Phi_{ijl} - D_1(T_{ijl})\right)^2 + \Sigma_{i,j,l,m} |Q_{ijlm}| \left(\cos \Phi_{ijlm} - D_1(Q_{ijlm})\right)^2}{\Sigma_{i,j,l} T_{ijl} + \Sigma_{i,j,l,m} |Q_{ijlm}|},$$

where

$$D_1(x) = I_1(x)/I_0(x)$$

is the ratio of the modified Bessel functions of order 1 and 0 respectively. Since triplets and quartets are functions of the individual phases, $R(\phi)$ is also implicitly defined as a function of them.

$R(\phi)$ is expected to have a global minimum, provided that the number of phases involved is sufficiently large, when all the phases are equal to their true values for some choice of origin and enantiomorph. Thus, the phasing problem reduces to that of finding the global minimum of $R(\phi)$ (the *minimum principle*). Let us consider now the value of $R(\phi)$, say $R_T(\phi)$, when the phase invariants

Third generation direct methods programs

assume values equal to their expected values. The triplet contribution to $R_T(\phi)$ will then be

$$(R_T(\phi))_{\text{triplets}} \simeq \frac{\Sigma_{i,j,l} T_{ijl} \left(\langle \cos^2 \Phi_{ijl} \rangle + D_1^2(T_{ijl}) - 2 \cos \Phi_{ijl} D_1(T_{ijl}) \right)}{\Sigma_{i,j,l} T_{ijl}}.$$

Since $\cos^2 \Phi = (1 + \cos 2\Phi)/2$, in accordance with (5.A.5), the above expression reduces to

$$(R_T(\phi))_{\text{triplets}} = \frac{\Sigma_{i,j,l} T_{ijl} \left(0.5 + 0.5 D_2(T_{ijl}) - D_1^2(T_{ijl}) \right)}{\Sigma_{i,j,l} T_{ijl}}.$$

$$= 0.5 + \frac{\Sigma_{i,j,l} T_{ijl} \left(0.5 D_2(T_{ijl}) - D_1^2(T_{ijl}) \right)}{\Sigma_{i,j,l} T_{ijl}}.$$

Since $0.5 D_2(x) < D_1^2(x)$, the following inequality will hold:

$$(R_T(\phi))_{\text{triplets}} < 1/2.$$

A similar expression will be obtained if we consider just the quartet contribution. On considering both triplets and quartets we have,

$$R_T(\phi) = 0.5 + \frac{\Sigma_{i,j,l} T_{ijl} \left(0.5 D_2(T_{ijl}) - D_1^2(T_{ijl}) \right) + \Sigma_{i,j,l,m} |Q_{ijlm}| \left(0.5 D_2(Q_{ijlm}) - D_1^2(Q_{ijlm}) \right)}{\Sigma_{i,j,l} T_{ijl} + |Q_{ijlm}|} < 1/2.$$

This may be considered as a guideline for recognizing the correct solution among numerous trials. A trial for which $R_T(\phi)$ is much larger than 0.5 may be considered to be an unreliable solution.

The six-step shake and bake procedure may be described as follows.

1. The number of reflections to phase is chosen to be close to $10t$, and the total number of invariants is chosen to be close to $100t$ (t being the number of atoms in the asymmetric unit). Triplet and negative quartet invariants are generated. Often, only triplets are used. If both triplets and quartets are employed, the number of quartets is chosen so as to satisfy the relation $\Sigma_{i,j,l} T_{ijl} \simeq \Sigma_{i,j,l,m} Q_{ijlm}$. This usually means that the number of negative quartets is somewhat larger than the number of triplets.
2. A trial structure of randomly positioned atoms (no two atoms to be closer than a minimum distance, say 1.2 Å) is generated.
3. Normalized structure factors are calculated from the trial coordinates.
4. The phase values are perturbed by the parameter-shift method, in which $R(\phi)$ is minimized as follows. The phase set is ordered in decreasing value of $|E|$ and evaluated in the same order. When considering a given phase ϕ_i, the value of $R(\phi)$ is initially evaluated three times: once at ϕ_i, once at $\phi_i + 90°$, and once at $\phi - 90°$. If ϕ_i yields the minimum of the three values of $R(\phi)$, then the phase is accepted and the procedure then examines phase ϕ_{i+1}. Suppose instead (see Fig. 6.14) that $\phi_i - 90°$ yields the minimum value of $R(\phi)$. Then the phase value is updated at $\phi_i - 90°$, but a further shift is checked (phase value at $\phi_i = 180°$), for which $R(\phi)$ is again evaluated. The final phase value chosen for ϕ_i will be that yielding the minimum of $R(\phi)$. For cs. space groups or for symmetry-restricted phases in n.cs. space groups, only a 180° shift has to be tested for each phase.

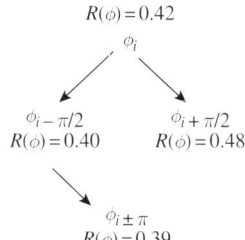

Fig. 6.14
An example of the parameter shift method used in *shake and bake*.

5. As soon as all of the phases have been evaluated, an E-map is calculated.
6. The electron density map is filtered, that is, the largest peaks in the Fourier map are selected (their number corresponds to about 0.8 of the number of expected atoms).

The procedure then returns to step 3 for cyclic improvement of the model. If the procedure converges to a false solution, a new trial model is generated (step 2) and the process is repeated.

Shake and bake (see Fig. 6.15) consists of a *shake* step (phase refinement) and a *bake* step (electron density modification), this second step aiming at imposing phase constraints implicit in real space. Accordingly the program requires two Fourier transforms per cycle, and numerous cycles. Thus, it can be time consuming and is not competitive with other direct methods for crystal solution of small molecules. However, it has introduced into the field extremely useful intensive computations for the direct solution of complex crystal structures. Indeed, shake and bake has succeeded in solving quite complex structures which only a few years ago were considered to be outside the range of applicability for direct methods. We note here the small protein structures solved so far by shake and bake. t is the number of non-hydrogen atoms in the asymmetric unit, d is the data resolution:

Structure	Space group	t	d (Å)
Vancomycin	$P4_32_12$	255	0.9
Gramicidin A	$P2_12_12_1$	317	0.86
Er-1 pheromone	C2	325	1.0
Crambin	$P2_1$	400	0.83
Alpha-1peptide	P1	450	0.92
Rubredoxin	$P2_1$	500	1.0
Tox II	$P2_12_12_1$	624	0.96
14 peptide	I4	289	1.1

It seems that data resolution has to be equal to or better than 1.1 Å. This condition is currently essential for success; further efforts will be necessary in order to make such a condition less severe. It may be instructive to check the predictive power of the z-test by applying it to the above list of structures. The $Fr(z)$ distribution for representatives of such a list is shown in Fig. 6.16. The figure

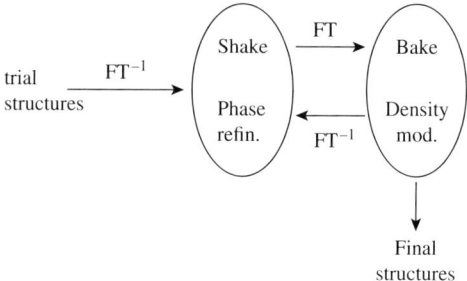

Fig. 6.15
The shake and bake scheme.

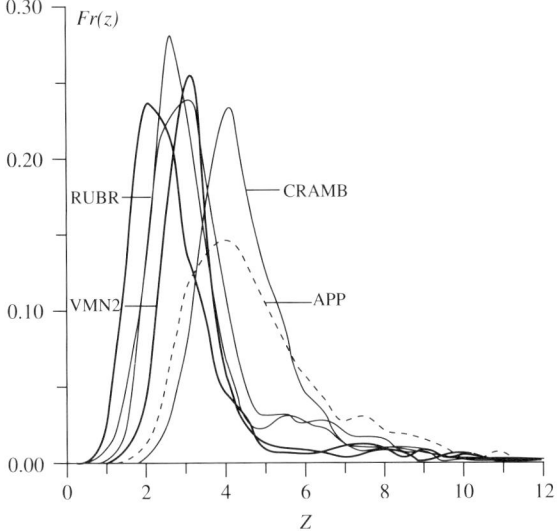

Fig. 6.16
$P(z)$ test for some of the structures solved by *shake and bake*.

indicates that each diffraction set contains enough information to allow crystal structure solution via the tangent formula or a related approach.

6.4.2 The half-bake approach

Due to Sheldrick (1997), *half-bake* does most of its work in direct space. Random phases are generated, to which the tangent formula is applied. E-maps are calculated, in which $1.3t$ peaks (t is the expected number of non-hydrogen atoms in the asymmetric unit) are searched.

Let E_o and E_c be the observed and the calculated (from the current model) normalized structure factors.

A number of peaks are eliminated subject to the condition that $\Sigma |E_c|(|E_0|^2 - 1)$ remains as large as possible (only reflections with $|E_0| > |E_{\min}|$ are involved, where $|E_{\min}| \simeq 1.4$). The phases of a suitable subset of reflections are then used as input for a tangent expansion. Then, an E-map is calculated from which peaks are selected; these are submitted to the elimination procedure.

Typically, from 5 to 20 cycles of this internal loop are performed. Then, the correlation coefficient (CC) between $|E_0|$ and $|E_c|$ is calculated for all the data. If CC is good (i.e. larger than a given threshold), then a new loop is performed; a new E-map is obtained, from which a list of peaks is selected which are all submitted to the elimination procedure. The criterion now is the value of CC, which is calculated for all the reflections. Typically from two to five cycles of this external loop are performed.

The program works indefinitely, restarting from random atoms until interrupted. It may work either applying the true spaces group symmetry or after having expanded the data to P1.

Half-bake has proved to be more efficient than *shake and bake*: it has led to 2000 non-hydrogen atoms in the asymmetric unit as the size limit for a structure solvable by direct methods. The procedure is now part of *SHELX-D* (Sheldrick, 1998).

6.4.3 The SIR2000-N approach

Unlike *shake and bake* and *half-bake*, in *SIR2000-N* (Burla et al., 2001) and in most of the subsequent programs of the *SIR* suite, the tangent refinement section is followed by a real-space phasing process without alternation. The tangent formula is first applied, within a multisolution random phase approach; the final phases are then submitted, in the order fixed by a figure of merit, to the real-space procedure, the moduli of which may be described as follows:

(a) Electron density modification (EDM). In repeated EDM cycles, only a fraction of the electron density map (say 2.5%) is used in the map inversion, the rest is set to zero (see Shiono and Woolfson, 1992, for a related method).
(b) Periodically, after some EDM cycles, the heaviest atomic species is associated with the selected peaks, with an occupancy factor taking into account peak height, site occupancy, and chemical connectivity. If heavy atoms are present, the displacement parameters are submitted to least squares.
(c) Peaks are labelled in terms of atomic species. The atomic positions are submitted to automatic diagonal-matrix least-squares cycles; combined with 2Fo-Fc syntheses, the procedure allows us to cyclically refine the structural model.

To avoid a waste of computing time, the crystal structures are divided into four classes: small (up to 80 atoms per asymmetric unit), medium (up to 200), large (e.g. proteins), and resistant. The number of cycles of each modulus is reduced according to the structure size.

The applications of *SIR2000-N* to a wide set of test structures has shown some interesting general trends which suggest how and when the synergy between real-space and reciprocal space procedures is essential for success:

 (i) Tangent procedures alone are able to solve the phase problem for small molecules, by driving random phase sets to the correct result, with quite small average phase error (say 10–25°).
 (ii) The efficiency of the tangent formula decreases with structural complexity. For medium-sized structures, the final phase error may be non-negligible (e.g. in the interval 20–50° or more).
 (iii) The final average phase error may be larger than 50° for larger structures and the best solution may not be distinguishable from the others.
 (iv) In a multisolution approach, the phase problem for small molecules may also be solved by real-space procedures alone, without the help of the tangent step. On the conversely, the tangent formula is still useful for medium and large-sized molecules, since it provides sets of phases which are more favourable (for subsequent real-space refinement) than trial random sets.
 (v) When applied to proteins, *SIR2000-N*, as well as its more efficient heir *SIR2011* (Burla et al., 2012a) and all the third generation programs, all

require diffraction data at atomic or nearly atomic resolution. Mooers and Matthews (2006) extended to about 2300 non-H atoms in the asymmetric unit as the size limit for structures solvable by direct methods, by phasing via *SIR2002* (Burla et al., 2003a) the previously unknown bacteriophage P22 lysozyme.

A last remark may be useful. All of the above described methods are less efficient than Patterson techniques when the unit cell contains heavy atoms. In such cases, Patterson deconvolution methods (implemented in *SIR2011*; see Caliandro et al., 2008a) bring the size limit to about 8000 non-H atoms in the asymmetric unit and the resolution limit to about 2 Å (see Section 10.4).

APPENDIX 6.A FINDING QUARTETS

In order to select quartets with large values of C (as given by equation (5.17)), the quartet invariants are usually searched among the subsets $\{S\}$ of reflections with $R > R_{Th}$, where R_{Th} is a suitable threshold. Two algorithms are commonly used.

Algorithm 1

1. The indices of the reflections symmetry equivalent to the unique reflections in $\{S\}$ are calculated, so obtaining the expanded list.
2. For a given $\boldsymbol{h}_1$ all the triplets $\boldsymbol{h}_2$, $\boldsymbol{h}_3$, $\overline{\boldsymbol{h}_1 + \boldsymbol{h}_2 + \boldsymbol{h}_3}$ with $R > R_{Th}$ are found. The search is made by letting $\boldsymbol{h}_1$ vary over the subset of standard reflections while $\boldsymbol{h}_2$, $\boldsymbol{h}_3$, and $\overline{\boldsymbol{h}_1 + \boldsymbol{h}_2 + \boldsymbol{h}_3}$ can span over the expanded list. In practice, all the sets of four vectors

$$\boldsymbol{h}_1, \boldsymbol{h}_2 \mathbf{R}_i, \boldsymbol{h}_3 \mathbf{R}_j \ \overline{\boldsymbol{h}_1 \pm \boldsymbol{h}_2 \mathbf{R}_i \pm \boldsymbol{h}_3 \mathbf{R}_j}$$

are found, where $\mathbf{R}_i$ and $\mathbf{R}_j$ vary over all the rotation matrices of the space group.
3. The cross-vectors are calculated and each quarter is estimated. This algorithm simultaneously finds positive and negative quartets, according to the specific values of the cross-magnitudes. Statistically speaking, the percentage of positive estimated quartets is, remarkably, larger than that of the negative ones. Indeed, $1 + \varepsilon_5 + \varepsilon_6 + \varepsilon_7$ is more probably positive than negative.

Algorithm 2

1. Triplets are calculated, and quartets are obtained as the sum of two triplets. For example, from the two triplets

$$\phi_{\boldsymbol{h}_1} + \phi_{\boldsymbol{h}_2} - \phi_{\boldsymbol{h}_5} \ (\boldsymbol{h}_1 + \boldsymbol{h}_2 - \boldsymbol{h}_5 = 0) \qquad (6.A.1a)$$

$$\phi_{\boldsymbol{h}_3} + \phi_{\boldsymbol{h}_4} + \phi_{\boldsymbol{h}_5} \ (\boldsymbol{h}_3 + \boldsymbol{h}_4 + \boldsymbol{h}_5 = 0). \qquad (6.A.1b)$$

Table 6.A.1 Number of triplets (NTRIP) and quartets (NQUAR) found among n reflections for *CEPHAL* [C2, $C_{18} H_{21} N O_4$, $Z = 8$] and *MUNICH* [C2, $C_{20} H_{16}$, $Z = 8$]

	n	NTRIP	NQUAR
CEPHAL	50	6	298
	100	113	4631
	200	836	76 092
	334	3751	645 213
MUNICH	50	34	317
	150	535	27 964
	310	3704	509 781

the quartet

$$\phi_{h_1} + \phi_{h_2} + \phi_{h_3} + \phi_{h_4} \; (h_1 + h_2 + h_3 + h_4 = 0)$$

is obtained. The method simultaneously provides one cross-magnitude of the quartet, i.e. R_{h_5}.

2. The other cross-vectors are calculated and each quartet is estimated.

If one is interested in negative quartets only, algorithm 2 is preferable. In this case, the two triplets (6.A.1) should be *psi-zero triplets* (see Section 6.3.4), and h_5 is the vectorial index of the weak reflection. Algorithm 2 in this case directly selects quartets having at least one small cross-magnitude, thus providing large percentages of negative estimated quartets.

Algorithm 2 is also preferable if one is only interested in positive quartets. In this case, the two triplets (6.A.1) should belong to the Σ_2 list and h_5 is the vectorial index of a strong reflection.

The number of quartets which can be found among a given subset $\{S\}$ is much larger than the corresponding number of triplets. In order to give a practical insight, in Table 6.A.1 we show the number of triplets (NTRIP) and quartets (NQUAR) found via algorithm 1 among a subset of n reflections. It is seen that NQUAR quickly increases with n; it is not infrequent that several millions of quartets can be found in high-symmetry space groups among only 400–500 standard reflections.

However, most of these quartets will have small reliability parameters (we should not forget that quartets are phase relationships of order N^{-1}).

Joint probability distribution functions when a model is available: Fourier syntheses

7.1 Introduction

The title of this chapter may seem a little strange; it relates Fourier syntheses, an algebraic method for calculating electron densities, to the joint probability distribution functions of structure factors, which are devoted to the probabilistic estimate of s.i.s and s.s.s. We will see that the two topics are strictly related, and that optimization of the Fourier syntheses requires previous knowledge and the use of joint probability distributions.

The distributions used in Chapters 4 to 6 are able to estimate s.i. or s.s. by exploiting the information contained in the experimental diffraction moduli of the *target structure* (the structure one wants to phase). An important tool for such distributions are the theories of *neighbourhoods* and of *representations*, which allow us to arrange, for each invariant or seminvariant Φ, the set of amplitudes in a sequence of shells, each contained within the subsequent shell, with the property that any s.i. or s.s. may be estimated via the magnitudes constituting any shell. The resulting conditional distributions were of the type,

$$P(\Phi|\{R\}), \qquad (7.1)$$

where $\{R\}$ represents the chosen phasing shell for the observed magnitudes. The more information contained within the set of observed moduli $\{R\}$, the better will be the Φ estimate.

By definition, conditional distributions (7.1) cannot change during the phasing process because prior information (i.e. the observed moduli) does not change; equation (7.1) maintains the same identical algebraic form. However, during any phasing process, various *model structures* progressively become available, with different degrees of correlation with the target structure. Such models are a source of supplementary information (e.g. the current *model phases*) which, in principle, can be exploited *during* the phasing procedure. If this observation is accepted, the method of joint probability distribution, as described so far, should be suitably modified. In a symbolic way, we should look for deriving conditional distributions

$$P\left(\Phi|\{R\},\{R_p\}\right), \qquad (7.2)$$

rather than (7.1), where $\{R_p\}$ represents a suitable subset of the amplitudes of the model structure factors. Such an approach modifies the traditional phasing strategy described in the preceding chapters; indeed, the set $\{R_p\}$ will change during the phasing process in conjunction with the model changes, which will continuously modify the probabilities (7.2). We will study in this chapter two invariants, $\Phi = (\phi_\mathbf{h} - \phi_{p\mathbf{h}})$ and $\Phi = (\phi_\mathbf{h} + \phi_\mathbf{k} - \phi_{\mathbf{h+k}})$, given suitable subsets of observed and calculated diffraction amplitudes.

The conditional distributions (7.2) are also able to optimize the *Fourier syntheses*, a fundamental tool in modern crystallography, devoted to the calculation of various types of electron density maps. Such syntheses may be computed only if a model structure is available: to obtain optimized syntheses, their coefficients should reflect the reliability of the model, by including a weight depending on model quality. This is the main reason why Fourier syntheses and joint probability distribution functions of structure factors are treated together in this chapter.

In Section 7.2, we will describe the probability distribution of the s.i., $\Phi = (\phi_\mathbf{h} - \phi_{p\mathbf{h}})$. In Section 7.3, we will analyse the different types of electron densities, i.e. the so-called *observed*, *difference*, and *hybrid Fourier syntheses*, their main properties, and the most useful weighting schemes, strictly connected with the distributions detailed in Section 7.2. In Section 7.4, the parameters affecting the quality of the electron density maps are discussed. In Section 7.5, a formula estimating triplet phase invariants from observed and calculated amplitudes is described.

7.2 Estimation of the two-phase structure invariant $(\phi_\mathbf{h} - \phi_{p\mathbf{h}})$

Model and target structures are often denoted as isomorphous. The usual condition is that they have the same unit cell, the same space group, and their crystal structures are sufficiently correlated. We will suppose, as a working hypothesis, that the model structure arose during the phasing approach and that it is correlated with the target structure. Thus, model and target are isomorphous.

According to Section 3.3, $\Phi_\mathbf{h} = (\phi_\mathbf{h} - \phi_{p\mathbf{h}})$ is a two-phase s.i., where $\phi_\mathbf{h}$ is the phase of the target, and $\phi_{p\mathbf{h}}$ is the phase of the model structure. $\Phi_\mathbf{h}$ may be estimated via the method of joint probability distributions, but now two isomorphous structures are involved. This implies a strong modification of the mathematical approach established in Chapters 4 to 6, where only one structure was considered. In that approach, the atomic position $\mathbf{r}_j$ of the target structure was assumed to be the primitive random variables, spanning uniformly over the unit cell. In this new approach, the atomic positions of the model structure also span the unit cell and therefore they should also be involved in the mathematical modelling. However, the target and the model structures are supposed to be isomorphous, and therefore their atomic positions can no longer be considered to be statistically independent.

In the following, we will assume that N is the number of atoms in the target and N_p the number of atoms in the model structure: $\mathbf{r}_j, j = 1, \ldots, N$

and $\mathbf{r}_{pj}$, $j = 1, \ldots, N_p$, with (usually) $N_p < N$, are the atomic positions of the target and of the model structure, respectively. The simplest mathematical approach was suggested by Sim (1959a,b). He made the following hypotheses:

(a) The vectors $\mathbf{r}_j$, $j = 1, \ldots, N$ are the primitive random variables, assumed to be uniformly distributed in the unit cell.
(b) $\mathbf{r}_{pj} = \mathbf{r}_j$, for $j = 1, \ldots, N_p$, with $N_p < N$.

The situation is as depicted in Fig. 7.1, for a non-realistic one-dimensional structure: two 'triangular' atoms of the model perfectly coincide with two atoms of the target.

In accordance with the above scheme, the normalized structure factors are modelled as follows:

$$E_\mathbf{h} = R_\mathbf{h} \exp(i\phi_\mathbf{h}) = \left\{ \sum_{j=1}^{N} f_j \exp(2\pi i \mathbf{h} \cdot \mathbf{r}_j) \right\} \Big/ (\varepsilon_\mathbf{h} \Sigma_N)^{1/2},$$

$$E_{p\mathbf{h}} = R_{p\mathbf{h}} \exp(i\phi_{p\mathbf{h}}) = \sum_{j=1}^{N_p} f_j \exp 2\pi i \mathbf{h} \cdot \mathbf{r}_{pj} \Big/ (\varepsilon_\mathbf{h} \Sigma_{N_p})^{1/2},$$

where $\varepsilon_\mathbf{h}$ is the correction factor for expected intensities arising from Wilson statistics, and

$$\Sigma_N = \sum_{j=1}^{N} f_j^2, \quad \Sigma_{N_p} = \sum_{j=1}^{N_p} f_j^2.$$

The Sim model is often not realistic; it is very rare that the atoms of a model structure occupy the same positions as in target structures. Srinivasan and Ramachandran (1965a,b) suggested less strict hypotheses:

(a) $\mathbf{r}_{pj}, j = 1, \ldots, N_p$, are fixed known parameters;
(b) $\mathbf{r}_j = \mathbf{r}_{pj} + \Delta \mathbf{r}_j, j = 1, \ldots, p$. The variables $\mathbf{r}_j$ are riding variables, which are variables correlated with the corresponding $\mathbf{r}_{pj}$s through local positional errors $\Delta \mathbf{r}_j$s. A schematic example of target and model structures is shown in Fig. 7.2, where $N = 4$ and $N_p = 2$; the $\Delta \mathbf{r}_j$ moduli have been chosen sufficiently small so as to secure, at least at low resolution, isomorphism between the model and target structures. In the absence of any information on their distribution or on their mutual correlation, it was assumed that the $\Delta \mathbf{r}_j$s are independent of each other and uniformly distributed around zero. In many practical situations, this condition is violated, e.g. when molecular fragments of the model are translated with respect to the correct position or are rotated with respect to the correct orientation; then, the $\Delta \mathbf{r}_j$s are equal to each other in the first case, or are referred to each other by a rotation in the second case. In the absence of any supplementary information, the independence of the $\Delta \mathbf{r}_j$s is a less demanding hypothesis one can apply.
(c) $\mathbf{r}_j, j = p+1, \ldots, N$ are primitive random variables, assumed to be uniformly distributed within the unit cell.

According to the preceding hypotheses, we can model the normalized structure factors as follows:

$$E_\mathbf{h} = R_\mathbf{h} \exp(i\phi_\mathbf{h}) = \left\{ \sum_{j=1}^{N_p} f_j \exp[2\pi i \mathbf{h} \cdot (\mathbf{r}_{pj} + \Delta \mathbf{r}_j)] + \sum_{j=N_p+1}^{N} f_j \exp(2\pi i \mathbf{h} \cdot \mathbf{r}_j) \right\} \Big/ (\varepsilon_\mathbf{h} \Sigma_N)^{1/2}$$

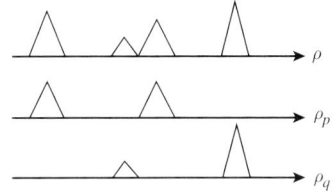

Fig. 7.1
Non-realistic one-dimensional four-atom target structure (ρ) and a two-atom model structure (say ρ_p), the atomic positions of which perfectly coincide with two atoms of the target structure (parameter $D = 1$).

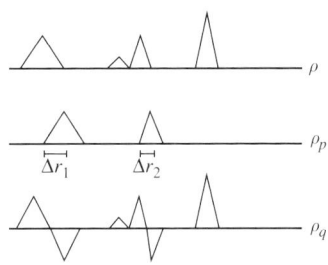

Fig. 7.2
Schematic target and model structures for $N = 4$, $N_p = 2$. Model and target structures partially overlap; parameter D is in the interval (0,1).

$$E_{p\mathbf{h}} = R_{p\mathbf{h}} \exp(i\phi_{p\mathbf{h}}) = \sum_{j=1}^{p} f_j \exp(2\pi i \mathbf{h} \cdot \mathbf{r}_{pj}) \Big/ (\varepsilon_{\mathbf{h}} \Sigma_{N_p})^{1/2}.$$

A further generalization was suggested by Caliandro et al. (2005c), who included the measurement error into the mathematical formalism. As for Srinivasan and Ramachandran, the $\mathbf{r}_{pj}$, $j = 1, \ldots, N_p$, are fixed known parameters, the $\mathbf{r}_j = \mathbf{r}_{pj} + \Delta\mathbf{r}_j$, $j = 1, \ldots, p$, are riding variables, and $\mathbf{r}_j, j = p + 1, \ldots, N$ are primitive random variables. The measurement error is parameterized through the following structure factor model:

$$E_{\mathbf{h}} = \left\{ \sum_{j=1}^{N_p} f_j \exp[2\pi i \mathbf{h}(\mathbf{r}_{pj} + \Delta\mathbf{r}_j)] + \sum_{j=N_p+1}^{N} f_j \exp(2\pi i \mathbf{h}\mathbf{r}_j) + |\mu|\exp(i\vartheta) \right\} \Big/ (\varepsilon \Sigma_N)^{1/2},$$

$$E_{p\mathbf{h}} = \sum_{j=1}^{p} f_j \exp 2\pi i \mathbf{h}\mathbf{r}_{pj} \Big/ (\varepsilon \Sigma_{N_p})^{1/2},$$

where $|\mu|\exp(i\vartheta)$ is the (complex) measurement error. $|\mu|$ and θ are two supplementary primitive random variables arising from the experimental uncertainty on the observed structure factor ($<|\mu|>$ is related to the standard deviation of the measured diffraction amplitude). The conclusive joint probability distribution function is

$$P(R, R_p, \phi, \phi_p) = RR_p \, \pi^{-2} (e - \sigma_A^2)^{-1}$$

$$\exp\left\{ -\frac{1}{(e - \sigma_A^2)} \left[R^2 + eR_p^2 - 2\sigma_A RR_p \cos(\phi - \phi_p) \right] \right\}, \quad (7.3)$$

where

$$\sigma_A = D\sqrt{\Sigma_{N_p}/\Sigma_N} \quad (7.4)$$

$$D = <\cos(2\pi \mathbf{h} \cdot \Delta \mathbf{r})> \quad (7.5)$$

$$e = (1 + \sigma_R^2), \quad \text{and} \quad \sigma_R^2 = <|\mu|^2>/\Sigma_N.$$

From (7.3), the conditional distribution

$$P(\Phi|R, R_p) = [2\pi I_0(X)]^{-1} \exp(X \cos \Phi) \quad (7.6)$$

is obtained, where $\Phi = \phi - \phi_p$ and

$$X = \frac{2\sigma_A RR_p}{(e - \sigma_A^2)}. \quad (7.7)$$

As for all von Mises distributions,

$$m = <\cos(\phi - \phi_p)> = I_1(X)/I_0(X). \quad (7.8)$$

Equation (7.6) suggests that if X is large enough, then $\phi \approx \phi_p$ m and is the weight to associate with such relation. The reliability parameter X depends not only on the product RR_p, but also on $\sqrt{\Sigma_{N_p}/\Sigma_N}$ (a factor which estimates the incompleteness of the model structure), and on the parameter D (a factor which estimates the similarity between the N_p-atom model and the corresponding N_p-atom substructure in the target structure). D (standing for $D_\mathbf{h}$), for the fixed

h reflection, is obtained through an average performed over the different $\Delta \mathbf{r}_j$, $j = 1, \ldots, N_p$. Values will be $D = 1$ when the N_p atomic positions of the model coincide with N_p atomic positions in the target structure, $D = 0$ when the model atomic positions are completely uncorrelated with the target atoms. Obviously, D is not a measurable quantity; it may only be estimated statistically during the phasing process, but it allows us to define a very important parameter, σ_A, quite central to distribution (7.6).

The mean value of σ_A (say $<\sigma_A>$; the average is calculated over all reflections) is a good estimate of the correlation between model and target structures. Indeed, if $\sigma_A = 1$, then $\Sigma_{N_p} = \Sigma_N$ and $D = 1$; if $\sigma_A = 0$, the two structures are uncorrelated. We will see in Appendix 7.A how σ_A may be estimated statistically.

For the centric case, the following distribution is obtained:

$$P(E, E_p) = \frac{1}{(2\pi)\sqrt{(e - \sigma_A^2)}} \exp\left\{-\frac{1}{2(e - \sigma_A^2)}\left[eE_p^2 + E^2 - 2\sigma_A E E_p\right]\right\}.$$

It can easily be seen that the Sim model is a particular case of the Srinivasan and Ramachandran model, obtained by fixing $\Delta \mathbf{r}_j = 0$, $j = 1, \ldots, N_p$. Furthermore, the classical Srinivasan and Ramachandran distribution is obtained from (7.3) by assuming that $e = 1$.

If equation (7.3) is integrated over ϕ and ϕ_p, the marginal distribution $P(R, R_p)$ is obtained, from which the conditional distributions, $P(R|R_p)$ and $P(R_p|R)$ may be obtained. The distribution $P(R|R_p)$ is important for maximum likelihood approaches (the reader is referred to McCoy (2004) for a recap, to Murshudov et al. (1997) and de la Fortelle and Bricogne (1997) for refinement, and to Read (2001, 2003a,b) for molecular replacement) and for *Free lunch* techniques (see Caliandro et al. (2005a,b) and Section 8.2). The conditional distributions are quoted in Appendix 7.C.

7.3 Electron density maps

As the reader may know, Fourier syntheses are a fundamental tool in the daily work of a crystallographer; if X-ray data have been collected and a model is available, such syntheses allow us to guess about the distribution of the electrons in the unit cell. The study of the properties of electron density maps started with W. H. Bragg in 1915 and continued with Bragg and West (1930), Booth (1946, 1947), Cruickshank (1949), Cochran (1951), and Cruickshank and Rollett (1953).

The properties of Fourier syntheses are also an important tool for solution of the phase problem, particularly when only an imperfect model structure is available and one wants to recover the target structure. Ramachandran and Raman (1959), Srinivasan (1961), and Main (1979), analysed the various types of Fourier synthesis and studied their convergence properties. In more recent years, syntheses and their possible applications were reconsidered by Read (1986), Ursby and Bourgeois (1997), Cowtan (1999), Lunin et al. (2002),

156 | Joint probability distribution functions when a model is available

Giacovazzo and Mazzone (2011, 2012), Giacovazzo, Mazzone, and Comunale (2011), and Altomare et al. (2008a,b; 2009c, 2010).

In Section 7.3.1 we will discuss the main features of an ideal Fourier synthesis, calculated via equation (1.18) when correct phases and amplitudes are available. In Sections 7.3.2 to 7.3.4, we will describe the most popular Fourier syntheses and their main properties.

7.3.1 The ideal Fourier synthesis and its properties

Let us recall equation (1.18):

$$\rho(\mathbf{r}) = \frac{1}{V} \sum_{\mathbf{h}} |F_{\mathbf{h}}| \exp(i\phi_{\mathbf{h}}) \exp(-2\pi i \mathbf{h} \cdot \mathbf{r}), \qquad (7.9)$$

where $\mathbf{r}$ is a generic point of the unit cell. $\rho(\mathbf{r})$ is a non-negative function whose maxima are expected to coincide with the atomic positions. We will describe the properties of synthesis (7.9), as a function of various parameters, when diffraction amplitudes and phases are known without error.

1. *Temperature factor*. Thermal displacement modifies the electron density of the atoms, by enlarging the volume frequented by the atomic electrons; consequently the atomic scattering factor decreases more rapidly with $\sin\theta/\lambda$. The larger the temperature factor, B, the smaller the corresponding peak intensity in the electron density map, and the larger its full width at half maximum. This behaviour may be verified in Fig. 7.3, where we show, in a one-dimensional example, the electron density for C atoms at different temperature factors. An important effect of the thermal displacement is the modification of RES; proteins are usually characterized by high values of B and consequently the experimental data resolution is often non-atomic (see point 4 in this section).

2. *Chemical occupancy*. If an atom is located at $\mathbf{r}$ and we reduce the chemical occupancy in that position, the corresponding peak in the electron density

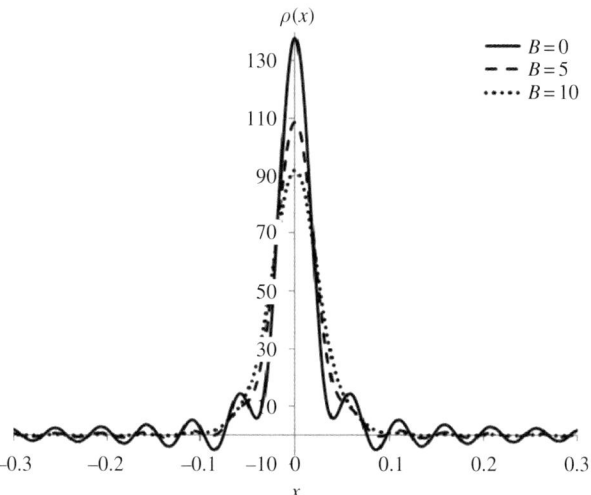

Fig. 7.3
One-dimensional space: the electron density of C atoms at different temperature factors (for the ripples, see point 4 of this section).

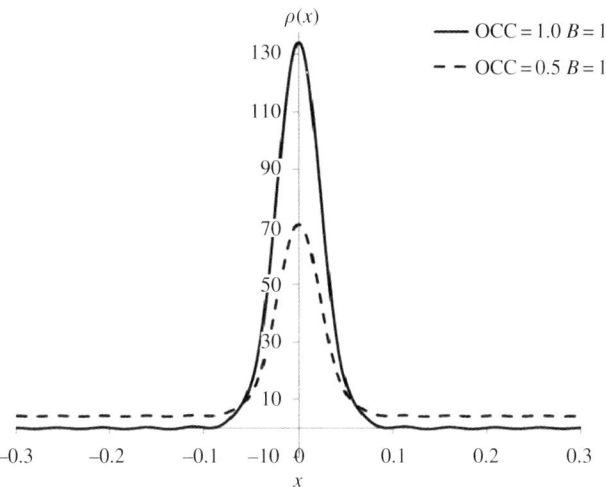

Fig. 7.4
One-dimensional example: reduction of the chemical occupancy from 1.0 to 0.5 implies a modification of the electron density profile.

will, similarly, be smaller (see Fig. 7.4). The reader may wonder whether an increase in the thermal factor may be distinguished from a reduction in the chemical occupancy. In Fig. 7.5, we show two C atoms, the first with full occupancy and $B = 30$ and the second with chemical occupancy 0.5 and $B = 2.2$. The intensities of the two peaks are equal, but the full width at half maximum of the second peak is sharper. The two profiles are indeed different, but in real cases it may be difficult to distinguish between occupancy and temperature displacement effects.

3. *Disorder*. The overall effect of disorder is a modification of RES. For example, fewer diffraction amplitudes can be measurable experimentally if the disorder involves a large fraction of the unit cell; electron density maps are then less informative. In some cases, disorder involves only specific regions of the electron density map (not unusual for both small and

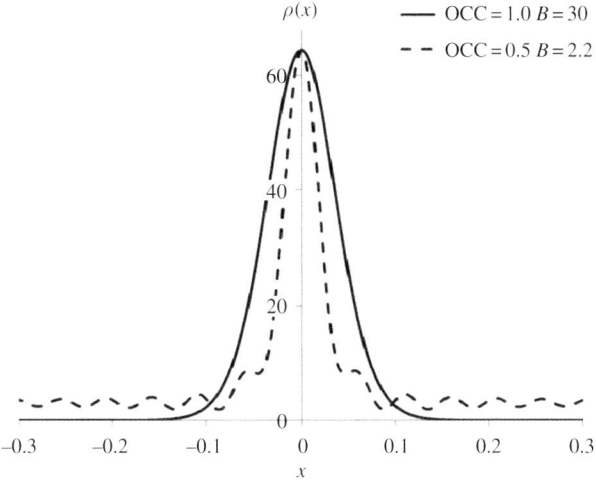

Fig. 7.5
Two electron density profiles of a C atom are superimposed, the first with full chemical occupancy and $B = 30$ and the second with chemical occupancy 0.5 and $B = 2.2$. The intensities of the two peaks are equal but the full width at half maximum of the second peak is sharper.

macro-molecules; e.g. peripheral groups of atoms which may rotate about one bond). In this case the electron density, in that region, will show a mean structure, resulting from an average over all of the different possible configurations. Correct interpretation of an electron density map therefore requires a good disorder model.

4. *Resolution bias*. The summation on the right-hand side of equation (7.9) goes over an infinite number of reflections. Because of the diffraction limits, only a limited number of measurable reflections are available; thermal displacement effects may make RES even worse. If vanishing diffraction data only are omitted from the right-hand side of (7.9), then all of the information contained in the structure factors is transferred to the electron density. If a significant signal is still present from the reflections used in the Fourier summation, then the Fourier transform has not been performed correctly. Owing to reflection truncation, the following effects will arise in the electron density map: the geometrical form of the peak profile is modified; peaks are broader; oscillating ripples are present close to each atomic peak, which interfere with the ripples of the closest peaks; the peaks positions themselves are modified. All of these effects fall under the heading resolution bias.

To provide a short mathematical treatment of truncation effects, let us assume that

$$\rho(\mathbf{r}) = \sum_{j=1}^{N} \rho_j(\mathbf{r} - \mathbf{r}_j)$$

is the electron density, expressed as sum of atomic electron densities $\rho_j(\mathbf{r})$ centred on atomic positions $\mathbf{r}_j$. $\rho(\mathbf{r})$ is a non-negative definite function (i.e. $\rho(\mathbf{r}) \geq 0$, at any point in the unit cell). In a practical diffraction experiment, the summation is limited to the measured domain of the reciprocal space, represented by the shape function $\Phi(\mathbf{r}^*)$ ($\Phi(\mathbf{r}^*) = 1$ inside the measured domain; $\Phi(\mathbf{r}^*) = 0$ outside). Correspondingly, the electron density map available in practice, say $\rho'(\mathbf{r})$, is based on the structure factors,

$$F'_\mathbf{h} = F_\mathbf{h} \Phi(\mathbf{r}^*),$$

rather than on the $F_\mathbf{h}$'s. Therefore,

$$\rho'(\mathbf{r}) = \rho(\mathbf{r}) \otimes T[\Phi(\mathbf{r}^*)] = \rho(\mathbf{r}) \otimes \zeta(\mathbf{r}) = \sum_{j=1}^{N} \rho_j(\mathbf{r} - \mathbf{r}_j) \otimes \zeta(\mathbf{r}), \quad (7.10)$$

where, $\otimes$ represents the convolution operation. $\zeta(\mathbf{r})$ is the Fourier transform of $\Phi(\mathbf{r}^*)$ (which may be deduced from the experiment), and primarily depends on RES; if Φ is a sphere, $\zeta(\mathbf{r})$ will have a spherical symmetry. In this case we have,

$$\zeta(\mathbf{r}) = \frac{4}{3} \pi r^{*3}_{max} \, y,$$

where

$$y = 3 \, \frac{\sin(2\pi r^*_{max} r) - (2\pi r^*_{max} r) \cos(2\pi r^*_{max} r)}{(2\pi r^*_{max} r)^3}$$

and $r^*_{max} = RES^{-1}$.

$\zeta(\mathbf{r})$ is an oscillating function, non-positive definite (see Fig. 7.6) and, consequently, $\rho'(\mathbf{r})$ is negative in more or less extended regions of the unit cell (while, by definition, the density should be non-negative at all points of the unit cell), the atomic peaks are broadened, and they are surrounded by a series of negative and positive ripples of gradually decreasing amplitude. Since the ripples may overlap with other ripples and with atomic peaks, the density maxima may be misplaced with respect to the correct positions.

In a one-dimensional case, $\Phi(x^*)$ corresponds to a *Heaviside unit step function*, and $\zeta(x) = \frac{\sin(2h_{\max}+1)\pi x}{\sin \pi x}$, where $h_{\max}$ is the maximum index used in the electron density calculation. The maximum value of $\zeta(x)$ is $(2h_{\max}+1)$, attained at the origin.

In Figs. 7.7 to 7.11, we depict what should be observed in one-dimensional electron density maps (say $\rho'(x)$) for the corresponding different schematic $\rho(x)$ structures, as the effect of resolution bias. In particular we show that:

(i) the ripple intensities are proportional to the intensity of the main peak (see Fig. 7.7);
(ii) the temperature factor modifies both the main peak and the ripple profiles (see Fig. 7.8);
(iii) for a one-point atom structure, when RES improves, the main peak becomes sharper and its intensity increases; the ripple frequency increases also (see Fig. 7.9);
(iv) for a one-dimensional two-point atom structure, the ripples corresponding with the two main peaks interact with each other and also interact with the main peaks, which may then move from their correct positions (see Fig. 7.10);
(v) ripples can significantly distort the main peak profile (see Fig. 7.11).

It is a common belief that resolution bias is unavoidable; it is generally considered to be an intrinsic characteristic of electron density maps, generated by the physics of the diffraction experiment. However, returning from $\rho'(\mathbf{r})$ to $\rho(\mathbf{r})$ via deconvolution procedures offers several advantages: peaks should move

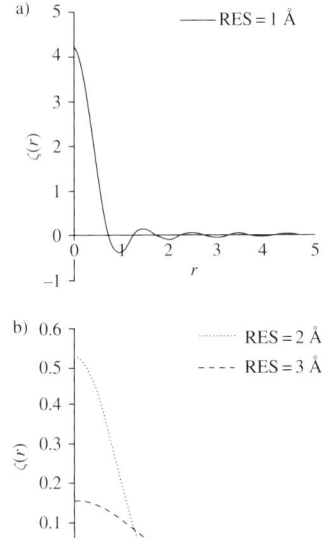

Fig. 7.6
The Fourier transform $\zeta(r)$ of the spherical reciprocal domain $\Phi(\mathbf{r}^*)$ as accessible for the experiment. Its form depends primarily on the experimental resolution, RES = $1/r^*_{\max} = d_{\min}$. a) $\zeta(r)$ for RES = 1 Å; b) $\zeta(r)$ for RES = 2 Å and 3 Å. The reader should notice how the amplitude and the width of the main peak and of the oscillations vary with RES.

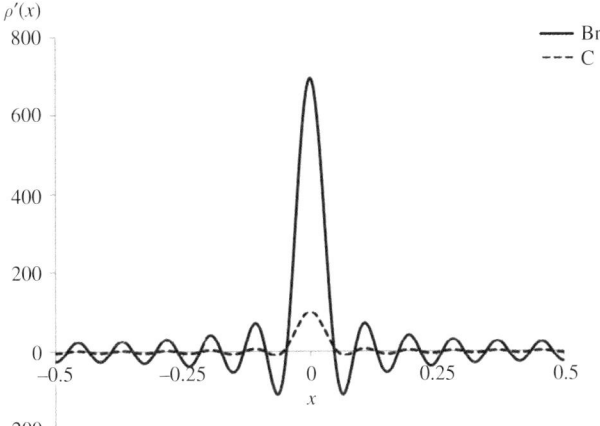

Fig. 7.7
$\rho'(x)$ of Br and C, at rest, when located at the origin of the unit cell, at RES = 1.8 Å (continuous line for Br, dashed line for C). The figure shows that ripple intensities are proportional to the intensity of the main peak.

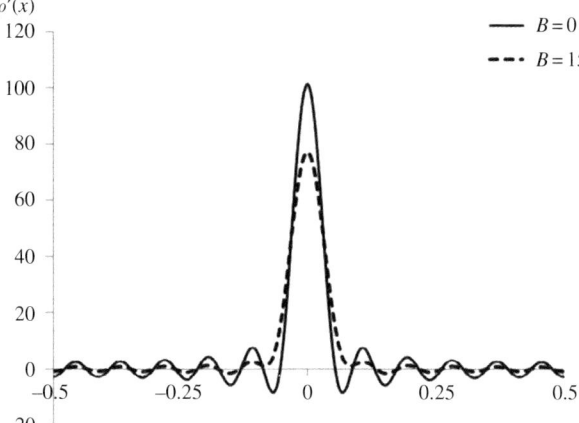

Fig. 7.8
$\rho'(x)$ at RES = 1.8 Å for a C atom located at the origin of the unit cell, with two different values of the temperature factor B (continuous line, $B = 0$; dashed line, $B = 15$). The figure shows how main peak and ripple profiles are modified by the temperature factor.

into more correct positions, false peaks due to limited experimental data resolution may be eliminated, peak broadening may be reduced. Some examples will be useful:

(a) In traditional Patterson search methods (see Chapter 8), the ripples generated by the origin peak are often suppressed, just by calculating $(|E|^2 - 1)$ Patterson maps.

(b) Altomare et al. (2008a,b, 2010) described a general algorithm for correcting resolution bias in E-maps, obtained by applying direct methods to powder data (see Chapter 12 for details).

(c) Burla et al. (2006a) described an algorithm allowing elimination of bias in a region around the heavy-atom peak in a protein electron density map. The algorithm allowed an extension from 2268 (a value attained by Mooers and Matthews, 2006) to around 6300 non-hydrogen atoms in the asymmetric unit, as the complexity limit for proteins solvable *ab initio* (provided that data resolution is better than 1.2 Å).

(d) An additional method to reduce the resolution bias (called *free lunch*) is described in Section 8.2.

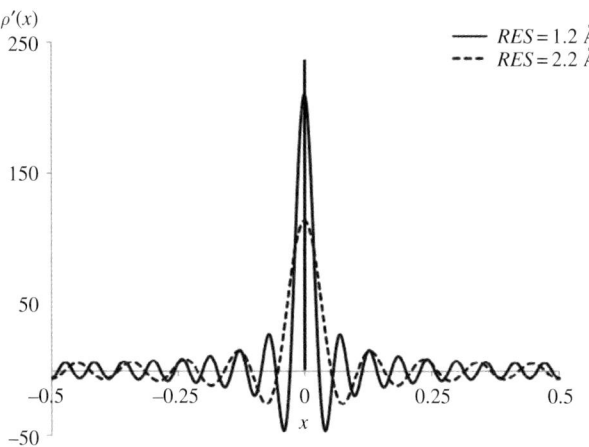

Fig. 7.9
A Dirac delta function, $\rho(x) = 6\delta(x)$ has been located at the origin of the unit cell (represented by a vertical bar). $\rho'(x)$, calculated by constant structure factors (these are the Fourier transform of a point atom) is shown at RES = 1.2 Å (continuous line) and RES = 2.2 Å (dashed line). In both cases we should observe, $\rho'(x) = 6\zeta(x)$, which is no longer a δ function. When RES diminishes, the sharpness of the main peak increases and the ripple frequency increases.

Electron density maps 161

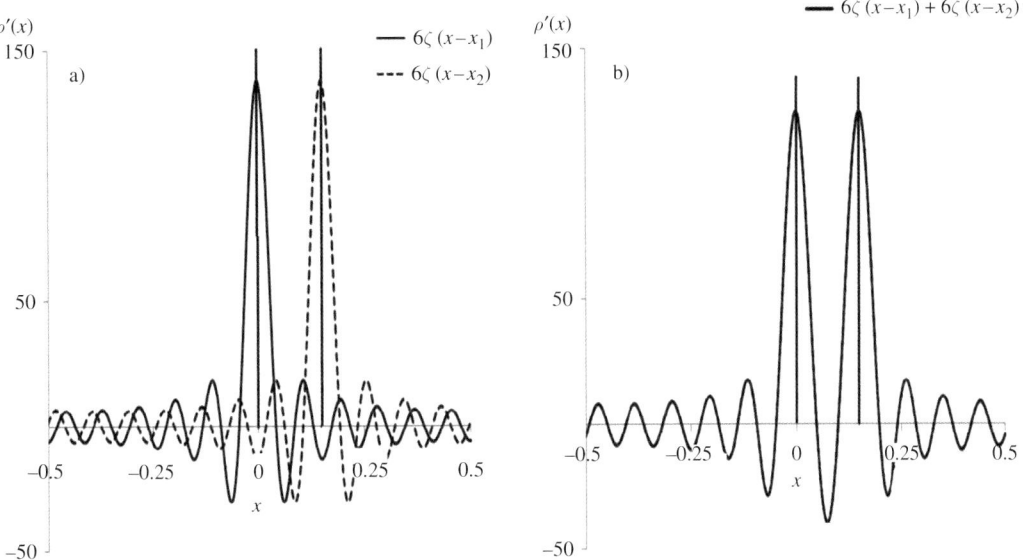

Fig. 7.10
Two Dirac δ functions [$6\delta(x-x_1)$ and $6\delta(x-x_2)$] were located at $x_1 = 0.0$, and $x_2 = 0.15$, respectively; they are represented by two vertical bars. a) $6\zeta(x-x_1)$ (continuous line) and $6\zeta(x-x_2)$ (dashed line) are reported at RES = 1.8 Å. b) $\rho'(x) = c_1\zeta(x-x_1) + c_2\zeta(x-x_2)$ (black line) at RES = 1.8 Å. As an effect of the ripple interaction, the two main maxima are now centred at $x'_1 = 0.006$ and $x'_2 = 0.145$.

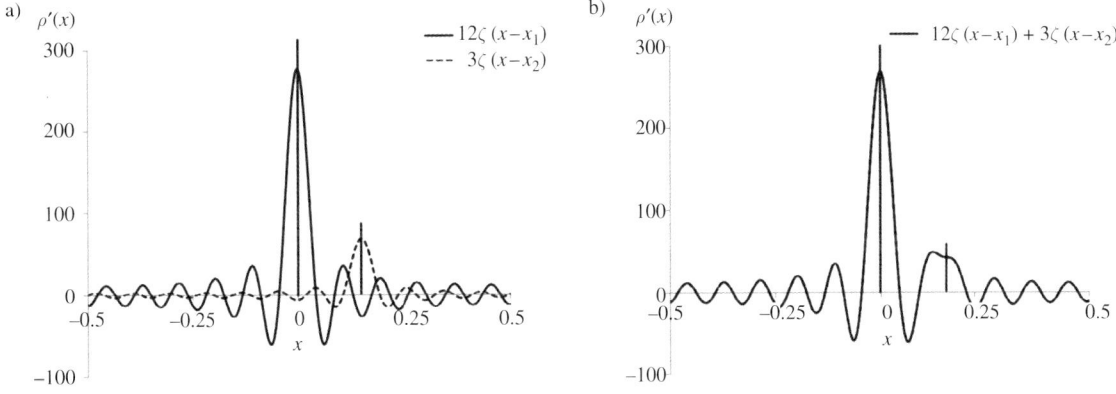

Fig. 7.11
Two-point atom structure: two δ functions, $12\delta(x-x_1)$ and $3\delta(x-x_2)$, were located at $x_1 = 0.0$ and at $x_2 = 0.15$. They are represented by vertical bars in Fig. 7.11a. a) functions $12\zeta(x-x_1)$ (continuous line) and $3\zeta(x-x_2)$ (dashed line) at RES = 1.8 Å. b) $\rho'(x) = 12\zeta(x-x_1) + 3\zeta(x-x_2)$ (black line). The two main maxima of $\rho'(x)$ are now centred at $x'_1 = 0.0$ and $x'_2 = 0.125$; a strong distortion of the second peak is clearly visible.

(5) *The grid*. Nowadays, Fourier syntheses are calculated by *FFT* (a *Fast Fourier Transform* algorithm, first described by Cooley and Tukey, 1965). A very popular example has been the program written by Ten Eyck (1973); only recently surpassed in performance by the code written by Frigo and Johnson (2005).

In order to reveal the atomic positions, the unit cell is sampled upon a three-dimensional grid. Peak search routines are then used to locate electron density maxima and to print these as a list in order of peak intensity (this last operation is very frequent for small to medium-sized molecules, less frequent for proteins due to lack of resolution). The location of the maxima is achieved using interpolation techniques (usually 19 grid points are used to centre each maximum). The spacing, Δ (in Å), of the grid should be about the same for each of the unit cell axes; e.g. a unit cell with a, b, c equal to 100, 50, 100, respectively, should be divided into $2n, n, 2n$ grid points. If the grid is too coarse, atoms lying between the grid points may show little effect on the surrounding grid points; then, the interpolation necessary to find the maximum electron density may be inaccurate. If the grid is too fine, computing time is wasted.

According to the *Nyquist theorem*, the optimal grid spacing should be chosen as a function of RES; it should be at least equal to RES/2 to preserve the information in the Fourier transform. In practice, spacings in the range RES/2–RES/3 are frequently used. The reader should notice that, for large structures, the number of grid points on which the electron density map is calculated may be extremely large; e.g. for a protein with cubic unit cell with a = 50 Å and $\Delta = 0.33$ Å, the number of points on which the Fourier map is sampled is about 3.37×10^6.

(6) *Contour levels.* To avoid the drawing of non-useful background features, a threshold is frequently applied when graphical programs illustrate electron density. A popular threshold is the standard deviation of the map,

$$\sigma_d = \frac{1}{n}\left\{\sum_{j=1}^{n}[\rho(\mathbf{r}_j) - <\rho>]^2\right\}^{1/2},$$

where n is the number of grid points on which the density is calculated and $<\rho>$ is the average value of the density. If we choose a *k-sigma map* we ask the program to represent only the map features for which $\rho(\mathbf{r}) > k\sigma_d$. For small values of k (e.g. $k < 1$), background details are also illustrated; for large values of k, only the strongest features of the map are revealed. As an example, we show in Fig. 7.12a), b), and c), the electron densities in a region of the unit cell of the protein *1kf3*. The density was obtained by molecular replacement, with an average phase error of 35°; RES is equal to 1.05 Å. Figure 7.12a), b), and c) correspond to $k = 2, 4$, and 5; they show that the continuity of the electron density map progressively breaks down as k increases. At $k = 5$, only a very small fraction of the molecule is included into the density. Conversely, the reader can imagine that if $k = 1$ were chosen, densities without structural meaning might also be present in the figure, which could lead to confusion.

The reader should not, however, forget that σ_d is only the standard deviation of the function 'map'; we will see in Section 7.4 that σ_d is not related to the variance of the map, which has to be calculated using other mathematical tools.

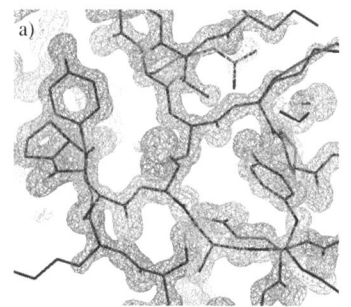
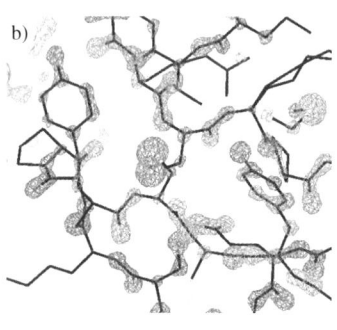
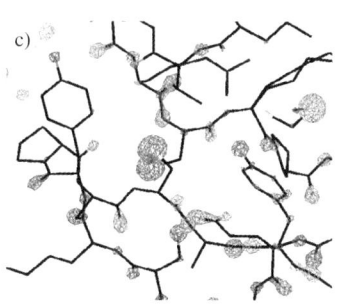

Fig. 7.12
1kf3: electron density corresponding to an average phase error of 35° (RES = 1.05 Å). a) by using the condition $\rho(\mathbf{r}) > 2\sigma_d$; b) by using the condition $\rho(\mathbf{r}) > 4\sigma_d$; c) by using the condition $\rho(\mathbf{r}) > 5\sigma_d$.

7.3.2 The observed Fourier synthesis

The ideal Fourier synthesis (7.9) may be calculated only at the end of a successful structure refinement process, which usually follows the phasing step.

During the phasing procedure, only a model structure is available, say $\rho_p(\mathbf{r})$, of which we know the calculated structure factor moduli (say $|F_{p\mathbf{h}}|$) and the corresponding phases (say $\phi_{p\mathbf{h}}$). So, how do we get information on $\rho(\mathbf{r})$, given the model density $\rho_p(\mathbf{r})$?

The classical recipe is to calculate the synthesis,

$$\rho_{obs}(\mathbf{r}) = \frac{1}{V} \sum_{\mathbf{h}} |F_{\mathbf{h}}| \exp(i\phi_{p\mathbf{h}}) \exp(-2\pi i\mathbf{h} \cdot \mathbf{r}), \quad (7.11)$$

where (7.11) is called the *observed Fourier synthesis*; its coefficients are the observed structure factors of the target structure $|F_{\mathbf{h}}|$ (which depend on all the atomic positions in the target, and therefore can provide information on these), and its phases are the model phases (the best phases we have at that stage of the phasing process). Because of the above properties, it may be expected that ρ_{obs} may provide some additional information on the target atoms, not included into the model structure (i.e. missed atoms). The maxima of ρ_{obs} should indicate the positions of the atoms, including the missed atoms; heavy atoms should correspond to the largest positive peaks and atoms affected by large thermal motion are expected to show broad but less intense peaks. Luzzati (1953) showed that peaks at the sites of missed atoms are at (about) only half the weight of the known peaks.

Since model phases do not coincide with target phases, when the equivalence $\phi = \phi_p$ is accepted, a systematic bias (*model bias*) is introduced in the map. In general, the density (7.11) tends to confirm the model from which phases were calculated, even if the model is only partially correct. It is usual to say that *observed electron density maps are biased towards the model*.

The theory developed in Section 7.2 allows us to design a better synthesis by replacing $\exp(i\phi_{\mathbf{h}})$ with its expected value. According to (7.6) and (7.8),

$$<\exp(i\phi_{\mathbf{h}})> = m\exp(i\phi_{p\mathbf{h}});$$

then (7.11) may be replaced by the *weighted observed Fourier synthesis*,

$$\rho_{obs}(\mathbf{r}) = \frac{1}{V} \sum_{\mathbf{h}} m_{\mathbf{h}} |F_{\mathbf{h}}| \exp(i\phi_{p\mathbf{h}}) \exp(-2\pi i\mathbf{h} \cdot \mathbf{r}). \quad (7.12)$$

A weighted synthesis (7.12) allows a better treatment of the model bias. However, phase errors that are too large do not allow additional information on the target structure to be recovered (e.g. the cyclic application of (7.12) does not lead from random to correct phases), while a moderate average phase error, suitably weighted, allows us to recover unknown atomic positions of the target. For a more practical insight into the use of (7.12), we show in Fig. 7.13a), b), for the small structure *GRA4*, two observed electron densities, the first calculated using unweighted coefficients (see equation (7.11)) and the second by using weighted coefficients (see equation (7.12)). *GRA4* crystallizes in $P\bar{1}$ with two molecules in the asymmetric unit, equivalent by non-crystallographic symmetry. In Fig. 7.13 we show the electron density on the second molecule, given prior knowledge on the atomic positions of the first. The quality of the map (7.12) is clearly higher than that of map (7.11).

If the model is poor, it may be difficult to recover the complete target structure via a single observed electron density map. Cyclic procedures

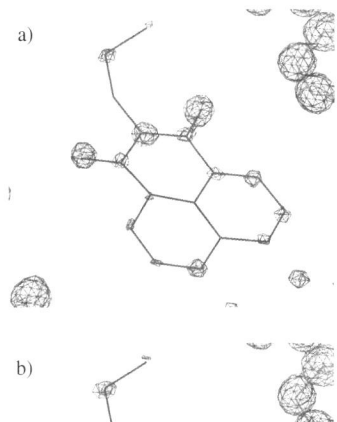

Fig. 7.13
GRA4: electron density on molecule B when the model consists of molecule A (equivalent to B by non-crystallographic symmetry and not shown in figure). a) unweighted coefficients are used (see equation (7.11)); b) weighted coefficients are used (see equation (7.12)).

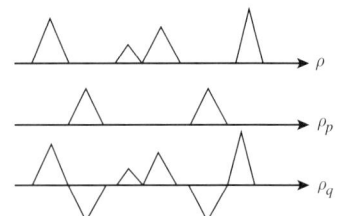

Fig. 7.14
A schematic representation of ρ, ρ_p, and ρ_q when the model is uncorrelated with the target structure (then, $D = 0$).

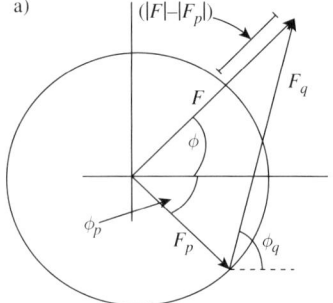

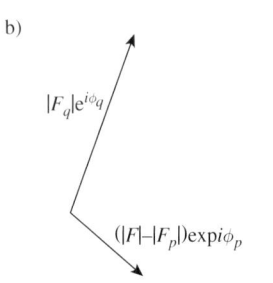

Fig. 7.15
a) A didactical case: F, F_p and F_q, with $|F| > |F_p|$ and ϕ remarkably different from ϕ_p. F_q is the coefficient to use for computing the ideal difference Fourier synthesis (7.14); it is unknown in practice, because ϕ is unknown. b) $(|F| - |F_p|)\exp(i\phi_p)$ is the coefficient to use for computing the synthesis (7.16). The two coefficients $||F_q|\exp(i\phi_q)|$ and $(|F| - |F_p|)\exp(i\phi_p)$ do not fit each other.

would be necessary; these are called EDM (*electron density modification*), and are described in Chapter 8. Reciprocal space approaches are also used (see Section 6.3.6), but are today less popular.

7.3.3 The difference Fourier synthesis

Let us now consider the *ideal difference map*,

$$\rho_q(\mathbf{r}) = \rho(\mathbf{r}) - \rho_p(\mathbf{r}). \tag{7.13}$$

By definition, $\rho_q(\mathbf{r})$ is the density, which summed to $\rho_p(\mathbf{r})$ gives $\rho(\mathbf{r})$, no matter the quality of $\rho_p(\mathbf{r})$. $\rho_q(\mathbf{r})$ may show positive and negative peaks, according to circumstances. In Fig. 7.1, we showed, schematically, a case in which the atomic positions of the model coincide with the corresponding positions of the target structure; no negative peak can be found in the $\rho_q(\mathbf{r})$ map. In Fig. 7.2, we showed a case in which the atomic positions of the model nearly coincide with the corresponding positions of the target structure; $\rho_q(\mathbf{r})$ shows positive and negative peaks. In Fig. 7.14, we show the case in which model and target structures are uncorrelated; again, positive and negative peaks are present in $\rho_q(\mathbf{r})$.

The ideal difference map may be calculated via the synthesis (7.14):

$$\rho_q(\mathbf{r}) = \frac{1}{V} \sum_{\mathbf{h}} |F_{q\mathbf{h}}| \exp(i\phi_{q\mathbf{h}}) \exp(-2\pi i \mathbf{h} \cdot \mathbf{r}), \tag{7.14}$$

where (see Fig. 7.15),

$$|F_{q\mathbf{h}}| \exp(i\phi_{q\mathbf{h}}) = \{|F_{\mathbf{h}}| \exp(i\phi_{\mathbf{h}}) - |F_{p\mathbf{h}}| \exp(i\phi_{p\mathbf{h}})\} \tag{7.15}$$

In vectorial notation, equation (7.15) is equivalent to

$$F_{q\mathbf{h}} = F_{\mathbf{h}} - F_{p\mathbf{h}}.$$

However, if the phases of the target structure are unknown, as is usual during the phasing process, synthesis (7.15) cannot be calculated. If only a model structure is available, as is usual during the phasing process, how do we approximate $\rho_q(\mathbf{r})$? The canonical recipe is to replace $\phi_{\mathbf{h}}$ by its best estimate $\phi_{p\mathbf{h}}$, and then calculate the difference synthesis,

$$\Delta \rho(\mathbf{r}) = \frac{1}{V} \sum_{\mathbf{h}} \left\{ [|F_{\mathbf{h}}| - |F_{p\mathbf{h}}|] \exp(i\phi_{p\mathbf{h}}) \right\} \exp(-2\pi i \mathbf{h} \cdot \mathbf{r}). \tag{7.16}$$

One can deduce from Fig. 7.15 how bad this approximation may be when $\phi_{\mathbf{h}}$ and $\phi_{p\mathbf{h}}$ are far from each other. The approximation is much better if $\phi_{\mathbf{h}}$ and $\phi_{p\mathbf{h}}$ are close (see Fig. 7.16). If this condition is satisfied for most of the reflections, then $\Delta \rho(\mathbf{r})$ will be a good approximation to $\rho_q(\mathbf{r})$; this occurs when model and target structure are well correlated. The first conclusion is:

the difference synthesis (7.16) is a good approximation of the ideal difference synthesis if the quality of the model is sufficiently high.

Let us now study synthesis (7.16) in more detail. $\Delta\rho(\mathbf{r})$ coefficients may be positive or negative, according to whether $|F_\mathbf{h}|$ is larger or smaller than $|F_{p\mathbf{h}}|$. It may be noticed that equation (7.16) is equivalent to (7.17):

$$\Delta\rho(\mathbf{r}) = \frac{1}{V}\sum_\mathbf{h}\left\{||F_\mathbf{h}|-|F_{p\mathbf{h}}||\exp i(\phi_{p\mathbf{h}}+s\pi)\right\}\exp(-2\pi i\mathbf{h}\cdot\mathbf{r}), \quad (7.17)$$

where we set $s = 0$ if $|F_\mathbf{h}| > |F_{p\mathbf{h}}|$, $s = 1$ if $|F_\mathbf{h}| < |F_{p\mathbf{h}}|$. In equation (7.17), the Fourier coefficients are always positive, but their phases are $\phi_{p\mathbf{h}}$ or $\phi_{p\mathbf{h}} + \pi$, according to whether $|F_\mathbf{h}|$ is larger or smaller than $|F_{p\mathbf{h}}|$.

As with the observed Fourier synthesis, the difference synthesis (7.16) is also affected by *model bias*; indeed, the phases used on the right-hand side of (7.17) are ϕ_p or $\phi_p + \pi$, according to circumstances. However, the occurrence of phases $\phi_p + \pi$, in opposition with the model phases ϕ_p, weakens the model bias.

The theory developed in Section 7.2 suggests a more efficient difference synthesis (Read, 1986):

$$\Delta\rho(\mathbf{r}) = \frac{1}{V}\sum_\mathbf{h}\left\{[|mF_\mathbf{h}|-D|F_{p\mathbf{h}}|]\exp i(\phi_{p\mathbf{h}})\right\}\exp(-2\pi i\mathbf{h}\cdot\mathbf{r}). \quad (7.18)$$

With respect to (7.17), two modifications have been introduced: $\exp(i\phi_\mathbf{h})$ has been replaced by its expected value,

$$<\exp(i\phi_\mathbf{h})> \,=\, m\exp(i\phi_{p\mathbf{h}}),$$

and the parameter D has been associated with $|F_{p\mathbf{h}}|$, in order to take into account the misfit between the atomic positions of the current model and the corresponding positions in the target structure ($D = 1$ if the atomic positions of the model perfectly coincide with corresponding atomic positions of the target structure). In terms of normalized structure factors, (7.18) becomes

$$\Delta\rho(\mathbf{r}) = \frac{1}{V}\sum_\mathbf{h}\left\{[|mE_\mathbf{h}|-\sigma_A|E_{p\mathbf{h}}|]\exp i(\phi_{p\mathbf{h}})\right\}\exp(-2\pi i\mathbf{h}\cdot\mathbf{r}). \quad (7.19)$$

So, which type of information provides a difference Fourier synthesis? Positive peaks should correspond to target atoms not included into the model, negative peaks to model atoms wrongly positioned. The quality of the map, however, depends, critically, on the quality of the model; a good map $\Delta\rho(\mathbf{r})$ may be obtained only if the model is sufficiently good. In Fig. 7.17a), b), we show for *BCDIMP* [C_{55} H_{76} N_4 O_{37}, space group $P2_1$, RES = 0.82 Å] two difference syntheses, obtained for average phase errors $<|\Delta\phi|>$, equal to 65° and 22°, respectively. The relative inefficiency of the first map is evident: several positive peaks are present where there is no missed atom present.

Nowadays, there is a new type of difference electron density available, which requires a probabilistic background different from that described in Section 7.2. This synthesis is the basis of the phasing algorithm *VLD* (*vive la difference*) and will be described in the Section 9.3.

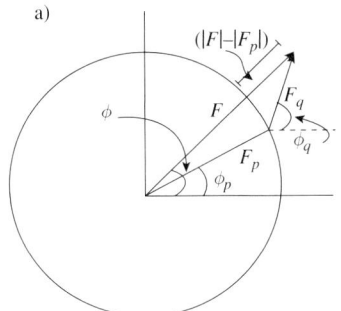

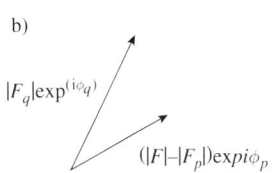

Fig. 7.16
a) A didactical case: F, F_p, and F_q with $|F|>|F_p|$ and ϕ close to ϕ_p. F_q is the coefficient to use for computing the ideal difference Fourier synthesis (7.14): it is unknown in practice, because ϕ is unknown. b) $(|F|-|F_p|)\exp(i\phi_p)$ is the coefficient to use for computing the synthesis (7.16). The two coefficients, $|F_q|\exp(i\phi_q)$ and $(|F|-|F_p|)\exp(i\phi_p)$, fit each other sufficiently well.

7.3.4 Hybrid Fourier syntheses

Let us consider, for any pair of rational numbers τ and ω, the following combination of Fourier syntheses:

$$\rho_Q(\mathbf{r}) = \tau\rho(\mathbf{r}) - \omega\rho_p(\mathbf{r})$$
$$= \frac{1}{V}\sum_{\mathbf{h}}[\tau|F_{\mathbf{h}}|\exp(i\phi_{\mathbf{h}}) - \omega|F_{p\mathbf{h}}|\exp(i\phi_{p\mathbf{h}})]\exp(-2\pi i\mathbf{h}\cdot\mathbf{r}). \quad (7.20)$$

ρ_Q is the *ideal hybrid synthesis*, i.e. the density which, by definition, summed to $\omega\rho_p$, gives $\tau\rho$, no matter the quality of ρ_p. It may be decomposed in different ways in a pair of components maps. For example, as schematized in equation (7.20), the first component may be $\tau\rho$ and the second component may be $\omega\rho_p$, but the decomposition may also be made according to the following rule:

$$\rho_Q = (\tau - \omega)\rho + \omega(\rho - \rho_p) = \omega\rho_q + (\tau - \omega)\rho, \quad \text{if } \tau > \omega, \quad (7.21a)$$
$$\rho_Q = (\tau - \omega)\rho_p + \tau(\rho - \rho_p) = \tau\rho_q - (\omega - \tau)\rho_p, \quad \text{if } \tau < \omega. \quad (7.21b)$$

The equations (7.21) suggest that ρ_Q is a linear combination of the three components ρ, ρ_p, and ρ_q; in particular, ρ_q may have a pre-eminent role when $\omega > \tau$. A schematic example is shown in Fig. 7.18; while the ideal $2\rho - \rho_p$ has strong positive peaks at the target atomic positions not belonging to the model, the ideal $\rho - 2\rho_p$ has strong negative peaks at wrongly positioned model atomic positions.

ρ_Q cannot be computed if the phases of the target structure are unknown, but if a model structure is available, it may be approximated by the *hybrid Fourier synthesis*,

$$\rho_{\tau\omega}(\mathbf{r}) = \frac{1}{V}\sum_{\mathbf{h}}[\tau|F_{\mathbf{h}}| - \omega|F_{p\mathbf{h}}|]\exp(i\phi_{p\mathbf{h}})\exp(-2\pi i\mathbf{h}\cdot\mathbf{r}). \quad (7.22)$$

Equation (7.22) is equivalent to

$$\Delta\rho(\mathbf{r}) = \frac{1}{V}\sum_{\mathbf{h}}\left\{||\tau F_{\mathbf{h}}| - \omega|F_{p\mathbf{h}}||\exp i(\phi_{p\mathbf{h}} + s\pi)\right\}\exp(-2\pi i\mathbf{h}\cdot\mathbf{r}) \quad (7.23)$$

where we set $s = 0$ if $\tau|F_{\mathbf{h}}| > \omega|F_{p\mathbf{h}}|$, $s = 1$ if $\tau|F_{\mathbf{h}}| < \omega|F_{p\mathbf{h}}|$. In equation (7.23), the Fourier coefficients are always positive, but their phases are $\phi_{p\mathbf{h}}$ or $\phi_{p\mathbf{h}} + \pi$, according to whether $\tau|F_{\mathbf{h}}|$ is larger or smaller than $\omega|F_{p\mathbf{h}}|$. As for the other Fourier syntheses, hybrid syntheses also suffer through model bias because the phases may only assume values $\phi_{p\mathbf{h}}$ or $\phi_{p\mathbf{h}} + \pi$. This bias is however different from that present in the observed syntheses, because of the occurrence of phase values $\phi_{p\mathbf{h}} + \pi$ in opposition to the model phases $\phi_{p\mathbf{h}}$.

We observe that:

1. for $\tau = 1$ and $\omega = 0$, $\rho_Q = \rho$, as given by (7.9), and $\rho_{10} = \rho_{obs}$, as given by (7.11);
2. for $\tau = 1$ and $\omega = 1$, $\rho_Q = \rho_q$, as given by (7.14), and $\rho_{11} = \Delta\rho$, as given by (7.18);
3. for $\tau = 2$ and $\omega = 1$, $\rho_Q = \rho + (\rho - \rho_p)$, and $\rho_{21} = \rho_{obs} + \Delta\rho$. Thus, ρ_{21} should show combined features of an observed and of a difference electron density. This interesting property is widely used in EDM approaches (see Chapter 8);

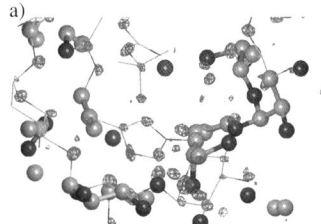

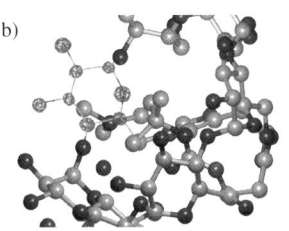

Fig. 7.17
BCDIMP. The model structure is represented by sticks and balls, the missed atoms in simple plotting style. The difference electron density, shown for $\rho(\mathbf{r}) > 3\sigma_d$, is calculated when $<|\Delta\phi|> = 65°1$; The difference electron density, shown for $\rho(\mathbf{r}) > 3\sigma_d$, is calculated when $<|\Delta\phi|> = 22°$.

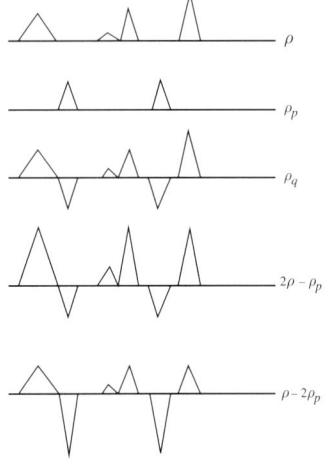

Fig. 7.18
A schematic example showing ρ, ρ_p, ρ_q (first three lines), ρ_Q when $\tau = 2$ and $\omega = 1$ (line 4), and ρ_Q when $\tau = 1$ and $\omega = 2$ (line 5).

4. for $\tau = 1$ and $\omega = 2$, $\rho_Q = -\rho_p + (\rho - \rho_p)$, and $\rho_{12} = -\rho_p + \Delta\rho$. This hybrid synthesis shows features belonging to a difference map plus density in opposition to the model.

We will see in Chapter 9 that hybrid syntheses may be conveniently used in the phasing process when the *VLD* algorithm is applied.

As for observed and difference syntheses, probabilistic criteria may be used to optimize the efficiency of hybrid syntheses. The coefficient suggested by Read (2001) in terms of normalized structure factors is

$$(\tau m R - \omega \sigma_A R_p) \exp(i\phi_p),$$

according to which,

$$\rho_{\tau\omega}(\mathbf{r}) = \frac{1}{V} \sum_{\mathbf{h}} [(\tau m |F| - \omega D |F_p|] \exp(i\phi_{p\mathbf{h}}) \exp(-2\pi i \mathbf{h} \cdot \mathbf{r}) \quad (7.24)$$

Finally, it may be useful to notice that:

(i) $\rho_{k\tau,k\omega}(\mathbf{r}) \equiv \rho_{\tau,\omega}(\mathbf{r})$. Accordingly, the properties of hybrid syntheses are fixed by the ratio τ/ω, rather than by the values of τ and ω; therefore, the most interesting hybrid syntheses are those of low order (i.e. defined by small values of τ and ω).

(ii) While in an observed Fourier synthesis always $\phi_\mathbf{h} \approx \phi_{p\mathbf{h}}$, in a hybrid Fourier synthesis there is always a subset of reflections for which it has been assumed that $\phi_Q = \phi_p + \pi$ (see Table 7.1). Therefore, hybrid and observed Fourier syntheses should never be confused.

(iii) With increasing values of the ratio τ/ω, hybrid Fourier syntheses become more and more similar to the observed synthesis; indeed, the percentage of reflections for which $\phi_\mathbf{h} \approx \phi_{p\mathbf{h}} + \pi$ diminishes with the ratio τ/ω (see Table 7.1). If the model is badly correlated with the target, $\rho_{\tau\omega}$ will also be badly correlated with ρ_Q.

(iv) For a hybrid Fourier synthesis with $\tau < \omega$, the number of reflections for which $\phi_\mathbf{h} \approx \phi_{p\mathbf{h}} + \pi$ is larger than for a hybrid Fourier synthesis with $\tau > \omega$ (indeed, the number of reflections for which $|F| < \frac{\omega}{\tau m} D|F_p|$ or $|F| < \frac{\omega}{\tau} D|F_p|$ is larger, see Table 7.1). As a consequence, a hybrid density map calculated with $\tau < \omega$ will differ from an observed synthesis more than a map with $\tau > \omega$.

Table 7.1 The assigned phase value (say ϕ) for each type of hybrid Fourier synthesis

Type	ϕ
$m|F| \exp(i\phi_p)$	$\phi = \phi_p$ always
$(m|F| - D|F_p|) \exp(i\phi_p)$	$\phi = \phi_p + \pi$ if $|F| < D|F_p|/m$, otherwise $\phi = \phi_p$
$(2m|F| - D|F_p|) \exp(i\phi_p)$	$\phi = \phi_p + \pi$ if $|F| < D|F_p|/(2m)$, otherwise $\phi = \phi_p$
$(\tau m|F| - \omega D|F_p|) \exp(i\phi_p)$	$\phi = \phi_p + \pi$ if $|F| < \frac{\omega}{\tau m} D|F_p|$, otherwise $\phi = \phi_p$

There is a new type of hybrid Fourier synthesis available today, which requires a probabilistic background different from that described in Section 7.2. We will consider this new synthesis and its applications in Section 9.3.

7.4 Variance and covariance for electron density maps

The density values at the grid points of an electron density map do not confirm confidence in the quality of the map; in other words, simple inspection of the map does not indicate whether the peak distribution fits the target structure. Indeed, a trivial change in scale or weight of the structure factors may increase or decrease the density values, without improving the map. As specified in point 6 of Section 7.3.1, it is usual, particularly in protein crystallography, to calculate the standard deviation σ_d of the pixel intensity distribution; it is tacitly assumed that pixels with $\rho(\mathbf{r}) > n\sigma_d$, and n sufficiently large, provide the most reliable information on atomic positions in the unit cell. The σ_d criterion, however, is unable to provide absolute confidence in the map; i.e. choosing peaks with $\rho(\mathbf{r}) > 5\sigma_d$ does not necessarily imply that such peaks coincide with the target atomic positions, if the model map is uncorrelated or poorly correlated with the target map. Therefore, the search for an absolute indicator has to be made among other tools.

The expected variance of an electron density map was first estimated by Cruickshank (1949) as

$$\sigma^2(\rho) = \frac{1}{V^2} \sum_{\mathbf{h}} \sigma^2(|F_{\mathbf{h}}|), \qquad (7.25)$$

where $\sigma^2(|F_{\mathbf{h}}|)$ is the variance of the observed amplitude. Equation (7.25) provides a global error, constant for any point on the map, depending only on the measurement error, and completely disregarding phase error. More interesting results were obtained by Coppens and Hamilton (1968), who obtained a variance expression, variable according to the point on the map:

$$\sigma^2\{\rho(\mathbf{r})\} = \frac{4}{V^2} \sum_{\mathbf{h}>0} \sigma^2(|F_{\mathbf{h}}|) \cos^2(2\pi \mathbf{h} \cdot \mathbf{r} - \phi_{\mathbf{h}}). \qquad (7.26)$$

Both equations (7.25) and (7.26) uniquely depend on measurement error. The reason for this is as follows. The authors were mainly interested in maps obtained in the final stages of crystal structure refinement (when the phases are considered to be perfectly determined, and therefore fixed), in order to assess the reliability of the conclusive structural parameters. During the phasing process, it is much more interesting to calculate the variance at a point in the map when the phases are uncertain; in other words, when a model, well or badly correlated with the target function, is available. Giacovazzo and Mazzone (2011) and Giacovazzo et al. (2011) calculated, for any space group, the variance of any point in an electron density map (observed, difference, or hybrid) by assuming that each phase $\phi_{\mathbf{h}}$ is distributed around $\phi_{p\mathbf{h}}$ according to the von Mises distribution (7.6). This assumption may be applied over quite a wide range, from the case in which the model is uncorrelated with the structure (which is equivalent to assuming $X \approx 0$ for most reflections), to the limit case in which the model coincides with the structure (X quite large for almost all of the reflections).

Because of the orthogonality between phase uncertainty (depending on the model) and measurement error (depending on the experiment), unique expressions of the variance, taking into account both phase and measurement errors, have been derived for any space group. The final expressions of the variance are shown in Appendix 7.C.

Giacovazzo and Mazzone (2012) showed how the *FFT* algorithm may be used for calculation of the variance and demonstrated that the ratio

$$\rho(\mathbf{r})/\sigma_\rho(\mathbf{r}), \qquad (7.27)$$

where $\sigma_\rho(\mathbf{r}) = [\mathrm{var}\rho(\mathbf{r})]^{1/2}$, is an absolute figure for estimating the confidence one should have in the density at a given point $\mathbf{r}$ on the map. This may be considered to be a 'signal to noise' ratio; when it is large at a point, the corresponding density may be considered a real feature of the target structure. In Fig. 7.19, for *BCDIMP*, the distribution of the ratio (7.27) is shown for two different model structures: for the first model, $<|\Delta\phi|> = 75°$, for the second, $<|\Delta\phi|> = 22°$. The curves suggest that, when the better model is used, (7.27) attains large values for larger percentages of map pixels (indeed, the corresponding map is more reliable).

Let us now deal with the covariance between two points on an electron density map. The crystal structure solution process usually ends with a set of atomic parameters, published together with the corresponding standard deviations, which fix the accuracy of the structural results. Least squares are considered to be a fundamental tool in crystallography, just because they provide optimal values of the structural parameters, of their variances, and of their covariances; least squares are usually inefficient if covariance between parameters is high.

Covariance between two points on an electron density map is a signal to investigate whether the electron density at one point of the map is amplified or depressed as an effect of the electron density at another point on the map. For example, pseudotranslational symmetry is a frequent source of covariance; if the pseudotranslational vector is equal to $\mathbf{u} = (\mathbf{a} + \mathbf{b})/2$, we should observe a

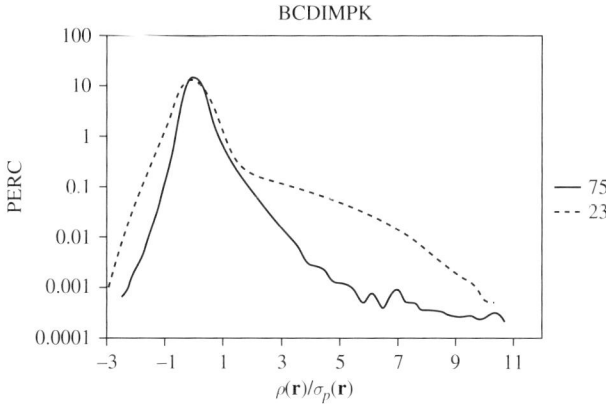

Fig. 7.19
BCDIMP: pixel percentage (*PERC*) with a given $\rho(\mathbf{r})/\sigma_\rho(\mathbf{r})$ value for two different average phase errors, $<|\Delta\phi|>$.

particularly strong correlation between any point **r** on the electron density map and the position **r** + **u**, if only (or most) reflections with h + l = even are used to compute the map.

An unavoidable covariance effect concerns peaks and their associated ripples (see point 4 of Section 7.3.1); indeed, the larger the peak, the larger will be the ripples. Mathematical formulas for calculating covariances between peak densities at any step of the phasing process have recently been described by Altomare et al. (2012) and are detailed in Appendix 7.B. In some way, covariances in the map are the counterpart in real space of least squares covariance calculations; large covariances should indicate undesired features in the map.

A figure normalized to the interval (0, 1) is the correlation coefficient between the densities in two points on the map:

$$C(\mathbf{r}_A, \mathbf{r}_B) = \frac{\text{cov}(\mathbf{r}_A, \mathbf{r}_B)}{\text{var}^{1/2}[\rho(\mathbf{r}_A)]\text{var}^{1/2}[\rho(\mathbf{r}_B)]}.$$

7.5 Triplet phase estimate when a model is available

In Section 7.2 we derived the distribution $P(E, E_p)$ for when a model structure is available. Here, we will derive a more complicated distribution, say

$$P(E_\mathbf{h}, E_\mathbf{k}, E_{\mathbf{h}+\mathbf{k}}, E_{p\mathbf{h}}, E_{p\mathbf{k}}, E_{p,-\mathbf{h}-\mathbf{k}}),$$

in order to estimate the triplet invariant reliability when a model is available.

In accordance with Burla et al. (2012b), we will suppose that a model structure is available, and that:

(i) $\mathbf{r}_{pj}$, $j = 1, \ldots, N_p$ are the model atomic positions. They are considered to be primitive random variables;
(ii) $\mathbf{r}_j = \mathbf{r}_{pj} + \Delta \mathbf{r}_j$, $j = 1, \ldots, N_p$ are the corresponding atomic positions in the target structure. They are considered to be random variables riding over the corresponding $\mathbf{r}_{pj}$;
(iii) The rest of the target atomic positions, say $\mathbf{r}_j$, $j = N_p + 1, \ldots, N$, are primitive random variables.

Then,

$$E = \left\{ \sum_{j=1}^{N_p} f_j \exp[2\pi \mathbf{h} \cdot (\mathbf{r}_{pj} + \Delta \mathbf{r}_j)] + \sum_{j=N_p+1}^{N} f_j \exp(2\pi \mathbf{h} \cdot \mathbf{r}_j) \right.$$
$$\left. + |\mu| \cos \vartheta) \right\} \bigg/ (\varepsilon \Sigma_N)^{1/2},$$

$$E_p = \sum_{j=1}^{N_p} f_j \exp(2\pi \mathbf{h} \cdot \mathbf{r}_{pj}) \bigg/ (\varepsilon \Sigma_p)^{1/2},$$

where $|\mu| \exp(i\vartheta)$ takes into account measurement errors.

To simplify the resulting formulas, we will use the following notation:
$E_i = E_\mathbf{h}, E_\mathbf{k}, E_{-\mathbf{h}-\mathbf{k}}$, respectively, for $i = 1, \ldots, 3$,
$E_{pi} = E_{p\mathbf{h}}, E_{p\mathbf{k}}, E_{p,-\mathbf{h}-\mathbf{k}}$, respectively, for $i = 1, \ldots, 3$;

$$R_i = |E_i|, \quad R_{pi} = |E_{pi}|, \quad i = 1, \ldots 3.$$

The required joint probability distribution is

$$P(R_i, R_{pi}, \phi_i, \phi_{pi}, i=1,\ldots,3) = \prod_{i=1}^{3}\left[\frac{R_i R_{pi}}{(e_i - \sigma_{Ai}^2)}\right]$$

$$\exp\left\{-\sum_{i=1}^{3}\frac{1}{(e_i - \sigma_{Ai}^2)}[R_i^2 + e_i R_{pi}^2 - 2\sigma_{Ai}R_i R_{pi}\cos(\phi_i - \phi_{pi})]\right\}$$

$$\exp\left\{\frac{2}{\sqrt{N_{p,eq}}}R_{p1}R_{p2}R_{p3}\cos(\phi_{p1}+\phi_{p2}+\phi_{p3})\right.$$
$$+ 2\beta[R_1 R_2 R_3 \cos(\phi_1+\phi_2+\phi_3) - \sigma_{A3}R_1 R_2 R_{p3}\cos(\phi_1+\phi_2+\phi_{p3})$$
$$- \sigma_{A2}R_1 R_{p2} R_3 \cos(\phi_1+\phi_{p2}+\phi_3) - \sigma_{A1}R_{p1}R_2 R_3 \cos(\phi_{p1}+\phi_2+\phi_3)$$
$$+ \sigma_{A2}\sigma_{A3}R_1 R_{p2}R_{p3}\cos(\phi_1+\phi_{p2}+\phi_{p3})$$
$$+ \sigma_{A1}\sigma_{A3}R_{p1}R_2 R_{p3}\cos(\phi_{p1}+\phi_2+\phi_{p3})$$
$$+ \sigma_{A1}\sigma_{A2}R_{p1}R_{p2}R_3\cos(\phi_{p1}+\phi_{p2}+\phi_3)$$
$$\left.- \sigma_{A1}\sigma_{A2}\sigma_{A3}R_{p1}R_{p2}R_{p3}\cos(\phi_{p1}+\phi_{p2}+\phi_{p3})]\right\}, \quad (7.28)$$

where

$$\beta = \frac{1}{(e_1 - \sigma_{A1}^2)(e_2 - \sigma_{A2}^2)(e_3 - \sigma_{A3}^2)}\frac{1}{\sqrt{N_{q,eq}}}$$

and

$$\frac{1}{\sqrt{N_{q,eq}}} = \left(\frac{1}{\sqrt{N_{eq}}} - \frac{1}{\sqrt{N_{p,eq}}}\sigma_{A1}\sigma_{A2}\sigma_{A3}\right).$$

$N_{q,eq}$ may be considered to be the *equivalent number of atoms in the difference structure*.

Equation (7.28) may be usefully compared with equation (5.3); while this only exploits the triplet $\Phi = (\phi_1 + \phi_2 + \phi_3)$, equation (7.28) is able to obtain information from the eight triplets belonging to the first representation of the invariant (see Section 4.4).

From equation (7.28), standard mathematical techniques lead to the conditional distribution,

$$P(\Phi|R_1, R_2, R_3, R_{p1}, R_{p2}, R_{p3}) \approx [2\pi I_0(G)]^{-1}\exp[G\cos\Phi] \quad (7.29)$$

where

$$G = \frac{2}{\sqrt{N_{p,eq}}}m_1 m_2 m_3 R_{p1}R_{p2}R_{p3} \quad (7.30)$$
$$+ 2\beta\left[(R_1 - \sigma_{A1}m_1 R_{p1})(R_2 - \sigma_{A2}m_2 R_{p2})(R_3 - \sigma_{A3}m_3 R_{p3})\right].$$

Equations (7.29) and (7.30) suggest the following conclusion: if a model is available which is (weakly or strongly) correlated with the target structure, the triplet phase Φ is no longer expected to always be distributed around zero, as suggested by the Cochran formula. The percentage of triplet phases close to π depends on the correlation between target and model structure.

To better understand equations (7.29) and (7.30), we simplify them by assuming $e_i = 1$ for $i = 1,\ldots,3$; this is a reasonable assumption when the diffraction amplitudes are sufficiently large. We will consider two extreme cases.

1. Target and model structures are completely uncorrelated. In this case, $D_i = 0$ for $i = 1,2,3$,

$$\sigma_{Ai} \approx m_i \approx 0, \text{ for i} = 1,\ldots,3, \quad \beta = \frac{1}{\sqrt{N_{eq}}}, \text{ and } G = \frac{2}{\sqrt{N_{eq}}} R_1 R_2 R_3.$$

In simple terms, and in accordance with expectations, an uncorrelated model does not provide any supplementary information to that contained in the Cochran formula.

2. Target and model structures are well correlated; i.e. the N_p atomic positions of the model perfectly coincide with the corresponding atomic positions in the target structure and the scattering power of the model structure is non negligible with respect to that of the target. Then, for most of the reflections (i.e. those with largest intensity), $D_i = m_i \sim 1$ for $i = 1,2,3$. In this case, the number of atoms in the difference structure is properly defined by $N_q = N - N_p$. Furthermore, the scattering power, $\sum_q$, of the difference structure reduces to $\sum_{qi} = \sum_{Ni} - \sum_{pi}$, for $i = 1,\ldots,3$. Under these conditions, we obtain

$$\sigma_{Ai} = \left(\sum\nolimits_{pi} / \sum\nolimits_{Ni}\right)^{1/2}, \quad \beta = \frac{1}{\sqrt{N_{q,eq}}} \left(\frac{\sum_{N1} \sum_{N2} \sum_{N3}}{\sum_{q1} \sum_{q2} \sum_{q3}}\right)^{1/2}$$

and

$$G = \frac{2}{\sqrt{N_{p,eq}}} R_{p1} R_{p2} R_{p3} + \frac{2}{\sqrt{N_{q,eq}}} \left[\frac{(|F_1| - |F_{p1}|)(|F_2| - |F_{p2}|)(|F_3| - |F_{p3}|)}{(\sum_{q1} \sum_{q2} \sum_{q3})^{1/2}}\right]. \tag{7.31}$$

Equation (7.31) suggests that the second terms on the right-hand side may be dominant with respect to the first term. Thus, triplet phase estimates should not be based on the Cochran formula, but may benefit from the supplementary information provided by the current model, even if it is weakly correlated with the target. Expression (7.30) is therefore the general relation to use; it requires prior knowledge of the σ_{Ai} parameters, which in turn may be estimated by standard statistical methods (see Appendix 7.A).

From distribution (7.28), the conditional distribution

$$P\left(\phi_1 | R_1, R_2, R_3, R_{p1}, R_{p2}, R_{p3}, \phi_2, \phi_3, \phi_{p1}, \phi_{p2}, \phi_{p3}\right)$$

may be obtained by standard techniques. We obtain,

$$P(\phi_1 | \ldots\ldots) = [2\pi I_0(G_1)]^{-1} \exp(G_1 \cos\phi_1 - \xi_1) \tag{7.32}$$

where $G_1^2 = a_1^2 + a_2^2$,

$$a_1 = 2R_1 \left\{ \frac{\sigma_{A1}}{e_1 - \sigma_{A1}^2} R_{p1} \cos\phi_{p1} + \beta[R_2 R_3 \cos(\phi_2 + \phi_3) \right.$$
$$\left. - \sigma_{A3} R_2 R_{p3} \cos(\phi_2 + \phi_{p3}) - \sigma_{A2} R_{p2} R_3 \cos(\phi_{p2} + \phi_3) \right.$$
$$\left. + \sigma_{A2}\sigma_{A3} R_{p2} R_{p3} \cos(\phi_{p2} + \phi_{p3})] \right\}$$

$$a_2 = 2R_1 \left\{ \frac{\sigma_{A1}}{e_1 - \sigma_{A1}^2} R_{p1} \sin\phi_{p1} + \beta[-R_2R_3 \sin(\phi_2 + \phi_3) \right.$$
$$+ \sigma_{A3}R_2R_{p3}\sin(\phi_2 + \phi_{p3}) + \sigma_{A2}R_{p2}R_3\sin(\phi_{p2} + \phi_3)$$
$$\left. - \sigma_{A2}\sigma_{A3}R_{p2}R_{p3}\sin(\phi_{p2} + \phi_{p3})] \right\},$$
$$\tan\xi_1 = a_2/a_1.$$

If model and target structures are strongly correlated, distribution (7.32) reduces to that found by Giacovazzo (1983a) and applied by Burla et al. (1989b). If model and target structure are uncorrelated (i.e. $\sigma_{Ai} \approx m_i \approx 0$, for $i = 1,\ldots,3$), then

$$\beta = \frac{1}{\sqrt{N_{eq}}}, \quad a_1 = \frac{2}{\sqrt{N_{eq}}} R_1R_2R_3 \cos(\phi_2 + \phi_3),$$
$$a_2 = -\frac{2}{\sqrt{N_{eq}}} R_1R_2R_3 \sin(\phi_2 + \phi_3),$$

in agreement with the Cochran formula. The expressions for a_1 and a_2 are much more useful if correlation between target and model structures is not vanishing.

APPENDIX 7.A   ESTIMATION OF σ_A

The parameter D as given by equation (7.5) is resolution dependent; consequently, σ_A is also resolution dependent. Luzzati (1952), Lunin and Urzhumtsev (1984), and Read (1986) contributed to clarification of the role of this parameter in the phasing process and to the design of an algorithm for its estimation. Caliandro et al. (2005c) showed that σ_A, in an acentric space group, may be estimated from the average

$$<R^2R_p^2> = \int_0^\infty \int_0^\infty R^2R_p^2 P(R,R_p)dRdR_p = (e + \sigma_A^2),$$

from which,

$$\sigma_A^2 = <R^2R_p^2> - e \qquad (7.A.1)$$

Giacovazzo et al. (2011) stated that any moment, $<R^s R_p^t>$, may be used for estimating σ_A, with s and t being any pair of integer numbers; for example, in acentric space groups, by choosing $s = t = 1$,

$$<RR_p> = \frac{\pi}{4}\sqrt{e}\left[1 + \left(\frac{4}{\pi} - 1\right)\frac{\sigma_A^2}{e}\right]$$

is derived. Different formulas are obtained if the space group is centric, where, for example,

$$<R^2R_p^2> = e + 2\sigma_A^2,$$

from which,

$$\sigma_A^2 = \frac{1}{2}\left(<R^2R_p^2> - e\right). \qquad (7.A.2)$$

The parameter σ_A has quite interesting statistical properties. Let us consider equation (7.A.1) under the assumption (valid in most cases of interest) that e is very close to unity and that the observed and calculated diffraction amplitudes satisfy Wilson distributions. Then,

(i) if $\sigma_A = 1$, model and target structures coincide, and $<R^2 R_p^2> = <R^4> = 2$, the value expected by Wilson statistics for acentric crystals;
(ii) if $\sigma_A = 0$, then $P(E)$ and $P(E_p)$ are uncorrelated, and $<R^2 R_p^2> = <R^2><R_p^2> = 1$.

The same properties hold when the structure is centric and equation (7.A.2) is taken into consideration. Then, one has only to remember that, when $\sigma_A = 1$,

$$<R^2 R_p^2> = <R^4> = 3.$$

The following conclusion arises: σ_A is a useful statistical indicator for estimating how much the current model is correlated with the target structure. Its values may be estimated statistically in a very simple way (routines are in the most popular packages, such as *CCP4*, *PHENIX*, *SIR2011*); measured reflections are partitioned in resolution shells, and for each shell, the joint moments, $<R^s R_p^t>$, are calculated. Very frequently the locally normalized quantities

$$\frac{<|FF_p|^2>}{<|F|^2><|F_p|^2>} \quad \text{or} \quad \frac{<R^2 R_p^2>}{<R^2><R_p^2>} \tag{7.A.3}$$

are taken into account, instead of the simple moments, $<|FF_p|^2>$ or $<R^2 R_p^2>$.

Very recently (Carrozzini et al., 2013b), a more satisfactory statistical interpretation of σ_A^2 has been suggested. σ_A^2 is nothing else but the correlation factor C between the R^2 and the R_p^2 sets:

$$\sigma_A^2 = C(R^2, R_p^2) = \frac{<R^2 R_p^2> - <R^2><R_p^2>}{\left(<R^4> - <R^2>^2\right)^{1/2}\left(<R_p^4> - <R_p^2>^2\right)^{1/2}} \tag{7.A.4}$$

Indeed:

(a) in accordance with its original definition in direct space, σ_A^2 always lies in the interval (0,1).
(b) equation (7.A.4) does not change with the centric or acentric nature of the crystal, and reduces to the equations (7.A.1) or (7.A.2) if the structure factor amplitudes satisfy the Wilson distribution.
(c) rescaling of the observed and/or the calculated amplitudes shell by shell, as described by equation (7.A.3), is no longer necessary because the correlation coefficient is scale independent.

APPENDIX 7.B VARIANCE AND COVARIANCE EXPRESSIONS FOR ELECTRON DENSITY MAPS

The variance expression obtained by Giacovazzo et al. (2011) may be divided into three components:

Variance and covariance expressions for electron density maps

(i) a term (say TH_1) which does not vary from point to point, but depends on the global misfit between model and target structures. It fixes the average variance of the map.

(ii) a second component (say TH_2), varying from point to point, strictly connected with the *implication transformations* (a quite basic tool for Patterson deconvolution: see Chapter 10). In some way, TH_2 'knows' the theory of the implication transformation.

(iii) A third term (say TD), varying from point to point, depending on the model phases and on the observed amplitudes. Its main task is to fix, via weights depending on the misfit between model and target structures, the variance at points related to the model structure.

The final expression is

$$\text{var}\,\rho(\mathbf{r}) = TH_1 + TH_2(\mathbf{r}) + TD(\mathbf{r}),$$

where

$$TH_1 = \frac{2}{V^2} \sum_{\mathbf{h}>0} (1 - m_{\mathbf{h}}^2)[|F_{\mathbf{h}}|^2 + \sigma^2(|F_{\mathbf{h}}|)]$$

$$TH_2(\mathbf{r}) = \frac{2}{V^2} \sum_{\mathbf{h},ind} [|F_{\mathbf{h}}|^2(1 - m_{\mathbf{h}}^2) + \sigma^2(|F_{\mathbf{h}}|)] \sum_{s \neq q=1}^{n} \cos\{2\pi \mathbf{h}[(\mathbf{C}_s - \mathbf{C}_q)\mathbf{r}]\}$$

$$TD(\mathbf{r}) = -\frac{2}{V^2} \sum_{\mathbf{h},ind} \Big\{ [|F_{\mathbf{h}}|^2[m_{\mathbf{h}}^2 - D_2(X_{\mathbf{h}})] - \sigma^2(|F_{\mathbf{h}}|)]$$

$$\sum_{s,q=1}^{n} \cos[2\phi_p(\mathbf{h}) - 2\pi \mathbf{h}(\mathbf{C}_s + \mathbf{C}_q)\mathbf{r}] \Big\}.$$

$\mathbf{C}_s$, $s = 1, \ldots, n$ are the symmetry operators of the space group; the subscript *ind* in the symbol of summation indicates that $\mathbf{h}$ varies over the set of symmetry-independent reflections.

For centric space groups, the corresponding expression is

$$\text{var}\,\rho(\mathbf{r}) = \frac{1}{V^2} \sum_{\mathbf{h},ind} [|F_{\mathbf{h}}|^2(1 - m_{c\mathbf{h}}^2) + \sigma^2(|F_{\mathbf{h}}|)] \sum_{s,q=1}^{n} \cos[2\pi \mathbf{h}(\mathbf{C}_s - \mathbf{C}_q)\mathbf{r}].$$

Giacovazzo and Mazzone (2012) described an algorithm for calculation of the variance via FFT.

We list here, the covariance expression between the two points $\mathbf{r}_A$ and $\mathbf{r}_B$:

$$<\rho(\mathbf{r}_A)\rho(\mathbf{r}_B)> = T_{obs} + T_{1a} + T_{2a},$$

where

$$T_{obs}(\mathbf{r}_A, \mathbf{r}_B) = \rho_{obs}(\mathbf{r}_A)\rho_{obs}(\mathbf{r}_B),$$

$$T_{1a}(\mathbf{r}_A, \mathbf{r}_B) = 2V^{-2} \sum_{\mathbf{h},ind} |F_{\mathbf{h}}|^2 (1 - m_{\mathbf{h}}^2) \sum_{s,q=1}^{n} \cos[2\pi \mathbf{h}(\mathbf{C}_s \mathbf{r}_A - \mathbf{C}_q \mathbf{r}_B)],$$

$$T_{2a}(\mathbf{r}_A, \mathbf{r}_B) = -2V^{-2} \sum_{\mathbf{h},ind} |F_{\mathbf{h}}|^2 [m_{\mathbf{h}}^2 - D_2(X_{\mathbf{h}})] \sum_{s,q=1}^{n} \cos[2\phi_p(\mathbf{h}) - 2\pi \mathbf{h}(\mathbf{C}_s \mathbf{r}_A + \mathbf{C}_q \mathbf{r}_B)].$$

APPENDIX 7.C SOME MARGINAL AND CONDITIONAL PROBABILITIES OF $P(R, R_p, \phi, \phi_p)$

By integrating (7.3) over ϕ and ϕ_p, the following marginal distribution is derived:

$$P(R, R_p) = \frac{4RR_p}{(e - \sigma_A^2)} \exp\left\{-\frac{1}{(e - \sigma_A^2)}\left[R^2 + eR_p^2\right]\right\} I_0[X], \qquad (7.C.1)$$

where I_0 is the modified Bessel function of order zero and X is defined by equation (7.7). From (7.C.1), the conditional distributions,

$$P(R|R_p) = \frac{2R}{(e - \sigma_A^2)} \exp\left\{-\frac{\left[R^2 + \sigma_A^2 R_p^2\right]}{(e - \sigma_A^2)}\right\} I_0[X], \qquad (7.C.2)$$

and

$$P(R_p|R) = \frac{2eR_p}{(e - \sigma_A^2)} \exp\left\{-\frac{\left[e^2 R_p^2 + \sigma_A^2 R^2\right]}{e(e - \sigma_A^2)}\right\} I_0[X] \qquad (7.C.3)$$

may be derived. Distribution (7.C.2) is useful when one wants to estimate R given R_p and distribution (7.C.3) may help when one wants to estimate R_p given R.

Phase improvement and extension

8

8.1 Introduction

The descriptions of the various types of Fourier synthesis (observed, difference, hybrid) and of their properties, given in Chapter 7, suggest that electron density maps are not only a tool for depicting the distribution of the electrons in the target structure, but also a source of information which may be continuously exploited during the phasing process, no matter whether *ab initio* or non-*ab initio* methods were used for deriving the initial model. Here, we will describe two important techniques based on the properties of electron density maps.

(i) The recursive approach for phase extension and refinement called EDM (*electron density modification*). Such techniques have dramatically improved the efficiency of phasing procedures, which usually end with a limited percentage of phased reflections and non-negligible phase errors. EDM techniques allow us to extend phase assignment and to improve phase quality. The author is firmly convinced that practical solution of the phase problem for structures with N_{asym} up to 200 atoms in the asymmetric unit may be jointly ascribed to direct methods and to EDM techniques.

(ii) The AMB (*automated model building*) procedures; these may be considered to be partly EDM techniques and they are used for automatic building of molecular models from electron density maps. Essentially, we will refer to proteins; the procedures used for small to medium-sized molecules have already been described in Section 6.3.5.

Two new ab initio phasing approaches, *charge flipping* and *VLD*, essentially based on the properties of the Fourier transform, belong to the EDM category, and since they require a special treatment, they will be described later in Chapter 9.

8.2 Phase extension and refinement via direct space procedures: EDM techniques

Phase extension and refinement may be performed in reciprocal and in direct space. We described the former in Section 6.3.6; here, we are just interested in direct space procedures, the so-called EDM (electron density modification) techniques. Such procedures are based on the following hypothesis: a poor electron density map, ρ, may be modified by a suitable function, f, to obtain a new map, say ρ_{mod}, which better approximates the true map:

$$\rho_{\text{mod}}(\mathbf{r}) = f[\rho(\mathbf{r})]. \tag{8.1}$$

If function f is chosen properly, more accurate phases can be obtained by Fourier inversion of ρ_{mod}, which may in turn be used to calculate a new electron density map. This map may be submitted to the function f, and so on, cyclically. Such a procedure should allow a model map to converge to the target structure, provided that f is chosen in a suitable way.

The reader should have noticed the first important EDM feature: modified electron density maps are Fourier inverted, a molecular model is not needed. Accordingly, the method may also work even if RES is far from the atomic resolution.

This approach was first proposed by Hoppe and Gassmann (1968) (see also Hoppe et al., 1970; Gassmann and Zechmeister, 1972; Gassmann, 1975, 1976, 1977; Simonov, 1976) who suggested the function,

$$\rho_{\text{mod}}(\mathbf{r}) = f[\rho(\mathbf{r})] = a\rho(\mathbf{r}) + b\rho^2(\mathbf{r}) + c\rho^3(\mathbf{r}), \tag{8.2}$$

where the parameters a, b, and c are chosen to sharpen strong peaks and weaken the smallest ones. The modification (8.2) is no longer in use today; more complicated, and often non-analytical, functions are necessary to face modern challenges.

In order to identify a better approach, we should answer the following question: what do we really want to obtain when we apply (8.1)? The practical answer should be: we want to introduce a virtuous distortion of the current electron density which, after Fourier inversion, should lead to improved phase values. To obtain such a result we should introduce into ρ some features which we know are present in high quality electron density maps; when better features are introduced, more information may be exploited. Below in this section, we will list and discuss the 'good features' exploited by modern EDM techniques.

Here, we address the reader's attention to another important capability of EDM procedures: *phase extension*. Let us suppose that only a subset of structure factors has been phased by some phasing procedure (e.g. if direct methods are used, only the reflections with the largest R values are phased; if molecular replacement has been employed, only structure factors up to a given resolution limit are usually phased; etc.) and that ρ_{obs} is the corresponding observed electron density map (e.g. calculated via equation (7.12)). If ρ_{obs} is Fourier inverted (without any modification), the structure factors F_{inv} are obtained, where

$$|F_\mathbf{h}|_{inv} = |F_\mathbf{h}|, \ \phi_{\mathbf{h}inv} = \phi_\mathbf{h} \text{ if } \mathbf{h} \text{ was used for the calculation of } \rho_{obs}; \tag{8.3a}$$

$$|F_\mathbf{h}|_{inv} \equiv 0, \text{ if } \mathbf{h} \text{ was not used for the calculation of } \rho_{obs}. \tag{8.3b}$$

In relation (8.3a), $|F_\mathbf{h}|$ is the observed amplitude of the target structure and $\phi_\mathbf{h}$ is the phase value used for computing the observed Fourier synthesis. The results (8.3) arise from intrinsic properties of Fourier transforms.

However, if ρ_{obs} is modified before being Fourier inverted, (8.3a) and (8.3b) are no longer valid. In this case, phases may be assigned to higher resolution observed reflections, in the molecular replacement case, or associated with small Rs in the case of direct methods. The additional phases gained via the phase extension process provide supplementary information used to improve subsequent electron density maps.

A typical flow diagram for a modern EDM procedure is shown in Fig. 8.1. $\{|F|\}$ are the observed amplitudes of the target structure and $\{\phi\}$ the best phase values available at a given step of the phasing process; $\rho(\mathbf{r})$ is the corresponding electron density map. ρ_{mod} is the modified (according to 8.1) density map, whose Fourier inversion provides the set $\{|F_{mod}|, \phi_{mod}\}$. In turn, set $\{\phi_{mod}\}$ is combined with set $\{\phi\}$ to yield set $\{\phi_{comb}\}$, which is combined with set $\{|F|\}$ for the calculation of a new electron density $\rho(\mathbf{r})$. The procedure may be repeated cyclically.

To clarify the rationale behind the various steps in Fig. 8.1, we observe:

(i) $\rho(\mathbf{r})$ is usually an observed Fourier synthesis, calculated using weighted coefficients, but may also be a hybrid synthesis. As described in Section 7.3, weights are essential for computing optimized electron density maps; they are calculated according to the probabilistic estimates of the cosine invariant $(\phi - \phi_p)$, described in Section 7.2.

(ii) Let us suppose that $\rho(\mathbf{r})$ was obtained by some non-ab initio technique [e.g. by molecular replacement (MR, see Chapter 13), isomorphous replacement (SIR-MIR, see Chapter 14), or anomalous dispersion technique (SAD-MAD, see Chapter 15)], and that $\{\phi\}$ are the corresponding phases. Such phases were obtained by exploiting prior information on the model molecule (in the case of MR), or via the diffraction experiment (in SIR-MIR or SAD-MAD cases). This information is statistically independent from the information used for modification of the electron density maps; therefore set $\{\phi_{mod}\}$ is usually combined with set $\{\phi\}$ to secure a smooth approach to the correct phases. The weights with which ϕ and ϕ_{mod} are associated may vary during the EDM process, in accordance with the confidence that the user has in the progressive improvement of the ϕ values.

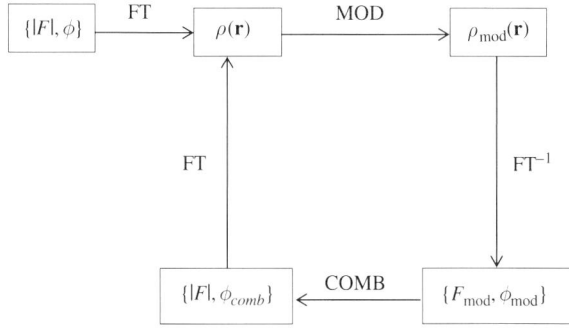

Fig. 8.1

Schematic EDM cycle. $\{|F|\}$ is the set of observed reflections, $\{\phi\}$ is the set of phases available at a certain step of the phasing process. FT and FT^{-1} indicate the direct and the inverse Fourier transform, respectively. MOD is the function used to modify the electron density $\rho(\mathbf{r})$. $|F|_{mod}$ and ϕ_{mod} are the structure factor amplitudes and phases obtained by Fourier inversion of ρ_{mod}. ϕ_{comb} arises from the combination of ϕ_{mod} with ϕ.

It may be useful to recall that, not only observed electron densities, but also difference and hybrid syntheses can be used in EDM procedures. For example, Read (1986) showed that, when strong coordinate errors affect the model, then ρ_{21} (see the symbols adopted in Section 7.3.4) may show a good correlation with the target map without presenting too high a correlation with the model. Since a high correlation with the model can hinder reduction of the model bias, Read suggested the use of ρ_{21} rather than ρ_{obs} in EDM procedures. The reason for this is easy to understand: the ρ_{21} map is the sum of an observed and a difference Fourier synthesis, and therefore it combines the features of both maps. ρ_{21} converges to the target map if a virtuous EDM technique is applied (then, F_p, the structure factor of the model structure, tends to F, and both m and D tend to unity; see again Section 7.3.4).

Let us now describe the expected positive features of electron density maps that are exploited by modern EDM programs.

1. *Flatness of the solvent region* (only in macromolecular crystallography). The electron density map of a highly refined protein structure is rather flat in the solvent region, because of the liquid nature of the solvent (the reader should not forget, however, that more static solvent molecules are found in the protein molecules, or as a monolayer or as double layer at their surface). If the map is calculated by using a model map poorly correlated with the target, then noise peaks will be present in the solvent region. Removing these would lead to better electron density maps, and therefore, by Fourier inversion, to better phases.

   The method is effective if the molecular boundary is known to a good approximation. The first techniques were manual (Hendrickson et al. (1975), Schevitz et al. (1981)), but were soon replaced by automated techniques (Bhat and Blow (1982), Wang (1985), Leslie (1987), Veilleux et al. (1995), Abrahams and Leslie (1996), Cowtan (1994, 1999), Abrahams (1997), Giacovazzo and Siliqi (1997)). In Appendix 8.A, the most popular methods for determining the solvent content and for flattening the corresponding region in the unit cell are described; some information on bulk solvent modelling is also given. An alternative approach is the sphere of influence algorithm, proposed by Sheldrick (2002).

Once the molecular envelope has been determined, the electron density in the solvent region is set to a constant value and a positivity constraint is applied to the electron density in the protein region. The structure factors are calculated by back Fourier transform and the resulting phases are combined with the corresponding ϕ values. The procedure is cyclically repeated.

In Fig. 8.2a we show, as an example, some details of the electron density of the protein *2sar*, as obtained by molecular replacement techniques: the published model in the same region is superimposed. In Fig. 8.2b, the same region of the map is shown as obtained after application of a solvent flattening procedure. The quality of the map (continuity of the electron density, interpretability, etc.) is highly increased.

2. *Distribution of density in a map*. The *actual histogram* of an electron density map is the frequency distribution of the electron density values;

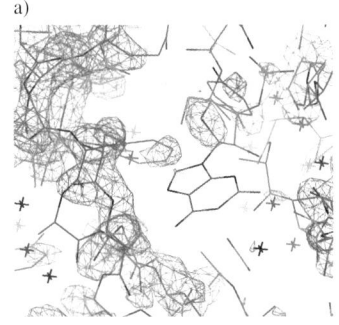

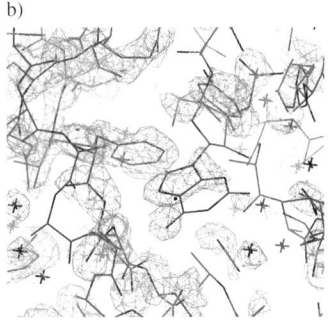

Fig. 8.2
2sar. a) Details of the electron density of the protein as obtained by molecular replacement; the published model in the same region is superimposed. b) The same region of the map after application of a solvent flattening procedure. In both figures, stars indicate the positions of the waters, as located in the published model.

Phase extension and refinement via direct space procedures

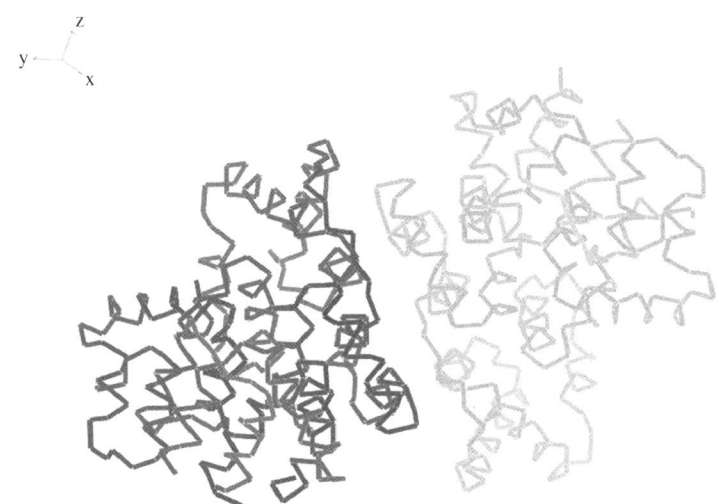

Fig. 8.3
1ycn. Two monomers in the asymmetric unit, referenced by non-crystallographic symmetry. Only the backbones are reported.

the *standard histogram* for an electron density map is the expected distribution of the density values. For light atom structures like proteins it is almost independent of the structure itself. Therefore, the histogram of a well-refined known structure may be used to predict the standard histogram of an unknown structure. Equivalently, the density values of a given trial map may be modified to fit the distribution expected for the ideal map. The fitting implies a modification of the current electron density map and therefore a constraint on the phases obtained by back inversion of the modified map. The technique used for fitting the current density histogram to the standard one is called *histogram matching*; the larger the difference between the two histograms, the larger the chance of an improvement by the process. Several approaches for the application of electron density histograms to the phase problem have been proposed [Luzzati et al. (1988); Lunin (1988); Harrison (1989); Zhang and Main (1990a); Lunin (1993); Gu et al. (1996); Refaat et al. (1996a)]. The technique is described in Appendix 8.B.

3. ***Non-crystallographic symmetry***. Crystallographic symmetry is valid for the full crystal, non-crystallographic symmetry (NCS) has local validity; therefore, five-, seven-, and higher-fold axes are permitted. NCS occurs if the asymmetric unit contains more chemically identical molecules (see Fig. 8.3), or if the unique molecule has its own symmetry axis (see Fig. 8.4).

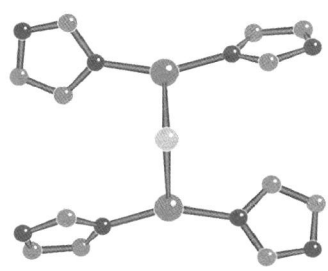

Fig. 8.4
Molecule with a local twofold symmetry axis perpendicular to the drawing plane.

In this section, we will only describe how NCS may be exploited in EDM procedures; the problem of how to discover and characterize NCS in molecular replacement procedures is described in Appendix 13.B.

The electron density is expected to be similar in corresponding regions, related by NCS (however, different contacts with neighbours may cause deviations from exact equality). The technique imposing equal density on sites equivalent by NCS is called *molecular averaging*. Equal densities of the molecules generate constraints on the protein structure factors, and consequently on the phase values. Rossmann and Blow (1963), Main (1967), and Crowther (1969) proposed reciprocal space methods for deriving the consequent phase relationships. Techniques working in reciprocal space were not successful; the real-space method by Bricogne (1974, 1976) soon became popular. It consists of the following steps.

(a) The envelope of the molecules in the asymmetric unit is found (it coincides with the molecular model, if the map has been obtained by molecular replacement). If only an initial electron density map is available, without any other information, the mask can be determined by using the local density correlation function (Veilleux et al., 1995) to distinguish volumes of the crystal which map onto similar density under transformation by the NCS operator.
(b) Equivalent (by NCS) densities are superimposed and the average is calculated; the noise will tend to be averaged out, and the protein density will improve.
(c) The solvent region is flattened.
(d) The thus modified electron density is back-transformed to obtain new phase values.
(e) The new phase values are combined with previous phase information.
  Steps (a) to (e) are cyclically repeated and eventually used for phase extension.

For virus structures, molecular averaging is dramatically effective; because of the large number of equivalent molecules related by non-crystallographic symmetry, molecular averaging produces phases practically without error. As a consequence, even diffraction data at 4 Å resolution gives a map that is easily interpretable. Rossmann and co-workers (Rossmann, 1990; McKenna et al., 1992) solved several spherical viruses using this method.

For common proteins, the situation is less favourable; however, even in the most difficult cases (e.g. via twofold averaging), remarkable reduction in noise and improvement in electron density are achieved, provided that good molecular envelopes are defined and the geometrical relationship between molecules is carefully identified. This last condition is not always easy to fulfill. A good way to determine intermolecular transformation is to look at the heavy atom sites; equivalent molecules have equivalent heavy atom positions.

4. *Continuity of the electron density*. Continuous chains of density are to be expected in a good macromolecular electron density. Locally, the density should reflect the shapes and the interatomic spacings common to macromolecules. A complete and economical depiction of density connectivity in large volumes is provided by the *skeletonization* techniques (Greer (1974), Johnson (1978), Williams (1982), Swanson (1979, 1994), Jones and Thirup (1986)), which use the fact that proteins consist of connected linear chains of atoms and provide corresponding one-dimensional paths passing through high-density regions of the electron density (see Figs. 8.5 and 8.6). The

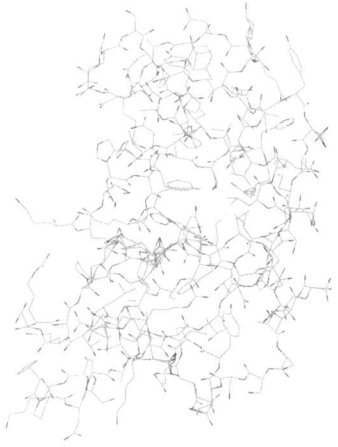

Fig. 8.5
1kf3. Wire model of the protein molecule.

continuity of the electron density in the current map is then a good quality indicator, and may therefore be used as a tool for improving phases (we will see that main chain identification is one of the tasks of automatic model-building programs, see Section 8.3). The strategy (Wilson and Agard (1993), Bystroff et al. (1993), Baker et al. (1993)) is based on the following observation: the information that proteins are linearly connected strings of atoms can be applied even in absence of an atomic model. Its application does not require any use of detailed stereochemistry rules, but only connectivity criteria. Therefore, a trial skeleton of the electron density may be forced to adopt 'protein-like' characteristics without possessing an implicit atomic model. This is equivalent to introducing additional strong constraints on the electron density map. The procedure may be schematized as follows:

(a) A trial skeleton is constructed from an original electron density map.
(b) A new map is created with density falling off smoothly with distance from the skeleton.
(c) The new map is Fourier transformed to obtain new structure factors, which are combined with starting phase information and experimental factor amplitudes.
(d) Steps (a), (b), and (c) are iterated.
The complementarity between solvent flattening and skeletonization techniques is easily understood. Skeletonization enforces density constraints, but it is susceptible to large errors due to the unavoidable inefficiency of the algorithm and to ignorance of the solvent. Integration of solvent flattening with skeletonization may benefit from both types of information.

Fig. 8.6
1kf3. Ribbon model of the protein molecule.

5. *Minimization of the resolution bias.* As stated at the beginning of this section, new structure factor estimates may be obtained by Fourier inversion of modified electron density maps. The aim is to estimate the phases of those measured reflections not involved in the preceding steps of the phasing process (we called this procedure *phase extension*); e.g. high resolution reflections unphased by MR, SIR-MIR or SAD-MAD procedures, or weak reflections unphased by direct methods or other phasing approaches. Often, very low resolution measured reflections are excluded from the phasing process because their amplitudes are affected by the solvent; phase extension may be applied to these reflections too. In all of the above cases, even if phase extension is limited to experimentally measured reflections, it substantially improves the quality of the electron density maps.

The question is now, can phase extension be applied to non-observed structure factors, that is, to structure factors with resolution better than the experimental one? The interest in such a question arises from the following observation: the quality of an electron density map could improve if non-observed structure factors are additionally used as coefficients for the synthesis, provided that their modulus and phase are extrapolated with sufficient accuracy.

A high quality map is usually a map with high correlation with the refined model of the corresponding structure. Low resolution structure factors provide the main contribution to such a correlation, but do not secure full interpretability of the map, for which the contribution of higher resolution terms is necessary. In this context, the unobserved higher resolution structure factors, suitably extrapolated, may significantly improve both the correlation of the current electron density map with the refined model map, and its interpretability. Obviously, the extrapolation process may be usefully extended to non-measured reflections with resolution worse than RES (various experimental reasons can make the dataset incomplete, even at very low resolution). Such reflections can contribute strongly to improving correlation of the current map with the 'true' case.

Caliandro et al. (2005a,b) first extrapolated unobserved reflections in modulus and phase, and applied them in a recursive way to obtain better structural models (see also Yao et al., 2006). They proposed two types of application:

(i) In EDM procedures, the additional use of extrapolated reflections produces a general improvement in the electron density maps. Indeed, RES is artificially improved, with beneficial effects on the sharpness of the peaks and on their location. Usually, the average phase error of the extrapolated reflections is not much larger than the current average phase error for those observed. On the other hand, amplitude extrapolation is very difficult; a posteriori analyses show that the correlation between extrapolated and calculated (from refined molecular models) amplitudes is very low. This technique was named *free lunch*, owing to the fact that it contributes to improvement in models without the need for additional spending on the experiment.

(ii) In ab initio phasing procedures the active use of extrapolated reflections allows a dramatic improvement in efficiency by leading to solution protein structures resistant to any other procedure. This success may be explained by the fact that phase and amplitude extrapolation progressively improves during the phasing procedure, and in the last stages, the extrapolated values are sufficiently close to the correct ones.

Two further observations may be useful. Firstly, extrapolated amplitudes and phases are largely dependent on the model. The risk is that their use may reinforce the model, even if it is very poor. Secondly, extrapolation may be performed up to very high resolution, thus involving a number of supplementary reflections in the phasing process, often much larger than the number of observed ones. This may occur, for example, if RES is 2.2 Å or worse, and extrapolation is performed up to 1.2 Å. To prevent extrapolated reflections from taking a dominant role, suitably reduced weights should be associated with them.

8.3 Automatic model building

A protein may be represented by long, highly flexible chains (the *backbone*), produced through repetition of identical units (i.e. the *peptide units*) to which small well-defined units (i.e. the 20 natural amino acids, the *side chains*) are attached. Building a molecular model in an electron density means to: (a) reconstruct the backbone; (b) locate the side chains.

In the early days this task, the last step of the phasing process, was performed manually and was time consuming; sufficient data resolution, good quality of the phases, and extensive experience of the crystallographer were necessary ingredients for the success of the building process. In recent years, semi-automated model-building (S-AMB) and fully automated model-building (AMB) procedures have been devised to speed up map interpretation and macromolecular structure determination; the historical trend has been analogous to that occurring in the small molecule field (see Section 6.3.5). However, accomplishing the task for macromolecules has been proved more difficult, because of the larger size of the problem and the poorer quality of available electron density maps. Among the earlier contributions to automation, we recall the work by Greer (1985), who suggested a fast method for tracing the polypeptide chain inside the region of high electron density, followed by Jones and Thirup (1986), Swanson (1994), and many others. More recently, Oldfield (2002) proposed starting the automated model-building process by identifying helices and strands. Cowtan (1998, 2001) and Terwilliger (2001) (see also Pavelcik et al., 2002) described a technique for identifying helices, strands, and other characteristic structures by template matching.

Today, the model-building process is fully automatized; computer graphic programs, pattern recognition techniques, real-space methods for fitting the model into the electron density, automated reciprocal space refinement of positional parameters, have all been combined into efficient automated model-building packages, capable of producing molecular models from scratch. The basic principle for most model-building algorithms uses the connectivity of the polypeptide chain and the presence of well-defined structures (e.g. helices, β-strands, etc.) in the chain. Model-building programs usually start by tracing the backbone of the protein via a careful analysis of the electron density map. Once the C_α backbone is traced, the next step is *sequence docking*: i.e., assign the protein sequence (usually known) to the main-chain fragments, previously located. This is usually done by checking the electron density at the coordinates of the atoms of the main chain.

Although model building may be accomplished in many different ways, some important features are common in any building program, that is the use of databases of protein fragments. A database example is associated with *Buccaneer* (Cowtan, 2012); it uses 500 well-refined protein structures (Lovell et al., 2003), which provide 106,295 amino acids in 1327 continuous fragments. For each amino acid, the residue type is specified and the coordinates of the N, C_α, and C atoms are stored. The full database is recorded as a single list of amino acids.

Such databases may be used for various purposes; they are frequently used to find all fragments whose C_α overlap, at least approximately, with a given (even discontinuous) set of C_α. The length of a search fragment may, in principle, vary, but libraries of fragments including 12 or more residues seem to be doubtful (Cowtan, 2012). Among the possible database applications, we recall the following:

(i) Jones and Thirup (1986) used pentapeptides for reconstructing a main-chain trace from C_α positions.
(ii) Kleywegt and Jones (1996) suggested the use of pentapeptide fragments for the validation of a protein backbone trace.
(iii) Terwilliger (2003) used a library of tripeptide fragments to extend fragments of protein chains, whereas Sheldrick (2010) used that to find initial protein fragments.
(iv) Cowtan (2012) used an eight-residue search fragment to find a missing loop of four residues (two residues before the missing loop, four null residues corresponding to the missing loop, and two residues after the missing loop).

Among the S-AMB programs, *Coot* deserves a special mention (Emsley and Cowtan, 2004; Emsley et al., 2010), an interactive molecular-graphics application for model building and validation. It may be used for:

(a) displaying electron density map and related atomic models;
(b) completing models generated by automatic model-building programs or for starting model building when no model is available. In this framework, *Coot* may generate rest of the main-chain atoms when other atoms were previously located, may find secondary structure elements such as α-helices and β-strands, may fit an α-helix or a β-strand in a given region of the unit cell, or may rebuild, regularize, and refine by real-space refinement an atomic model against an electron density map;
(c) improving the fit of the model with the density and simultaneously obtaining a high-quality Ramachandran plot (Ramachandran et al., 1963);
(d) validating the current or the final model against the electron density. Also, geometrical checks are used, such as the identification of incorrect chiral volumes, a peptide analysis, a rotamer analysis, the Ramachandran plot, the Kleywegt plot (useful for highlighting differences between chains referred by non-crystallographic symmetry) (Kleywegt and Jones,1996), etc.;
(e) providing interfaces to external validation tools such as *MolProbity* (Davis et al., 2007) and *REFMAC* (Murshudov et al., 1997, 2011) refinement software.

AMB may be accomplished in many different ways, as is testified by the variety of computer programs available. Among these, we quote *ARP/wARP* (Lamzin and Wilson, 1993; Perrakis, et al., 1999; Morris et al., 2002), *RESOLVE* (Terwilliger, 2004), *TEXTAL* (Gopal et al., 2007), *MAID* (Levitt, 2001), *BUCCANEER* (Cowtan, 2006). The most popular program is *ARP/wARP*. Its approach is described in Appendix 8.C.

AMB frequently starts after the application of EDM techniques, implies a further modification of the electron density map and ends with a substantial improvement in the crystallographic phases; which is the reason why model building may be considered to be included in the EDM methods.

Automatic model building

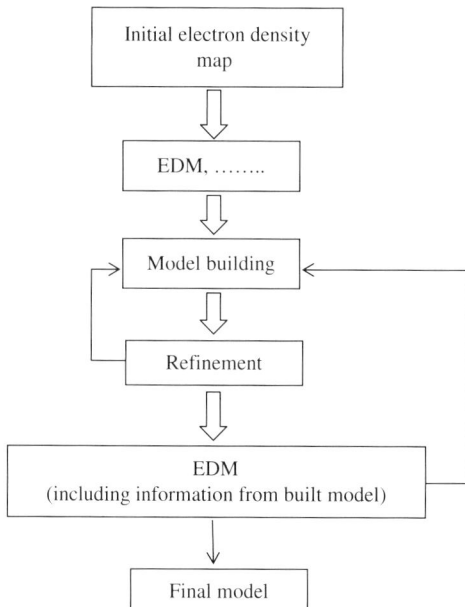

Fig. 8.7
Typical outline of an AMB program.

The general outline of a typical AMB procedure is shown in Fig. 8.7, which may be clarified by a short description of the main steps.

(i) An observed electron density map (regardless of technique used to obtaining it) is submitted to EDM, to improve phases and increase the map interpretability. If an ab initio approach is used, the starting observed density corresponds to the best available map. If molecular replacement techniques (MR) are used, the starting density may be obtained as follows: first submit to restrained least-squares refinement the model structure as oriented and positioned by MR techniques, and then use the resulting phases and the observes amplitudes to calculate the observed map to submit to EDM procedures.

(ii) Cycles of model building are performed, aimed to find C_α atoms and define protein fragments. Fragments are grown by adding flanking residues at either end, using the electron density and the Ramachandran plot as guide.

(iii) Fragments may be joined in longer fragments when they overlap.

(iv) Fragments may be linked by inserting one or more residues. As anticipated at the beginning of this section, pentapeptides (Jones et al., 1991; Kleywegt, 1997; Esnouf, 1997; Joosten et al., 2008) may be used to add an additional C_α (see Fig. 8.8); indeed, if a tetrapeptide belongs to the terminal part of the main-chain fragment, correctly positioned, they may predict the position of the fifth C_α and so extend the fragment. In case of a gap involving more C_α, the above process may be iterated until the gap is filled.

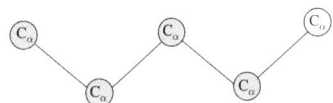

Fig. 8.8
Pentapeptide. The black tetrapeptide belongs to a main-chain fragment, correctly positioned; the position of the fifth may be predicted from it.

(v) The sequence is assigned. Residues non-docked into the sequence or lying in a poor density region are eliminated or pruned. As anticipated at the beginning of this section, rotamers may be used for locating and improving side chains; for this purpose, a rotamer library (Ponder and Richards, 1987; Dunbrack and Cohen, 1997; Lovell et al., 2000) is usually contained in S-AMB and AMB programs. It may be worthwhile recalling that a rotamer is defined by specific torsion angles (not all the angles are permitted, in the sense that there exists a rotational energy barrier that needs to be overcome to convert one rotamer to another). For each amino acid side chain, up to 50 rotamers may be found; the list includes side chain torsion (chi) and probability values for each rotamer. For a particular side chain, a set of the most likely rotamers is used. Each test rotamer is rigid body refined for best fit with the electron density; a score is assigned according to the fit of the map. The best fit rotamer is chosen, and the old coordinates of the side chain atoms are replaced by the new ones.

(vi) Various protein fragments are submitted to restrained refinement cycles (e.g. via *REFMAC*), which are interspersed among the building cycles. The model refinement improves the built model and the corresponding phases.

A successful application of AMB programs frequently ends with a large percentage of built residues (P_{built}), a large percentage of residues assigned to the sequence (P_{docked}) and a small number of chains. Small values of P_{docked} correspond to failures. If the number of chains is too high, the chain connectivity is broken in many places.

8.4 Applications

EDM procedures are usually applied after any phasing approach: traditional and non-traditional direct methods, Patterson deconvolution techniques, MR, SIR-MIR, SAD-MAD, etc. In a way they are an obligatory route to phase extension and refinement. Probably the reader is interested in verification of how automatic application of a chain of programs, consisting of a phasing approach followed by some phase refinement procedures, may improve the starting set of phases and lead to an electron density map automatically interpretable by an AMB program. We will give two examples.

1. Protein *1pm2*: two chains, 750 residues, data resolution equal to 1.80 Å. The target symmetry-independent molecules are depicted in Fig. 8.9a. There are 14 Hg and 4 Yb in the asymmetric unit; thus, ab initio Patterson methods (see Chapter 10) were applied to obtain a preliminary electron density map, which, submitted to *ARP/wARP*, provided the structural model reported in Fig. 8.9b. EDM procedures, via *SOLOMON*, were applied to the electron density map provided by Patterson techniques; phases improved significantly, so that a next application of *ARP/wARP* led to the more complete

Applications

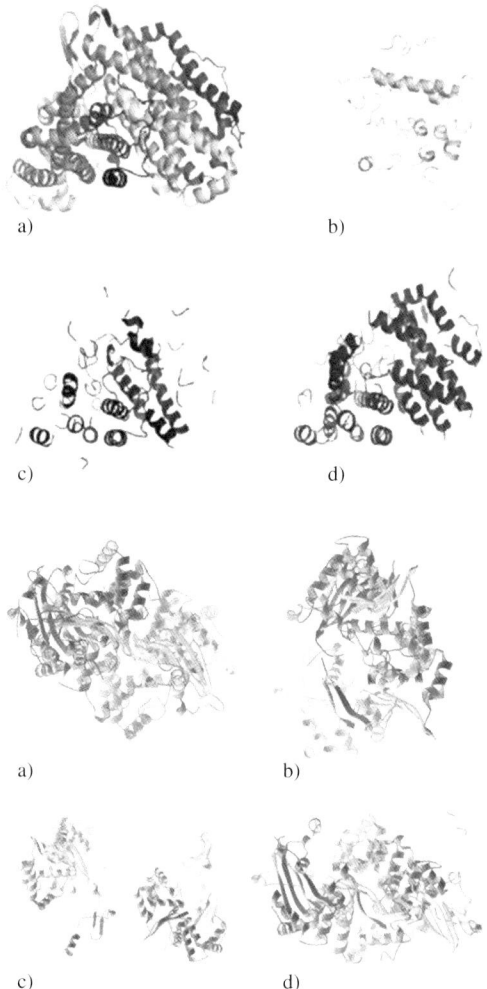

Fig. 8.9
1pm2. Structural models available by means of the chain *Patterson deconvolution + Solomon + Free Lunch + ARP/wARP*. (a) Published model; (b) model available after the application of *ARP/wARP* to the electron density, provided by a Patterson deconvolution program (in symbols, by *PATTERSON + ARP/wARP*); (c) model available by *PATTERSON + SOLOMON + ARP/wARP*; (d) model available by *PATTERSON + SOLOMON + FREE LUNCH + ARP/wARP*.

Fig. 8.10
1e3u. Structural models available by means of the chain *Patterson deconvolution + Solomon + Free Lunch + ARP/wARP*. (a) Published model; (b) model available after the application of *ARP/wARP* to the electron density, provided by a Patterson deconvolution program (in symbols, by *PATTERSON + ARP/wARP*); (c) model available by *PATTERSON + SOLOMON + ARP/wARP*; (d) model available by PATTERSON + SOLOMON + FREE LUNCH + *ARP/wARP*.

model depicted in Fig. 8.9c. The last electron density map provided by *SOLOMON* was submitted to cycles of *FREE LUNCH*. The final model is shown in Fig. 8.9d.

2. *1e3u*, four chains, 984 residues, data resolution equal to 1.65 Å. The asymmetric unit is shown in Fig. 8.10a; it contains 8 Au, and therefore the structure is ideal for the application of ab initio Patterson deconvolution techniques. The electron density map so obtained was submitted to *ARP/wARP* : the resulting structural model is shown in Fig. 8.10b. EDM procedures, via *SOLOMON*, were applied to the electron density map provided by the Patterson deconvolution techniques; phase improvement is not evident, so application of *ARP/wARP* does not lead to a more complete model (see Fig. 8.10c). The last electron density map provided by Solomon was submitted to cycles of *FREE LUNCH*. The final, more complete model, is shown in Fig. 8.10d.

APPENDIX 8.A SOLVENT CONTENT, ENVELOPE DEFINITION, AND SOLVENT MODELLING

A solvent flattening procedure may be described, schematically, by the following steps.

1. Estimate
   V'_p = (volume of the solvent)/(volume of the unit cell).
2. Find which region of the unit cell is occupied by the solvent and flatten the solvent space to obtain better phases.

Points 1 and 2 are described in Sections 8.A.1 and 8.A.2, respectively. In Section 8.A.3, we describe possible model solvents, useful at various stages of the crystal structure solution process.

8.A.1 Solvent content according to Matthews

Matthews (1968) surveyed 116 different crystalline proteins with known molecular weights and found that for many protein crystals the ratio

$$V_M = V/(ZM) = \text{unit cell volume/(total molecular weight)}$$

is between 1.7 and 3.5 Å³/Da, with most values being around 2.15 Å³/Da. Z is the number of molecules in the unit cell, M the weight of a molecule in Da. V_M is also called *Matthews number*; it may be used for estimating Z.

For example, suppose that the space group is C2, $V = 320\,000$ Å³ and $M = 32\,000$. Given $Z = 2, 4, 8$, we obtain $V_M = 5, 2.5, 1.25$ Å³/Da, the first and the third exceeding the allowed range. Thus the correct Z value is 4, equivalent to assuming that there is one molecule per asymmetric unit.

This result may usefully be compared with the practical rule suggested in Section 2.9 for small molecules, according to which V/N (the volume per atom) should be approximately equal to 18 Å³. If we assume that the small structure is mainly composed of C atoms, the rule is equivalent to the following: ratio $V/(ZM)$ is expected to be about 1.5 Å³/Da. The larger volume per protein atom mainly arises from the solvent.

Matthews found that the fraction of a crystal usually occupied by protein (say $V'_p =$) is about 57% of the unit cell volume and may vary between 30% and 75% (values correlated with the size of the molecule); similarly, the solvent volume ranges from 70% to 25%. Since

$$V'_p = V_{\text{prot}}/V,$$

where V_{prot} is the volume occupied by the protein per unit cell, we have.

$$V'_p = \left(V_{\text{prot}}/M_{\text{prot}}\right)\left(M_{\text{prot}}/V\right) = \left(V_{\text{prot}}/M_{\text{prot}}\right)/V_M, \quad (8.A.1)$$

where $M_{\text{prot}} = MZ$. Since M, and therefore M_{prot}, are expressed in Daltons and V_{prot} is in Å³, (8.A.1) reduces to

$$V'_p = 1.6604/\left(d_{\text{prot}} V_M\right),$$

where d_{prot} is the protein density (g cm^{-3}). If this is assumed equal to 1.35, then

$$V'_p = 1.23/V_M, V'_{solv} = 1 - V'_p. \qquad (8.A.2)$$

8.A.2 Envelope definition

Wang (1985) proposed an automatic cyclic procedure for defining a mask, which should separate the map into solvent and molecule space, in accordance with the ratio fixed by the Matthews criterion. The Wang procedure may be described as follows:

(a) The current electron density map (no matter if it has been obtained by ab initio or non-ab initio techniques) is truncated according to:

$$\rho_{trunc}(\mathbf{r}) = \rho(\mathbf{r}) \text{ if } \rho(\mathbf{r}) > \rho_{solv}, \quad \rho_{trunc}(\mathbf{r}) = 0 \text{ if } \rho(\mathbf{r}) \leq \rho_{solv},$$

where the threshold, ρ_{solv}, is chosen to meet the expected solvent content.

(b) ρ_{trunc} is smoothed (into $\rho_{sm}(\mathbf{r})$) by associating, at each point $\mathbf{r}$ of ρ_{trunc}, the weighted average density over the points included in an encompassing sphere of radius R (between 8 and 4 Å, according to the resolution or, also, to the quality of the structure):

$$\rho_{sm}(\mathbf{r}) = \sum_{r} w(\mathbf{r} - \mathbf{r}') \rho_{trunc}(\mathbf{r}'), \qquad (8.A.3)$$

where

$$w(\mathbf{r} - \mathbf{r}') = 1 - d(\mathbf{r}')/R \text{ for } d < R, \quad w(\mathbf{r} - \mathbf{r}') = 0 \text{ for } d > R.$$

$d = |\mathbf{r} - \mathbf{r}'|$ is the distance between points $\mathbf{r}$ and $\mathbf{r}'$.

(c) A cut-off value, ρ_{cut}, is calculated, which divides the unit cell into two regions, solvent and protein; solvent pixels are marked by the condition $\rho_{sm}(\mathbf{r}) \leq \rho_{cut}$, voids internal to the molecular envelope are polished.

(d) A solvent corrected map is obtained by setting all the values outside the protein envelope to a low constant value; the electron density values inside the molecular envelope are set to the current values (say, to the values defined at point a).

(e) New phases are obtained by Fourier inversion of the solvent corrected map. Often such phases are combined with the experimental phases (those obtained via SAD-MAD, SIR-MIR, or MR techniques). The corresponding electron density map is the new basis for the application of point a.

Two years later, Leslie (1987) observed that (8.A.3) is a convolution, and that flattening may be more easily performed via its Fourier transform:

$$T[\rho_{sm}(\mathbf{r})] = F_{sm}(\mathbf{h}) = T[w] T[\rho_{trunc}] = g(s) \cdot F_{trunc},$$

where F_{trunc} is readily calculated by Fourier inversion of the truncated map. $g(s)$ is the Fourier transform of the weight function, sum of two components, the first of which is the Fourier transform of a sphere. According to James (1962),

$$g(s) = Y(uR) - Z(uR),$$

where $s = 2\sin\theta/\lambda$, $u = 2\pi s$,

$$Y(x) = 3(\sin x - x\cos x)/x^3,$$

and

$$Z(x) = 3[2x\sin x - (x^2 - 2)\cos x - 2]/x^4.$$

Leslie's procedure improved the efficiency of the flattening technique and dramatically reduced the computing time.

It may be useful to mention that several attempts have been made to estimate the protein envelope at very low resolution (say, about 8 Å or worse). The necessary prior information consists of unit cell parameters, space group, high quality diffraction data, complete up to a fixed resolution, and a rough estimate of the solvent fraction. Attempts began with Kraut (1958). Somewhat later, different algorithms were proposed, summarized as follows: the *histogram method* (Luzzati et al., 1988; Mariani et al., 1988; Lunin et al., 1990), the *condensing protocol* (Subbiah, 1991, 1993; David and Subbiah, 1994), *the one sphere method* (Harris, 1995), *FAM* (Lunin et al., 1995; Moras et al., 1983; Urzhumtsev et al., 1996). The general idea at the basis of all these algorithms was to define at very low resolution a rough envelope, which may be easier than at high resolution. Once a model envelope is obtained, phase extension at higher resolution should be performed, mainly via solvent flattening, histogram matching, etc. to progressively improve identification of the solvent region, and then allow solution of the protein structure.

The above methods were able to find good (even if rough) envelope models, but their weak point was the phase extension, quite difficult from very low resolution. In recent years these methods have been shelved, however it may be that in the future their appeal will again increase.

8.A.3 Models for the bulk solvent

The narrow boundary region (within a 7 Å boundary layer) between the protein and the solvent exhibits an ordered structure of strongly bonded water molecules. As a rule of thumb, about one water molecule per residue belongs to such an ordered substructure (Kleywegt and Jones, 1997a). The solvent is disordered beyond this shell, and solvent flattening techniques use this characteristic property to improve the protein phases. Since the bulk solvent may significantly contribute to the structure factors, taking into account its contribution may improve agreement between calculated and observed structure factors; this may be useful both in the refinement step, and in the phasing step itself (e.g. in the translation step of molecular replacement).

The effect of the solvent on structure factors may be understood as follows: cancelling the solvent contribution from the calculated structure factor is equivalent to setting the electron density of the bulk solvent to zero. This implies an infinitely sharp contrast between protein surface and solvent, with an overestimate of the low resolution structure factor amplitudes. We will quote two models for the solvent:

1. *Exponential bulk solvent*. This is based on the Babinet principle, according to which, if the unit cell is divided in two parts, one relative to the disordered solvent and the other to the protein molecule, then $F_{solv} = -F_P$, where the first structure factor refers to the solvent volume and the second to the protein volume. To understand the above relationship, we notice that, by a property of the Fourier transform, a unit cell with constant electron density will show vanishing structure factor amplitudes, except for F_{000}. The contribution of the solvent bulk is therefore opposite to that of the protein volume, and will tend to weaken the amplitudes of the latter. An approximation to solvent scattering may be achieved by placing atoms with very high temperature factor (e.g. 200 Å^2) in the solvent region. The effects of the above model may be represented by calculating the total structure factor, F_t, as (Glikos and Kokkinidis, 2000b)

$$F_t(\mathbf{h}) = F_P(\mathbf{h})[1 - k_{sol} \exp(-B_{sol} s^2/4)],$$

where $s = 2\sin\theta/\lambda$ and k_{sol} is the ratio between the mean electron densities of the solvent and of the protein. Since:

   (i) the electron density of water is about 0.334 e^-/Å^3, and that for a salt solution may be estimated at around 0.40 e^-/Å^3;
   (ii) the protein density may be estimated as close to 0.439 e^-/Å^3,
   then k_{sol} may be approximated to 0.76. If we choose $B_{sol} \approx 200$ Å^2, the effects of the solvent will disappear rapidly at higher resolution.

2. *Flat bulk solvent*. A flat mask is used as the solvent model; it is located into the solvent region, at a distance of about 1.4 Å from the van der Waals surface of the protein. The bulk solvent region is then uniformly filled by a continuous electron density, which contributes to the total structure factor, in accordance with (Jiang and Brunger, 1994),

$$F_t(\mathbf{h}) = F_P(\mathbf{h}) + k_{sol} F_{sol}(\mathbf{h}) \exp(-B_{sol} s^2/4)] \qquad (8.\text{A}.4)$$

The residual between the observed and the solvent corrected structure factors, F_t, provides optimal values for the parameters k_{sol} and B_{sol} (typical values are, $k_{sol} \approx 0.4$ and $B_{sol} \approx 45$ Å^2). This kind of bulk solvent correction is implemented in several refinement programs and is also used to improve the efficiency of the translation step in MR programs.

APPENDIX 8.B HISTOGRAM MATCHING

This technique is widely used in image processing; it aims to improve the image quality by fitting the density distribution of an image with the ideal distribution. From this point of view, the electron density is an image of the crystal structure, the quality of which should be improved by fitting the density frequency with standard distributions.

The actual form of a histogram depends on several parameters, among which are:

1. the fraction of the unit cell volume occupied by the solvent;
2. the resolution at which the diagram is calculated;

3. the mean phase error associated with the structure factors;
4. the overall temperature factor.

To circumvent the effects of the temperature parameter, histogram matching procedures remove the overall temperature factor from all the |F|s. This allows simplification of the method, since it is not necessary to use different standard histograms for different temperature factors. Accordingly, the standard histogram, which is relative to the frequency distribution of the density in the protein region, may be treated as a function of the resolution only. It may be obtained from the electron density map of a similar known structure or from a formula. Main (1990a,b) (see also Lunin and Skovoroda, 1991) has developed a six-parameter formula which produces useful histograms over a range of resolutions from 4.5 to 0.9 Å. Histograms are calculated by considering the densities only within the molecular envelope. We note that:

1. The flatness of the histogram increases with the average phase error. In Fig. 8.B.1 we overlap the histogram corresponding to the refined structure with one obtained from approximate phase values.
2. Histograms are asymmetric; the asymmetry is a consequence of the positivity of the electron density (negative density values are less frequent than positive ones) and may be used as a criterion for phase correction (Podjarny and Yonath, 1977). On the other hand, the negative regions must be present in the histograms because they are generated by unavoidable series termination errors. *Skewness*, say,

$$\gamma = \left\langle \left(\frac{\rho - \langle \rho \rangle}{\sigma_d} \right)^3 \right\rangle \quad \text{with} \quad \sigma_d = \langle (\rho - \langle \rho \rangle)^2 \rangle^{1/2},$$

is usually calculated to evaluate the asymmetry; it can be positive or negative, or undefined. Negative skewness values indicate that the tail on the left-hand side of the probability density function is longer than on the right-hand side; a positive skewness indicates that the tail on the right-hand side is longer than on the left-hand side; a zero value indicates that the values are relatively evenly distributed on both sides of the mean. In our case, skewness is expected to be positive.
3. The histogram changes with the data resolution (see Fig. 8.B.2). The histogram for high resolution maps has its maximum close to $\rho = 0$; for low resolution maps, the maximum shifts to higher values of ρ, and the peak is broader. The peak of the histogram lowers to a minimum at about 3 Å resolution; as the resolution decreases, the peak rises again, moves towards higher density, and becomes broader. Long tails towards high density are present in high resolution maps.
4. The histogram matching technique may be applied as follows (Zhang and Main, 1990a,b):

   (i) From a given set of *B*-parameter corrected structure factors, the Fourier synthesis and the corresponding histogram are calculated. The latter is compared with the standard histogram.
   (ii) The electron density histogram of the actual map is divided into smaller areas with boundaries, ρ_i, $i = 1, \ldots, n$ ($n \sim 100$) (see Fig. 8.B.3a). The

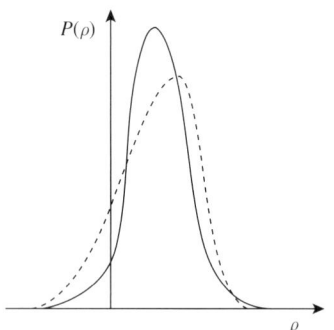

Fig. 8.B.1
Electron density histograms obtained from refined phases (——) and from approximated phases (- - -).

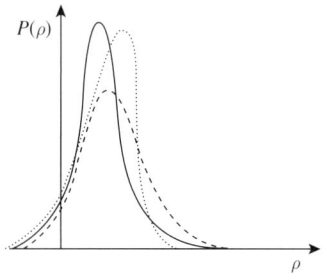

Fig. 8.B.2
Electron density profile variation with resolution.

standard histogram is also divided into smaller areas, with boundaries, $\rho'_i, i = 1, \ldots, n$ (see Fig. 8.B.3b).

(iii) Scale factors a_i and shifts b_i are calculated to map ρ into ρ' for the ith interval:

$$\rho' = a_i \rho + b_i \qquad (8.B.1)$$

where

$$a_i = \frac{\rho'_{i+1} + \rho'_i}{\rho_{i+1} - \rho_i}, \quad b_i = \frac{\rho'_i \rho_{i+1} - \rho'_{i+1} \rho_i}{\rho_{i+1} - \rho_i}.$$

For example, if only a scale factor k relates the two maps (e.g. $\rho' = k\rho$), then

$$a_i = k \text{ and } b_i = 0 \text{ for any } i.$$

If only a shift relates the two maps (e.g. $\rho' = \rho + b$), then

$$a_i = 1 \text{ and } b_i = k \text{ for any } i.$$

(iv) The operation (8.B.1) is applied to the actual map for each interval; the new map will show the same density distribution as that expected.
(v) A new set of structure factors is calculated from the modified electron density, whose phases are employed for a next cycle.

A more intuitive approach is to let $P(\rho)$ and $P_s(\rho)$ be the current and the standard reference density histograms, respectively (both sum to unity), and $N(\rho)$ and $N_s(\rho)$ the corresponding cumulative distributions. The transformation of $P(\rho)$ into $P_s(\rho)$ is made as follows. For any density value $P(\rho)$, the corresponding point in $N(\rho)$ is calculated; this is mapped in $N_s(\rho)$, and the desired modified value in the standard distribution is obtained by inverting the cumulative standard distribution:

$$\rho = N_s^{-1}[N_s(\rho)].$$

Histogram matching is usefully combined with solvent flattening techniques as follows:

(a) The molecular envelope is obtained.
(b) The solvent region is flattened, while the density within the molecular envelope is matched with the expected histogram. Obviously, histogram matching efficiency is high when the solvent region is a small percentage of the unit cell. When the reverse condition occurs, solvent flattening effects are dominant.
(c) Structure factors are calculated from the above modified map and their phases are (eventually) combined with experimental phases. If a phase extension process is started, the extended phases are accepted at the calculated values.
(d) A new map is calculated using data obtained at step (c), and the procedure is repeated from step (a) until convergence is obtained.

It is obvious that histogram matching and solvent flattening procedures are not able to suppress all false peaks from an experimental electron density map and/or generate all the supplementary peaks to complete the structure. Indeed,

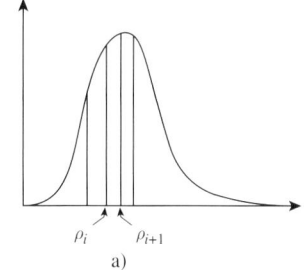
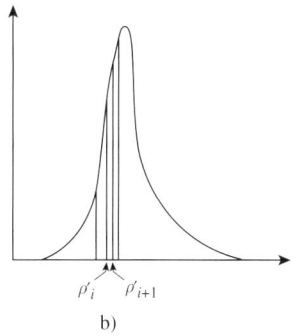

Fig. 8.B.3
(a) Electron density histogram for the actual electron density model; (b) standard electron density histogram.

false density inside the envelope tends to remain, while molecular density outside the envelope may remain strongly depressed. However, these procedures are able to produce remarkable improvements in the maps.

APPENDIX 8.C A BRIEF OUTLINE OF THE *ARP/wARP* PROCEDURE

The *automatic refinement procedure* (*ARP*) is based on a free atom approach. A set of dummy atoms is created, new atoms are added and old ones are deleted to create new models which, cyclically re-evaluated, should end in a final model describing electron density and target protein. The procedure may be described schematically as follows.

Dummy atoms (of the same atomic species, say O) are located in the high density regions of the best available electron density map; a fine grid of about 0.25 Å is used. The initial model is gradually expanded; the density threshold is gradually lowered and additional atoms are located at bonding distances from existing atoms. The model is completed when the number of dummy atoms is about 3 x *NAT*, where *NAT* is the expected number of atoms. At the end of the updating process (see below) the number of atoms is reduced to about 1.2 x *NAT*.

The model is updated as follows.

  (i) *Atom rejection*. Hybrid electron densities of type $3F_o–2F_c$ are calculated; an atom is removed on the basis of the density at the atomic centre, on shape criteria (e.g. sphericity), and distance criteria (e.g. too close to accepted atoms).
 (ii) *Atom addition*. The $F_o–F_c$ synthesis is calculated; the grid point with the highest density value is selected as a new atomic position, provided that this satisfies defined distance constraints in relation to other positioned atoms. Grid points at small distance from this added atom are rejected and the next higher grid point is selected.
(iii) *Model refinement*. This may be performed by a cyclic procedure based on unrestrained least squares or maximum likelihood refinement (in both the cases the procedure aims at matching calculated to observed structure factors), and/or by real-space refinement (an atom may be moved from the peak position on the basis of a density shape analysis around it). Usually, the reciprocal space refinement is performed by *REFMAC* (Murshudov et al., 1996); it needs a number of observations, greater than the number of model parameters, which then sets the resolution limit of *ARP/wARP* to about 2.5 Å.

So far no attempt has been made to establish a chemical sense to the atoms, in terms of atomic species, bond distances, bond angles, protein secondary structures, etc.; typically, free atoms lie within 0.5–0.6 Å of the corresponding positions in the correct structure.

Following this, *model reconstruction* starts; its task is to discard atoms in false positions, assign atomic species to the well-located atoms, and to establish their connectivity. Only when atomic species, bonds, and angles for a group of atoms have been defined (see below), will stereochemical restraints be applied in *restrained refinements*; this will improve the ratio of observations to parameters, and will increase the efficiency of the least squares refinement (which will then become *hybrid*, because free and restrained atomic positions will coexist).

Model reconstruction starts with identification of the main-chain atoms. Every C_α atom should stay at 3.8 Å from at least one other candidate C_α atom, which may be connected to the first one by a *forward (outgoing) directionality* $(-C(=O)-N-C_\alpha)$ or by a *incoming (backward) directionality* $[N-C(=O)-C_\alpha]$. If two atoms i and j are C_α candidates (see Fig. 8.C.1), then a peptide unit plane is placed among the candidates and rotated about the i–j axis. If, for a given rotation angle, the interpolated electron density at the peptide atomic positions is larger than a given threshold, atoms i and j are flagged as C_α atoms.

The electron density maps to which *AMB* algorithms are applied usually show non-negligible phase errors; therefore, the condition according to which two consecutive C_α atoms should lie at 3.8 Å apart must be replaced by a more permissive condition, say, the distance should lie in a range (e.g. 3.8 ± 1 Å). The result is that many candidates may be connected by more than one incoming and one outgoing connection, with the consequent combinatorial explosion of the possible chains. *ARP/wARP* solves the problem by dividing each candidate chain into small structural subunits and by evaluating, by stereochemical arguments, the probability of each subunit being the correct one. Subunits consisting of four consecutive C_α atoms are used, say $C_\alpha(n) - C_\alpha(n+1) - C_\alpha(n+2) - C_\alpha(n+3)$, and the two-dimensional frequency distributions of the angle $C_\alpha(n) - C_\alpha(n+1) - C_\alpha(n+2)$ and of the dihedral angle $C_\alpha(n) - C_\alpha(n+1) - C_\alpha(n+2) - C_\alpha(n+3)$ are tested against the distribution derived from database analysis (Oldfield and Hubbard, 1994; Kleywegt, 1997). This information is of a three-dimensional nature, and may be used to obtain a score for the subunit (of length four) parameters. The main chain is built by overlapping the last three atoms of one subunit with the first three of the following. The chain scores are then obtained by summation of the subunit scores.

Limited data resolution and quality of the phases, combined with the natural conformational flexibility of the chain, may not allow recovery of a full continuous chain; several main-chain fragments may be obtained, separated by gaps, and some chain fragments may be wrongly identified. The lower the quality of the starting electron density map and data resolution, the larger the probability of having a large number of gaps.

Once one or more main-chain fragments have been correctly identified, side chains may be built by taking into account the C_α positions, the density distribution in the map, and connectivity criteria; the aim is to dock the polypeptide fragments into the sequence (assumed to be known). A score is associated with each possible docking position, so that the chain would have the most probable side chain conformation.

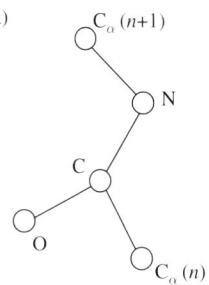

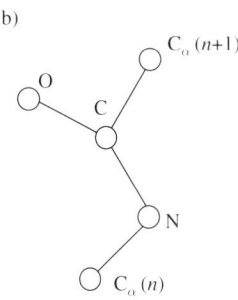

Fig. 8.C.1
Connection between two candidate C_α atoms. The peptide plane is located between these atoms and rotates about the axis $C_\alpha(n) - C_\alpha(n+1)$. The two candidates are flagged as C_α atoms if, for a given orientation of the plane, the interpolated density values at the atomic positions are larger than a given threshold. (a) Forward directionality; (b) backward directionality.

9 Charge flipping and *VLD* (*vive la difference*)

9.1 Introduction

Direct methods procedures (see Chapter 6) or Patterson techniques (see Chapter 10), primarily the former, have been methods of choice for crystal structure solution of small- to medium-sized molecules from diffraction data. Over the last 30 years, several new phasing algorithms have been proposed, not requiring the use of triplet and quartet invariants, but based only on the properties of Fourier transforms. These were not competitive with direct methods and have never became popular, but they contain a nucleus for further advances. Among these we mention:

 (i) Bhat (1990) proposed a *Metropolis* technique (Metropolis et al., 1953; Kirkpatrick et al., 1983; Press et al., 1992), also known as *simulated annealing* (the reader is referred to Section 12.9 for details on the algorithm). From a random set of phases, an electron density map is calculated, modified, and inverted. The corresponding phases are altered according to the simulated annealing algorithm, and then used to calculate a new electron density map. The procedure is cyclic.
(ii) A strictly related simulated annealing procedure has been proposed by Su (1995). The objective function to minimize was

$$R = \sum_{\mathbf{h}} (S|F_{\mathbf{h}}|_{calc} - |F_{\mathbf{h}}|_{obs})^2,$$

where S is the scale factor. The scheme is as follows: random atomic positions are generated and in succession shifted; the simulated annealing algorithm is applied to accept or reject atomic shifts. At the end, a new atomic structure is generated, whose positions are shifted in succession, and so on in a cyclic way.
(iii) *The forced coalescence method* (*FCP*) was proposed by Drendel et al. (1995). Hybrid electron density maps (see Section 7.3.4) were actively used with different values of τ and ω.

Even if never popular, the above algorithms opened the way to two other methods which are much more efficient, *charge flipping* and *VLD* (*vive la difference*), to which this chapter is dedicated. Both are based on the properties of the Fourier transform; they do not require the explicit use of structure invariants and seminvariants, or a deep knowledge of their properties. The reader should not, however, conclude that the invariance and seminvariance concepts are not necessary in the handling of these approaches, on the contrary, understanding these basic concepts is essential to the appreciation of these new methods.

To be more clear, when an electron density is modified, a model map is simultaneously identified; and when the model map is Fourier inverted, model structure factors with modulus $|F_p|$ and phase ϕ_p are obtained. The reliability of the new phases is usually calculated via the distribution $P(R, R_p, \phi, \phi_p)$, described in Section 7.2, which involves estimation of the two-phase structure invariants, $(\phi_\mathbf{h} - \phi_{p\mathbf{h}})$.

9.2 The charge flipping algorithm

Charge flipping was developed by Oszlányi and Suto (2004, 2005, 2008) and has been successfully applied to small molecules (Wu et al., 2004; Palatinus and Chapuis, 2007), modulated structures (Palatinus et al., 2006), powder data (Baerlocher et al., 2007a,b), high resolution protein data (Dumas and van der Lee, 2008). We describe the algorithm step by step (see Fig. 9.1):

1. The list of unique reflections, as fixed by the space group symmetry, is expanded in P1 to produce a complete list of reflections; Friedel pairs, if present, are merged.
2. Random starting phases are assigned to the expanded list of reflections.
3. An electron density map is calculated over a grid with spacing adjusted to RES/2. It may be seen from Fig. 9.2a that, at least at high resolution, large density values are restricted to a small percentage of pixels, which therefore carry almost all of the structural information. Figure 9.2a is a different way of representing the density distributions shown previously in Figs. 8.B.1 and 8.B.2.
4. The electron density is modified so that all the pixels with density smaller than a given positive threshold δ (see Fig. 9.2b) are submitted to flipping (i.e. their density is multiplied by -1). In Fig. 9.1, the modified map is called ρ_{mod}.
5. The inverse Fourier transform of ρ_{mod} is treated as follows: for large amplitudes (about 80% of the total number), calculated phases are associated with observed amplitudes, for weak reflections, the calculated modulus

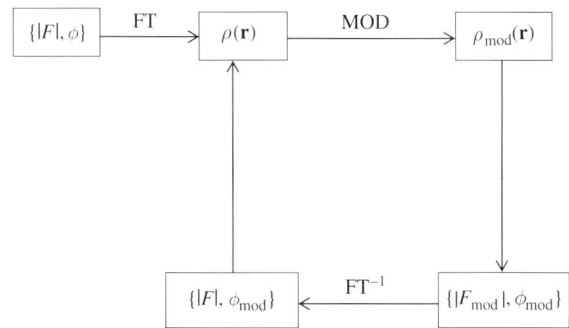

Fig. 9.1

Charge flipping algorithm. $\{|F|\}$ is the set of observed reflections, $\{\phi\}$ is the set of random phases. FT and FT^{-1} indicate the direct and the inverse Fourier transforms, respectively, MOD is the function used to modify the electron density, $\rho(\mathbf{r})$. $\{|F|_{\mathrm{mod}}\}$ and $\{\phi_{\mathrm{mod}}\}$ are the structure factor amplitudes and phases obtained by Fourier inversion of ρ_{mod}.

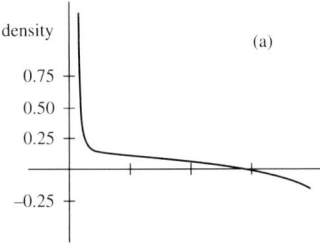

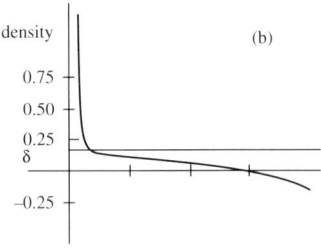

Fig. 9.2
Typical high-resolution electron density distribution, sorted in descending order. The number of pixels is in the abscissa.

is retained and the phase shifted by $\pi/2$. A new electron density map is calculated and the cycle starts again.

Let us first explain the source of the algorithm name. The total charge in a map is assumed to be

$$c_{tot} = \sum_i \rho_i,$$

where i varies over all grid points and the flipped charge is defined as

$$c_{flip} = \sum_{\rho_i < \delta} |\rho_i|,$$

where the summation goes over all points satisfying the condition $\rho_i < \delta$. The various applications showed that, for proper values of δ, the ratio c_{flip}/c_{tot} should lie at around 0.9. As a rule of thumb, it should roughly correspond to inverting the low density pixels shown in Fig. 9.2b. Flipping the density in this region modifies the electron density distribution and allows us to explore the phase space efficiently. Giacovazzo and Mazzone (2011) observed that the flipped region corresponds to that with the largest values of electron density variance. As in other EDM techniques, the algorithm modifies a model without destroying it; in this way the region to reverse is not part of the solution, otherwise the convergence would never be reached. When a good model is obtained, reversing the sign of the density for high variance pixels provides negligible perturbation of the model, which cannot be destroyed.

δ is the most critical parameter; it usually changes during the phasing process and sometimes has to be tuned to lead the algorithm to succeed.

During each charge flipping cycle, the crystallographic residual and the *skewness* coefficient of the density map (see Appendix 8.B) are calculated. The convergence is assumed to be reached when a sharp increase in relative skewness occurs, accompanied by a drop in the crystallographic residual. The phasing process may usually be subdivided into an initial transient step, a long stagnation period where the phase space is extensively explored, a stable state where a sharp increase in the relative skewness occurs, accompanied by a drop in the crystallographic residual. Such a sharp improvement in the figure of merit denotes that convergence has been reached.

Charge flipping solves the structures in P1. This trick was first applied by Sheldrick and Gould (1995), when solution in the correct space group was not being successful; it was later adopted by other authors. The main advantage of solving a structure in P1 is that the restraints imposed by symmetry on the phase values are relaxed and the phases may sometimes converge smoothly to the correct values. However, the use of P1 remained infrequent for direct methods; indeed, symmetry is important prior information which should not generally be suppressed. Charge flipping, however, renounces this information; it is not clear why, but its efficiency decreases dramatically when phasing is attempted using the correct space group symmetry.

In accordance with the above observations, the charge flipping crystal structure solution step is followed by a second step, restating the correct space group symmetry; i.e. it locates the space group symmetry elements in the P1 density map. A technique is therefore necessary to automatically find the shift between the origin of the P1 map and the conventional origin of the space group. This process has to be accompanied by density averaging over the symmetry equivalent (in the correct space group) grid points. The algorithm used for returning to the correct space group is similar to the *RELAX* procedure developed by Burla et al. (2000), and later improved by Caliandro et al. (2007a). Since *RELAX* also plays an important role in the VLD approach, we describe it in Appendix 9.B.

9.3 The *VLD* phasing method

In Section 7.2, we assumed that a model structure is available; to deal in an optimal way with the phasing problem, we calculated the joint probability distribution $P(R, R_p, \phi, \phi_p)$ (see equation (7.3)). In Section 7.3, we showed its extraordinary usefulness for optimization of some widely used crystallographic tools and phasing procedures; we refer in particular to the observed Fourier synthesis via the use of the weight m, and to the difference Fourier synthesis via the use of the Read coefficients, $mE - \sigma_A E_p$.

Let us now consider ρ, ρ_p, ρ_q; these are the target, the model, and their ideal difference structure; $\rho_q = \rho - \rho_p$ has the property that, summed to ρ_p, it provides ρ, no matter what is the quality of ρ_p. Let $R, R_p, R_q, \phi, \phi_p, \phi_q$ be the corresponding normalized diffraction amplitudes and phases. Would the distribution,

$$P(R, R_p, R_q, \phi, \phi_p, \phi_q), \tag{9.1}$$

be more useful than (7.3)? The hope is that including into the probabilistic approach the additional variate E_q could lead to more accurate conditional distributions, estimating phases given three rather than two magnitudes.

Distribution (9.1) (studied by Burla et al., 2010a) is the theoretical basis of the *VLD* (*vive la différence*) algorithm; for the interested reader, some details are quoted in Appendix 9.A.1, together with the conditional distributions which support *VLD*. The *VLD* algorithm (Burla et al., 2010b, 2011a,b) as an ab initio phasing technique is described in Sections 9.3.1 to 9.3.2; its applications to ab initio phasing are summarized in Section 9.3.3. We delay until Section 10.4 some *VLD* applications in combination with Patterson deconvolution techniques: *VLD* combination with molecular replacement is described in Section 13.10.

9.3.1 The algorithm

Distribution (9.1) is practicable only if:

(i) measurement errors are included in the mathematical model;
(ii) the parameter σ_A (calculated between the model and the target structure) is not unity.

Indeed, according to the definition of ρ_q, if condition (i) is violated, then $F_q = F - F_p$ is determined perfectly by the other two variates, and cannot be introduced as a third variable in equation (9.1). If condition (ii) is violated (*i.e.*, $\sigma_A = 1$), then $\rho_p \equiv \rho$ and $\rho_q \equiv 0$; then it is not necessary to calculate a six-variate distribution (indeed, F_q will be identically equal to zero).

These circumstances oblige us to adopt the following mathematical model for the normalized structure factors:

$$E_{\mathbf{h}} = \left\{ \sum_{j=1}^{N_p} f_j \exp[2\pi i \mathbf{h}(\mathbf{r}_{pj} + \Delta \mathbf{r}_j)] \right.$$
$$\left. + \sum_{j=N_p+1}^{N} f_j \exp(2\pi i \mathbf{h} \mathbf{r}_j) + |\mu| \exp(i\vartheta) \right\} \Big/ (\varepsilon \Sigma_N)^{1/2} \quad (9.2a)$$

$$E_{p\mathbf{h}} = \sum_{j=1}^{p} f_j \exp 2\pi i \mathbf{h} \mathbf{r}_{pj} \Big/ (\varepsilon \Sigma_{N_p})^{1/2} \quad (9.2b)$$

$$E_{q\mathbf{h}} = \left\{ \sum_{j=1}^{N_p} f_j \exp[2\pi i \mathbf{h}(\mathbf{r}_{pj} + \Delta \mathbf{r}_j)] \right.$$
$$\left. + \sum_{j=N_p+1}^{N} f_j \exp(2\pi i \mathbf{h} \mathbf{r}_j) - \sum_{j=1}^{N_p} f_j \exp(2\pi i \mathbf{h} \mathbf{r}_{pj}) \right\} \Big/ (\varepsilon \Sigma_q)^{1/2}, \quad (9.2c)$$

where (compare with the corresponding assumptions in Section 7.2):

(a) $\mathbf{r}_{pj}, j = 1, \ldots, N_p$, are known fixed parameters;
(b) $\mathbf{r}_j = \mathbf{r}_{pj} + \Delta \mathbf{r}_j, j = 1, \ldots, p$. The variables $\mathbf{r}_j$ are riding variables, correlated with the corresponding $\mathbf{r}_{pj}$s through local positional errors, $\Delta \mathbf{r}_j$s;
(c) $\mathbf{r}_j, j = p + 1, \ldots, N$ are primitive random variables, assumed to be uniformly distributed within the unit cell;
(d) $|\mu| \exp(i\vartheta)$ is the (complex) measurement error, considered as a two-dimensional primitive random variable;
(e) All of the primitive random variables are assumed to be statistically independent of each other.

The distribution $P(R, R_p, R_q, \phi, \phi_p, \phi_q)$ obtained under the above assumptions and the conditional distributions $P(\phi_q | R, R_p, R_q, \phi_p)$ and $P(\phi | R, R_p, R_q, \phi_p, \phi_q)$ are given in Appendix 9.A. The first conditional distribution is used to calculate the best estimates of $\rho_q(\mathbf{r})$ given $\rho(\mathbf{r})$ and $\rho_p(\mathbf{r})$, the second is used to calculate a new $\rho(\mathbf{r})$ map, given $\rho_p(\mathbf{r})$ and the estimate of $\rho_q(\mathbf{r})$.

The *VLD* algorithm may be briefly schematized as follows (Burla et al., 2011b):

Let $\rho_i(\mathbf{r})$ and $\rho_{pi}(\mathbf{r})$ be the current target and model map estimates at step i; then, $\rho_{qi}(\mathbf{r})$ is calculated via Fourier coefficients suggested by $P(\phi_{qi} | R, R_{pi}, R_{qi}, \phi_{pi})$ and it is suitably modified. The new target map is obtained via the equation,

$$\rho_{i+1}(\mathbf{r}) = \rho_{pi}(\mathbf{r}) + \rho_{q \bmod i}(\mathbf{r}).$$

The procedure is cyclic.

In greater detail, the *VLD* algorithm, theoretically justified in Appendix 9.A, is described below, step by step (see Fig. 9.3 for a flow chart of the algorithm):

1. Random phases, ϕ, are assigned to the observed structure factors; then the electron density, $\rho_{obs}(\mathbf{r})$, (using $R_{obs} \exp(i\phi)$ as coefficients) is calculated.
2. The electron density, $\rho_p(\mathbf{r})$, is obtained by a simple modification of $\rho_{obs}(\mathbf{r})$; i.e. 2.5% of the pixels (those with the largest positive intensity) are accepted unchanged, the rest are set to zero. γ (about ten) EDM cycles, not shown in the figure for brevity, are applied to improve the electron density map; for each cycle, the σ_A parameter is calculated. At the end of the EDM cycles, a new pair $[\rho_{obs}(\mathbf{r}), \rho_p(\mathbf{r})]$ is obtained (see Appendix 9.A).

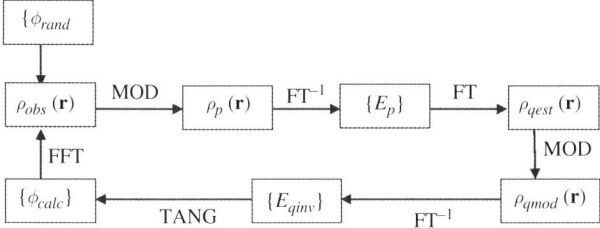

Fig. 9.3
Typical cycle of the standard VLD algorithm.

3. The difference electron density, $\rho_{qest}(\mathbf{r})$, an estimate of the ideal difference synthesis, $\rho_q(\mathbf{r})$, is calculated via coefficients, $(mR - R_p)\exp(i\phi_p)$, suggested by distribution (9.1) (see Appendix 9.A.1). It is suitably modified in $\rho_{qmod}(\mathbf{r})$, as follows: 4% of the pixels with positive density and 4% of the pixels with negative density (those with the largest absolute value of the density) are accepted unchanged, the rest are set to zero.

4. E_{qinv} are obtained by Fourier inversion of $\rho_{qmod}(\mathbf{r})$ and by subsequent histogram matching with a Wilson distribution; they are then summed to E_p (say, $E_{calc} \sim E_p + wE_{qinv}$ via the tangent formula (9.A.8)) to obtain new ϕ estimates (say, ϕ_{calc}).

5. An updated electron density, $\rho_{obs}(\mathbf{r})$, is calculated and the cycle starts again.

Three remarks will be useful here:

(i) According to the Read formulation, if the model is uncorrelated with the target (e.g. in the case of a random model), then the best difference electron density is calculated via the coefficients

$$E_{qest} = mR - \sigma_A R_p \equiv 0. \tag{9.3a}$$

Since all the E_q's are expected to vanish (indeed, m and σ_A are expected to be zero), the best difference electron density should vanish everywhere, because no information is available on the rest of the structure from a random model; consequently, the best difference electron density is

$$\rho_{qest}(\mathbf{r}) \equiv 0. \tag{9.3b}$$

In these conditions, the best target electron density map is

$$\rho(\mathbf{r}) \approx \rho_p(\mathbf{r}) + \rho_{qest}(\mathbf{r}) = \rho_p(\mathbf{r}); \tag{9.3c}$$

In other words, according to the Read formulation, the best target electron density coincides with the model, evidently in contradiction with the assumption that the model is casual.

Distribution (9.1) suggests a different logic: under the condition $(\Sigma_p \approx \Sigma_N)$, the best coefficient for calculating the difference Fourier synthesis is

$$E_{qest} = mR - R_p \equiv -R_p, \tag{9.4a}$$

according to which the best difference electron density is

$$\rho_{qest}(\mathbf{r}) = -\rho_p(\mathbf{r}) \neq 0, \tag{9.4b}$$

and the best target electron density is

$$\rho(\mathbf{r}) = \rho_p(\mathbf{r}) + \rho_{qest}(\mathbf{r}) = 0 \tag{9.4c}$$

The logical result suggested by the above relations is that if the model is random, the best target electron density is that which identically vanishes at all points in the map. In this case, the best $\rho(\mathbf{r})$ is no longer biased towards the model, which was defined as casual. This result agrees well with the expectation that the electron density should not show any structural features, because the model is casual and does not provide any information on $\rho(\mathbf{r})$.

The fact that $\rho_{qest}(\mathbf{r})$ is not vanishing everywhere allows it to be modified; if this is sensible, $\rho_{q\,\text{mod}}(\mathbf{r})$ may better fit the ideal difference electron density $\rho_q(\mathbf{r})$, and, when summed to $\rho_p(\mathbf{r})$, may provide a new, better electron density $\rho(\mathbf{r})$.

(ii) At step 2 of the algorithm, we specified that the σ_A parameter is calculated at each EDM cycle. Since σ_A tends to increase during the EDM cycles even if the phases do not improve, the restraint

$$s_A = (<\sigma_A> - <\sigma_{Ainit}>)/(1 - <\sigma_{Ainit}>)$$

is calculated, where $<\sigma_A>$ is the average (over the resolution shells) σ_A value for the current *VLD* cycle, $<\sigma_{Ainit}>$ is the corresponding experimental value for the initial *VLD* random model. s_A is an estimate of the path fraction covered by $<\sigma_A>$ during the phasing process towards the correct solution (it is assumed that the correct solution is attained when $\sigma_A \approx 1$). It is suggested that s_A should not be larger than

$$r = \left(R_{cryst_{init}} - R_{cryst}\right) / \left(R_{cryst_{init}} - 0.20\right),$$

which is an estimate of the path fraction covered by R_{cryst} towards the correct solution (it is assumed that the correct solution is found when $R_{cryst} = 0.20$).

(iii) As stated in Section 9.2, Sheldrick and Gould (1995) discovered that, when the crystal structure solution is attempted via direct or Patterson methods, the phasing process may occasionally succeed in P1 and fail in the correct space group. It was then clear that symmetry elements can trap phases in false minima, while the absence of symmetry elements allows a softer phasing moving towards their correct values.

The drawbacks of the P1 procedure were essentially twofold. Firstly, lowering the symmetry corresponds to deleting some prior information. Secondly, supplementary work is necessary after crystal structure solution, to find the translation locating the origin in the correct space group.

A different way of overcoming the occasional perverse effects of symmetry operators was suggested by Burla et al. (2000) and made more efficient by Caliandro et al. (2007a). This is the so-called *RELAX* procedure (see Appendix 9.B), which has recently been integrated into the *VLD* algorithm (Burla et al., 2011b). *VLD* phasing is always attempted in the correct space group; a jump in P1 is only made during the cycles in which the figures of merit attain local maxima. In P1 phases are relaxed, but soon the procedure automatically returns to the correct space group, where the phasing process is continued. The procedure is based on the observation that often direct and

Patterson methods provide molecular fragments, correctly oriented but incorrectly located. *RELAX* is the algorithm encharged with the origin shift; its use has allowed a dramatic improvement in *VLD* efficiency.

9.3.2 VLD and hybrid Fourier syntheses

The main properties of hybrid Fourier syntheses were discussed in Section 7.3.4. In the literature, many papers have been dedicated to the subject (Ramachandran and Srinivasan, 1970; Dodson and Vijayan, 1971; Main, 1979; Vijayan, 1980). According to Read (1986), they may be estimated via the coefficients

$$(\tau m|F| - \omega D|F_p|)\exp(i\phi_p), \qquad (9.5)$$

where τ and ω are any pair of real numbers. For brevity, in the following we will denote the hybrid synthesis obtained via τ and ω parameters as the $(\tau - \omega)$ – synthesis; e.g. the usual difference and observed Fourier syntheses coincide with the (1–1)- and (1–0)-syntheses, respectively. Among applications, we quote:

(i) (2–1)-syntheses are employed both in powder (Altomare et al., 2009c) and in macromolecular crystallography (Read, 1986) for improvement of the model structure. Indeed (2–1)-syntheses suffer from model bias much less than (1–0)-syntheses, and may thus allow the model to evolve towards the target structure more easily;
(ii) (4–3)- and (2–1)-syntheses were used by Drendel et al. (1995) in the *forced coalescence algorithm* for ab initio phasing. Even if theoretically interesting, the limited efficiency of the algorithm did not make the method a popular choice.

$(\tau - \omega)$-syntheses with $\tau < \omega$ are less used in crystallographic methods, probably because they are the difference between an ω difference Fourier synthesis and an $\omega - \tau$ observed Fourier synthesis, with a pre-eminent role for the difference component (see Section 7.3.4).

Quite recently (Burla et al., 2011c), study of the joint probability distribution $P(E, E_p, E_Q)$ suggested that, under the condition $\Sigma_p \approx \Sigma_N$, the best coefficient for a hybrid Fourier synthesis is

$$\Delta E = (\tau mR - \omega R_p)\exp(i\varphi_p). \qquad (9.6)$$

The above scenario opened up new perspectives for the *VLD* algorithm, as it may be combined with a $(\tau - \omega)$-synthesis instead of only the (1–1)-synthesis. This combination is then analogous to the use of different types of structure invariants and seminvariants in direct methods phasing procedures. More explicitly, while direct methods may use invariants of different order (i.e. triplet and quartet invariants and/or structure seminvariants), *VLD* can use phase relationships arising from different hybrid electron densities; from the (1–1)-synthesis to any $(\tau - \omega)$-hybrid density.

9.3.3 VLD applications to ab initio phasing

In Section 9.2, we specified the numerous applications for the charge flipping algorithm. *VLD* is a very new approach, and has, so far, only been applied to two fields: to ab initio phasing and combined with Patterson and molecular replacement techniques. Here, we will only deal with the ab initio approach; we postpone to Section 10.4 description of some *VLD* applications in combination with Patterson deconvolution techniques, and to Section 13.10, *VLD* combination with molecular replacement.

To provide the reader with an idea of the efficiency of *VLD*, we remember that Burla et al. (2011c) used 100 small (up to 80 non-hydrogen atoms in the asymmetric unit) structures to check how effective the *VLD* algorithm is when combined with (1–1), (1–2), (1–3), and (2–1) syntheses (four protocols). The set of test structures contained minerals as well as organic and metal-organic compounds; some structures showed strong pseudo-translational symmetry and 42 of them were centrosymmetric.

The results were as follows: all structures were solved in default, no matter which protocol was used; the average number of random models necessary to solve the structures (say $<n_{\text{seed}}>$) always lay between 1.1 and 1.3 (if $<n_{\text{seed}}>$ equal to 1, the correct solution would be obtained from any random model), $<|\Delta\phi|>$ was between 8° and 12°, $<t>$ sec between 26 and 33 (data for current desk computers). Thus, at least for small molecules, the *VLD* user does not need to choose among the various protocols.

Burla et al. (2011c) checked the various protocols against 86 medium-sized (up to about 350 non-H atoms in the asymmetric unit) test structures: 19 were centric, 12 had more than 200 non-H atoms in the asymmetric unit, and for 40 of them, F was the heaviest chemical species. The results were as follows: (i) the *VLD* combinations with (1–1) and (1–2) syntheses were the most effective; (ii) in default conditions, neither syntheses could solve two of the structures; (iii) for the solved ones $<n_{\text{seed}}> = 1.95$, $<t> = 10.7$ min for the (1–1) protocol, $<n_{\text{seed}}> = 2.13$ and $<t> = 6.3$ min for the (1–2) protocol.

It should be stressed that the above result does not imply that *VLD* is unable to solve the missed structures, which may be achieved by increasing the number of random models (by default set to a maximum of 20), or by using non-default parameters.

Protocols (1–1) and (1–2) were also applied to a set of 35 protein test structures at atomic resolution. Results were as follows: (i) in default conditions, protocol (1–1) was unable to solve 13 of the 35 structures. For the solved ones $<n_{\text{seed}}> = 2.18$; (ii) protocol (1–2) in default did not solve 11 proteins, with $<n_{\text{seed}}> = 5.04$.

APPENDIX 9.A ABOUT *VLD* JOINT PROBABILITY DISTRIBUTIONS

9.A.1 The *VLD* algorithm based on difference Fourier synthesis

Under the assumptions stated in Section 9.3, the following joint probability distribution has been found (Burla et al., 2010a):

$$P(R, R_p, R_q, \phi, \phi_p, \phi_q) \cong \pi^{-3} e^{-1} L^{-1} R R_p R_q \exp\{-[\lambda_{11} R^2 + \lambda_{22} R_p^2 + \lambda_{33} R_q^2 \\ + 2\lambda_{12} R R_p \cos(\phi - \phi_p) + 2\lambda_{13} R R_q \cos(\phi - \phi_q) \\ + 2\lambda_{23} R_p R_q \cos(\phi_p - \phi_q)]\},$$

(9.A.1)

where, in accordance with definitions given in Section 7.2, model and target structures consist of N_p and N atoms, respectively,

$$\sigma_A = D\sqrt{\sum\nolimits_{N_p} \Big/ \sum\nolimits_N}$$

$$D = <\cos(2\pi \mathbf{h} \cdot \Delta \mathbf{r})>$$

$$e = (1 + \sigma_R^2), \quad \text{and} \quad \sigma_R^2 = <|\mu|^2>/\Sigma_N.$$

Furthermore,

$$L = \frac{(e-1)(1-\sigma_A^2)\Sigma_N}{e\Sigma_q},$$

$$\lambda_{11} = \frac{1}{(e-1)}, \quad \lambda_{22} = \frac{\Sigma_q}{\Sigma_N}\frac{1}{1-\sigma_A^2} + \frac{\Sigma_p}{\Sigma_N}\frac{1}{e-1},$$

$$\lambda_{33} = \frac{\Sigma_q}{\Sigma_N}\left[\frac{1}{e-1} + \frac{1}{1-\sigma_A^2}\right], \quad \lambda_{12} = -\left(\frac{\Sigma_p}{\Sigma_N}\right)^{1/2}\frac{1}{e-1},$$

$$\lambda_{13} = -\left(\frac{\Sigma_q}{\Sigma_N}\right)^{1/2}\frac{1}{e-1},$$

$$\lambda_{23} = \frac{\left(\Sigma_p \Sigma_q\right)^{1/2}}{\Sigma_N}\left[\frac{e-\sigma_A^2}{(e-1)(1-\sigma_A^2)}\right] - \left(\frac{\Sigma_q}{\Sigma_N}\right)^{1/2}\frac{\sigma_A}{1-\sigma_A^2}.$$

Σ_q is a D-dependent parameter, as the following relationship suggests:

$$\Sigma_q = <|F_q|^2> = \Sigma_p(1-2D) + \Sigma_N.$$

Accordingly, Σ_q depends on the quality of the model; it tends to $\Sigma_N - \Sigma_p$ when $D = 1$ and it tends to $\Sigma_N + \Sigma_p$ when ρ_p progressively loses (up to $D = 0$) its isomorphism with ρ. This behaviour is confirmed in Figs. 9.A.1 to 9.A.3.

From equation (9.A.1), the conditional distribution

$$P(\phi_q|R, R_p, R_q, \phi_p) \cong [2\pi I_0(G_q)]^{-1} \exp\left\{G_q \cos(\phi_q - \phi_p)\right\} \qquad (9.A.2)$$

is obtained, where

$$G_q = \frac{2R'_q}{e-1}\left\{(mR - \sigma_A R_p) - R'_p(1-D)\left(\frac{e-\sigma_A^2}{1-\sigma_A^2}\right)\right\} \qquad (9.A.3)$$

where $R'_p = R_p(\Sigma_p/\Sigma_N)^{1/2}$ and $R'_q = R_q(\Sigma_q/\Sigma_N)^{1/2}$.

In terms of structure factors, equation (9.A.3) becomes,

$$G_q = \frac{2|F_q|}{(e-1)\Sigma_N}\left\{(m|F| - D|F_p|) - |F_p|(1-D)\left(\frac{e-\sigma_A^2}{1-\sigma_A^2}\right)\right\} \qquad (9.A.4)$$

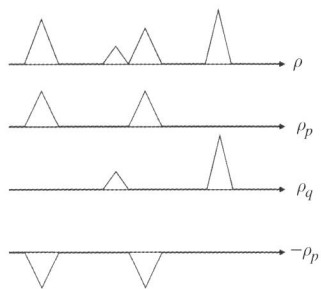

Fig. 9.A.1
Non-realistic one-dimensional four-atom target structure (ρ) and a two-atom model (say, ρ_p), the atomic positions of which perfectly coincide with two atoms of the target structure (parameter $D = 1$). In this case, ρ_q is everywhere positive and has a reduced scattering power (compared with that in Figs. 9.A.2 and 9.A.3).

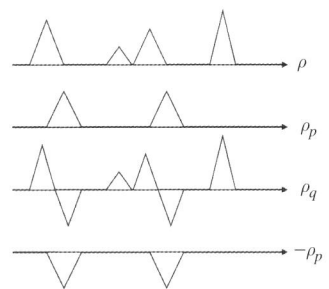

Fig. 9.A.2
Schematic target and model structures for $N = 4$, $N_p = 2$. Model and target structures partially overlap; parameter D in the interval (0,1).

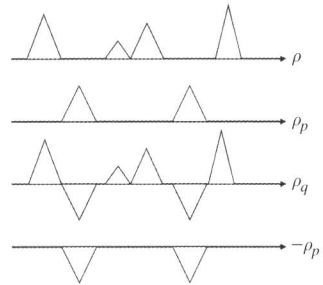

Fig. 9.A.3
Schematic representation of ρ, ρ_p, and ρ_q when the model is uncorrelated with the target structure (then, $D = 0$). ρ_q is not everywhere positive and has very large scattering power.

Equations (9.A.3) and (9.A.4) suggest the following considerations:

1. The phase relationship used for the difference Fourier synthesis (7.18), which is

$$\phi_q \approx \phi_p \text{ if } m|F| > D|F_p|, \phi_q \approx \phi_p + \pi \text{ if } m|F| < D|F_p|,$$

is no longer supported. Indeed, according to equation (9.A.4), ϕ_q is expected to be close to ϕ_p, or close to $\phi_p + \pi$, according to whether G_q is positive or negative.

2. The sign of G_q does not always coincide with the sign of $(m|F| - D|F_p|)$. Indeed, the right-hand side of equation (9.A.4) is the sum of two contributions, the first depending on the value of $(m|F| - D|F_p|)$, called the *difference term*; the second (say, $|F_p|(1-D)\left(\frac{e-\sigma_A^2}{1-\sigma_A^2}\right)$), is called the *flipping term*, and is always negative and proportional (via a positive factor) to $-|F_p|$. Its contribution depends on the quality of the model structure, and increases with the poorness of the model. It is dominant when the model is very poor (then $(m|F| - D|F_p|) = 0$) and tends to vanish for high quality structural models.

3. A new difference Fourier coefficient may be conjectured, equal to

$$\Delta E = \left[(mR - \sigma_A R_p) - R'_p(1-D)\left(\frac{e-\sigma_A^2}{1-\sigma_A^2}\right)\right]\exp(i\phi_p). \quad (9.A.5)$$

The corresponding difference Fourier map will show three different types of peaks, the properties of which are determined by the quality of the model. For a very poor model, the map will show the following peaks:

(i) very strong negative peaks where model atoms do not overlap with target atoms. In this case, both the *difference* and *flipping* terms will generate negative electron density;
(ii) medium intensity negative peaks, where model and target atoms overlap. In this case, the *difference term* does not provide any contribution to the electron density, while the *flipping term* will generate negative electron density;
(iii) medium intensity positive peaks, where target atoms do not overlap with model atoms. In this case, the *difference term* provides a positive electron density, while the *flipping term* does not provide any contribution to the electron density. The intensity ratio between peaks (i) and peaks (ii) and (iii) will decrease when the model becomes a better approximation of the target structure;
(iv) estimates of ϕ_q are available no matter what is the quality of the model structure. For example, even in the limit case in which σ_A and D are zero (model and target structures completely uncorrelated, as in Fig. 9.A.3), the parameter G_q may be large and therefore the ϕ_q estimate may be reliable. That is equivalent to the following statement:

it is possible to obtain a meaningful estimate of ρ_q, even when ρ_p and ρ are completely uncorrelated (e.g. when ρ_p is randomly fixed). (9.A.6)

This property is not shared by the classical synthesis (7.18): when $CORR \sim 0$, then $<|F_q|> = |m|F|-D|F_p||$ is vanishing, and consequently the intensity of any pixel of the difference Fourier synthesis, calculated via those coefficients, is expected to vanish. In practice, no information in, no information out.

Statement (9.A.6) seems to be without foundation, however, the correctness of the approach becomes clear if one considers Fig. 9.A.3, where simplified models for ρ, ρ_p, and ρ_q are schematized, when $\sigma_A = 0$. On assuming that ρ is unknown and that ρ_p is uncorrelated with ρ (they have no peak in common), then ρ_q consists of N positive and N_p negative peaks. It can easily be seen that ρ_q and $-\rho_p$ are positively correlated, because they have p negative peaks in common (corresponding to the wrongly located atoms in the ρ_p structure). Equivalently, ρ_p and ρ_q are anticorrelated, and it is just the *flipping term* in equations (9.A.5) and (9.A.6) which guarantees the anticorrelation. Accordingly, an estimate of ρ_q is always possible, even when the atoms of the model structure are randomly located.

Similar observations also hold for Fig. 9.A.2, where σ_A and D do not coincide neither with 0 nor with 1. In this case, ρ_q consists of $N - N_p$ positive peaks and by N_p electron density residuals, consisting of pairs of positive and negative peaks. Now, ρ_q and ρ_p are weakly anticorrelated; such a relation is taken fully into account when both the difference and the flipping term are used.

If we consider Fig. 9.A.1, where the case $D = 1$ is schematized, it can be seen that ρ_q and ρ_p are uncorrelated. Accordingly, the flipping term vanishes and structure completion may only occur through the difference term.

To check the extraordinary property of coefficients (9.A.5), we selected (see Table 9.A.1) a number of protein test structures for which models are available, with different degrees of correlation with the target structure (*CORR* in the table). We calculated the difference electron density by using both Read coefficients and (9.A.5) coefficients. Correlations of the corresponding maps with the ideal difference density are denoted by $CORRq_{READ}$ and $CORRq_{VLD}$. The table shows that, even when *CORR* is close to zero, the difference map calculated with *VLD* coefficients is well correlated with the ideal difference density.

A second conditional probability may be derived from equation (9.A.1), say, $P(\phi|R, R_p, R_q, \phi_p, \phi_q)$, which assumes that the moduli R, R_p, R_q and the phases ϕ_p, ϕ_q are known:

$$P(\phi|R, R_p, R_q, \phi_p, \phi_q) = [2\pi I_0(Q)]^{-1} \exp[Q\cos(\phi - \vartheta)], \quad (9.A.7)$$

where ϑ is the most probable value of ϕ, given by

$$\tan\vartheta = \frac{R'_p \sin\phi_p + R'_q \sin\phi_q}{R'_p \cos\phi_p + R'_q \cos\phi_q} = \frac{Q_T}{Q_B}, \quad (9.A.8)$$

and

$$Q = 2R(e-1)^{-1}(Q_T^2 + Q_B^2)^{1/2}$$
$$= 2R(e-1)^{-1}\left[R'^2_p + R'^2_q + 2R'_p R'_q \cos(\phi_p - \phi_q)\right]^{1/2} \quad (9.A.9)$$

is its reliability factor.

Equation (9.A.7) suggests that, when an estimate for ϕ_q is obtained (even from a model uncorrelated with the model structure) via distribution (9.A.2),

Table 9.A.1 For some protein test structures (PDB is the protein data bank code) we show: (1) The correlation (*CORR*) of the protein electron density with the model structure (not reported): values of *CORR* close to unity indicate good models, values close to 0 indicate uncorrelated models. (2) $CORR_{qRead}$ is the correlation between the difference electron density calculated by Read coefficients and the ideal difference structure. $<\Delta\phi_q>_{Read}$ is the corresponding average phase error. When the model is uncorrelated with the target, $CORR_{qRead}$ is close to zero. (3) $CORR_{VLD}$ is the correlation between the difference electron density calculated by coefficients (9.A.5) and the ideal difference structure. $<\Delta\phi_q>_{VLD}$ is the corresponding average phase error. When the model is uncorrelated, $CORR_{VLD}$ is particularly high

PDB	CORR	$CORR_{qRead}$	$<\Delta\phi_q>_{READ}$	$CORR_{VLD}$	$<\Delta\phi_q>_{VLD}$
1kf3	0.93	0.51	57	0.62	53
6rhn	0.89	0.40	63	0.62	54
1zs0	0.82	0.38	64	0.61	54
2p0g	0.72	0.30	66	0.60	59
2sar	0.70	0.36	62	0.59	55
1cgn	0.39	0.19	65	0.62	53
2iff	0.36	0.18	71	0.51	67
9pti	0.02	0.01	73	0.62	54
6ebx'	0.01	0.00	70	0.66	46
9pti'	0.01	−0.02	74	0.66	50

it may be combined with ϕ_p to provide a new phase estimate for the target structure. This result is the basis of the *VLD* algorithm.

The parameters involved in distribution (9.A.1), and therefore in the conditional distributions (9.A.2) and (9.A.7), seem to be complicated to use. We will now see how their expressions may be simplified. The *VLD* algorithm works on maps, therefore it does not need a molecular model, a model map being sufficient. Accordingly, model structure factors may be calculated by inversion of some percentage of the electron density map. In the present version of the *VLD* algorithm, it has been assumed that the percentage is large enough to contain the quasi-totality of the electrons; in this case, $\Sigma_{N_p} \approx \Sigma_N$. This approximation simplifies the parameters as follows:

(i) Since $R'_p = R_p(\Sigma_p/\Sigma_N)^{1/2}$, R'_p reduces to R_p.
(ii) Since $R'_q = R_q(\Sigma_q/\Sigma_N)^{1/2}$, $D = \sqrt{\frac{\Sigma_N}{\Sigma_p}}\sigma_A$ and $\Sigma_q = \Sigma_p(1-2D) + \Sigma_N$, we have,

$$\Sigma_q/\Sigma_N = 2(1-\sigma_A) \quad \text{and} \quad R'_q = R_q\sqrt{2(1-\sigma_A)}.$$

(iii) If, in addition, it is assumed that e is very close to unity then,

$$\Delta E \approx (mR - R_p)\exp(i\phi_p) \qquad (9.A.10)$$

is the coefficient to use for calculation of the difference Fourier synthesis, and the tangent formula (9.A.8) may be rewritten as

$$\tan\vartheta = \frac{R_p\sin\phi_p + w_qR_q\sin\phi_q}{R_p\cos\phi_p + w_qR_q\cos\phi_q}, \qquad (9.A.11)$$

where

$$w_q = \sqrt{2(1-\sigma_A)}.$$

Equation (9.A.11) makes clear the important role of E_q in the first steps of the phasing process, when σ_A is close to zero and $w_q \approx \sqrt{2}$, and its marginal contribution in the last steps, when σ_A is frequently close to unity and $w_q \approx 0$. In the first case, E_q contributes strongly to completion or modification of the crystal structure model, in the second, the crystal structure refinement may only rely on the efficiency of the EDM procedures.

9.A.2 The VLD algorithm based on hybrid Fourier syntheses

Quite recently (see Burla et al., 2011c), the joint probability distribution

$$P(E, E_p, E_Q) \qquad (9.A.12)$$

was studied, where E_Q is the normalized structure factor of the hybrid map, $\rho_Q = \tau\rho - \omega\rho_p$.

The Fourier coefficient suggested by (9A.12) for the $(\tau - \omega)-$ synthesis is

$$\left\{(\tau mR - \omega\sigma_A R_p) - R'_p \left[\omega(1-D) + (\omega - \tau D)\frac{(e-1)}{1-\sigma_A^2}\right]\right\} \exp(i\phi_p), \qquad (9.A.13)$$

or, in terms of structure factors,

$$\left\{(\tau m|F| - \omega D|F_p|) - |F_p| \left[\omega(1-D) + (\omega - \tau D)\frac{(e-1)}{1-\sigma_A^2}\right]\right\} \exp(i\phi_p). \qquad (9.A.14)$$

Again, Read and flipping coefficients are simultaneously present in expressions (9A.13) and (9A.14). If $\omega = \tau = 1$, equations (9.A.13) and (9.A.14) reduce to the difference Fourier coefficients given in Section 9.A.1.

Under the condition $\sum_p \approx \sum_N$, the best coefficient for a hybrid Fourier synthesis is

$$\Delta E = (\tau mR - \omega R_p) \exp(i\phi_p), \qquad (9.A.15)$$

and the VLD algorithm may be schematized as follows:

(a) The chosen hybrid synthesis is calculated via the coefficient ΔE, given by (9A.15); it is conveniently modified and inverted, no matter what is the quality of the model. Let ϕ_Q be the phase obtained by Fourier inversion, and R_Q, the corresponding normalized amplitudes (obtained by submitting the moduli obtained by Fourier inversion to histogram matching with the Wilson distribution).

(b) The corresponding Fourier coefficients, E_Q, are combined with the normalized structure factors of the model structure, through the tangent formula,

$$\tan\phi = \frac{\omega R_p \sin\phi_p + w_Q R_Q \sin\phi_Q}{\omega R_p \cos\phi_p + w_Q R_Q \cos\phi_Q}, \qquad (9A.16)$$

where

$$w_Q = [\tau^2 + \omega^2 - 2\omega\tau\sigma_A]^{1/2},$$

and
$$\xi_Q = 2R\tau^{-1}\left[\omega^2 R_p^2 + w_Q^2 R_Q^2 + 2\omega w_Q R_p R_Q \cos(\phi_p - \phi_Q)\right]^{1/2} \quad (9A.17)$$
is the reliability parameter of the phase indication.

(c) The observed Fourier synthesis (using the phases ϕ, defined by equation (9.A.16)) is calculated and submitted to cycles of EDM. At the end, a new model structure is obtained and the procedure returns to (a).

APPENDIX 9.B THE *RELAX* ALGORITHM

The algorithm may be described as follows:

1. Structure solution is attempted in the correct space group; the phases ϕ are assigned.
2. The reflections are expanded in P1. So far the space group symmetry is still fulfilled; indeed, reflections that are symmetry equivalent in the correct space group have, in P1, the same amplitudes and symmetry related phases.
3. Some EDM cycles are performed to relax the phases. After this EDM stage, the space group symmetry is not fully obeyed; while the reflections that are symmetry equivalent in the correct space group maintain in P1 the same observed amplitudes, the phases of the corresponding calculated structure factors usually do not reflect the space group symmetry, because they were refined in the EDM step without any constraint.
4. A permissible origin for the correct space group is searched. To do that, the Cheshire cell (Hirshfeld, 1968) is explored according to a grid of 0.3 Å. Each node of the grid is defined by the generic vector $\mathbf{X}_j$ and for each node the following figure of merit is calculated:

$$S_2 = \sum w_\mathbf{h} w_\mathbf{hR} \cos(\phi'_\mathbf{h} - \phi'_\mathbf{hR} + 2\pi \mathbf{hT})/\sum w_\mathbf{h} w_\mathbf{hR}$$
$$- \sum w_\mathbf{h} w_\mathbf{hR} |\sin(\phi'_\mathbf{h} - \phi'_\mathbf{hR} + 2\pi \mathbf{hT})|/\sum w_\mathbf{h} w_\mathbf{hR},$$

where $\mathbf{R}$ and $\mathbf{T}$ are the rotational and the translational matrices of the generic symmetry operator, respectively. The summation covers all phased reflections. Furthermore,

$$\phi'_\mathbf{h} = \phi_\mathbf{h} - 2\pi \mathbf{hX}_j \text{ and } \phi'_\mathbf{hR} = \phi_\mathbf{hR} - 2\pi \mathbf{hRX}_j,$$

are the phase values to assign to $F_\mathbf{h}$ and to $F_\mathbf{hR}$, respectively, when the origin has been translated by the vector $\mathbf{X}_j$. $w_\mathbf{h} = |F_\mathbf{h}|_{calc}$ is the weight assigned to the reflection $\mathbf{h}$.
Since, in the correct space group,

$$\phi_\mathbf{hR} = \phi_\mathbf{h} - 2\pi \mathbf{hT},$$

a good translation should show $(\phi'_\mathbf{h} - \phi'_\mathbf{hR} - 2\pi \mathbf{hT})$ values close to zero. Accordingly, the maximum of S_2 should identify the suitable origin translation.

5. The grid point for which S_2 is a maximum fixes the most probable origin translation. Let $\mathbf{X}_0$ be such a grid point; then the P1 phases $\phi_\mathbf{h}$ are modified according to the relation

$$\phi'_\mathbf{h} = \phi_\mathbf{h} - 2\pi \mathbf{hX}_0.$$

6. Since the various ϕ_{hR} (obtained by varying the **R** matrix) were refined in P1 without constraints, at this stage they are not expected to strictly satisfy the space group symmetry. To re-establish this, for each set of equivalent reflections, a unique reflection has to be selected, to which a suitable phase value should be assigned. Then symmetry related phase values should be assigned to the symmetry equivalent reflections.
7. Let $\mathbf{h}_a$ be the unique reflection, and $\mathbf{h}_a\mathbf{R}_j$ be the vectorial index of its jth symmetry equivalent reflection. The reflection $F_{\mathbf{h}_a}$ is assigned the phase $\phi_{\mathbf{h}_a}$ defined by

$$\tan\phi'_{\mathbf{h}_a} = \sum_j w_j \sin(\phi'_{\mathbf{h}_a})_j \bigg/ \sum_j w_j \cos(\phi'_{\mathbf{h}_a})_j = T/B.$$

The summation is over the m symmetry equivalent reflections, and

$$(\phi'_{\mathbf{h}_a})_j = \phi'_{\mathbf{h}_a\mathbf{R}_j} + 2\pi\mathbf{h}_a\mathbf{T}_j$$

is the phase value which one should associate with the reflection $\mathbf{h}_a$ if $\phi'_{\mathbf{h}_a\mathbf{R}_j}$ was determined correctly.
8. Eventually, other phase refinement cycles are performed to improve the structure model.

Unfortunately, *RELAX*, as defined by the above algorithm, is very time consuming; indeed its application requires the calculation of S_2 for all of the grid points in the Cheshire cell. As a consequence, *RELAX*, in the above version, was only ever used in difficult cases and for a very small number of selected trial solutions.

In order to simplify the *ab initio* crystal structure solution of proteins, Caliandro et al. (2007a) combined *RELAX* with Patterson deconvolution methods; this time *RELAX* was calculated via FFT, and led to a reduction in cpu time of more than one order of magnitude.

FFT calculations allowed more extended applications of *RELAX*. Thus direct methods and the *VLD* approach may also benefit; indeed the use of *RELAX* allows us to obtain the correct solution from a misplaced electron density, which is not possible with traditional approaches.

However, with the *VLD* procedure, the use of *RELAX* must be judicious; indeed, thousands of trial models may be generated, and the trivial combination *VLD-RELAX* may dramatically slacken the phasing process. Luckily, only some of the models are appropriate to be treated using *RELAX*, and these may be preliminarily identified via suitable figures of merit.

10 Patterson methods and direct space properties

10.1 Introduction

According to the basic principles of structural crystallography, stated in Section 1.6:

(i) it is logically possible to recover the structure from experimental diffraction moduli;
(ii) the necessary information lies in the diffraction amplitudes themselves, because they depend on interatomic vectors.

The first systematic approach to structure determination based on the above principle was developed by Patterson (1934a,b). In the small molecule field related techniques, even if computerized (Mighell and Jacobson, 1963; Nordman and Nakatsu, 1963), were relegated to niche by the advent of direct methods. Conversely, in macromolecular crystallography, they survived and are still widely used today. Nowadays, Patterson techniques have been reborn as a general phasing approach, valid for small-, medium-, and large-sized molecules.

The bases of Patterson methods are described in Section 10.2; in Section 10.3 some methods for Patterson deconvolution (i.e. for passing from the Patterson map to the correct electron density map) are described, and in Section 10.4 some applications to ab initio phasing are summarized. The use of Patterson methods in non-ab initio approaches like MR, SAD-MAD, or SIR-MIR are deferred to Chapters 13 to 15.

We do not want to leave this chapter without mentioning some fundamental relations between direct space properties and reciprocal space phase relationships. Patterson, unlike direct methods, seek their phasing way in direct space; conversely, DM are the counterpart, in reciprocal space, of some direct space properties (positivity, atomicity, etc.). One may wonder if, by Fourier transform, it is possible to immediately derive phase information from such properties, without the heavy probabilistic machinery. In Appendix 10.A, we show some of many relations between electron density properties and phase relationships, and in Appendix 10.B, we summarize some relations between Patterson space and phase relationships.

Patterson (1949) defined a second synthesis, known as the *Patterson synthesis of the second kind*. Even if theoretically interesting, it is of limited use in practice. We provide information on this in Appendix 10.C.

10.2 The Patterson function

10.2.1 Mathematical background

Let $\rho(\mathbf{r})$ be the electron density of our target structure, and

$$F_\mathbf{h} = \sum_{j=1}^{N} f_j \exp(2\pi i \mathbf{h} \cdot \mathbf{r}_j) = T[\rho(\mathbf{r})] = \int_V \rho(\mathbf{r}) \exp(2\pi i \mathbf{h} \cdot \mathbf{r}) \, d\mathbf{r}$$

its generic structure factor. We express $F_\mathbf{h}$ in different ways to emphasize both its practical and its basic definitions. T indicates the Fourier transform, and V is the volume of the unit cell. Conversely, the electron density may be obtained via Fourier transform of $F_\mathbf{h}$ (assumed to be known in modulus and phase):

$$T^{-1}[F_\mathbf{h}] = \rho(\mathbf{r}) = \frac{1}{V} \sum_\mathbf{h} F_\mathbf{h} \exp(-2\pi i \mathbf{h} \cdot \mathbf{r}). \quad (10.1)$$

Let us now consider the squared amplitude $|F_\mathbf{h}|^2$; we wonder which information the inverse Fourier transform of $|F_\mathbf{h}|^2$ should provide us, say,

$$T^{-1}\left[|F_\mathbf{h}|^2\right] = \frac{1}{V} \sum_\mathbf{h} |F_\mathbf{h}|^2 \exp(-2\pi i \mathbf{h} \cdot \mathbf{r}). \quad (10.2)$$

This question is very important because equation (10.2) may be directly calculated from the diffraction amplitudes, without any phase information. Since $|F_\mathbf{h}|^2 = F_\mathbf{h} \cdot F_{-\mathbf{h}}$, we have, because of the convolution theorem,

$$T^{-1}\left[|F_\mathbf{h}|^2\right] = T^{-1}[F_\mathbf{h} \cdot F_{-\mathbf{h}}] = \rho(\mathbf{r}) \otimes \rho(-\mathbf{r}) = \int_V \rho(\mathbf{r}) \, \rho(\mathbf{r}+\mathbf{u}) d\mathbf{r}. \quad (10.3)$$

The integral on the right-hand side of (10.3) is the *autoconvolution* of the function $\rho(\mathbf{r})$, called a *Patterson function* in crystallography, and indicated by $P(\mathbf{u})$. We can rewrite (10.3) as

$$P(\mathbf{u}) = \frac{1}{V} \sum_\mathbf{h} |F_\mathbf{h}|^2 \exp(-2\pi i \mathbf{h} \cdot \mathbf{u}) = \frac{2}{V} \sum_{\mathbf{h}>0} |F_\mathbf{h}|^2 \cos(2\pi \mathbf{h} \cdot \mathbf{u}), \quad (10.4)$$

where the symbol $\mathbf{h} > 0$ means that the summation goes over half the reciprocal space (the Friedel opposites are included within the summation). The result is that the integral on the right-hand side of (10.3) may be calculated via the Fourier synthesis on the right-hand side of (10.4),

$$P(\mathbf{u}) = \int_V \rho(\mathbf{r}) \, \rho(\mathbf{r}+\mathbf{u}) d\mathbf{r} = \frac{2}{V} \sum_{\mathbf{h}>0} |F_\mathbf{h}|^2 \cos(2\pi \mathbf{h} \cdot \mathbf{u}) \quad (10.5)$$

We now have to discover what information is provided by $P(\mathbf{u})$.

10.2.2 About interatomic vectors

According to (10.5):

1. The periodicity of $P(\mathbf{u})$ is the same as that of $\rho(\mathbf{r})$; both functions are characterized by the same unit (primitive or centred) cell.
2. $P(\mathbf{u}) = 0$ in $\mathbf{u}$, unless at least two atoms with interatomic vector $\mathbf{u}$ exist (see the medium term in equation (10.5)). Indeed, if only two atoms exist, one in $\mathbf{r}_i$ with density $\rho(\mathbf{r}_i)$ and one in $\mathbf{r}_i + \mathbf{u}$ with density $\rho(\mathbf{r}_i + \mathbf{u})$, then $P(\mathbf{u}) \approx \rho(\mathbf{r}_i) \cdot \rho(\mathbf{r}_i + \mathbf{u})$. If more pairs exist in the unit cell, all with the same interatomic vector $\mathbf{u}$, each jth pair contributes to $P(\mathbf{u})$ with a quantity proportional to $\rho(\mathbf{r}_j)\rho(\mathbf{r}_j + \mathbf{u})$. In this case, $P(\mathbf{u}) \approx \sum_{pairs} \rho(\mathbf{r}_j)\rho(\mathbf{r}_j + \mathbf{u})$.
3. If N atoms are in the unit cell, there are N-1 vectors from any given atom to the others, and one (null) vector from itself. Accordingly, we will have N^2 interatomic vectors, say, $\mathbf{r}_i - \mathbf{r}_j, i, j = 1, \ldots, N$. When $i = j$ the vectors correspond to N zero-distances of each atom with itself; they superimpose at the origin of the Patterson map, which is by far the most intense peak on the map. The remaining $N \cdot (N-1)$ vectors are distributed within the unit cell.
4. $P(\mathbf{u})$ is always cs., even if ρ is n.cs. (indeed, if $\mathbf{r}_i - \mathbf{r}_j$ exists, $\mathbf{r}_j - \mathbf{r}_i$ also exists). As an example, in Fig. 10.1a, a one-dimensional very simple electron density is shown, consisting of three atoms with different atomic weights, denoted 1, 2, and 3. In Fig. 10.1b, the corresponding Patterson map is shown; each peak is labelled with two numbers, corresponding to the atom labels in the electron density. The largest peak is in $\mathbf{u} = 0$, where the zero-distances of each atom with itself overlap. The intensities of the other peaks are proportional to the products of the corresponding atomic numbers, say, Z_3Z_2, Z_3Z_1, Z_2Z_1.
5. Since the number of interatomic vectors is much larger than the number of atomic positions, the Patterson function shows a strong peak overlapping (e.g. for 1000 atoms in the unit cell there are 10^6 interatomic vectors). In particular, in the origin, N zero-distance vectors overlap; the overall intensity is proportional to $\sum_{j=1}^{N} Z_j^2$.

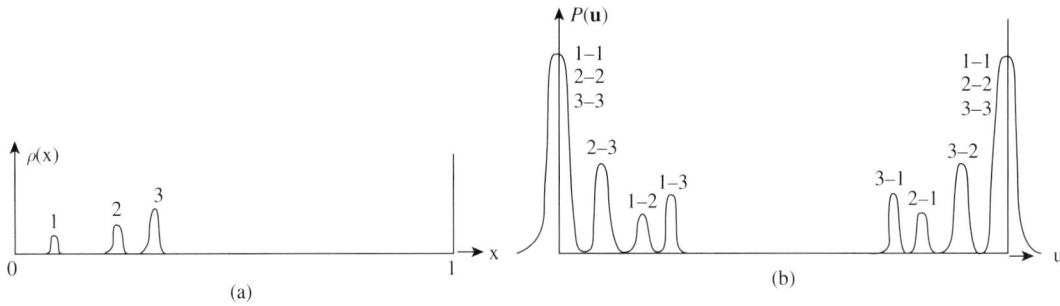

Fig. 10.1
a) A simple one-dimensional structure composed of three atoms with different atomic number Z. Z is assumed to be proportional to the peak label; b) the corresponding Patterson map.

6. A Patterson peak in **u** has a width approximately equal to the sum of the widths of the corresponding electron density peaks in $\mathbf{r}_i$ and in $\mathbf{r}_i + \mathbf{u}$. This is a consequence of the fact that $P(\mathbf{u})$ is the autoconvolution of the electron density. Indeed, if we approximate the electron density of two atoms by two Gaussians functions, with standard deviations σ_A and σ_B, the convolution will still be Gaussian, and will have standard deviation equal to $(\sigma_A^2 + \sigma_B^2)^{1/2}$. This behaviour is an additional reason for the peak overlapping; as a consequence, the resulting Patterson map may be almost featureless, unless some heavy atoms are present.

7. To reduce the overlapping problem, it is convenient to sharpen the map, by using $|E|^2$ or $(|E|^3|F|)^{1/2}$ as Fourier coefficients; we should not forget, however, that any sharpening increases the series truncation errors and the consequent ripples (see Section 7.3.1).

Figure 10.1 suggests that a three-dimensional Patterson map, in ideal conditions (that is in the absence of peak overlapping and devoid of resolution effects), provides the information contained in the following matrix:

$$\begin{vmatrix} Z_1Z_1, \mathbf{r}_1 - \mathbf{r}_1 & Z_1Z_2, \mathbf{r}_1 - \mathbf{r}_2 & \cdots & Z_1Z_N, \mathbf{r}_1 - \mathbf{r}_N \\ Z_2Z_1, \mathbf{r}_2 - \mathbf{r}_1 & Z_2Z_2, \mathbf{r}_2 - \mathbf{r}_2 & \cdots & Z_2Z_N, \mathbf{r}_2 - \mathbf{r}_N \\ \cdots & \cdots & \cdots & \cdots \\ Z_NZ_1, \mathbf{r}_N - \mathbf{r}_1 & Z_NZ_2, \mathbf{r}_N - \mathbf{r}_2 & \cdots & Z_NZ_N, \mathbf{r}_N - \mathbf{r}_N \end{vmatrix}.$$

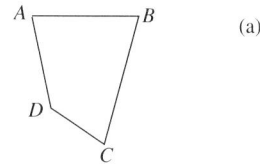

Each element of the matrix contains information on the peak intensity and on its location; the diagonal entries correspond with zero-distance peaks, all overlapping in the origin. It may also be noted that each *j*th line (or column) provides an image of the structure as seen from the *j*th atom. Accordingly, a Patterson map may be thought of as a sum of images (Wrinch, 1939; Buerger, 1946, 1959; Clastre and Gray, 1950; Garrido, 1950). In Fig. 10.2a, a schematic four-atom structure is shown, and in Fig. 10.2b, the corresponding vector set is marked; the vector set as the sum of structure images is clearly recognizable. This property will be exploited in the Sections 10.3.3 and 10.3.4 to solve the structural problem.

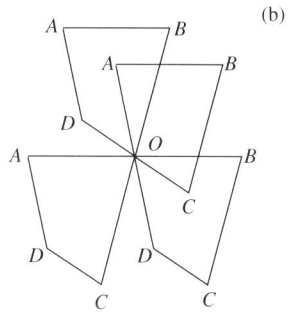

Properties (1) to (7), listed above, can be easily recognized in Figs. 10.3a and b; in (a) we show a schematic two-dimensional unit cell, space group *pg* (equivalent positions (x,y), (−x, y + 0.5)) with three symmetry independent atoms (one S and two O); in (b) the corresponding Patterson map is shown. Because of the periodic nature of the Patterson map, the peaks lying on the edges of the unit cell are repeated after a period.

10.2.3 About Patterson symmetry

The electron density,

$$\rho(\mathbf{r}) = \frac{1}{V} \sum_{\mathbf{h}} F_{\mathbf{h}} \exp(-2\pi i \mathbf{h} \cdot \mathbf{r}), \quad (10.6)$$

shows space group symmetry because the summation on the right-hand side of (10.6) embraces all reflections, symmetry equivalents included; each $\bar{\mathbf{h}}R$

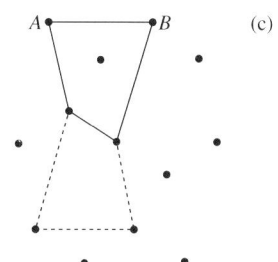

Fig. 10.2
a) Schematic four-atom molecule. b) Corresponding vector set. c) Two images of the molecule obtained by superimposing on the map in b) the same map shifted by the vector OC.

218 Patterson methods and direct space properties

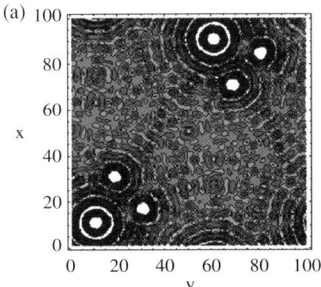

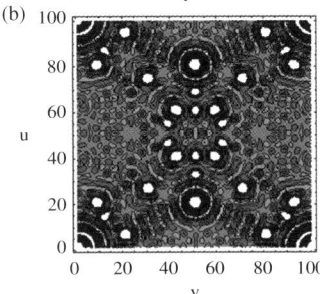

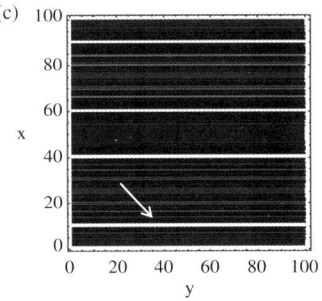

Fig. 10.3
a) Schematic two-dimensional structure, plane group *pg*, consisting of three symmetry-independent atomic positions (one S and two O). b) Patterson map of the structure in a). c) The $I_s(\mathbf{r})$ function, for $s = 2$, as defined by equation (10.8). By courtesy of Rocco Caliandro.

reflection, symmetry equivalent to **h**, enters into equation (10.6) as a complex number defined by

$$F_{\bar{\mathbf{h}}R} = F_{\mathbf{h}} \exp(-2\pi i \bar{\mathbf{h}} \mathbf{T}).$$

The Fourier coefficient of the Patterson function (10.5) is $|F_{\bar{\mathbf{h}}R}|^2$; this satisfies the rotational symmetry of the electron density space group, but is not sensitive to the symmetry translational components. Accordingly, the Patterson map symmetry must be symmorphic; since the Patterson function is always cs., its symmetry group will coincide with the Laue group of the $\rho(\mathbf{r})$ space group.

We stressed in Section 10.2.2 that the unit cell parameters and type (primitive or centred) do not change, passing from $\rho(\mathbf{r})$ to $P(\mathbf{u})$. Accordingly, the Patterson maps will show one of 24 symmetry groups; e.g.

$$P2, P2_1, \ldots, P2_1/c \text{ all reduce to } P2/m;$$

$$C2, Cc, \ldots, C2/c \text{ all reduce to } C2/m;$$

$$P2_12_12_1, Pca2_1, \ldots, Pbca \text{ all reduce to } Pmmm.$$

The reader will easily verify that, for the two-dimensional electron density map shown in Fig. 10.3a (plane group *pg*), the Patterson symmetry is *p2mm*.

An important consequence of the symmetry is the existence of interatomic vectors between the atom in $\mathbf{r}_i$ and its symmetry equivalents, say, the vectors $\mathbf{r}_j - \mathbf{C}_s \mathbf{r}_j = (\mathbf{I} - \mathbf{C}_s)\mathbf{r}_j = (\mathbf{I} - \mathbf{R}_s)\mathbf{r}_j - \mathbf{T}_s$. These are called *Harker vectors* (Harker, 1936), which lie in special sections of the unit cell called *Harker sections*, and may therefore be easily recognized. In Table 10.1, we show, for some space groups, the relations between positional vectors and Harker vectors.

By generalizing the data in the last column of Table 10.1, the reader can easily verify that, if the rotational component of **C** represents symmetry operators $\bar{1}, \bar{3}, \bar{4}, \bar{6}$, then the three coordinates of **r** may be derived from **u** without ambiguity; if it represents symmetry operators 2, 3, 4, 6 (or corresponding screw axes), one component of **r** remains undefined; if **C** represents the operator $\bar{2}$ (or corresponding glide planes), two coordinates remain undefined. In space groups like P2$_1$/c or P2$_1$2$_1$2$_1$, where more Harker sections coexist and intersect with each other, the $(\mathbf{r}_j)_{est}$ can be combined, so providing an unambiguous definition of $\mathbf{r}_j$.

Harker sections and Harker peaks are therefore an important key to passing from the Patterson map to the electron density map. They also have another important property worthy of mention: the Fourier transform of a Harker section provides a direct estimate of one-phase s.s.s (Ardito et al., 1985; Cascarano et al., 1987c; see Appendix 10.B).

10.3 Deconvolution of Patterson functions

Shortly, we will describe some methods for Patterson deconvolution: the heavy-atom method, the method of implication transformations, and superposition techniques.

Table 10.1 In a given space group (Sp.Gr), $\mathbf{r}_j$ and $\mathbf{Cr}_j$ are the positional vectors of the jth atom and of its symmetry equivalent in the unit cell, (u_j, v_j, w_j) are the coordinates of the corresponding Harker vector, $\mathbf{u}_j = (\mathbf{r}_j - \mathbf{Cr}_j) = (\mathbf{I} - \mathbf{C})\mathbf{r}_j = (\mathbf{I} - \mathbf{R})\mathbf{r}_j - \mathbf{T})$ in $P(\mathbf{u})$, HS is the corresponding Harker section. In the last column, $(\mathbf{r}_j)_{est}$ is the $\mathbf{r}_j$ estimate which may be obtained from $\mathbf{u}_j$; the bar indicates that no information on the corresponding coordinate is obtained

Sp. Gr.	$\mathbf{r}_j$	$\mathbf{Cr}_j$	$\mathbf{u}_j$	HS	$(\mathbf{r}_j)_{est}$
$P\bar{1}$	x_j, y_j, z_j	$\bar{x}_j, \bar{y}_j, \bar{z}_j$	$2x_j, 2y_j, 2z_j$	u, v, w	$u_j/2, v_j/2, w_j/2$
$P2$	x_j, y_j, z_j	$\bar{x}_j, y_j, \bar{z}_j$	$2x_j, 0, 2z_j$	$u, 0, w$	$u_j/2, -, w_j/2$
$P2_1$	x_j, y_j, z_j	$\bar{x}_j, y_j+1/2, \bar{z}_j$	$2x_j, 1/2, 2z_j$	$u, 1/2, w$	$u_j/2, -, w_j/2$
Pc	x_j, y_j, z_j	$x_j, \bar{y}_j, z_j+1/2$	$0, 2y_j, 1/2$	$0, v, 1/2$	$-, v_j/2, -$
$P2_1/c$	x_j, y_j, z_j	$\bar{x}_j, y_j+1/2, \bar{z}_j+1/2$	$2x_j, 1/2, 2z_j+1/2$	$u, 1/2, w$	$u_j/2, -, \left(\frac{w_j}{2} - \frac{1}{4}\right)$
		$\bar{x}_j, \bar{y}_j, \bar{z}_j$	$2x_j, 2y_j, 2z_j$	u, v, w	$u_j/2, v_j/2, w_j/2$
		$x_j, \bar{y}_j+1/2, z_j+1/2$	$0, 2y_j+1/2, 1/2$	$0, v, 1/2$	$-, \left(\frac{v_j}{2} - \frac{1}{4}\right), -$
$P2_12_12_1$	x_j, y_j, z_j	$\bar{x}+1/2, \bar{y}_j, z_j+1/2$	$2x_j+1/2, 2y_j, 1/2$	$u, v, 1/2$	$\left(\frac{u_j}{2} - \frac{1}{4}\right), \frac{v_j}{2}, -$
		$\bar{x}_j, y_j+1/2, \bar{z}_j+1/2$	$2x_j, 1/2, 2z_j+1/2$	$u, 1/2, w$	$\frac{u_j}{2}, -, \left(\frac{w_j}{2} - \frac{1}{4}\right)$
		$x_j+1/2, \bar{y}_j+1/2, \bar{z}_j$	$1/2, 2y_j+1/2, 2z_j$	$1/2, v, w$	$-, \left(\frac{v_j}{2} - \frac{1}{4}\right), \frac{w_j}{2}$

10.3.1 The traditional heavy-atom method

In Table 10.1, we showed how to construct the Harker vector $\mathbf{u}_j = (\mathbf{I} - \mathbf{C})\mathbf{r}_j \equiv (u_j, v_j, w_j)$ given the atomic position (x_j, y_j, z_j). The question we answer in this section is: given the Harker vector $\mathbf{u}_j$ corresponding to the distance between two symmetry equivalent atoms, can we then return to (x_j, y_j, z_j)? In the last column of Table 10.1, we showed that, from $\mathbf{u}_j$, only $(\mathbf{r}_j)_{est}$ can be derived, for which, often, one or two coordinates are not fixed. We will show how such an ambiguity may be overcome by giving some examples.

In Section 10.2.2, we stated that the height of the Patterson peak in $\mathbf{u}$ is proportional to the product of the atomic numbers of the two atoms i and j with interatomic vector $\mathbf{u}$:

$$P(\mathbf{u}) \propto Z_i Z_j.$$

If both i and j correspond to heavy atoms, then $P(\mathbf{u})$ may be much larger than other Patterson peaks. The same conclusion holds if i and j are heavy atoms and are symmetry equivalent. Suppose now that in the unit cell, space group $P\bar{1}$, there are two bromine atoms ($Z = 35$), referred by an inversion centre, and many O and C atoms. The Harker peak Br–Br will be proportional to 1225, while peaks Br–O, Br–C, O–O, and O–C will be proportional to 280, 210, 64, and 36, respectively. In this case the Harker peak Br–Br would be easily recognized; from its coordinates (u, v, w), the Br position is completely defined. Indeed, according to the last column of Table 10.1, the Br coordinates in the electron density map will be $(u/2, v/2, w/2)$.

In P2$_1$, the Harker peak Br–Br will have coordinates of type (u, 1/2, w), from which the x and z Br coordinates may be derived (see the last column of Table 10.1): y, however, can lie everywhere between 0 and 1. Since the origin of the unit cell may float freely along y, we can arbitrarily fix $y = 0$, thereby simultaneously fixing the Br position without ambiguity. The same should occur in Pc, where we cannot derive, from the Harker peak at (0, v, 1/2), the Br x and z coordinates. Luckily, in Pc, the origin can float freely in the plane x,z and may be arbitrarily fixed; accordingly, the Br position may be fixed without ambiguity.

In P1, the coordinates (x, y, z) can float freely along the three directions; therefore the first atomic position can be arbitrarily fixed (for example, in (0, 0, 0)).

In the above cases, fixing the free Br coordinates simultaneously fixes the origin; therefore, if a second heavy atom is present in the structure, the choice to arbitrarily fix its floating coordinates is not allowed. They should be referred to the origin fixed by the first Br through the use of interatomic vectors.

The procedure so far described works in the following way: look for the outstanding Harker peaks and derive the heavy-atom coordinates. Some automatic Patterson search programs work in a reverse way. They consider each grid point in the asymmetric unit of the crystal to be a possible heavy-atom site, and compute a score, based on the corresponding Harker peak densities (the Harker peaks may lie on different Harker sections if the symmetry is sufficiently rich). The value of the calculated score is associated with the grid point: at the end of the process a map similar to an electron density map is obtained; this can then be used to locate the heavy atoms. When more than one heavy atom needs to be located, Patterson interatomic vectors should be used to define the heavy-atom substructure.

In both procedures, once the heavy-atom substructure model has been fixed, EDM procedures (like those described in Chapter 8) are started, to recover the full structure.

10.3.2 Heavy-atom search by translation functions

The basic technique is part of a *molecular replacement* approach and is described in Section 13.7. In a naive way, a single heavy atom is translated throughout the asymmetric unit; for each point, the structure factor of the heavy atom is calculated (say F_p) and the correlation coefficient between $|F|^2$ and $|F_p|^2$ is computed (we denote this as $C(|F|^2,|F_p|^2)$, but also, $C(|E|^2,|E_p|^2)$ may be used if normalized structure factors are preferred). It may be expected that the target function $C(|F|^2,|F_p|^2)$ is a maximum for the correct position of the heavy atom.

In practice, such calculations will be very time consuming for macromolecules. Luckily, $C(|F|^2,|F_p|^2)$ and $C(|E|^2,|E_p|^2)$ may be calculated using fast Fourier transform (Fujinaga and Read, 1987; Navaza and Vernoslova, 1995) in a few seconds, even in the case of large proteins and large numbers of heavy-atom sites (Grosse-Kunstleve and Brunger, 1999; Vagin and Tepliakov, 1998). If more than one heavy atom is to be located, the technique does not change significantly; the only difference being that the calculated structure factor includes a contribution from the previously accepted sites.

If the space group is P1, in accordance with the observations made in Section 10.3.1, the first heavy atom may be arbitrarily located, and the heavy-atom search starts from the second.

10.3.3 The method of implication transformations

The naive heavy-atom approach described in Section 10.3.1 is not normally sufficient for solution of complex structures. The *implication transformation approach* (Buerger, 1946; Beevers and Robertson, 1950; Clastre and Gray, 1950; Garrido, 1950), strictly in relation to heavy-atom techniques, is more efficient and allows complete automation of the Patterson deconvolution procedure.

In Section 10.2.3, we showed that any generic point $\mathbf{r}$ of the unit cell may be transformed into a Harker vector $\mathbf{u}$, by applying the transformation

$$\mathbf{u} = (\mathbf{r} - \mathbf{Cr}) = (\mathbf{I} - \mathbf{C})\mathbf{r} = (\mathbf{I} - \mathbf{R})\mathbf{r} - \mathbf{T}. \tag{10.7}$$

For example, in $P2_1$, where

$$\mathbf{I} - \mathbf{R} = \begin{vmatrix} 2 & 0 & 0 \\ 0 & 0 & 0 \\ 0 & 0 & 2 \end{vmatrix}, \quad \mathbf{T} = \begin{vmatrix} 0 \\ 1/2 \\ 0 \end{vmatrix},$$

the vector $\mathbf{r} = (0.12, 0.28, 0.15)$ is transformed by (10.7) into $\mathbf{u} = (0.24, 0.5, 0.30)$. The transformation is from many to one; indeed, all of the points $\mathbf{r} = (0.12, y, 0.15)$ are transformed into the same Harker point, $\mathbf{u} = (0.24, 0.5, 0.30)$. Conversely, from $\mathbf{u}$ it is possible to return to points $\mathbf{r}$ (see the last column in Table 10.1), even if with ambiguity about the y position; this is what in crystallography is called an *implication transformation*. More specifically, an *implication transformation*, $I_s(\mathbf{r})$, is a function which transforms the Patterson density in $\mathbf{u} = \mathbf{r} - \mathbf{C}_s \mathbf{r}$, belonging to the Harker section defined by the symmetry operator $\mathbf{C}_s$, into electron densities for appropriate points $\mathbf{r}$, defined by (10.8):

$$I_s(\mathbf{r}) = P(\mathbf{r} - \mathbf{C}_s \mathbf{r})/n_s, \tag{10.8}$$

where n_s is the multiplicity of the Harker vector, i.e. the number of symmetry operators that generate the given Harker vector.

In practice, an implication transformation tries to reconstruct the electron density in the unit cell, starting from the densities in the Harker sections. To give a practical example, if the target structure space group is $P2_1$, and if a Harker peak lies in $\mathbf{u} = (0.24, 0.5, 0.30)$, the implication transformation associates the same Harker intensity with the set of points $\mathbf{r} = (0.12, y, 0.15)$, a column parallel to the y-axis.

In order to fully exploit the symmetry, the concept of *symmetry minimum function (SMF)* is introduced. Suppose that in P222, three Harker peaks (belonging to three different Harker sections) lie at

$$\mathbf{u}_1 = (0.0, 0.44, 0.30), \quad \mathbf{u}_2 = (0.24, 0.0, 0.30), \quad \mathbf{u}_3 = (0.24, 0.44, 0.0).$$

The corresponding intensities are transferred to the three columnar sets,

$$\mathbf{r}_1 = (x, 0.22, 0.15), \quad \mathbf{r}_2 = (0.12, y, 0.15), \quad \mathbf{r}_3 = (0.12, 0.22, 0.0) \quad (10.9)$$

respectively. Let us now apply the *symmetry minimum function*,

$$SMF(\mathbf{r}) = \operatorname*{Min}_{s=1}^{\overline{m}}[I_s(\mathbf{r})], \quad (10.10)$$

where the minimum operator, *Min*, indicates that the lowest value among the $\overline{m}$ functions $I_s(\mathbf{r})$ has been chosen, and $\overline{m}$ is the number of independent Harker domains in the Patterson map.

The reader will notice that, in the case where no symmetry is present (i.e. space group P1), the *SMF* map coincides with the Patterson map itself.

If we apply equation (10.10) to the above P222 example, *SMF* is different from zero only in $\mathbf{r} = (0.12, 0.22, 0.15)$; all of the other points belonging to sets (10.9) have vanishing intensity.

It seems, from the previous results, that the implication transformation approach can definitively solve the phase problem, at least in those favourable cases in which Patterson peaks do not overlap and more Harker sections exist. Unfortunately, there are several supplementary sources of ambiguity. The first is trivial: in the above P222 example, the peak $\mathbf{r} = (0.12, 0.22, 0.15)$ will be correctly identified only if the Harker vectors are not affected by resolution bias (see Section 7.3.1) or by peak overlapping effects; both may shift the Harker peaks from the ideal positions. The second source of ambiguity concerns the cs. nature of the Patterson. When the space group is acentric, how can the enantiomorph be automatically defined from the (centric) Patterson map? Evidently, in a *SMF* map, there are residual cs. features which should be cancelled by suitable filtering approaches.

The third source of ambiguity arises from the following question: *a structure is defined with respect to a given origin. How may the diffraction amplitudes used to calculate the Patterson map, which are structure invariants, automatically determine the origin?* Evidently, the *SMF* must present an ambiguity on the origin (see Chapter 3). The reader can immediately verify that in $P2_1$, if a Harker vector lies at $\mathbf{u} = (0.24, 0.5, 0.30)$, not only is the column, $\mathbf{r} = (0.12, y, 0.15)$, but also the columns,

$$\mathbf{r} = (0.12 + 1/2, y, 0.15 + 1/2), \quad \mathbf{r} = (0.12, y, 0.15 + 1/2), \quad \mathbf{r} = (0.12 + 1/2, y, 0.15),$$

are compatible with the Harker vector $\mathbf{u}$. This ambiguity arises from the set of permissible origins defined in Section 3.4. In short, the electron density map created by the function $I_s(\mathbf{r})$ in $P2_1$ is of a columnar type, and shows a pseudotranslational symmetry in the plane perpendicular to y, fixed by the allowed origin translations.

The origin ambiguity may be arbitrarily resolved by the crystallographer through choosing one of the allowed origins; but this only holds for the position of the first atom.

To represent the above mentioned problems graphically, we reconsider the schematic two-dimensional structure shown in Fig. 10.3a, with unit cell parameters a = 20 Å, b = 20 Å, plane group *pg* (equivalent positions (x,y), (−x, y + 0.5)). There are three independent atomic positions: the sulphur S at (0.1, 0.1), the oxygen O_1 at (0.16, 0.30), and the oxygen O_2 at (0.30, 0.18).

For the Patterson map shown in Fig. 10.3b, the Harker vectors are of type **u** = (2x, 0.5), and therefore the Harker section is the line v = 0.5. The heaviest maximum in the Harker section corresponds to the S–S interatomic vector, $\mathbf{u}_{S-S}$ = (0.2, 0.5) (and to its cs. Harker vector with coordinates (0.8, 0.5); the Harker peaks corresponding to $\mathbf{u}_{O_1-O_1}$ = (0.32, 0.50) and to $\mathbf{u}_{O_2-O_2}$ = (0.6, 0.5) (and to their cs. vectors) show lower density.

According to Table 10.1, $(\mathbf{r}_S)_{est}$ = (0.1, y); this corresponds to a strong two-dimensional column marked by an arrow in Fig. 10.3c. Of the other three high intensity columns, one is cs. to the column (0.1, y), the other two are due to origin ambiguity. It is also easy to identify in Fig. 10.3c the columns corresponding to $(\mathbf{r}_{O_1})_{est}$ and to $(\mathbf{r}_{O_2})_{est}$, and the supplementary columns related to them. The columnar nature of the $I(\mathbf{r})$ map, as defined by equation (10.8), is now clearly understandable.

The position of the S atom may be arbitrarily chosen at any point in the four strong columns; this choice simultaneously defines the origin of the model electron density map. The O positions remain confined to suitable columns of the $I(\mathbf{r})$ map, but they cannot be arbitrarily assigned; non-Harker peaks relating the oxygens to the sulphur should be considered.

In conclusion, the *SMF* map contains an image of the target structure, but it is frequently immersed in huge noise, created by:

(i) the intrinsic nature of the implication transformation (from one Harker vector **u** to many points **r**);
(ii) the residual Patterson symmetry;
(iii) Harker peak intensities (frequently affected by peak overlapping) transferred without filtering to the *SMF* peaks;
(iv) resolution bias and peak overlapping effects.

To reduce the noise and recover the target structure, supplementary tools may be necessary, one of which may be identified in the superposition techniques.

10.3.4 Patterson superposition methods

We will now describe the overlapping of shifted Patterson maps with the *SMF* function. To simplify our treatment, let us assume that the target structure contains N atoms, NH of which are heavy atoms, defined by the positional vectors, $\mathbf{r}_{Hi}$, $i = 1, \ldots, NH$; $\mathbf{r}_{lv}$, for $v = NH + 1, \ldots, N$ are the positional vectors of the light atoms. For ideal diffraction data, Patterson peak positions will be the union of the following three sets:

$$\{\mathbf{r}_{Hi} - \mathbf{r}_{Hj}, \, i,j = 1, \ldots, NH\} \quad (10.11a)$$

$$\{\pm(\mathbf{r}_{Hi} - \mathbf{r}_{lv}) \, i = 1, \ldots, NH, \, v = NH + 1, \ldots, N\} \quad (10.11b)$$

$$\{\mathbf{r}_{lv} - \mathbf{r}_{l\mu}, \, v, \mu = NH + 1, \ldots, N\}. \quad (10.11c)$$

In turn, these correspond with heavy–heavy, heavy–light, and light–light atom distances.

Suppose now that the position, $\mathbf{r}_{Hq}$ of the qth heavy atom has been found by application of the *SMF* function. Let us now shift the Patterson map by a vector, $\mathbf{r}_{Hq}$; we will denote this shifted map as $P(\mathbf{u} - \mathbf{r}_{Hq})$. Such a shift should generate a noised image of the structure. Indeed the following peak sets arise:

$$\{\mathbf{r}_{Hi} - \mathbf{r}_{Hj} + \mathbf{r}_{Hq},\ i,j = 1,\ldots,NH\} \quad (10.12a)$$

$$\{\pm(\mathbf{r}_{Hi} - \mathbf{r}_{lv}) + \mathbf{r}_{Hq},\ i = 1,\ldots,NH,\ \nu = NH+1,\ldots,N\} \quad (10.12b)$$

$$\{\mathbf{r}_{l\nu} - \mathbf{r}_{l\mu} + \mathbf{r}_{Hq},\ \nu,\mu = NH+1,\ldots,N\}. \quad (10.12c)$$

Emphasizing the case $j = q$ for subset (10.12a) and the case $i = q$ for subset (10.12b), allows us to rewrite the peaks (10.12) as,

$$\{\mathbf{r}_{Hi}, i = 1,\ldots,NH,\ \mathbf{r}_{Hi} - \mathbf{r}_{Hj} + \mathbf{r}_{Hq},\ i,j = 1,\ldots,NH, j \ne q\} \quad (10.13a)$$

$$\{\mathbf{r}_{l\nu}, \nu = NH+1,\ldots,N,\ \pm(\mathbf{r}_{Hi} - \mathbf{r}_{lv}) + \mathbf{r}_{Hq},\ i = 1,\ldots,NH,\ \nu = NH+1,\ldots,N\}, \quad (10.13b)$$

$$\{\mathbf{r}_{l\nu} - \mathbf{r}_{l\mu} + \mathbf{r}_{Hq}, \nu,\mu = NH+1,\ldots,N\} \quad (10.13c)$$

Let us now overlap the peaks (10.13) with the *SMF* map (which, as we know, contains an image of the target structure) and take the minimum; this corresponds with the *minimum superposition function* (Buerger, 1951, 1953; Taylor, 1953),

$$MS(\mathbf{r}) = Min\left[P(\mathbf{u} - \mathbf{r}_q), SMF(\mathbf{r})\right]. \quad (10.14)$$

$MS(\mathbf{r})$ would provide: the heavy atom substructure, $\{\mathbf{r}_{Hi}, i = 1,\ldots,NH\}$ plus noise from peaks (10.13a), the light atom positions, $\{\mathbf{r}_{l\nu}, \nu = NH+1,\ldots,N\}$ plus noise from the set (10.13b), and only noise from the set (10.13c). As a result, the superposition approach leads to an image of the structure, say,

$$\{\mathbf{r}_{Hi}, i = 1,\ldots,NH\} \bigcup \{\mathbf{r}_{l\nu}, \nu = NH+1,\ldots,N\},$$

plus noise; the noise has to be minimized via suitable filtering procedures followed by EDM cycles.

Simple control of the above vectorial algebra is achieved, using the example illustrated in Fig. 10.3. Let us suppose that the S position has been fixed via *SMF*; without loss of generality we have located it at the correct position, say (0.1, 0.1). In Fig. 10.4a, we show the Patterson map shifted by the vector $\mathbf{r}_{Hq} = (0.1, 0.1)$, and in Fig. 10.4b, the minimum superposition function *MS*, as given by equation (10.14). The result may be summarized as follows: the target structure is located correctly, but some noise (e.g. spurious peaks) is present. Proper filtering algorithms and *EDM* techniques should be used to eliminate the noise.

Superposition techniques may be generalized in several ways (Raman and Lipscomb, 1961):

(i) By calculating the more general minimum superposition function,

$$MS(\mathbf{r}) = Min\left[P(\mathbf{u} - \mathbf{r}_q), P(\mathbf{u} - \mathbf{r}_p),\ldots, SMF(\mathbf{r})\right], \quad (10.15)$$

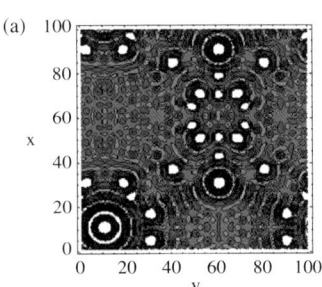

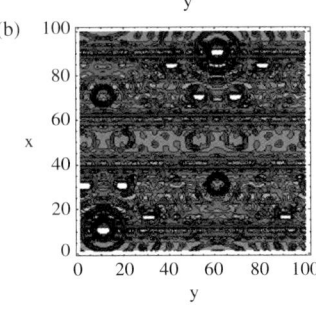

Fig. 10.4
For the example depicted in Fig. 10.3: a) Patterson map shifted by the vector $\mathbf{r}_{Hq} = (0.1, 0.1)$; b) superposition map according to equation (10.14) (in practice, obtained by overlapping Fig. 10.4a with Fig. 10.3c). By courtesy of Rocco Caliandro.

where $\mathbf{r}_q, \mathbf{r}_p, \ldots$, etc. are atomic positions, previously defined. The use of (10.15) may reduce the residual noise still present in (10.14), but presents an additional risk; the final expected signal (that is, the image of the structure) may be very imperfect, because Patterson peaks are often displaced from their correct positions, and perfect overlapping of the displaced Patterson peaks on the atomic sites is impossible. In short, $MS(\mathbf{r})$, as given by (10.15), may be too discriminating; it will vanish in $\mathbf{r}$ if one of the overlapping maps has, by chance, a very low peak density value in $\mathbf{r}$.

(ii) By calculating the *sum function*,

$$SS(\mathbf{r}) = SMF(\mathbf{r}) + \sum_i P(\mathbf{u} - \mathbf{r}_i), \qquad (10.16)$$

where the sum goes over the located atomic positions.

$SS(\mathbf{r})$ may easily be calculated by using the relation,

$$\sum_i P(\mathbf{u} - \mathbf{r}_i) = \frac{1}{V} \sum_{\mathbf{h}} |F_{\mathbf{h}}|^2 \exp[-2\pi i \mathbf{h} \cdot (\mathbf{u} - \mathbf{r}_i)]$$

$$= \frac{1}{V} \sum_{\mathbf{h}} |F_{\mathbf{h}}|^2 \exp -(2\pi i \mathbf{h} \cdot \mathbf{u}) \left[\sum_i \exp(2\pi i \mathbf{h} \cdot \mathbf{r}_i) \right].$$

If we rewrite

$$\sum_i \exp(2\pi i \mathbf{h} \cdot \mathbf{r}_i) = m \exp(i\psi_{\mathbf{h}}),$$

then,

$$\sum_i P(\mathbf{u} - \mathbf{r}_i) = \frac{1}{V} \sum_{\mathbf{h}} m_{\mathbf{h}} |F_{\mathbf{h}}|^2 \exp -(2\pi i \mathbf{h} \cdot \mathbf{u} + \psi_{\mathbf{h}}).$$

$SS(\mathbf{r})$ tends to have a low discriminatory power; if one of the maps has a high density value in $\mathbf{r}$, but the others are vanishing in $\mathbf{r}$, then the map may show a peak where there is none.

(iii) By calculating the *product function*,

$$PS(\mathbf{r}) = SMF(\mathbf{r}) \cdot \left[\sum_i P(\mathbf{u} - \mathbf{r}_i) \right]. \qquad (10.17)$$

$PS(\mathbf{r})$ is the most vigorous; the density at a point $\mathbf{r}$ is zero if one of the maps vanishes at that point, but it may show exceptional values if two maps are strong in $\mathbf{r}$.

$MS(\mathbf{r})$ is the most popular technique and is frequently used in modern Patterson deconvolution programs. The harmonic mean (reciprocal of the mean of the reciprocals of the density values) seems to be the medium choice amongst the three.

10.3.5 The C-map and superposition methods

Let us suppose that a model structure, $\rho_p(\mathbf{r})$, is known; one could be interested in the convolution,

$$C(\mathbf{u}) = \rho(\mathbf{r}) \otimes \rho_p(\mathbf{r}) = \int_S \rho(\mathbf{r}) \rho_p(\mathbf{r} + \mathbf{u}) d\mathbf{r}, \qquad (10.18)$$

which, by the properties of the Fourier transform, may be written as,

$$C(\mathbf{u}) = \frac{1}{V} \sum_{\mathbf{h}} |F_{\mathbf{h}} F_{p\mathbf{h}}| \exp i(\phi_{\mathbf{h}} - \phi_{p\mathbf{h}}) \exp(-2\pi i \mathbf{h} \cdot \mathbf{u}), \qquad (10.19)$$

where $F_\mathbf{h} = |F_\mathbf{h}|\exp(i\phi_\mathbf{h})$ and $F_{p\mathbf{h}} = |F_{p\mathbf{h}}|\exp(i\phi_{p\mathbf{h}})$ are the structure factors of $\rho(\mathbf{r})$ and $\rho_p(\mathbf{r})$, respectively. $|F_\mathbf{h} F_{p\mathbf{h}}|\exp i(\phi_\mathbf{h} - \phi_{p\mathbf{h}})$ is the Fourier coefficient of $C(\mathbf{u})$ and is a complex number; therefore $C(\mathbf{u})$ is acentric, it is centric only if both $\rho(\mathbf{r})$ and $\rho_p(\mathbf{r})$ are centric. It was shown by Carrozzini et al. (2010) that the space group of the C-function is the symmorphic variant of the space group of the target structure (e.g. P222, if the target space group is P2$_1$2$_1$2$_1$).

The map, $C(\mathbf{u})$, cannot be computed during the phasing process, essentially because the $\phi_\mathbf{h}$s are unknown. Fortunately, the approximating function $C'(\mathbf{u})$, given by

$$C'(\mathbf{u}) = \frac{1}{V}\sum_\mathbf{h} m_\mathbf{h}|F_\mathbf{h} F_{p\mathbf{h}}|\exp(-2\pi i\,\mathbf{h}\cdot\mathbf{u}), \qquad (10.20)$$

is easily computable, where,

$$m = <\cos(\phi - \phi_p)> = I_1(X)/I_0(X)$$

and (see Section 7.2)

$$X = 2\sigma_A|EE_p|/\left(1 - \sigma_A^2\right).$$

$C'(\mathbf{u})$ is a useful approximation to $C(\mathbf{u})$ if the m_h values are sufficiently large. The approximation has a principal consequence in that the Fourier coefficients of $C'(\mathbf{u})$ are real numbers, consequently, the space group of $C'(\mathbf{u})$ is centric, so coinciding with the Patterson space group (e.g. Pmmm if the space group of the target is P2$_1$2$_1$2$_1$).

Carrozzini et al. (2010) observed that both $\rho(\mathbf{r})$ and $\rho_p(\mathbf{r})$ are non-negative definite functions, therefore, $C(\mathbf{u})$ and $C'(\mathbf{u})$ are both non-negative definite. The map, $C'(\mathbf{u})$, may therefore be suitably modified and Fourier inverted, as in the usual EDM procedures (see Chapter 8), so leading to better estimates of the invariants $(\phi_\mathbf{h} - \phi_{p\mathbf{h}})$. Also (Caliandro et al., 2013), $C'(\mathbf{u})$ may be combined with vector superposition techniques, as follows.

Let us suppose that a model is available, consisting of one heavy atom located at $\mathbf{r}_{Hq}$ (e.g. obtained via the study of the *SMF* function). If m is a good approximation of $\cos(\phi - \phi_p)$ for a sufficiently large set of reflections (for the moment we will assume that this condition is satisfied), then the C' peaks will be located at (with the notation used in the Section 10.3.3)

$$\{\mathbf{r}_{Hi} - \mathbf{r}_{Hj}, i,j = 1,\ldots,NH\} \cup \{\pm(\mathbf{r}_{l\nu} - \mathbf{r}_{Hi}), j = 1,\ldots,NH, \nu = NH+1,\ldots,N\} \qquad (10.21)$$

By adding the heavy atom position, $\mathbf{r}_{Hq}$, to the interatomic vectors in (10.21), we obtain

$$\{\mathbf{r}_{Hi}, i = 1,\ldots,NH, \mathbf{r}_{Hi} - \mathbf{r}_{Hj} + \mathbf{r}_{Hq}, i,j = 1,\ldots,NH, j\neq q\}, \qquad (10.22a)$$

$$\{\mathbf{r}_{l\nu}, \nu = NH+1,\ldots,N;\ \pm(\mathbf{r}_{l\nu} - \mathbf{r}_{Hi}) + \mathbf{r}_{Hq}, i = 1,\ldots,NH, i\neq q, \nu = NH+1,\ldots,N\}. \qquad (10.22b)$$

Let us compare the sets (10.22), obtained by use of the C' map, with the sets (10.13), obtained using the P map:

(i) (10.22a) and (10.22b) coincide with (10.13a) and (10.13b). Therefore, both set (10.22a) and set (10.13a) provide the atomic positions of the full target structure, with the same amount of noise.

(ii) The noise (10.13c) has no correspondent if the C'-map is used. This property may be very important when the target structure contains a large number of light atoms, as frequently occurs in medium-sized structures and proteins.

This superior characteristic of the C' map requires that one condition should be satisfied: i.e. that C' is a good approximation of C, and this occurs only if the model is sufficiently good. In practice, m is only a statistical estimate of $\cos(\phi - \phi_p)$; therefore, the light atoms–light atoms peaks may be present in the C' map, but their intensity is expected to be smaller compared with the corresponding peaks in the Patterson map. A further property makes use of the C' map more interesting. While the Patterson map is invariant during the phasing process, the C' map changes with the current model. This characteristic is crucial for successful crystal structure determination.

10.4 Applications of Patterson techniques

Any description of a phasing approach should always be followed by some detail about its practical efficiency, in order to orient the potential user towards the most appropriate choice of phasing method. The short report below excludes examples of protein substructure solution in *SIR-MIR* and *SAD-MAD* cases, which are delayed until Chapters 14 and 15.

There has recently been a strong renewal of interest in Patterson deconvolution techniques, as described in Section 10.3; new approaches have been applied to a wide set of experimental data under different situations. For example, to powder diffraction data (Burla et al., 2007a), to ab initio crystal structure solution of proteins (Burla et al., 2006a; Caliandro et al., 2007a,b, 2008a,b), to protein substructure solution in *SAD-MAD* cases (Burla et al., 2007b). The reader is referred to the original papers for details of the algorithms and experimental results. Here, we will merely summarize some of the above applications, to emphasize the versatility of these methods.

1. *Ab initio applications of Patterson superposition techniques to proteins.*
   To collect the results in a few lines, we divide the test structures into suitable subsets, in order to study the potential of the techniques versus the size, resolution, and type of heavy atom included within the structure. For each subset we give the average *cpu* time ($<cpu>$) necessary to obtain the solution (naturally, the time refers to solved structures) using a common desk computer.

   (a) 23 test structures with RES <1.2 Å, number of non-H atoms in the asymmetric unit (N_{asym}) up to about 2000, an atomic species heavier than Ca. All structures solved, $<cpu> = 6.8$ min.
   (b) 31 test structures with RES <1.2 Å, N_{asym} up to about 2000, no atomic species heavier than Ca. All structures solved, $<cpu> = 16.6$ min.
   (c) 20 test structures with 1.2 Å < RES <1.7 Å, N_{asym} up to about 2000, heavy atoms from S to Br. Four structures, with RES between 1.3 and 1.7 remain unsolved.

(d) 20 test structures with RES <1.2 Å, 2000 <N_{asym} < 6300, and atomic species from Ca up. Three structures remained unsolved, all with Ca as heaviest atom.

(e) 32 test structures, with 1.5 Å < RES <1.7 Å, N_{asym} < 8000, and atomic species from Fe up. All 19 structures up to 1.7 Å resolution (including *1e3u*, a large protein with about 8000 atoms in the asymmetric unit and RES = 1.65 Å, and seven proteins with RES between 1.7 and 1.9 Å) could be solved.

2. *Patterson superposition techniques and superposition techniques combined with the C′ map.* Such a technique has recently been applied (Caliandro et al., 2013) to a total of 186 test structures, with up to about 350 atoms in the asymmetric unit. In Table 10.2, the test structures are divided into subsets, according to the number of non-hydrogen atoms in the asymmetric unit (N_{asym}) and to the heaviest atomic species (*L* stands for 'all light atoms', *H* for if some heavy atoms are present): for practicality, an atom is considered heavy if its atomic number is larger than that of Ca.

In Table 10.2, we also show the average phasing efficiencies of two deconvolution methods, the first C′-based (as described in Section 10.3.4), the second Patterson-based (as described in Section 10.3.3). For each structure subset, $E_{ffC'}$ and E_{ffP} are the ratio

(number of structures solved in default conditions)/(number of structures)

when the C′-based and the Patterson-based algorithms are used, respectively.

The deconvolution algorithm based on the C′ map is more efficient than that based on the Patterson map for structures having only light atoms, and in particular for difficult cases (i.e. for structures with N_{asym} >150). If heavy atoms are present, both algorithms work to full efficiency.

In Fig. 10.5, we show, for the C′-based algorithm and for *H* and *L* subsets, the cpu time necessary to reach a solution (*t* in minutes), versus the average values of N_{asym} (<N_{asym}>). The averages have been calculated over the test structures contained in each of the four subsets considered in Table 10.2. We observe that, for structures containing heavy atoms, the solution is very soon attained, indicating that the true heavy atom position is always found in

Table 10.2 The test structures are grouped according to the number of non-H atoms in the asymmetric unit (N_{asym}) and the presence or absence of atoms heavier than calcium (*H/L*). Entries correspond to: number of structures in each class (*Num*), phasing efficiencies of the C′-based ($Eff_{C'}$), and of the Patterson-based deconvolution (Eff_P) algorithms

	H			L		
	Num	$Eff_{C'}$	Eff_P	Num	$Eff_{C'}$	Eff_P
$N_{asym} \leq 20$	23	100	100	14	100	100
$20 < N_{asym} \leq 80$	21	100	100	45	100	96
$80 < N_{asym} \leq 150$	14	100	100	46	100	93
$N_{asym} > 150$	4	100	100	21	90	81

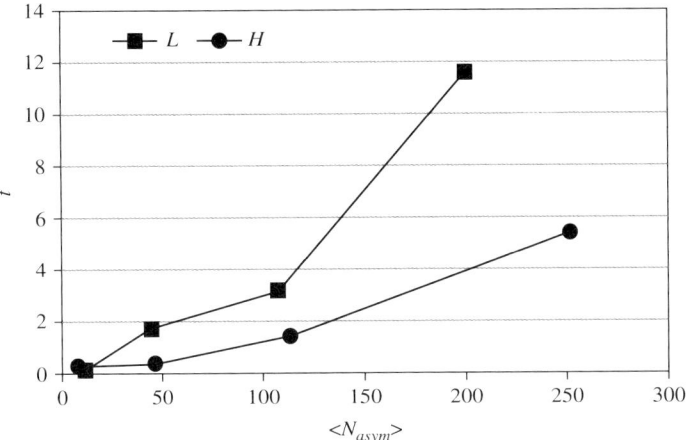

Fig. 10.5
For H- and L-structures we show, versus $<N_{asym}>$, the cpu time (t, in minutes) needed to reach the solution (see text for details).

the very early stages of the phasing procedure. In this case the efficiency of the deconvolution algorithm does not depend on the size of the structure.

For structures containing only light atoms, the cpu time increases strongly with structure complexity, much more rapidly than for heavy atoms, indicating that locating light atoms in their true positions by superposition techniques becomes more challenging as the size of the structure increases.

The subset of structures with $N_{asym} > 150$ contains various structures for which O is the heaviest atom. The success of superposition methods dramatically changes common judgement on the versatility of Patterson techniques. In the *International Tables for Crystallography*, volume B, Rossmann and Arnold (1993), write:

> The feasibility of structure solution by the heavy-atom method depends on a number of factors which include the relative size of the heavy atom and the extent and quality of the data. A useful rule of thumb is that the ratio
>
> $$r = \frac{\sum\limits_{heavy} Z^2}{\sum\limits_{light} Z^2}$$
>
> should be near unity if the heavy atom is to provide useful starting phase information (Z is the atomic number of an atom). The condition that r >1 normally guarantees interpretability of the Patterson function in terms of the heavy atom positions.

The authors also state that the rule is rather conservative and quote as an outstanding example, vitamin B_{12} with formula $C_{62} H_{88} Co O_{14} P$ (Hodgkin et al., 1957) which gave $r = 0.14$ for the cobalt atom alone.

The above results indicate that there has been great progress in Patterson deconvolution techniques over the last few years; heavy atoms are no longer strictly necessary to the success of Patterson or C' based procedures, which may also be successfully applied to structures with atoms not heavier than oxygen. Accordingly, Patterson and C'-map techniques are probably the

APPENDIX 10.A ELECTRON DENSITY AND PHASE RELATIONSHIPS

This subject will be treated in a very short digest, since most of the mentioned phase relationships are not particularly in use today; the interested reader will find a more substantial account in *Phasing in Crystallography*. We will discuss techniques as follows:

1. *Sayre–Hughes equation*. Let the electron density, $\rho(\mathbf{r})$, be composed of identical atoms which are fully resolved from one another. Then, $\rho^2(\mathbf{r})$ may be expanded in the Fourier series

$$\rho^2(\mathbf{r}) = \frac{1}{V} \sum_{\mathbf{h}=-\infty}^{+\infty} {}_2F_{\mathbf{h}} \exp(-2\pi i \, \mathbf{h}.\mathbf{r}), \qquad (10.A.1)$$

where ${}_2F_{\mathbf{h}}$ represents the Fourier coefficient of the expansion (10.A.1). If $\rho(\mathbf{r})$ is non-negative everywhere, $\rho(\mathbf{r})$ and $\rho^2(\mathbf{r})$ are very similar, but not equal (see Fig. 10.A.1); indeed, they show maxima at the same positions, but different peak shapes. If $s = 2\sin\theta/\lambda$, the following relationships hold:

$$F_{\mathbf{h}} = f_s \sum_{j=1}^{N} \exp(2\pi i \mathbf{h} \cdot \mathbf{r}_j), \quad {}_2F_{\mathbf{h}} = {}_2f_s \sum_{j=1}^{N} \exp(2\pi i \mathbf{h} \cdot \mathbf{r}_j),$$

where ${}_2f_s$ is the scattering factor of the squared atom. When the two equations are divided term by term,

$$ {}_2F_{\mathbf{h}} = \frac{{}_2f_s}{f_s} F_{\mathbf{h}} = \beta_{\mathbf{h}} F_{\mathbf{h}} \qquad (10.A.2)$$

is obtained, and $\beta_{\mathbf{h}}$ appears as a function only of the parameter s.

On the othe side, $\rho^2(\mathbf{r}) = \rho(\mathbf{r}) \cdot \rho(\mathbf{r})$, and its Fourier transform, because of the convolution theorem, will be

$$ {}_2F_{\mathbf{h}} = \frac{1}{V} \sum_{\mathbf{k}=-\infty}^{+\infty} F_{\mathbf{k}} F_{\mathbf{h}-\mathbf{k}}. \qquad (10.A.3)$$

Combining (10.A.2) and (10.A.3) gives

$$F_{\mathbf{h}} = \frac{1}{V\beta_{\mathbf{h}}} \sum_{\mathbf{k}=-\infty}^{+\infty} F_{\mathbf{k}} F_{\mathbf{h}-\mathbf{k}}. \qquad (10.A.4)$$

Equation (10.A.4) is the Sayre–Hughes equation (Sayre, 1952; Hughes, 1949), from which a tangent expression may be derived relating $\phi_{\mathbf{h}}$ to the phases $\phi_{\mathbf{k}} + \phi_{\mathbf{h}-\mathbf{k}}$. However, both (10.A.4) and the related tangent formula are asymptotical relationships, i.e. they may only be applied under the condition that the summations are carried out over an infinite number of reflections; no reliability parameter is provided. This requirement restricts the use of (10.A.4) in practical procedures, where $\phi_{\mathbf{h}}$ is usually evaluated from a limited number of known phases.

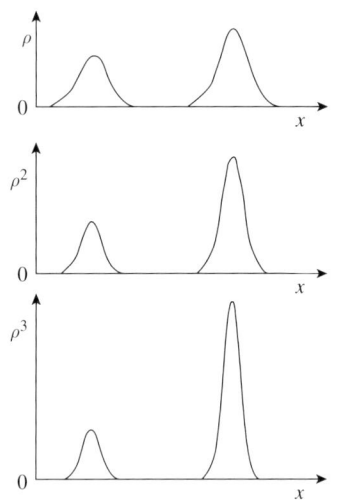

Fig. 10.A.1
A one-dimensional electron density distribution and its first powers.

2. *Von Eller polynomial method.* (von Eller, 1973) Let $\rho(\mathbf{r}) = \langle E_\mathbf{h} \exp(-2\pi i \mathbf{h}.\mathbf{r}) \rangle$ be the electron density, Z the atomic number of the unique atomic species present in the crystal and $\sigma_2 = \sum_{j=1}^{N} Z_j^2 = NZ^2$. Then the relation

$$\rho\left(\rho - Z/\sigma_2^{1/2}\right) = \rho\left(\rho - 1/\sqrt{N}\right) = 0 \qquad (10.A.5)$$

is identically satisfied for all values of $\mathbf{r}$; in fact, at any point in the unit cell at least one of the two factors on the left-hand side of (10.A.5) is zero. The Fourier transform of (10.A.5) gives

$$E_\mathbf{h} = \sqrt{N} \langle E_\mathbf{k} E_{\mathbf{h}-\mathbf{k}} \rangle,$$

strictly related to the Sayre–Hughes equation.

Equation (10.A.5) may easily be extended to the case in which there are n different atomic species via the polynomial,

$$\rho\left(\rho - Z_1/\sigma_2^{1/2}\right)\rho\left(\rho - Z_2/\sigma_2^{1/2}\right),\ldots\ldots\ldots,\rho\left(\rho - Z_n/\sigma_2^{1/2}\right) = 0 \qquad (10.A.6)$$

The Fourier transform of (10.A.6) involves triplets, quartets, etc.

3. *Cochran integral criterion.* Cochran (1952) argued that the integral

$$\int_V \rho^3(\mathbf{r})d\mathbf{r} = \frac{2}{V}\sum_{\mathbf{h}_1,\mathbf{h}_2} |F_{\mathbf{h}_1}F_{\mathbf{h}_2}F_{\mathbf{h}_1-\mathbf{h}_2}|\cos(\phi_{\mathbf{h}_1} - \phi_{\mathbf{h}_2} - \phi_{\mathbf{h}_1-\mathbf{h}_2}) \qquad (10.A.7)$$

should be a maximum for the correct structure; this occurs when, for large values of $|F_{\mathbf{h}_1}F_{\mathbf{h}_2}F_{\mathbf{h}_1-\mathbf{h}_2}|$, the triplet phase $\phi_{\mathbf{h}_1} - \phi_{\mathbf{h}_2} - \phi_{\mathbf{h}_1-\mathbf{h}_2}$ is close to zero. Clearly, (10.A.7) also leads to a tangent formula expression.

4. *Allegra and Colombo integral criterion.* Allegra and Colombo (1974) observed that the difference between the real and the squared structure must approach zero for the correct phase set obtained in a multisolution procedure. In formula,

$$\int_V \left(\frac{\rho^2(\mathbf{r})}{I_2} - \frac{\rho(\mathbf{r})}{I_1}\right)^2 d\mathbf{r} \approx \min, \qquad (10.A.8)$$

where

$$I_n = \int_V \rho^n(\mathbf{r})d\mathbf{r}.$$

Equation (10.A.8) leads to a tangent formula in which $\phi_{\mathbf{h}_1}$ is determined by means of triplet and quartet relationships.

5. *Rius and Miravittles criterion.* Rius and Miravittles (1989) (see also Giacovazzo, 1991) modified the Cochran criterion by maximizing the difference

$$\int_V \rho^3(\mathbf{r})d\mathbf{r} - \int_V \rho(\mathbf{r}+\mathbf{u}_0)\rho^2(\mathbf{r})d\mathbf{r}, \qquad (10.A.9)$$

where $\mathbf{u}_0$ represents a shift approximately equal to the average radius of the atomic peak ($\sim$0.4–0.5 Å). The second integral on the right-hand side of (10.A.9) is expected to be small.

6. *Karle and Hauptman determinants*. Karle and Hauptman (1950) stated that, as a consequence of the non-negativity property of the electron density, determinants exist satisfying the relation,

$$D_n = \begin{vmatrix} 1 & U_{12} & U_{13} & \ldots & U_{1n} \\ U_{21} & 1 & U_{23} & \ldots & U_{2n} \\ \ldots & \ldots & \ldots & \ldots & \ldots \\ U_{n1} & U_{n2} & U_{n3} & \ldots & 1 \end{vmatrix} \geq 0, \quad (10.A.10)$$

where $U_{pq} = U_{\mathbf{h}_p - \mathbf{h}_q}$ are the so-called *unitary structure factors*, defined by $U = F/(\sum_{j=1}^{N} f_j)$.

$D_n = 0$ when $n > N$. Since, on varying n from 1 to N, the positiveness of the Hermitian form (10.A.10) holds, it follows that all the principal minors of D_n must be positive; conversely, the positiveness of all the principal minors assures that D_n is positive. In conclusion, if the positiveness of the electron density function ensures the validity of (10.A.10), the discreteness of the points with $\rho(\mathbf{r}) \neq 0$ determines the maximum order of the positive determinants.

Karle Hauptman determinants have been the basis for a probabilistic approach proposed by Tsoucaris (1970) (see also Lajzerowicz and Lajzerowicz, 1966; de Rango et al., 1969, 1974) under the name the *maximum determinant rule*. Further advances were achieved by Vermin and de Graaff (1978), Taylor et al. (1979), de Gelder et al. (1990), and de Gelder (1992), who successfully applied such methods to the crystal structure solution of small molecules.

The reader will find in *Phasing in Crystallography*, Chapter 4, more details on all of the methods mentioned in this appendix.

APPENDIX 10.B PATTERSON FEATURES AND PHASE RELATIONSHIPS

Two methods will be sketched which relate Patterson features to phase relationships.

(a) *High-order Patterson syntheses*. An important contribution relating direct-space properties and reciprocal space relationships has been given by Vaughan (1958), who introduced the function,

$$P_n(\mathbf{u}_1, \mathbf{u}_1, \ldots, \mathbf{u}_n) = \int_V \rho(\mathbf{r})\rho(\mathbf{r} + \mathbf{u}_1) \ldots \rho(\mathbf{r} + \mathbf{u}_n) dV, \quad (10.B.1)$$

where V is the unit cell volume. When $n = 1$, function (10.B.1) coincides with the usual Patterson function and when $n = 2$ it coincides with the so-called *double Patterson* introduced by Sayre (1953) (see also Main, 1993). It may be shown that some sections of the double Patterson deal with modified Patterson functions, like

$$\int_V \rho^2(\mathbf{r})\rho(\mathbf{r} + \mathbf{u}) dV, \quad (10.B.2)$$

which, submitted to a Fourier transform, is related to the triplet invariants. The criterion (10.B.2) is exploited directly by a criterion proposed by Collins et al. (1996).

(b) *Estimate of one-phase structure seminvariants from Harker sections.* (Ardito et al., 1985; Cascarano et al., 1987c). In accordance with Section 4.3, the reflection with vectorial index $\mathbf{H} = \bar{\mathbf{h}}(\mathbf{I} - \mathbf{R}_s)$ is a s.s.; then,

$$F_\mathbf{H} = \sum_{j=1}^{N} f_j \exp[2\pi i \bar{\mathbf{h}}(\mathbf{I} - \mathbf{R}_s)\mathbf{r}_j] = \exp(2\pi i \bar{\mathbf{h}} \mathbf{T}_s) \sum_{j=1}^{N} f_j \exp[2\pi i \bar{\mathbf{h}}(\mathbf{I} - \mathbf{C}_s)\mathbf{r}_j]. \quad (10.B.3)$$

According to equation (10.6), the vectors $(\mathbf{I} - \mathbf{C}_s)\mathbf{r}_j$ are Harker vectors and lie on the sth Harker section, or, in other words, define the density on the sth Harker section. In a more general expression we can rewrite (10.B.3) as

$$F_\mathbf{H} = \frac{1}{L} \exp(2\pi i \mathbf{h} \cdot \mathbf{T}_s) \int_{HS} P(\mathbf{u}) \exp(2\pi i \mathbf{h} \cdot \mathbf{u}) d\mathbf{u}, \quad (10.B.4)$$

where $\mathbf{u}$ varies over the complete sth Harker section, and L is a constant which takes into account the dimensionality of HS. Relation (10.B.4) provides a phase estimate for the s.s., $F_\mathbf{H}$.

The reader will find more details on all the methods mentioned in this appendix in *Phasing in Crystallography*, Chapter 4.

(a) *Patterson synthesis of the second kind*. Patterson (1949) defined a second type of Patterson synthesis, say,

$$P_\pm(\mathbf{u}) = \int_V \rho(\mathbf{u}+\mathbf{r})\rho(\mathbf{u}-\mathbf{r})d\mathbf{r} = \frac{2}{V} \sum_{\mathbf{h}>0} |F_\mathbf{h}|^2 \cos(4\pi \mathbf{h} \cdot \mathbf{r} - 2\phi_\mathbf{h}).$$

$P_\pm(\mathbf{u})$ will show a large peak when the product of the densities at $\mathbf{u} + \mathbf{r}$ and $\mathbf{u} - \mathbf{r}$, integrated over all $\mathbf{r}$ values, is large. In this case, $\mathbf{u}$ is a pseudo-centre of symmetry for part (or all) of the electron density; the larger is $P_\pm(\mathbf{u})$, the larger is the percentage of electron density related by the inversion centre in $\mathbf{u}$. Since each atom is centrosymmetric in itself, small peaks will be present at the atomic positions. Small peaks will also be present at midpoints between atoms, with intensity proportional to the product of the centrosymmetrically shared part of the two atoms. $P_\pm(\mathbf{u})$ may be trivially used when a centric structure has been solved in P1 and one wants to recognize the positions of the inversion centres. It may also be used during a Patterson deconvolution process to check if the centrosymmetric features of the Patterson function have been eliminated from the model (Burla et al., 2006b). A large value of $P_\pm(\mathbf{u})$ is some $\mathbf{u}$ would suggest that the enantiomorph has not been well defined.

11 Phasing via electron and neutron diffraction data

11.1 Introduction

Among the statistics freely available on the webpage of the *Cambridge Structural Database,* there is a detail of interest for this chapter: of the 596 910 crystal structures deposited up to 1 January 2012, only 1534 were solved by neutron data (see Table 1.11). No information is provided on the number of structures solved by electron data because it is negligible (organic samples are soon damaged by the electron beams).

A statistical search of the *Inorganic Crystal Structure Database* (*ICSD,* Ver. 2012–1, about 150 000 entries; by courtesy of Thomas Weirich) on structures that have been solved by means of electron diffraction, eventually in combination with other techniques, indicates a total of about 0.7%.

In spite of limited impact on the databases, electron and neutron diffraction play a fundamental role in materials science and in crystallography. The main reason is that they provide alternative techniques to X-rays. Let us first consider electron diffraction (ED) techniques.

The study of crystalline samples at the nanometer scale is mandatory for many industrial applications; indeed, physical properties depend on the crystal structure. Unfortunately it is not unusual for compounds to only exist in the nanocrystalline state; then, traditional X-ray diffraction techniques for atomic structure determination cannot be applied, because of the weak interactions between X-rays and matter. As a consequence, such structures remain unknown, in spite of their technological importance. This limits the contribution of X-ray crystallography to nanoscience, a growing scientific area, crucial to many fields, from semiconductors to pharmaceuticals and proteins. The result is a lack of knowledge on the underlying structure–property relationships, which often retards further research and development.

Structure analysis by electron diffraction began as early as the 1930s (in particular, by Rigamonti, in 1936), but the interest of the crystallographic community in such a technique soon faded, mostly because electron diffraction intensities are not routinely transferable into kinematical $|F|^2$. In spite of this limitation, the technique has been used for investigating the structure of many inorganic, organic, and metallo-organic crystals, biological structures, and various minerals, especially layer silicates. We will therefore, in this chapter, describe briefly (see Sections 11.2 to 11.7 and Appendix 11.A) the specific features of electron scattering, in order to allow the reader to understand the special problems one has to face when direct phasing by electrons data is attempted. For further reading, the reader is addressed to the IUCr monograph, *Electron Diffraction Techniques*, volumes I and II (Cowley, 1992), with special emphasis on Chapter 6 by Vainshtein, Zuyagin, and Avilov, also, to a monograph by Dorset (1995) and to the *International Tables for Crystallography*, Vol. B (1993) and Vol. C (1992). Probably the most updated and complete presentation of electron crystallography is the monograph by Zou, Hovmöller, and Oleynikov (2011).

Let us now consider neutron diffraction. This is advised for when crystal structure solution is attempted in order to obtain details which are not available via X-ray crystallography (e.g. the accurate positions of H atoms in organic, inorganic, and biological molecules). Phasing via neutron data is today common practice, mostly when powder diffraction techniques are used (see Chapter 12); here, only the aspects concerning single crystals will be described. Since the application of appropriate phasing techniques requires a prior knowledge of the neutron scattering mechanism, we briefly recall this topic in Section 11.8. In Section 11.9, possible violation of the positivity postulate, which may occur when neutron radiation is used, will be discussed.

11.2 Electron scattering

Electrons are produced in an electron gun by a filament a few micrometers in size and they are accelerated through a potential difference of E volts. Their divergence is restricted to 10^{-4} rad or less (smaller than for conventional X-ray sources) by electromagnetic lenses and the spread of wavelengths is small (10^{-5} or less). The wavelength may be calculated as

$$\lambda = \frac{12.3}{(E + 10^{-6}E^2)^{1/2}}.$$

If high energy electron diffraction (HEED) is used, $E \geq 100kV$ (the range may be extended to 1 MeV) and $\lambda \leq 0.05$ Å. Since electrons are charged particles, they are strongly absorbed by matter; therefore, electron diffraction in transmission is applicable only to very thin layers of matter (10^{-7} to 10^{-5} cm). While the electron density distribution is responsible for X-ray scattering, for electron scattering it is the potential distribution which plays that role. Such a distribution is the sum of the field caused by the nucleus and the field caused by the electron cloud. Two processes contribute to electron scattering:

1. *Elastic scattering*: the electrons are scattered by the Coulombic potential due to the nucleus. Since the proton is much heavier than the electron, no energy transfer occurs.
2. *Inelastic scattering*: electrons of the primary beam interact with the atomic electrons and are scattered after having suffered a loss in energy. In a microscope, such electrons are focused at different positions and produce the so-called *chromatic aberration*, which causes a blurring of the image.

The strong scattering of electrons by matter implies some advantages but also serious hindrances to structure analysis work. If we look to the atomic scattering amplitudes, f_e, f_x, f_n (for electron, X-ray, and neutron scattering, respectively), on average $f_e \sim 10^{-8}$ cm, $f_x \sim 10^{-11}$ cm, and $f_n \sim 10^{-12}$ cm. In terms of intensity, I, the ratios will approximately satisfy the relation,

$$I_x : I_e : I_n = 1 : 10^6 : 10^{-2}.$$

Minimum specimen thickness for each diffraction technique may be roughly summarized as follows:

about 0.1 mm for neutron radiation (at SNS, Oak Ridge, USA);
about 0.1 mm for common laboratory X-ray diffractometers, about a few microns at the Grenoble synchrotron;
about 30 nm for electron diffraction.

The above data suggest that electron diffraction may reduce the most severe limitations of today's crystallographic research: crystal sample size and the necessity for a single chemical phase. For example, the chemical synthesis of thin films and coatings, superconductors, or improved materials for long-life batteries typically do not yield large single crystals, but rather produce small grain size multiphase powders. From a report based on data extracted from the *JCPDS-ICDD* database, 1997, for a small subset of technologically relevant substances (by courtesy of Thomas Weirich), it may be estimated that the fraction of unknown crystal structures is about 81% for pharmaceuticals, 65% for pigments, 67% for general organic compounds, and 33% for zeolites. It can be envisioned that these large fractions of materials with unknown crystal structure will decrease considerably if the phasing capacity of electron diffraction techniques improves.

11.3 Electron diffraction amplitudes

For electron diffraction, the structure factor may be written as

$$F_\mathbf{h}^B = \sum_{j=1}^{N} f_j^B \exp(2\pi i \mathbf{h} \cdot \mathbf{r}_j),$$

where

$$f^B(s) = 4\pi K \int_0^\infty \rho(r) r^2 \frac{\sin sr}{sr} dr, \quad (11.1)$$

$$s = 4\pi \sin\theta/\lambda, \quad K = \frac{2\pi me}{h^2},$$

h is the Planck constant
and $\rho(r)$ is the atomic potential distribution. $f^B(s)$ is related to the atomic scattering factor for X-rays, $f_x(s)$, by the Mott–Bethe formula,

$$f^B(s) = \frac{K\lambda^2}{\varepsilon_0}(Z - f_x(s))/\sin^2\theta, \quad (11.2)$$

where ε_0 is the permittivity of the vacuum. If λ is in Å, $f^B(s)$ in Å, and f_x in electron units, (11.2) reduces to

$$f^B(s) = 0.023934\lambda^2 (Z - f_x(s)) / \sin^2\theta.$$

At low values of s, the Mott–Bethe formula is less accurate; indeed, $(Z - f_x(s))$ vanishes for neutral atoms. In this case, the formula given by Ibers (1958),

$$f^B(0) = \frac{4\pi m e^2}{3h^2} Z \langle r^2 \rangle, \qquad (11.3)$$

may be used, where $\langle r^2 \rangle$ is the mean square atomic radius.

The $f^B(s)$ values (in Å) are listed in the *International Tables for Crystallography* (1992), Vol. C, Table 4.3.11 for all neutral atoms and most chemically significant ions. Most of the values were derived by Doyle and Turner (1968) using the relativistic Hartree–Fock atomic potential; for some atoms and ions, $f^B(s)$ has been derived using the Mott–Bethe formula (11.2) integrated with (11.3). Relativistic effects can be taken into account by multiplying the tabulated $f^B(s)$ by $m/m_0 = (1 - \beta^2)^{-1/2}$, where $\beta = v/c$ and v is the velocity of the electron. In order to obtain the Fourier coefficients of the potential distribution in volts, $f^B(s)$ values (and therefore F_h^B values) are usually multiplied by the ratio $47.87/V$, where V is the volume of the unit cell in Å^3.

The difference between $f_x(s)$ and $f^B(s)$ can be schematized as:

1. With increasing s value, $f^B(s)$ decreases more rapidly than $f_x(s)$.
2. While $f_x(0) = Z$ coincides with the electron shell charge, $f^B(0)$ is the 'full potential' of the atom. On average, $f^B(0) \simeq Z^{1/3}$, but for small atomic numbers, $f^B(0)$ decreases with increasing Z.
3. The scattering factor of ions may be markedly different from a neutral atom; for small $\sin\theta/\lambda$ ranges, f^B may also be negative (see Fig. 11.1).

The above features of f^B reflect the peculiarity of the potential distribution (Vainshtein, 1964):

(a) The peaks of the atomic potential (being related by Fourier transform to f^B) are more blurred than electron density peaks.
(b) The peak height (that is, the potential in the maximum) is not strongly dependent on the atomic number. Therefore, light atoms (hydrogen included) can be revealed in an easier way than via X-ray data. Typical peak height ratios are

$$H : C : O : Al : Cu = 35 : 165 : 215 : 330 : 750.$$

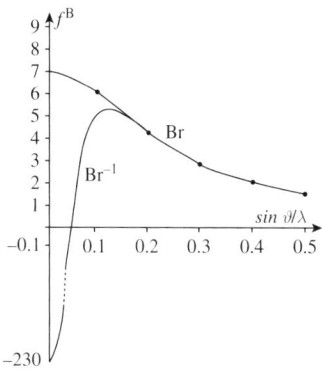

Fig. 11.1
Kinematic electron scattering factor f^B for Br and Br^{-1}.

11.4 Non-kinematical character of electron diffraction amplitudes

Crystal structure analysis via electron diffraction was initiated during the years 1937–8 by a group of crystallographers in the Soviet Union, led by Pinsker and Vainshtein. Ten years later, Vainshtein and Pinsker (1949) published the Fourier map of $Ba Cl_2 \cdot H_2O$. The same kinematic approach was used by

Cowley (1953a,b,c; 1955) for solving a small number of structures. Some years later, Cowley and Moodie (1957, 1959) described the *n-beam dynamical diffraction theory* which, through multislice calculations, more closely describes the physical phenomena involved in electron diffraction. Such models are highly successful in explaining the details of scattering but are, in part, structure and crystal shape dependent (the observed diffraction pattern does not contain direct information on the crystal shape); therefore, corrections eliminating the dynamical effects from the intensities are still not trivial.

The significance of dynamical effects, often predominant with respect to kinematical scattering, has reduced the interest in structure analysis via electron data. In the following, we summarize qualitatively the components of electron diffraction amplitudes.

Dynamic scattering. The transition from kinematic to dynamic scattering occurs when the thickness t of the crystal reaches a critical value for which

$$\lambda \left| \frac{<F_{\mathbf{h}}^B>}{V} \right| t \geq 1. \tag{11.4}$$

$F_{\mathbf{h}}^B$ is the structure factor amplitude (first Born approximation, see Appendix 12.A) and V is the volume of the unit cell. Since $<|F_{\mathbf{h}}^B|>$ is proportional to $Z^{0.8}$, condition (11.4) is soon violated for heavy atoms. Even a 50 Å thickness may be enough to produce dynamic diffraction effects for heavy atom materials. Condition (11.4) may be easily understood from Fig. 11.2, where the type of scattering is monitored as a function of thickness t. The number of unscattered electrons rapidly decreases with t; with increasing values of t, singly scattered electrons are scattered again.

The effects of dynamic scattering on the success of direct methods was evaluated by Dorset et al. (1979) (see also Tivol et al., 1993) by calculating n-beam diffraction data from crystals with increasing thickness and using different wavelengths. The tangent formula was then applied to the dynamic structure factors. Failures occurred for accelerating voltages too low with respect to crystal thickness. However, even if higher voltages are used, dynamic effects are always present in electron diffraction data. Formula (11.4) suggests that one should use very thin crystals, and also higher voltages since they generate smaller wavelengths.

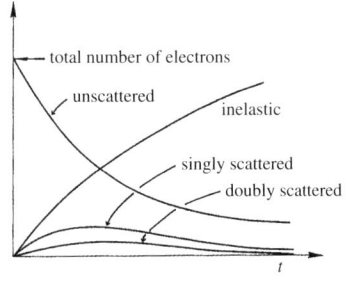

Fig. 11.2
Proportion of electrons which remain unscattered, singly scattered, and doubly scattered as a function of the thickness t. The proportion of inelastically scattered electrons is also shown.

Secondary scattering. A perturbation of the diffracted intensities occurs in thick layered crystals when strongly diffracted beams from upper layers, uncoupled from the lower crystalline regions owing to defects, act as the primary beam for the lower layers. The intensities can then deviate remarkably from the kinematic value. It is worthwhile stressing that the secondary scattering only involves superposition of the intensities of diffracted beams without any interference between coincident beams (such interference is present in dynamic scattering). Accordingly, one measures

$$I'_{h} = I_{h} + m_1 I_{h} \otimes I_{h} + m_2 I_{h} \otimes I_{h} \otimes I_{h} + \cdots, \tag{11.5}$$

where '$\otimes$' denotes the convolution operation.

Among other consequences, owing to (11.5), space group forbidden reflections (because of screw or glide planes) could not remain extinct under the convolution operation. Dorset (1995) applied a secondary scattering model to correct data for copper perchlorophthalocyanine, obtaining a good fit with the kinematic intensities.

Diffraction incoherence. An additional source of incoherent scattering is crystal bending. Even if its occurrence is easily recognizable, the influence on electron diffraction is often not taken into account. Cowley (1961) and Cowley and Goswami (1961) noted that bending is a non-negligible source of incoherent scattering; however, it is not easy to model its effects on diffraction intensities (Turner and Cowley, 1969; Moss and Dorset, 1983).

Radiation damage. Inelastic scattering often damages the crystal specimen and, generally, the resulting damage is of a chemical nature and it is different for different types of materials. The damage may influence the various diffraction intensities by different amounts.

The above considerations suggest that, in spite of great experimental and theoretical advances, the problem of deriving kinematic intensities from observed data is still not completely solved. Thus, one should be prepared to apply phasing methods to scrambled diffracted intensities, and to suffer an unavoidable loss of efficiency. In the case of success, phasing methods should provide phases which, coupled with the diffraction magnitudes, should provide approximated potential maps.

This situation has a counterpart in the final stages of crystal structure analysis, which usually end with a value of the crystallographic residual, R_{cryst}, larger than for X-ray single crystal data. R_{cryst} values close to 0.25–0.35 may be obtained, which reduces to 0.15–0.20 for data with larger kinematical nature.

11.5 A traditional experimental procedure for electron diffraction studies

The traditional experimental diffraction procedure may be summarized as follows:

1. The crystals are transferred to electron microscope grids; a relatively large but thin (see below) defect-free region of the sample is selected for which almost no bend contour is observed.
2. Based on the diffraction pattern of the initial zone, appropriate axes are chosen for *tilting,* to provide different zonal projections. Tilting is performed using the tilt holder in the goniometric stage of the electron microscope. Since the sample is extremely thin, the Fourier transform of the lattice function in the beam direction is not a delta function. In such conditions, tilt angles are not well defined and a tilt series is used to better approximate the tilt angle.

   Electron diffraction patterns usually provide a subset of the reflections within reciprocal space. This weakens the efficiency of direct methods, since a large percentage of strong reflections, and consequently of strong phase relationships, would be lost. A good example is the structure, CBNA (Voigt-Martin et al., 1995; see also Chapter 18), space group $P2_1/c$, for which the b^* axis was chosen as the tilt axis (see Fig. 11.3). Firstly,

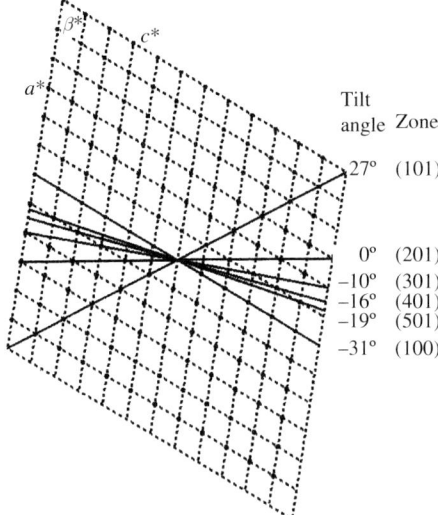

Fig. 11.3
CNBA: reciprocal space view down the tilt axis **b***. Diffraction data have been collected for the emphasized zones (courtesy of I. G. Voigt-Martin and U. Kolb).

zone (201) was recorded (0° tilt angle; $(hk\overline{2h})$ reflections recorded). Then measurements were made for the following zones:

Zone	Tilt angle	Reflection type
(101)	27°	$(hk\overline{h})$
(301)	−10°	$(hk\overline{3h})$
(401)	−16°	$(hk\overline{4h})$
(501)	−19°	$(hk\overline{5h})$
(100)	−31°	$(0kl)$

Tilting can be performed only over a range of 31°; therefore, the data so collected represent only part of the information within the reciprocal space.

It may be worthwhile mentioning two additional reasons for the lack of popularity of traditional ED techniques.

(i) Organic crystals are often subject to severe electron beam damage, and recording diffraction data from different zones of the same crystal is nearly impossible. The problem is made more acute by the low speed of the manual tilting and alignment of the crystal. As a consequence, data acquisition may be an extremely time-consuming task, inappropriate for routine investigations.

(ii) Geometrical restrictions on the specimen holder do not allow the use of high tilt angles with conventional transmission microscopes. This causes further severe limitations in data collection (*missing cone problem*) and therefore additional difficulties in the phasing step.

3. High accuracy in estimated cell constants is difficult to achieve for ED techniques, because data are confined to very small angles. Consequently, the accuracy of the unit cell parameters is usually low, and cells with higher

symmetry may frequently be simulated. A synergy with powder diffraction data is often practised; indeed, unit cell constants obtained by ED data are used as a starting point for indexing powder patterns with strong peak overlapping. Conversely, powder diffraction is used to improve the accuracy of unit cell parameters obtained via ED.

4. Electron diffraction intensities are quantified by application of specifically designed programs (Zou et al., 1993).
5. The space group is identified. This may be performed in three basic ways:

   (a) by exploiting the dynamical effects present in the convergent beam patterns (CBED). The resulting diffraction pattern consists of disks (rather than sharp spots) of diameter proportional to the chosen convergence angle. Interplanar spacing and angular information are obtained from the centres of the disks. The point group symmetry and often the space group symmetry may be derived from the fine structure of intensity variations within the disks. Not necessarily high quality CBED patterns may be obtained, and in any case, this technique requires further supplementary experimental work.

   (b) from a few zone axis microdiffraction patterns, i.e. at least the zero and first-order Laue zones should be recorded and analysed (Morniroli and Steeds, 1992; Morniroli et al., 2007). This approach implies a four-step procedure: identification of the crystal system, of the Bravais lattice, of the glide planes, and of the screw axes.

   (c) using a fully automatic approach (Camalli et al., 2012) based on analysis of the diffraction intensities; similarly to that used for X-ray data (see Section 2.6). This task is not trivial. Indeed, dynamical effects introduce discrepancies among expected symmetry equivalent reflections, and Laue groups belonging to the same crystal system can frequently not be clearly distinguished by checking equivalent intensities. Furthermore, symmetry forbidden reflections (i.e. reflections expected to be systematically absent) show non-vanishing intensity, so making it difficult to identify the diffraction symbol.

   (d) by direct inspection of the projected potential maps, obtained straight from the experiment (see Section 11.5). Indeed, each recorded planar image would show a specific planar symmetry compatible with the three-dimensional space group. For example, if the space group is $P6_3/mcm$, then [001], [100], and [1$\bar{1}$0] images should show projected $p6mm$, pmm, and pgg symmetry, respectively.

11.6 Electron microscopy, image processing, and phasing methods

There is an important supplementary advantage which may be exploited by electron crystallography. Electron diffracted beams can be focused by electromagnetic lenses (equivalent to inverse Fourier transform).

These images, however, are sensitive to focus, crystal thickness, orientation, and astigmatism. Indeed, only very thin and well-aligned crystals can provide images interpretable in terms of crystal structure projection, provided that the effects of other parameters have been corrected. Point-to-point resolution is between 1.5 and 2 Å for a conventional TEM operating at about 200–400kV. In complex structures, atoms overlap (nearly or exactly) in any projection, and therefore cannot be resolved in a single projection image. The way to overcome this problem is to collect several images from different directions and to then combine the images to provide a model structure. This was the basic contribution of de Rosier and Klug (1968), who described a method for the reconstruction of three-dimensional structures from a set of two-dimensional microscope images. Their work allowed the solution of hundreds of molecular structures, including membrane proteins and viruses. Later on, it was found that the contrast of a high-resolution electron microscopy (HREM) image changes with optical conditions and crystal thickness. The so-called contrast transfer function (CTF) plays a fundamental role in contrast changes. The interpretation of contrast became simpler when simulation computer programs (O'Keefe, 1973) using multislice methods, became available. It was then the custom to interpret experimental images via image simulation. This method was essentially a trial and error technique; a structural model is assumed, various optical parameters (thickness, defocus, etc.) are varied, and calculated images are compared with the experimental image. The structural model is modified, then simulation is started again.

The method is time consuming. Klug and his group revived the technique by application of the crystallographic image processing method, which proved capable of recovering the correct structure projection from each individual image. The method has been further improved by Hovmöller and his group in Stockolm (Hovmöller et al., 1984; Wang et al., 1988; Li and Hovmöller, 1988).

What is the accuracy and resolution with which the projection of a structure can be deduced from an image obtained via the back Fourier transform performed by electromagnetic lenses? There are two main factors limiting the immediate use of the electron micrograph:

1. The image does not represent the projection of the crystal potential, but instead, its convolution with the Fourier transform of the contrast transfer function. Therefore, a deconvolution operation is necessary to restore the desired image.
2. Widely scattered electrons are focused at positions other than those to which electrons travelling close to the lens axis (spherical aberration) are focused. As a consequence a point object is spread over length Δr in the image plane, so that the real resolution of an electron microscope is no less than about 100 times the electron wavelength. In practice, 1 Å resolution images can seldom be obtained by high-resolution microscopes; more often the image resolution for organic crystals is 2–3 Å or lower, and from 4 to 15 Å for a two-dimensional protein crystal.

Images with a resolution of 1 Å are only obtained for special inorganic structures. Then, atoms are resolved; since images are projections of the three-dimensional structure, peak overlapping could, even in this case, hinder correct three-dimensional location of the atoms. Several projections are therefore needed for a three-dimensional reconstruction of the structure (Wenk et al.,

1992). However, if the projection axis is short, packing considerations can lead to solution, even from a single projection.

The effectiveness of high-resolution images for crystal structure solution of macromolecules is limited by radiation sensitivity and poor crystal ordering. Membrane proteins are particularly suited to electron microscopy, because they often form two-dimensional crystals. A first significant result was the resolution (at 7 Å resolution) of the purple membrane (Hendersson and Unwin, 1975; Unwin and Hendersson, 1975); the model was further refined at 3.5 Å resolution by combining electron microscope images and electron diffraction intensities.

In spite of the above limitations, Fourier transformation of electron micrography is quite an important branch of electron crystallography devoted to crystal structure. However, the image intensities constitute a non-linear representation of the projected potential and depend on crystal specimen (e.g. thickness and orientation) and on instrumental parameters (e.g. aberration, alignment, defocusing, etc.). Interpretation of the image in terms of charge density distribution is meaningful only when all of the experimental parameters have been correctly adjusted, and/or when it is supported by the image calculated via many-beam dynamic diffraction theory. If this process turns out to be successful, the image may be quite useful for determination of crystallographic phases; it may also be employed as prior information towards extending the phasing process to higher resolution or to a different set of reflections.

Direct methods can play an essential role in this field. Among the various recent achievements we quote the following:

1. Image processing methods have been combined with direct methods (Fan et al., 1985; Hu et al., 1992) and maximum entropy methods (Bricogne 1984, 1988a,b, 1991; Dong et al., 1992; Gilmore et al., 1993; Voigt-Martin et al., 1995).
2. Structure factor statistics has been used to estimate crystal thickness under near-kinematic conditions (Tang et al., 1995).
3. Phases derived from a 10 Å resolution image of a two-dimensional *E. coli* Omp F outer membrane porin (space group P31*m*) have been expanded to 6 Å resolution by the tangent formula. The mean phase error for the 25 determined reflections was about 43° (Dorset, 1996).
4. Phases derived from 15 Å resolution images from bacteriorhodopsin have been extended by maximum entropy and likelihood procedures to the diffraction limit (Gilmore et al., 1993).
5. Useful results were obtained by Dorset (1996) for phasing ab initio, via tangent methods, the centrosymmetric projection of halorhodopsin to 6 Å resolution.
6. Maximum entropy and likelihood methods have been used for an ab initio phase determination (at about 6–10 Å resolution) for two membrane proteins, the Omp F porin from the outer membrane of *E. coli* and for halorhodopsin (Gilmore et al., 1996). Potential maps revealed the essential structural details of the macromolecules.
7. Three-dimensional reconstruction of ordered materials from diffraction images. Particularly interesting was the combination of 13 zone axes for

the structure of a very complex intermetallic compound, v-AlCrFe, with 129 atoms in the asymmetric unit (Zou et al., 2003).

11.7 New experimental approaches: precession and rotation cameras

In the preceding sections we have emphasized the limitations of the traditional experimental ED techniques:

(i) data resolution is limited (only low-index zones are recorded);
(ii) experimental procedures are very time consuming (crystal orientation is in itself time consuming and contributes to a deterioration of beam sensitive samples);
(iii) since conventional manual techniques allow collection of reflections from a few well-oriented zone axes, a diffraction experiment usually provides less than 30% of the full three-dimensional reciprocal space. Owing to the scarcity of observations phasing is difficult, the electron density maps are poor, and least squares refinement is not effective.
(iv) the diffraction intensities (which carry information about the crystal structure) are of poor quality due to multiple/dynamic scattering of the electrons.

The precession electron diffraction technique, recently developed by Vincent and Midgley (1994), allows us to significantly reduce the dynamic effects and improve data resolution. The technical bases of the precession camera are the following: the electron beam is tilted by a small angle, typically 1–3 degrees, and then rotated around the TEM optical axis. The precession movement of the reciprocal plane in diffraction allows only a small number of reflections to be excited at any time (which reduces the multiple/dynamic scattering). Furthermore, the movement integrates over the excitation error because a volume of reciprocal space is explored, not just a surface.

Although the precession technique curtails the problem of dynamic diffraction, another important problem remains: how to collect full 3D reflection data, or, in other words, how to collect data as in automated X-ray diffractometry? If such a technique should become available, ED would show an important advantage over X-ray diffraction and it may be extendable to nanocrystals.

A sequential electron diffraction data collection (*automated diffraction tomography*, *ADT*) and related data processing routines have been developed by Kolb et al. (2007a,b). The technique uses tilts around an arbitrary axis; the reciprocal space is sequentially sampled in fine steps, so that most of the reflections lying in the covered reciprocal space may be collected. The technique combines well with precession techniques; as stated above, better integration of the diffraction intensities may be performed because several cuts through the reflection body can be collected and, in this way, the true reflection intensities are more accurately reconstructed.

One of the most complex systems solved so far, via ADT + precession, is the mineral charoite (Rozhdestvenskaya et al., 2010; $V = 4500$ Å^3, 90 non-H atoms in the asymmetric unit), a silicate structurally close to a zeolite. The structure was solved by direct methods, as implemented in the program, *SIR2008*. Around 9000 reflections with 97% coverage up to 1.1 Å resolution were measured and used and the final crystallographic residual was 17%.

Electron rotation uses a technique rather similar to that employed by electron precession (Zhang et al., 2010), but the main difference is that, in the rotation technique the electron beam is tilted along a straight line, like a pendulum, whereas it is tilted around a circle in precession. Rotations up to 5° may be used and the line can be along the x or the y direction, or along any diagonal in between. Data are collected in small angular steps, in order to handle partially recorded reflections. Measurements can start from any orientation of the crystal, because there is no need to align it.

11.8 Neutron scattering

A neutron is a heavy particle with spin 1/2 and magnetic moment of 1.9132 nuclear magnetrons. The most common sources of neutrons suitable for scattering experiments are *nuclear reactors* and *spallation sources*. Nuclear reactors are based on a continuous fission reaction; fast neutrons are produced whose energy is reduced by collisions in a moderator of heavy water and graphite (thermalization process). The neutrons, thus retarded, obey the Maxwell distribution and the wavelength for the scattering experiment is selected by a monochromator, usually a single crystal of Ge, Cu, Zn, or Pb.

Neutrons are also produced by striking target nuclei (usually tungsten or uranium) with charged particles (protons, α-particles). These are accelerated in short pulses ($<1\,\mu$s) to 500–1000 MeV and cause, by impact with the target, the 'evaporation' of high-energy neutrons. Hydrogenous moderators (typically polyethylene) thermalize the fast neutrons, making them suitable for the scattering experiment.

There are two basic differences between the neutrons produced by a reactor and those from a spallation source: the neutron flux is pulsed when obtained from a spallation source, consequently the experiments must be performed by *time-of-flight* techniques; high intensities at short wavelength ($\lambda < 1$ Å) is a very significant characteristic of spallation sources.

The scattering of neutrons by atoms comprises interaction with the nucleus and interaction with the *magnetic moment of the neutron-magnetic moment of the atom*. This last effect mainly occurs in atoms with incompletely occupied outer electron shells; since the usefulness of phasing methods to diffraction effects caused by magnetic interaction is marginal, this topic will not be covered in this book.

Since the nuclear radius is of the order of 10^{-15} cm (several orders of magnitude less than the wavelength associated with the incident neutrons), the nucleus behaves like a point scatterer and its scattering factor, b_0, will be isotropic and not dependent on θ/λ. In a gas, the nucleus is free to recoil under the impact of the neutrons; then the *free-scattering length* should be calculated by

$$b_{\text{free}} = \frac{M}{m_n + M} b_0,$$

where m_n is the mass of the neutron and M is the nuclear mass. If the nucleus does not behave like an impenetrable sphere, a metastable *nucleus + neutron* system is created, which decays by re-emitting the neutron. For appropriate energies, a reasonable effect may occur; then,

$$b = b_0 + b' + ib'',$$

where b' and b'' are the real and imaginary parts of the resonance scattering. This occurs for relatively few nuclei; for the majority of them the imaginary component is small, so we will ignore it in direct methods applications.

The angular momentum of the nucleus, I, will influence the neutron–nucleus interaction. Indeed, I combines with the neutron spin in both a parallel and an antiparallel fashion, yielding two possible values,

$$J = I + 1/2$$

or

$$J = I - 1/2,$$

corresponding to scattering factors b_+ and b_-, respectively.

According to quantum mechanics, there are $2J + 1$ orientations in the space compatible with one spin of value J. The number of possible states is therefore

$$[2(I + 1/2) + 1] + [2(I - 1/2) + 1] = 2(2I + 1),$$

the fraction

$$w_+ = \frac{2(I + 1/2) + 1}{2(2I + 1)} = \frac{I + 1}{2I + 1}$$

of which corresponds to states with parallel spins; while the fraction

$$w_- = \frac{2(I - 1/2) + 1}{2(2I + 1)} = \frac{I}{2I + 1}$$

corresponds to states with antiparallel spins.

For a single element containing several isotopes, each isotope has its own characteristic scattering length. Accordingly, the mean value of the scattering length of the atom is obtained by averaging first over the two spin states of the isotope,

$$\langle b \rangle_{\text{isotope}} = w_+ b_+ + w_- b_-,$$

and then over all isotopes, taking into account their relative abundance. The final quantity gives the coherent scattering length of the atom which, for the sake of simplicity, is called b throughout the text.

However, incoherent scattering will also occur, with square amplitude given by

$$\left(w_+ b_+^2 + w_- b_-^2\right) - (w_+ b_+ + w_- b_-)^2,$$

which contributes to the background and does not provide diffraction effects. The incoherent scattering is particularly important for hydrogen, for which

$$I = 1/2, \quad b_+ = -1.04 \times 10^{-12} \text{ cm},$$

$$b_- = -4.7 \times 10^{-12} \text{cm}, \quad w_+ = 0.75, \quad w_- = 0.25.$$

Then, $b = -0.39$ and most of the scattering is incoherent.

From the above considerations, a number of important scattering features are obtained:

1. The interaction of neutrons with matter is weaker than for X-rays and electrons (see Section 11.2). Therefore, higher neutron fluxes or larger single crystals are needed in order to measure appreciable scattered intensities.
2. The b values do not vary monotonically with atomic number Z. This specific property allows us to distinguish between atoms having quite close values of Z and different values of b. Specifically, neutron diffraction data are particularly useful for localizing H atoms.
3. Isotopes of the same element have different values of b. Thus, in order to reduce incoherent scattering, H is often replaced by deuterium.
4. For some elements, $b < 0$. A property basic to direct methods may therefore be violated when neutron radiation is used. This topic is discussed in the next section.
5. Coherent scattering, giving rise to Bragg scattering, defines the structure factor,

$$F_h = \sum_{j=1}^{N} b_j \exp\left(2\pi h \cdot r_j\right) \exp\left(-B_j \sin^2 \theta / \lambda^2\right),$$

where b_j is a positive or negative value (b does not vary with $\sin(\theta/\lambda)$), and B_j is the temperature factor.

11.9 Violation of the positivity postulate

In Chapters 5 and 6, we underlined that positivity and atomicity of the electron density are basic conditions for the validity of the traditional direct methods. We want to show here that positivity is not an essential ingredient (Hauptman, 1976). There are several examples of crystal structures solved by direct methods (via data collected via neutron scattering) which involve atoms with negative scattering factors. Our treatment will follow closely the paper by Altomare et al. (1994b).

In order to understand the effect that the violation of the positivity condition has on triplet and quartet relationships we will consider three cases:

Case A: all the scattering factors are real non-negative functions of $\sin \theta / \lambda$ (as for standard X-ray or electron scattering).

Case B: all the scattering factors are real negative functions. Accordingly, ρ is supposed non-positive everywhere. This situation seldom occurs in neutron scattering.

Case C: some scattering factors are real positive functions, some others are real negative functions. Cases A and B are limiting situations of case C.

Case A. If the electron density, ρ, satisfies the relationship

$$\rho \simeq \rho^2, \tag{11.6a}$$

then the Sayre (1952) equation (10.A.4), the Cochran relationship (5.6), and the tangent formula (6.10) arise.

Then the relation

$$\rho \simeq \rho^3 \tag{11.6b}$$

also holds, and Hauptman and Giacovazzo formulas for quartet invariants follow (see Chapter 5).

Case B. Let us replace f_j by $-f_j$. In this case, the structure factor is given by

$$\sum_{j=1}^{N} (-f_j) \exp(2\pi i \mathbf{h} \cdot \mathbf{r}_j) = -F_{\mathbf{h}} = F_{\mathbf{h}} \exp(i\pi). \tag{11.7}$$

Relation (11.6a) is now replaced by

$$\rho \simeq -\rho^2 \tag{11.8}$$

and

$$F_{\mathbf{h}} = -\theta_{\mathbf{h}} \sum_{\mathbf{k}} F_{\mathbf{k}} F_{\mathbf{h}-\mathbf{k}}. \tag{11.9}$$

Comparison of (11.8) and (11.9) with (11.6a) and (10.A.4) shows that the Sayre equation does not hold when ρ is non-positive definite. However, since (11.8) and (11.6a) differ only by their sign, we will refer to (11.9) as a *modified Sayre equation*.

By application of the joint probability distribution method, the triplet phase distribution

$$P(\Phi_{\mathbf{h},\mathbf{k}}) \simeq (2\pi I_0 (-G_{\mathbf{h},\mathbf{k}}))^{-1} \exp(-G_{\mathbf{h},\mathbf{k}} \cos \Phi_{\mathbf{h},\mathbf{k}}) \tag{11.10}$$

is obtained. More generally,

$$\tan \phi_{\mathbf{h}} = \frac{\sum_{\mathbf{k}} -G_{\mathbf{h},\mathbf{k}} \sin(\phi_{\mathbf{k}} + \phi_{\mathbf{h}-\mathbf{k}})}{\left(\sum_{\mathbf{k}} -G_{\mathbf{h},\mathbf{k}} \cos(\phi_{\mathbf{k}} + \phi_{\mathbf{h}-\mathbf{k}})\right)}$$
$$= \frac{\sum_{\mathbf{k}} G_{\mathbf{h},\mathbf{k}} \sin(\phi_{\mathbf{k}} + \phi_{\mathbf{h}-\mathbf{k}} + \pi)}{\left(\sum_{\mathbf{k}} G_{\mathbf{h},\mathbf{k}} \cos(\phi_{\mathbf{k}} + \phi_{\mathbf{h}-\mathbf{k}} + \pi)\right)}, \tag{11.11}$$
$$= -T_{3,\mathbf{h}}/-B_{3,\mathbf{h}}$$

with reliability parameter (unmodified with respect to case A),

$$\alpha_{3,\mathbf{h}} \simeq \left(T_{3,\mathbf{h}}^2 + B_{3,\mathbf{h}}^2\right)^{1/2}. \tag{11.12}$$

A tangent routine which uses negative values of G (or equivalently, assumes positive values of G and adds π to the triplet phase) is expected to be as efficient in finding the correct positions of the negative 'atoms' as the usual tangent routine for the location of positive atoms. Since FOMs based on the distribution of the α moduli (psi-zero FOM included) are not sensitive to the sign of ρ, a structure with positive atomic scattering factors will be marked by the same FOMs as its negative image.

When we move from case A to case B, relation (11.6a) is replaced by (11.8), but relation (11.6b) still holds. Accordingly, the quartet formulas of Hauptman and Giacovazzo still hold. In short, the result is as follows: while triplet phase relationships depend on the 'sign' of the scattering matter, quartet relationships and relative FOMs are insensitive to it. Therefore, the set of phases one obtains at the end of a direct procedure based only on quartet relationships can lead to the correct solution, either via a collection of positive peaks, or equivalently, via a collection of negative peaks.

Case C. Now, neither the Sayre equation (10.A.4) nor the modified Sayre equation (11.9) hold. However, application of the joint probability distribution method brings us to useful formulas. In particular, the Cochran relationship is still valid, but $G_{h,k}$ must be calculated from the general expression (Hauptman, 1976; Altomare et al., 1994b).

$$G_{h,k} = 2C_{h,k}|E_h E_k E_{h-k}|, \quad (11.13)$$

where

$$C_{h,k} = \sum_{j=1}^{N} v_j(\boldsymbol{h}) v_j(\boldsymbol{k}) v_j(\boldsymbol{h}-\boldsymbol{k}) \simeq \sigma_3 \sigma_2^{-3/2}.$$

If all the f_j are positive (case A), then $C_{h,k} = 1/\sqrt{N_{eq}}$, as in the standard formula. If all the f_j are negative (case B), then $C_{h,k} \simeq -\left(1/\sqrt{N_{eq}}\right)$, in agreement with eqation (11.10). If some f_j are positive and others are negative, then $C_{h,k}$ contains contributions of opposite sign. If their balance is close to zero, then $C_{h,k}$ is close to zero too. On preserving the analogy with case A, one could then say that the apparent value of N_{eq} (i.e. $N_{eq} = 1/C_{h,k}^2$) becomes infinitely large (accordingly, N is no longer the number of scatterers in the unit cell, as in case A). In this situation the structure could hardly be solved via triplet invariants.

The equation $\rho \simeq \rho^3$ also holds in case C, and therefore the Hauptman and Giacovazzo formulas still hold. However, estimation of a quartet phase via the joint probability distribution method requires some caution.

With the assumption that $\sigma_4/\sigma_2^2 \simeq (\sigma_3/\sigma_2^{3/2})^2$, the classical quartet estimate is obtained, as described for cases A and B. If some f_j are positive and some are negative, σ_4/σ_2^2 may be quite different from σ_3^2/σ_2^3. In particular, if $\sigma_3 \simeq 0$, the cross-magnitudes do not influence the quartet estimate. Then this depends only on the term,

$$2\left(\sigma_4/\sigma_2^2\right)|E_h E_k E_l E_{h-k-l}| \quad (11.14)$$

and the quartet is always expected to be positive. In particular, for situations in which $\sigma_3 \approx 0$ while σ_4 is still large, the structure solution may be obtained more easily via the active use of quartet invariants rather than triplets.

APPENDIX 11.A ABOUT THE ELASTIC SCATTERING OF ELECTRONS: THE KINEMATICAL APPROXIMATION

We remember some of the properties of elastic scattering of electrons, useful in phasing approaches.

1. Elastic scattering may be described by the Schrödinger equation,

$$\nabla^2 \Psi(\mathbf{r}) + \frac{8\pi^2 me}{h^2}(E + \rho(\mathbf{r}))\Psi(\mathbf{r}) = 0, \quad (11.A.1)$$

where $e = |e|$ is the magnitude of the electronic charge, m is the mass of the electron, E is the accelerating voltage (then eE is the kinetic energy of the electron), and $\rho(\mathbf{r})$ is the (positive) crystal potential, assumed to be periodic as in a perfect crystal. We explicitly stress that the potential distribution, $\rho(\mathbf{r})$, does not coincide with the electron density distribution, even if we use the same symbol, ρ, for both. The kinematical approximation assumes that the electron experiences only a single (weak) scattering through the material, usually from a spherically symmetrical atom. In the Fraunhofer conditions (i.e. measurements of scattered radiation are made at a large distance from the crystal), the total scattered wave is made up of the incident plane wave plus a scattered spherical wave. After some simple approximations, the amplitude of the scattered spherical wave is $F^B(\mathbf{r}^*)$, as defined in Section 11.3.

2. The concept of the Ewald sphere may also be used to describe electron diffraction, but two important differences should be emphasized.

   (a) The short wavelengths imply that the scattering angles are much smaller than for X-ray or neutron radiation experiments. Most of the reflections have 2θ values between $0°$ and $4°$, so that in this region, $\sin\theta$ may be approximated by θ. Since the radius of the Ewald sphere ($= 1/\lambda$) is very large, the sphere is in practice reduced to a plane; thus, several points of the reciprocal lattice plane can be simultaneously in Bragg position (see Fig. 11.A.1). Thousands of diffracted beams can be simultaneously observed on a screen, or collected on a photographic plate for measurement of intensities. Thist is the physical reason for the popularity of a traditional experimental technique collecting diffracted intensities from reciprocal lattice zones (see Section 11.5).

   (b) Since electron radiation interacts strongly with the crystal, the specimen must be very thin (less than 100 Å). In this case, transformation of the shape of the specimen will affect the form and the dimensions of the reciprocal lattice points; each node of such a lattice (a point-like function only for an infinite crystal) is replaced by a distribution function which is non-zero over an extended domain. Accordingly, the wave intensity scattered by a crystal for a reflection $\mathbf{h}$ is given by,

$$I_\mathbf{h}(x^*, y^*, z^*) = I_0 |F_\mathbf{h}^B|^2 |D(x^*, y^*, z^*)|^2,$$

where I_0 is the primary beam intensity, D is the Fourier transform of the crystal shape, and x^*, y^*, z^* are the coordinates in reciprocal space around the $\mathbf{h}$ node.

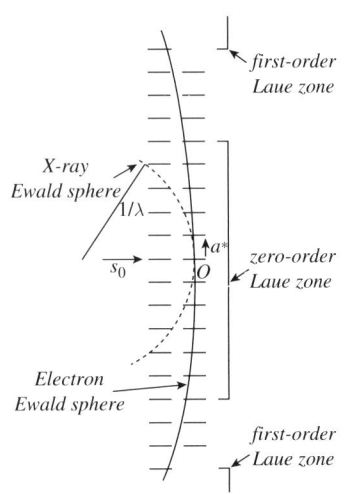

Fig. 11.A.1
Ewald sphere and Laue zones for electron diffraction.

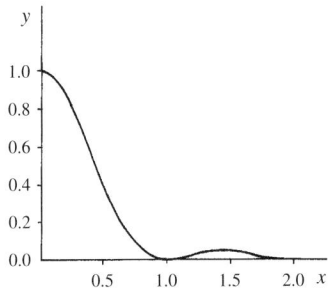

Fig. 11.A.2
$y = [\sin(\pi x)/(\pi x)]^2$ is plotted against x.

In order to give a simple example, let us consider a parallel-sided crystal plate of thickness t, belonging to the orthorhombic crystal system, with $a \simeq b \simeq 10$Å, $c \simeq 6$Å. Let us suppose that the incident electron beam has direction s_0, parallel to c, and wavelength $\lambda \simeq 0.03$Å. Then the radius of the Ewald sphere (see Fig. 11.A.1) is several orders of magnitude larger than the reciprocal lattice vectors. Let N_1, N_2, N_3 be the number of unit cells along $\boldsymbol{a}$, $\boldsymbol{b}$, and $\boldsymbol{c}$, respectively. If N_1 and N_2 are very large and N_3 is small (say $t \simeq 6c$), the reciprocal lattice spots (modelled by the crystal shape transform) are extended in the $\boldsymbol{c}$ direction, and can intersect the Ewald sphere even if the reciprocal lattice points are not on it. Then, the intensities of the $(\boldsymbol{h}, \boldsymbol{k}, 0)$ reflections are given by the expression,

$$I_{hk0} \simeq \frac{C}{r^2}|F^B_{hk0}|^2 \left(\frac{\sin(\pi t \zeta_{hk0})}{\pi t \zeta_{hk0}}\right)^2, \qquad (11.A.2)$$

where C is a constant and ζ is the *excitation error*, which measures the distance of the Ewald sphere from the reciprocal lattice point. For the reflection $(hk0)$,

$$\zeta_{hk0} = \left(h^2 a^{*2} + k^2 b^{*2}\right) \lambda/2.$$

The function $y = \left[\sin(\pi x)/\pi x\right]^2$, with $x = t\zeta_{hk0}$, is shown in Fig. 11.A.2. Its first zero value occurs at $x = 1$ or, in terms of ζ, at $\zeta_{hk0} = t^{-1}$. In practice, the decay of I with ζ, as suggested by (11.A.2), is rarely observed for several practical reasons (variation of thickness, bent crystals, etc.).

The overall result obtained from Fig. 11.A.1 is that several hundred $(hk0)$ spots will, simultaneously, intersect the Ewald sphere and so produce diffracted intensities (all of them belonging to the so-called zero-order Laue zone). From the same figure, it may be seen that a ring of spots belonging to the $(hk1)$ layer (first-order Laue zone) intersect the Ewald sphere and can therefore be collected on the same plate as used for the $(hk0)$ layer. Further rings, corresponding to $(hk2)$, $(hk3)$, etc. layers can also be measured (in favourable situations), which constitute the higher-order Laue zones.

12 Phasing methods for powder data

12.1 Introduction

Powder diffractometry plays (and will probably continue to play in the near future) a central role in research and technology, because it allows us to investigate materials which are not available as a single crystal of adequate size and quality. Therefore, recently, much effort has been devoted to the development of powder diffraction. Improvements include the design of better instruments (e.g. optimized synchrotron radiation lines, time-of-flight technology at pulsed neutron sources, optics, generators, detectors), as well as more sophisticated methods for data analysis. As a result, in favourable cases, high quality powder patterns of proteins may be collected which contain sufficient information to allow identification of the unit cell and of the space group, a result unthinkable 30 years ago. This has opened the way for qualitative analysis and study of the polymorphism of macromolecules (Margiolaki et al., 2005; Collings et al., 2010).

Advances in the experimental and the theoretical aspects of powder crystallography have been able to reduce losses of information from a powder pattern with respect to single crystal data, and have made ab initio crystal structure solution from powder experiments possible.

The reader may deduce the increasing popularity of powder techniques from:

(i) Table 1.11, where, among the *CSD (Cambridge Structural Database)*, entries on 1 January 2012, 2354 powder diffraction studies were counted;
(ii) Figure 12.1, where the cumulative statistics (up to the year 2006) on the number of structures solved via powder diffraction data is shown (*SDPD* database);
(iii) Figure 12.2, where the statistics on the number of studies in the *ICDD (Inorganic Crystal Structure Database)* (to the year 2005) for different types of data is given. For the powder case, 21 472 cases are counted for which powder data have been used, mostly for refinement purposes.

In this chapter, we will neglect experimental aspects, unless unrelated to the phasing problem. We will describe in Sections 12.2 to 12.5, the basic features

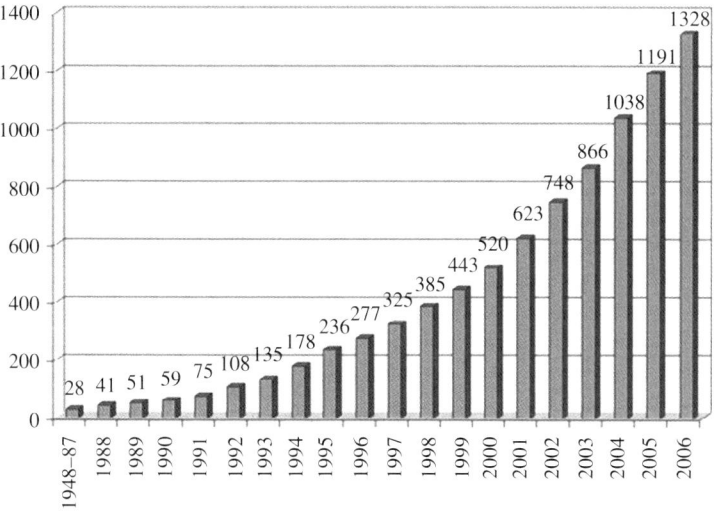

Fig. 12.1

From *Structure Determination from Powder Diffraction – Database* < http://sdpd.univ-lemans.fr/iniref.html>. By courtesy of Armel Le Bail.

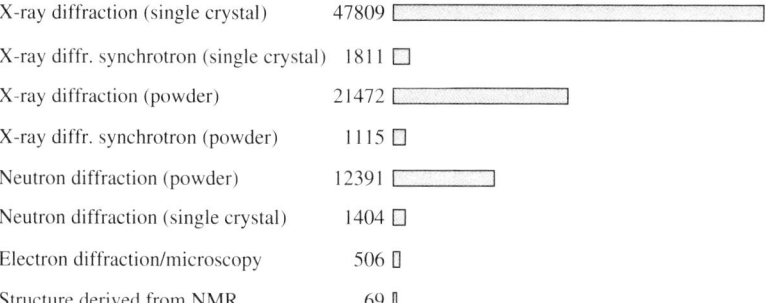

Fig. 12.2

Inorganic Crystal Structure Database (*ICDD*; Nist and Fachinformationszentrum Karlsruhe). Number of deposited crystal structures, solved or refined, per experimental technique. Data extracted from the 2005 release of *ICDD*; by courtesy of Thomas Weirich.

of powder pattern diagrams, and in Sections 12.6 and 12.7, the procedures for full pattern indexing and space group determination. Ab initio phasing will be treated in Section 12.8 and non-ab initio methods in Section 12.9. The combination of anomalous dispersion techniques with powder methods is postponed to Section 15.9.

For brevity, we will not mention numerous computer programs performing a wide range of powder data collection and analysis. Only a few will be reported, those strictly related to the phasing problem. The reader is referred to an extensive paper by Cranswick (2008) for a complete review.

12.2 About the diffraction pattern: peak overlapping

An ideal powder is an ensemble of a very large number of randomly oriented crystallites, usually of the order of a few microns. A reciprocal lattice may be associated with each crystallite; as a result, the various reciprocal

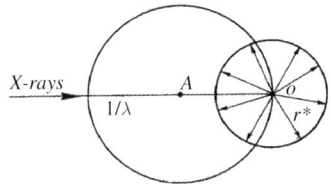

Fig. 12.3
Bragg conditions for ideal powders. The sphere with radius $1/\lambda$ and centre at A is the Ewald sphere. When the specimen is an aggregate of randomly oriented crystallites, the vector $r_\mathbf{h}^*$, corresponding to a fixed reciprocal vector $\mathbf{h}$, is found in all possible orientations.

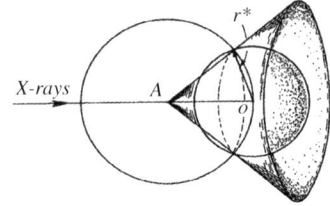

Fig. 12.4
The intersection of a sphere of radius $r_\mathbf{h}^* = h\mathbf{a}^* + k\mathbf{b}^* + l\mathbf{c}^*$ with the Ewald sphere is a circle which defines a diffraction cone (from A), fixing all of the possible directions in which diffraction may be observed.

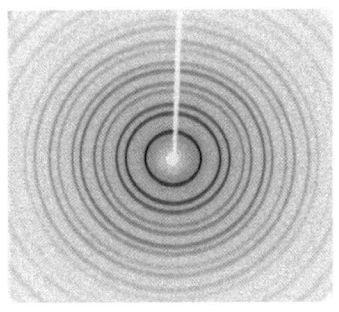

Fig. 12.5
Silicium powder pattern collected in the laboratory: Mo $K\alpha$ radiation, CCD detector perpendicular to the diffraction cones.

lattices, all with the same origin, are themselves randomly oriented. The corresponding ideal powder pattern may be explained by means of the Ewald sphere (see Fig. 12.3). For ideal powders, and for a given $\mathbf{h}$, the vector $r_\mathbf{h}^* = h\mathbf{a}^* + k\mathbf{b}^* + l\mathbf{c}^*$ is found in all possible orientations with respect to the incident X-ray (or neutron) beam; then we can say that the reciprocal lattice point defined by $r_\mathbf{h}^*$ degenerates in a sphere of radius $|r_\mathbf{h}^*|$ with its centre at the origin of the reciprocal space. In accordance with Chapter 1, diffraction will occur for such a reflection $\mathbf{h}$ when the sphere of radius $|r_\mathbf{h}^*|$ intersects the Ewald sphere (see Fig. 12.4). The locus of the intersection points of the two spheres is a circle which, together with point A (the centre of the Ewald sphere), defines a diffraction cone specifying all the possible directions in which diffraction may be observed.

Measurements may be made using a flat detector which intercepts all the diffraction cones (we will see a set of rings, as shown in Fig. 12.5), or in Bragg–Brentano or Debye–Scherrer geometry, which projects peaks on the 2θ axis (as shown in Fig. 12.6).

Figures 12.3 and 12.4 suggest that two reciprocal lattice points with the same modulus, $|r_\mathbf{h}^*|$, will give rise to the same diffraction cone and their diffraction intensities will completely overlap (see Fig. 12.7a); if the $|r_\mathbf{h}^*|$s are slightly different, the overlap will be partial (see Fig. 12.7b). In the first case we are not able to decompose the overall peak intensities in the reflection diffraction intensities; in the second, some information is still present, but additional work taking into account peak profiles is necessary in order to be able to extract the desired information. This is the well-known *overlapping problem* in powder crystallography.

The above considerations suggest the main differences between single crystal data and powder diffraction patterns. In a single crystal diffraction experiment all the measured reflection intensities are accessible at their respective points in the three-dimensional lattice, while in a polycrystalline sample experiment, the three-dimensional reciprocal lattice degrades into a one-dimensional projection, the observed pattern of powder peaks, each characterized by a Bragg angle and by non-vanishing width. If more Bragg reflections overlap, a more severe loss of information occurs.

We will distinguish three different types of peak overlapping, two of systematic type, one of occasional type:

(i) Symmetry-equivalent reflections share the same $|r^*|$ and therefore fully overlap (*systematic overlapping*). This case is however not a problem for phasing approaches. Since the reflection multiplicity m is known (provided that the Laue group has been previously defined, m is the number of symmetry-equivalent reflections), the intensity corresponding to the single reflection $\mathbf{h}$ may be found simply by dividing the overall intensity by m.

(ii) Two symmetry-independent reflections with occasionally equal or similar $|r^*|$ will fully or partially overlap, respectively. The overlap may be total or partial according to the misfit of the corresponding 2θ angles (*occasional overlapping*). Let us consider an example in which a special metric relationship generates occasional overlapping. In the orthorhombic

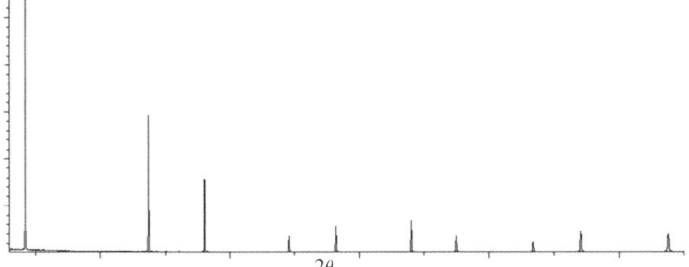

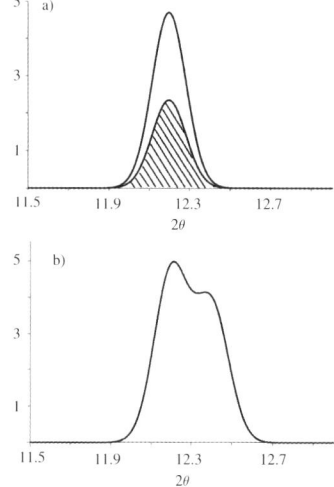

Fig. 12.6
Silicium powder pattern collected in the laboratory: Debye–Scherrer geometry, Cu $K\alpha$ radiation.

unit cell defined by $a = 10.00$ Å, $b = 5.77$ Å, $c = 14.32$ Å, we have, $b \approx a/\sqrt{3}$. Accordingly, the reflections (200) and (110) overlap, as well as (400) and (220), etc.

(iii) In high symmetry crystal systems (i.e. in trigonal, tetragonal, hexagonal, and cubic systems), where the lattice symmetry may be higher than the Laue symmetry (check Table 1.3), *systematic overlapping* may occur between reflections which are not symmetry equivalent. As an example, let $4/m$ be the Laue group of our crystal; the symmetry equivalent reflections are

$$(hkl), (\bar{h}\bar{k}l), (\bar{k}hl), (k\bar{h}l), (\bar{h}\bar{k}\bar{l}), (hk\bar{l}), (k\bar{h}\bar{l}), (\bar{k}h\bar{l}).$$

Since overlapping is ruled by the lattice symmetry, in our case $4/mmm$, the following reflections systematically overlap:

$$(hkl), (\bar{h}\bar{k}l), (\bar{k}hl), (k\bar{h}l), (\bar{h}\bar{k}\bar{l}), (hk\bar{l}), (k\bar{h}\bar{l}), (\bar{k}h\bar{l}),$$

$$(\bar{h}k\bar{l}), (h\bar{k}\bar{l}), (kh\bar{l}), (\bar{k}\bar{h}\bar{l}), (h\bar{k}l), (\bar{h}kl), (\bar{k}\bar{h}l), (khl).$$

In the Laue group $4/m$, the first eight are symmetry equivalent to (hkl), and the second eight to (khl). Since $|F_{hkl}|^2$ and $|F_{khl}|^2$ are uncorrelated, the measured overall intensity (summing the contributions of 16 reflections) cannot be reliably partitioned into $|F_{hkl}|^2$ and $|F_{khl}|^2$.

The quality of the diffraction pattern can determine the success or the failure of the phasing process: the larger the overlapping, the larger the loss of experimental information. Owing to several experimental factors, the typical diffraction pattern provided by a laboratory diffractometer shows peaks with large FWHM (*full width at half maximum*). In particular, unless a special monochromator is used, the peak shape is rather complicated because of the simultaneous presence of three wavelengths (see Fig. 12.8). Therefore, synchrotron radiation powder patterns are often used for crystal structure solution.

Synchrotrons are circular accelerators where charged particles (usually electrons) are accelerated to speeds very close to that of light. Through magnetic fields produced by the so-called bending magnets the electrons are maintained in a circular orbit, tangentially to which synchrotron radiation is emitted. Third generation synchrotrons also implement multiple magnets placed in ad hoc straight sections inside the ring. Inside these multiple magnets, the electrons perform multiple oscillations and emit synchrotron radiation after each oscillation. The result is enhanced radiation emission intensity. Multiple magnets can

Fig. 12.7
(a) Two diffraction peaks completely overlap; (b) two diffraction peaks partially overlap.

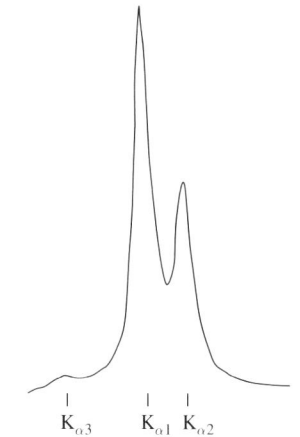

Fig. 12.8
Peak shape model using X-rays from tubes.

be undulators or wigglers, depending on whether coherent or incoherent superposition of these single emissions after each oscillation occurs. Synchrotron radiation is extremely intense, highly collimated (vertically, in the case of bending magnets and both vertically and horizontally in the case of undulators and to a lesser extent, wigglers), linearly polarized in the plane of the electron orbit, and has a continuous spectral distribution over a quite wide spectral range (See *Scientific Discussions* on <www.excels.us>).

X-ray powder diffraction measurements can be performed by using either point detectors or 1D- or 2D-display detectors. Point detectors record full diffraction patterns by spanning the detector in the 2-theta range step by step or continuously, whereas 1D and 2D detectors simultaneously record the full diffraction pattern, a single cut of the Debye–Scherrer rings or the entire rings, respectively.

A point detector with crystal analyser allows the highest angular resolution and therefore the smallest FWHM; their efficiency depends strongly on the photon beam monochromaticity and on its degree of collimation (which requires a considerable amount of photon intensity and explains why the highest FWHM resolutions are achieved at synchrotron facilities). Figure 12.9 shows the full diffraction pattern of *NAC* (chemical composition $Na_2Ca_3Al_2F_{14}$; Courbion and Ferey (1988)) recorded at the Swiss Light Source Materials Science (SLS-MS) beamline using a multicrystal analyser detector (Gozzo et al., 2006). Figure 12.10 shows the diffraction pattern of the same material recorded at the same beamline by the 1D solid state MYTHEN II detector (Bergamaschi et al., 2010). The difference in FWHM may be clearly appreciated from the magnified plots.

Neutron radiation patterns have a FWHM which is larger than for synchrotron data, but have other advantages with respect to X-ray radiation. The scattering power is still high at high $|r^*|$ values (indeed, the nuclei can be considered as point scatterers), and structure refinement is more straightforward when the crystal structure contains atoms of quite different atomic number

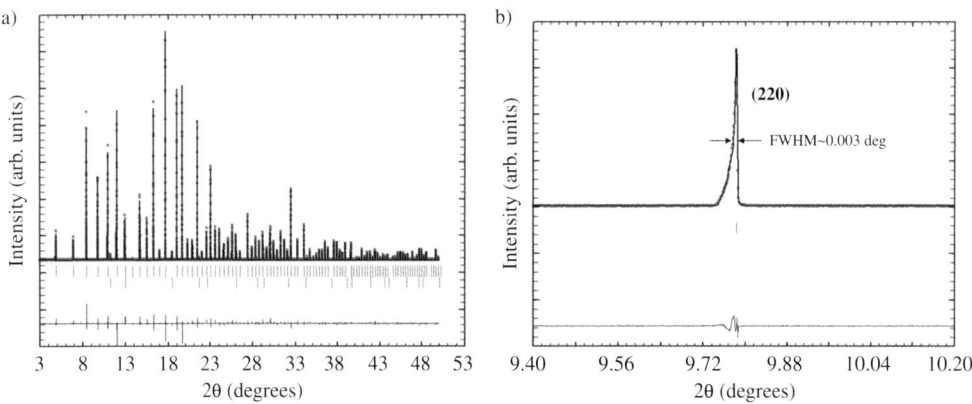

Fig. 12.9
NAC. (a) Diffraction pattern collected at SLS-MS (Villigen, Switzerland) at 20.1 keV by a multicrystal analyser detector: six hour acquisition time; (b) a zoom on the reflection (220). By courtesy of Fabia Gozzo.

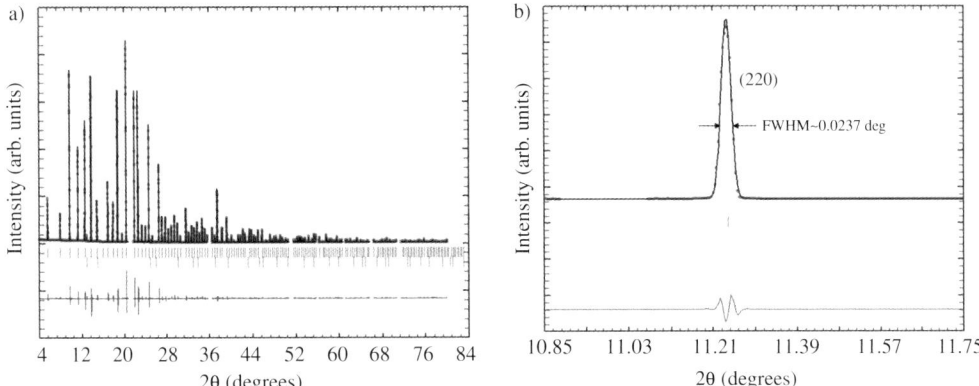

Fig. 12.10
NAC. (a) Diffraction pattern collected at SLS-MS (Villigen, Switzerland) at 17.5 keV by the 1D solid state detector MYTHEN II: 10 second acquisition time; (b) a zoom on the reflection (220). By courtesy of Fabia Gozzo.

(see Sections 11.8 and 11.9). The situation is complementary to that occurring with X-ray data, where location of some dominant atoms may be easy, but subsequent accurate location of light atoms may be much more difficult. It is therefore not infrequent that a partial solution is obtained via synchrotron data and then structure refinement is performed by exploiting both synchrotron and neutron data (Larson and von Dreele, 1987).

In Fig. 12.11a, typical neutron, synchrotron, and laboratory diffractometer patterns are shown for the same structure; in Fig. 12.11b, a high 2θ angle interval is selected to show the different profiles of the three patterns (high contrast peak–background for neutron data, still good contrast and small FWHM for synchrotron data, very noised for the conventional diffractometer).

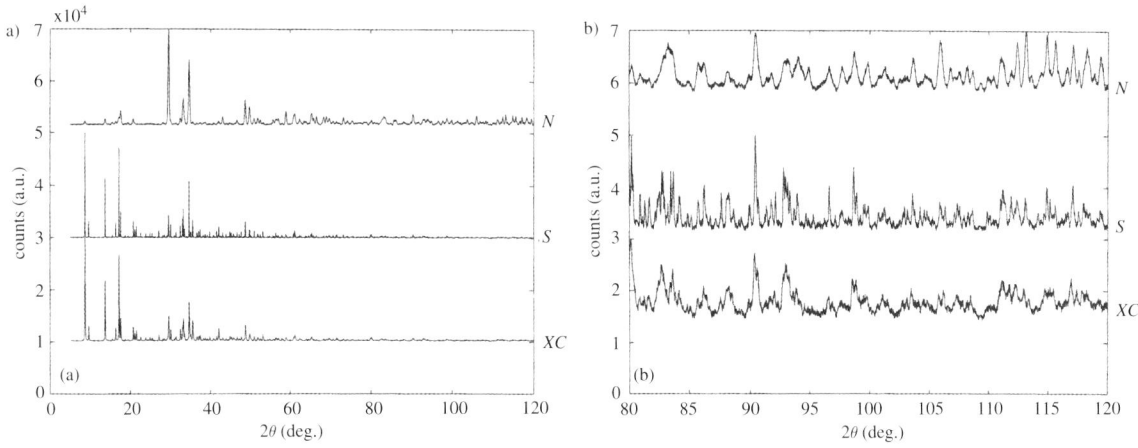

Fig. 12.11
RUCO [Ru(CO)$_4$]. (a) Simulated full pattern for neutrons (*N*), synchrotron (*S*), and conventional X-ray (*XC*). (b) High $\sin\theta/\lambda$ regions of the patterns.

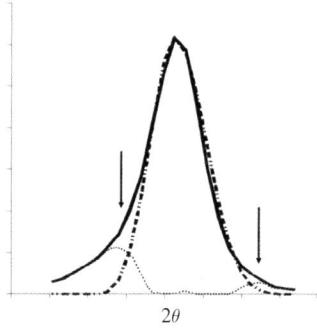

Fig. 12.12
Experimental profile (continuous line) and model profile (dashed line), both centred at the maximum count value. The dotted line is the difference profile between the experimental and the model peak profile: two false peaks (indicated by arrows) will be simulated (artifact) if the difference profile is explored.

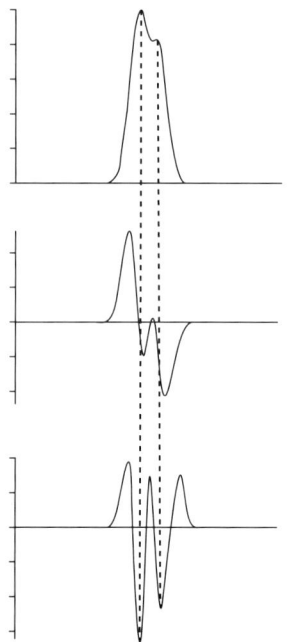

Fig. 12.13
Experimental peak profile, its first and second derivatives are shown in order (from top to bottom). Vertical bars indicate the locations of the two peaks.

12.3 Modelling the diffraction pattern

Modelling a diffraction pattern requires the analysis of all of its profile components. Schematically, this may be reduced to two main activities: analysis of the peak profile and background definition.

The background function is rather complicated because it is the result of numerous phenomena (Suortti and Jennings, 1977; Riello et al., 1995), e.g. air scattering, incoherent and diffuse scattering, amorphous contribution, etc.

Line profile analysis depends (and therefore may provide information) on several factors: size of crystalline domains, line and plane defects (e.g. dislocations and stacking faults, respectively), antiphase domains for materials undergoing disorder/order transformations, compositional fluctuations, etc. Such analysis is basic for the study of microstructure and lattice defects, and is not object of our attention here. The reader is referred to Scardi (2008) for general and advanced information and for a more complete bibliography.

We will deal here with a simplified use of profile analysis, which is, however, extended to the full diffraction pattern. We refer to *full pattern decomposition*, which involves the following steps: recognize peaks and find their locations, define their shape, and model the background. We briefly describe them below.

Peak search. This is the preliminary operation, not at all trivial, as small perturbations in the peak or in the background may simulate false peaks. These artifacts should be eliminated using procedures that are insensitive to noise. Among the most popular methods are peak profile fit (Huang and Parrish, 1975), the derivative method (Sonneveld and Visser, 1975; Huang, 1988), and their combination (Altomare et al., 2000a).

In Fig. 12.12 we show an example of an artifact. Let us suppose that a peak has been located at the maximum experimental count value, but the profile of the model peak is sharper than the experimental one. If the difference pattern is analysed to find supplementary peaks, two small peaks arise which are the cause of the profile misfit.

In Fig. 12.13, a two-peak profile is shown with its first and second derivatives. Overlapping of the two peaks does not allow automatic location of both peaks, the second derivative reveals their positions.

No matter which specific approach is used for the pattern decomposition, a typical computer program generally ends the automatic peak search step by marking the determined peak positions with bars (see Fig. 12.14: this figure is extracted from the *EXPO2009* (Altomare et al., 2009a) graphical output); the user can delete or add additional bars after visual review of the powder pattern.

Pattern modelling. This aims at defining the background and decomposing the diffraction pattern into individual profiles for single reflections. A simple case of profile modelling is that shown in Fig. 12.15, where the profiles of three reflections, weakly overlapping, are determined. Pattern modelling may be performed using two basic algorithms: one (Rietveld modelling), is based on the a posteriori use of analytical functions, the second (the convolution approach) is based on physical principles.

The Rietveld (1969) approach. The calculated diffraction intensity at step i may be written as

$$y_i = S \sum_{\mathbf{k}} m_{\mathbf{k}} L_{\mathbf{k}} |F_{\mathbf{k}}|^2 G(i, \mathbf{k}) A(i, \mathbf{k}) O_{\mathbf{k}} + y_{ib}, \qquad (12.1)$$

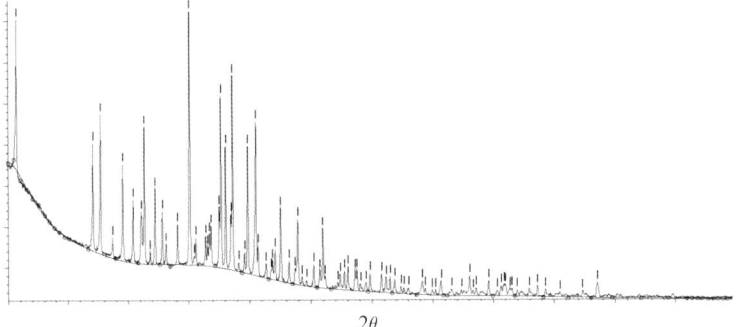

Fig. 12.14
Clomipramine hydrochloride (Florence, A. J. et al. (2005)). Data from laboratory diffractometer. Powder pattern after an automatic peak search step performed by *EXPO2009*. The estimated peak positions are marked by vertical bars.

where S is a scale factor, $m_\mathbf{k}$ is the multiplicity of the reflection $\mathbf{k}$, L is the Lorentz polarization factor, $F_\mathbf{k}$ is the structure factor, $A(i,\mathbf{k})$ is the profile asymmetry function for the reflection $\mathbf{k}$, calculated in i, $O_\mathbf{k}$ is the preferred orientation correction factor, $G(i,\mathbf{k})$ is the profile shape function associated with the reflection $\mathbf{k}$ and calculated at step i, and y_{ib} is the background value at step i. This is mainly due to insufficient shielding, diffuse scattering, incoherent scattering (often high for neutrons), and electron noise of the detector system. The background and its variation with 2θ is usually modelled through a polynomial of a given order (Young, 1993):

$$y_{ib} = \sum_{j=1}^{m} b_j (2\theta_i/2\theta_0 - 1)^{j-1},$$

where $2\theta_0$ is a constant (it corresponds to the 2θ position associated with the minimum observed count in the pattern).

The profile shape function, $G(i,\mathbf{k})$, is determined by several factors, such as the finite crystallite size, imperfections and inhomogeneities of the crystalline material, radiation-specific distributions of intensities and wavelengths, and geometrical and instrumental aberrations. There are many choices for the analytical peak-shape functions G. These include:

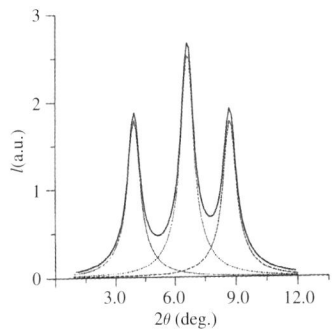

Fig. 12.15
A simple case: the overall profile corresponding to three weakly overlapping peaks (continuous line) is decomposed into component peaks (broken lines).

$$\frac{C_0^{1/2}}{\sqrt{\pi} H_k} \exp(-C_0 X_{ik}^2) \quad \text{(Gaussian)}$$

$$\frac{C_1^{1/2}}{\pi H_k} (1 + C_1 X_{ik}^2)^{-1} \quad \text{(Lorentzian)}$$

$$\frac{2 C_2^{1/2}}{\pi H_k} (1 + C_2 X_{ik}^2)^{-2} \quad \text{(modif. 1 Lorentzian)}$$

$$\frac{C_3^{1/2}}{2 H_k} (1 + C_3 X_{ik}^2)^{-1.5} \quad \text{(modif. 2 Lorentzian)}$$

$$\frac{\eta C_1^{1/2}}{\pi H_k} (1 + C_1 X_{ik}^2)^{-1} + (1-\eta) \frac{C_0^{1/2}}{\pi^{1/2} H_k} \exp(-C_0 X_{ik}^2),$$

with $0 \leq \eta \leq 1$ (pseudo-Voigt)

$$\frac{\Gamma(\beta)}{\Gamma(\beta - 0.5)} \left(\frac{C_4}{\pi}\right)^{1/2} \frac{2}{H_k} (1 + 4C_4 X_{ik}^2)^{-\beta}. \quad \text{(Pearson VII)}$$

where $C_0 = 4\ln 2, C_1 = 4, C_2 = 4(\sqrt{2} - 1), C_3 = 4(2^{2/3} - 1), C_4 = 2^{1/\beta} - 1$, $X_{ik} = \Delta\theta_{ik}/H_k$. H_k is the FWHM of the kth Bragg reflection, and Γ is the gamma function.

It is easy to see that the pseudo-Voigt function presents the mixing parameter η, which gives the percentage Lorentzian character of the profile ($1 - \eta$ is the percentage of the Gaussian component). When $\beta = 1, 2, \infty$, Pearson VII becomes the Lorentzian, modified Lorentzian, and Gaussian function, respectively. Also of use is the pure Voigt function which is the convolution of the Gaussian and Lorentzian forms.

The FWHM is usually considered to vary with scattering angle according to,

$$(\text{FWHM})_G = (U \tan^2 \theta + V \tan \theta + W)^{1/2}$$

for the Gaussian component, and according to

$$(\text{FWHM})_L = X \tan \theta + Y/\cos \theta$$

for the Lorentzian component. $U, V, W,$ and/or X, Y are variable parameters in the profile refinement.

The convolution approach (Wilson, 1963; Klug and Alexander, 1974; Prince, 1993; Cheary and Coelho, 1992; Ida and Toraya, 2002; Gozzo et al., 2006; Zuev, 2006). The calculated diffraction intensity at the angle 2θ is assumed to be the convolution of specific experimental functions (instrumental and sample dependent, taking into account the finiteness of the source, of the sample, of the slits, etc.), to which a background function may be added:

$$y(2\theta) = [w(2\theta) \otimes g(2\theta)] \otimes f(2\theta) + y_b(2\theta). \quad (12.2)$$

$f(2\theta)$ is a specimen-related function, $g(2\theta)$ represents the diffractometer optics, $w(2\theta)$ depends on the wavelength distribution of the incident radiation, and y_b is the background function. The convolution $w(2\theta) \otimes g(2\theta)$ is usually called *instrument function* and it is the result of six specific functions (see Fig. 12.16),

$$g = g_1 \otimes g_2 \otimes \ldots \otimes g_6,$$

where g_1 depends on the projected spot profile, g_2 on the displacement of the flat specimen surface from the focusing circle, g_3 on the axial divergence of the beam, g_4 on the specimen transparency, g_5 on the receiving slits, and g_6 on the possible misalignment of the experimental set-up.

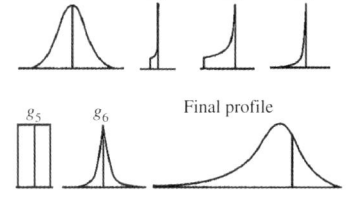

Fig. 12.16
Functions defining the instrument function g.

12.4 Recovering $|F_{hkl}|^2$ from powder patterns

Let us suppose that pattern indexing has been suitably performed. As soon as pattern modelling starts (i.e. in the case of Rietveld modelling, suitable functions for $A(i, \mathbf{k}), O_\mathbf{k}, G(i, \mathbf{k})$ and y_{ib} become fixed), estimates of the $|F_{hkl}|^2$s from the pattern profile become available (for each (hkl), $|F_{hkl}|^2$ is the integral of the profile function associated with that reflection). Recovering the $|F_{hkl}|^2$s is usually achieved by one of the two following procedures:

1. *The Pawley (1981) method*. This is essentially an iterative process which minimizes the sum of squares of the differences between the observed and the calculated profiles. In a standard Pawley refinement the set of parameters which are varied includes, besides cell parameters and those modelling background and peak shape, the integrated intensities themselves. The method suffers from severe problems of ill-conditioning, caused by the assumption that all reflections have independent variable intensities which often refine to negative values. Pawley himself was aware of this problem and introduced slack constraints in the least squares process to force the intensities to assume allowed values. The problem, however, was not solved. Some notable papers have tried to obviate this:

    (a) Since a strong ambiguity in intensities mostly occurs when two or more reflections strongly (or completely) overlap, peaks which fulfil this condition are treated as a single peak and the overall intensity is equipartitioned amongst separate reflections (Toraya, 1986; Jansen et al., 1992).
    (b) The Bayesian approach is used as a natural probabilistic method to impose a positivity constraint (Sivia and David, 1994). The method is consistent with that of French and Wilson (1978), but is generalized so as to include peak overlapping. The method provides meaningful results even when high overlapping occurs.

2. The method of Le Bail (Le Bail et al., 1988). In accordance with the original proposal of Rietveld (1969), single intensities are obtained by partitioning the total observed intensity according to the calculated values for the overlapping reflections. The method does not require inversion of least squares matrices, provides positive values for the intensities if the background is correctly estimated, is computationally efficient, but does not provide standard deviations for the intensity estimates (luckily this is not a real problem for direct methods applications).

    The Le Bail algorithm is quite simple and may be described as follows:

    (a) For each reflection, $\boldsymbol{h}$, an arbitrary value for the intensity is chosen, say $I_{0\boldsymbol{h}}$ (usually the same for all reflections).
    (b) The contribution to the pattern profile corresponding to each reflection $\boldsymbol{h}$ at the ith step, is modelled according to (12.1):

    $$y_{i,h} = L_h I_{0h} G(i, \boldsymbol{h}) A(i, \boldsymbol{h}).$$

    (c) New intensities $I_{1\boldsymbol{h}}$ are calculated by using the formula

    $$I_{1h} = \sum_i (y_i - y_{ib}) \frac{Y_{i,h}}{\Sigma_k y_{i,k}},$$

    where the summation over k goes over all the peaks overlapping with $\boldsymbol{h}$, and the summation over i covers the 2θ interval in which the reflection $\boldsymbol{h}$ contributes to the pattern.

(d) The values I_{1h} so obtained replace I_{0h} at step (b) to generate a new model.

The Le Bail algorithm converges rapidly to better intensities. Some least squares cycles are usually associated with the Le Bail algorithm to reduce the residual between the calculated and the experimental profiles. The parameters to refine are scale factor, background coefficients, 2θ zero shift, peak asymmetry, half-width at half-maximum, the parameters fixing the chosen profile function (i.e. Gaussian, Lorentzian, etc.), and unit cell parameters. Each point of the profile has a weight which is inversely proportional to the experimental intensity value associated with it. The above least squares cycles are followed by application of the Le Bail algorithm, and so on, cyclically.

Sometimes the quality of the powder diagram is not sufficient to allow good estimates of the structure factor amplitudes. Three physical properties may be used to improve the estimates: the positivity of the Patterson function, the anisotropic thermal expansion, and texture effects.

Positivity of the Patterson function. This property may be useful when peak overlapping is not severe (David, 1987; Estermann et al., 1992; Estermann and Gramlich, 1993; Altomare et al., 1998). It is a cyclic procedure which sets to zero the negative values of the Patterson function and back-transforms the resulting map. For completely overlapping reflections the diffraction intensities are shifted away from the original equipartitioned data and become closer to the correct values.

Anisotropic thermal expansion. This may significantly change peak overlapping (e.g. two peaks in strong overlapping at a given temperature may be separated at another temperature); this effect may be used to improve the quality of the extracted reflection intensities (Shankland et al., 1997a,b). As an example, the technique has been applied to an organic structure whose data were collected at ESRF, Grenoble, by using a 1.5 mm capillary and a multianalyser detector, by Brunelli et al. (2003). All attempts to solve the structure failed, despite the high quality of the data (FWHM $\sim 0.016°$ in 2θ). Data were recollected at five different temperatures and, suitably combined, led to the crystal structure solution.

Texture effects. If the powder is not ideal, the experimental data will be affected by texture effects (i.e. systematic distortion of the intensity ratios due to non-random orientation of the crystallites; Hedel et al., 1994; Matthies et al., 1997). Such a distortion may be eliminated by correcting the effects of the preferred orientation, so as to simulate ideal powder data; these techniques are described in Appendix 12.A. The preferred orientation may, however, be explicitly used as a source of information, for improving the pattern decomposition (Wessels et al., 1999; Baerlocher et al., 2004).

Let us specify equation (12.1) for the case in which the orientation distribution function has been experimentally determined as a function of the sample tilt angle, χ, and rotation angle, ϕ. By emphasizing in (12.1) the peak shape function and the preferred orientation correction we can rewrite it as,

$$y(2\theta, \chi, \phi) = \sum_{h,k,l} I_{hkl} O_{hkl}(\chi, \phi) G(2\theta - 2\theta_{hkl}). \tag{12.3}$$

If we now suppose that $y(2\theta, \chi, \phi)$ has been measured for several (χ,ϕ) settings and that $G(2\theta-2\theta_{hkl})$ has been determined by standard Rietveld procedures, then equation (12.3) may be solved for single-crystal reflection intensities I_{hkl}.

12.5 The amount of information in a powder diagram

In single crystal diffraction experiments, the NREFL measured intensities may be considered to be statistically independent of each other (strictly speaking, this is not really so, as triplet relationships suggest): the information provided by the experiment may then be considered to be proportional to NREFL. Let us now consider a powder diagram and let us suppose that NREFL reflections fall within the measured 2θ range. If two or more peaks overlap, their intensities should be correlated (Pawley, 1981; Young, 1993; Sivia and David, 1994). For example, if two or more peaks completely overlap, the sum of intensities must be equal to the clump overall intensity. As a consequence, individual intensities, given prior information on the value of the clump overall intensity, cannot be considered to be statistically independent of the other intensities belonging to the clump. The question now is: is it possible to transform NREFL (the number of correlated intensities) into N_{ind}, an equivalent number of statistically independent intensities?

This problem has several implications:

(a) phasing methods work well when the ratio,
 number of independent intensities/number of atoms to find is sufficiently high;
(b) usual crystallographic least squares consider reflections which are not symmetry equivalent as statistically independent observations; their efficiency is determined by the ratio
 number of independent intensities/number of atoms to refine;
(c) at high 2θ ranges, where the contrast peak/background intensity is extremely low, it may happen that peak overlapping is extremely high. In such a case, the 2θ interval is usually omitted from the calculations.

An algorithm has been proposed by Altomare et al. (1995), which, via a systematic study of peak overlapping, is able to provide an estimate of N_{ind}. Depending on data quality and structural complexity, the ratio N_{ind}/NREFL is often between 1/2 (this is the case for very high data quality, usually obtained by synchrotron radiation) and 1/6 (for laboratory data, relatively large structure). In the second case, crystal structure solution and refinement may be very difficult. Then, supplementary information is needed to solve the phase problem (see for example Section 12.4) and to refine the structure (e.g. via the introduction of restraints on some geometric or energy parameters (Immirzi, 1980; Pawley, 1981; Baerlocher, 1982; Elsenhans, 1990; Izumi, 1989), or by combining X-ray with neutron data (Larson and von Dreele, 1987)).

12.6 Indexing of diffraction patterns

When a single crystal diffraction experiment is performed the usual outcome is as follows:

1. to each diffraction spot a vectorial index **h** is routinely assigned;
2. the diffracted intensities are measured with reasonable accuracy;
3. unit cell and space group are unambiguously found (within the obvious limits of the diffraction laws).

From a powder diffraction experiment unit cell, reflection intensities, and space group are not directly available; some supplementary work is therefore necessary to recover these.

Indexing is the first step in the phasing process; its success is essential to the other steps. Accurate measurements of the peak positions are basic conditions for correct indexing; different types of error, random or systematic, may affect peak positions and shapes (Klug and Alexander, 1974; Wilson, 1963; Parrish, 1965) thus making indexing a difficult task. Among the systematic sources of error we quote: a wrong zero position in the 2θ circle, axial divergence of the incident beam (important at low angles), specimen surface displacement, specimen transparency, and receiving slit amplitude. Most of the errors may be corrected by ensuring perfect alignment of the diffraction instrument and by calibrating it by using standard reference materials.

Indexing is in general easier when synchrotron data are available (because of smallest peak overlapping) and is in general difficult for laboratory diffractometer data (because of the generally large width of the peaks). Shirley (1980) underlined that 'powder indexing works beautifully on good data, but with poor data it will usually not work at all'.

Let us recall the basic quadratic form relating indices and reciprocal cell parameters in the form most suitable for applications:

$$Q_{hkl} = h^2 A_{11} + k^2 A_{22} + l^2 A_{33} + hk A_{12} + hl A_{13} + kl A_{23} \qquad (12.4)$$

where

$$Q_{hkl} = \frac{10^4}{d_{hkl}^2}, \quad d_{hkl} = \frac{\lambda}{2 \sin \theta_{hkl}},$$

$$A_{11} = 10^4 \cdot a^{*2}, \; A_{22} = 10^4 \cdot b^{*2}, \; A_{33} = 10^4 \cdot c^{*2}, \; A_{12} = 10^4 \cdot 2\, a^* \, b^* \cos \gamma^*,$$

$$A_{13} = 10^4 \cdot 2\, a^* \, c^* \cos \beta^*, \; A_{23} = 10^4 \cdot 2\, b^* \, c^* \cos \alpha^*,$$

Since (12.4) is linear with respect to the unknown parameters A_{ij}, the right unit cell may be identified by associating correct indices to n interplanar distances d_{hkl}, where n depends on the lattice symmetry. As a minimum, n equals the number of independent unit cell parameters, therefore $n = 1$ for the cubic system, $n = 2$ for the tetragonal and hexagonal crystals, $n = 3,4,6$ for orthorhombic, monoclinic, and triclinic systems, respectively.

Indexing has historically been addressed using three programs, *ITO* (Visser, 1969), *TREOR90* (Werner et al., 1985), and *DICVOL91* (Boultif and Louër, 1991), with the following approaches:

ITO. Reciprocal lattice planes, defined *via* the origin itself and any two lattice points (zones), are searched. Pairs with a common row are identified, and the angles between each pair are used to describe the lattice.

TREOR90. This is a classical trial and error method. Starting from cubic symmetry, the unit cell search is extended step by step to lower symmetry crystal systems. For each investigated system 'basis diffraction lines' are selected to which tentative indices are associated; the linear equations above are then solved. Testing several different combinations of the basis lines enables us to find a correct solution, even in the case of error in one or more basis lines. The efficiency of *TREOR90* has been greatly improved by two modified versions of the original approach: *N-TREOR* (Altomare et al., 2000a) and *NTREOR-09* (Altomare et al., 2009b), both implemented in the *EXPO2009* code (Altomare et al., 2009a).

DICVOL91. Its systematic indexing approach is based on the *dichotomy method*. It is applied in direct space and relies on the variation, by finite increments, in length of cell edges and of the interaxial angles; the search is m-dimensional, where m is the number of unknown unit cell parameters. The search is performed from high to low symmetry crystal systems by using partitioning of the volume space.

Often, more than one feasible solution is provided, so that proper figures of merit (FOM) must be used to distinguish between bad and good solutions. The most popular has been M_{20} (de Wolff, 1961), defined as follows:

$$M_{20} = \frac{Q_{20}}{2 <\varepsilon> N_{20}},$$

where Q_{20} is the Q value in the case of the 20th observed and indexed peak, $<\varepsilon>$ is the average discrepancy between the observed and calculated Q values for the twenty indexed peaks, and N_{20} is the number of calculated reflections up to the d value corresponding to Q_{20}.

The progress in power and speed of modern computers suggests that new generation indexing programs will rely on the full pattern rather than just on the line positions. Solution may be described as follows: the best lattice $[a, b, c, \alpha, \beta, \gamma]$ is that providing the best agreement between the observed and calculated diffraction patterns. At the same time, the concept of effective figure of merit has evolved. For example, it should:

(a) be definable for any interval of the experimental pattern;
(b) take into account the peak intensities (e.g. correct indexing should not miss the strongest lines of the pattern);
(c) not be very sensible to the presence of impurity lines;
(d) take into account the number of generated peaks and their overlap (e.g. quite large wrong unit cells may easily index a pattern);
(f) be comprised in the interval (0,1), to be interpreted as the probability of the correct indexing.

An example of such an evolved FOM is *WRIP20*, introduced into the program *NTREOR-09*.

12.7 Space group identification

Unequivocal definition of the extinction symbol (and therefore of the space group) is often difficult from powder data, even when the unit cell parameters have been correctly defined. This is because: (a) the Laue group may be identified with difficulty when overlapping is severe. Indeed, the intensities of symmetry-independent reflections may be estimated with low accuracy when they overlap; (b) systematic absences, if present, can be obscured by other non-zero intensity reflections with which they overlap (see Fig. 12.17). Thus, the chosen space group must be carefully reconsidered if any attempt at solving the crystal structure turns out to be unsuccessful.

Two alternative automatic methods have been proposed, the first one, implemented into the package *DASH* (Markvardsen et al., 2001), and the second (Altomare et al., 2004, 2005, 2007), implemented into *EXPO2009*. Both are based on statistical analysis of the reflection integrated intensities and provide a quantitative estimate of the probabilities of the different extinction symbols that are compatible with the crystal system (that determined by the indexing process). The following preliminary step is necessary: the experimental powder diffraction diagram is decomposed via Le Bail or Pawley algorithms into single diffraction intensities in the space group with the largest Laue symmetry and no extinction conditions (e.g. P2/*m* for monoclinic, P2/*m*2/*m*2/*m* for orthorhombic, P4/*mmm* for tetragonal, P6/*mmm* for trigonal–hexagonal systems, and P*m*3*m* for the cubic system).

The *EXPO2009* approach is essentially that described in Section 2.6 for single crystal data. The normalized intensities, $z_\mathbf{h} = |E_\mathbf{h}|^2$, are calculated and submitted to statistical analysis for the determination of the space group symmetry. The analysis is based on the expectation that the mean value $<z_\mathbf{h}>$ calculated for the systematically absent reflections should be close to zero, while it should be close to unity when calculated for non-absent reflections.

At the end of the calculations, the probability value for each extinction symbol, compatible with the lattice symmetry established by the indexing procedure, is provided. The extinction symbol with the largest probability is preferred; the space groups compatible with it are the best candidates.

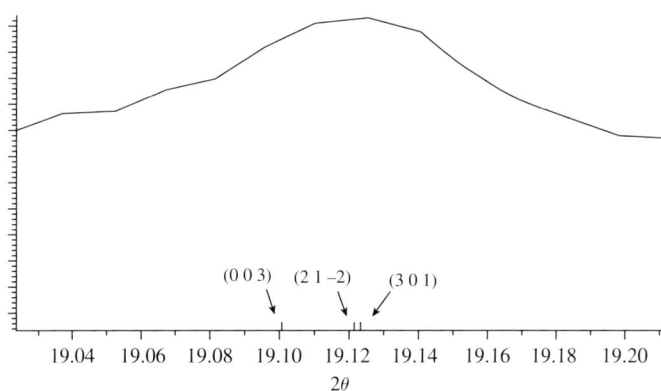

Fig. 12.17

Clomipramine hydrochloride, space group P2$_1$/c. The experimental peak profile arises from the overlapping of three peaks (the corresponding reflection indices are quoted in the figure), one of which [say (3 0 1)] belonging to the set of systematically absent reflections.

12.8 Ab initio phasing methods

Let us suppose that there is no information on the molecular geometry available. If the unit cell has been found, if the full pattern decomposition program has been successfully applied to obtain the $|F_\mathbf{h}|^2$ values for each reflection, and if the space group has been identified, then we have all we need to start ab initio phasing methods for single crystal data. Then, traditional direct Methods, Patterson techniques, charge flipping, or *VLD* may be applied: indeed, the phase information is hidden in the diffraction moduli (see Chapter 1) and these methods just aim to recover it. However, if moduli are estimated approximately, as usually occurs in powder crystallography, the phasing process will be less efficient. This is exactly what happens in practice: crystal structure solution is more difficult because the amount of information collected by a powder is smaller than for a single crystal experiment. This has been the reason for the vigorous development activity and for the success of non-ab initio methods, described in Section 12.9.

For ab initio phasing techniques it is useful to note the following:

(a) Application of the Wilson method (see Chapter 2) does not provide useful information on the centric or acentric nature of the compound (Cascarano et al., 1992c; Estermann and Gramlich, 1993). Different indications may be obtained according to which pattern decomposition algorithm is used. The Pawley technique, if not suitably modified, may provide negative intensities for some of the overlapping reflections; as compensation, single individual intensities larger than the clump overall intensity could be provided. This effect enlarges the variance of the intensity distribution and suggests centric groups, even when the structure is acentric. Conversely, the Le Bail algorithm intrinsically tends to equipartition the overall intensity among severely overlapping reflections. The general statistical consequence is that the dispersion of the structure factor amplitudes appears smaller than it really is, and acentric groups are suggested even when the real space group is centric.
(b) The Wilson scaling process sometimes ends with a negative overall temperature factor (Cascarano et al., 1992c). This result may be caused by truncation of experimental data to where the Debye effects are important (see Section 2.9) or by imperfect modelling of the background and the peak shape (Lutterotti and Scardi, 1990). Indeed at high-angle regions of the pattern the decay of the X-ray atomic scattering factors can make background definition critical; even small errors in the background modelling can produce non-negligible changes in amplitude estimation.
(c) The experimental data resolution RES may be non-atomic; indeed, owing to the strong overlapping, the pattern at high 2θ angles may be very noisy and therefore may be omitted from the phasing process. The overall effect is a shortage of reliable triplets in the case of direct methods, and, in the electron density maps, a strong resolution bias. An effective method for minimizing the resolution bias (see Section 8.2) has recently been described and successfully applied to powder data (Altomare et al., 2008a, 2008b, 2009c, 2010). The algorithm is based on a generalization of the concept of a Gaussian-like peak, which is replaced by a two-component

function, extending over the full unit cell, and consisting of the *main peak* and by the corresponding *ripples*. The theory models main peaks and corresponding ripples both in direct and in reciprocal space, and provides mathematical tools for minimizing the resolution bias of a given electron density map calculated at the experimental data resolution. Application of the algorithm is cyclic and, combined with EDM techniques, is able to move atoms into more correct positions, thus contributing to a full recovery of the structure.

(d) The least squares normally used to refine a given structural model are not very effective for powder data because of:

(i) the usually low ratio between number of reflections in the measured range and the number of parameters to refine;
(ii) the correlation between reflection intensities caused by peak overlapping;
(iii) low accuracy in the observed diffraction moduli (i.e. the estimated intensities provided by the pattern decomposition programs).

To overcome point (i), supplementary information is needed. This usually comes from restraints on some geometric (angles, distances, planarity, etc.) or energy parameters (Immirzi, 1980; Pawley, 1981; Baerlocher, 1982; Elsenhans, 1990; Izumi, 1989).

To reduce the effects of correlation among overlapping reflections, a special weight, w, may be used within the S function that least squares tries to minimize (Altomare et al., 2006):

$$S = \sum_{\mathbf{h}} w_{\mathbf{h}} (I_{o\mathbf{h}} - I_{c\mathbf{h}})^2;$$

$w_{\mathbf{h}}$ is also designed to take into account the reflection overlapping and $I_{o\mathbf{h}}$ and $I_{c\mathbf{h}}$ are the estimated observed and calculated intensities for the reflection **h**. A different approach is based on a non-diagonal weighted least squares procedure (Will, 1979). The function that is minimized is

$$\bar{D}WD,$$

where D is the vector of the residuals between observed and calculated intensities and W is a weight matrix which is proportional to the inverted variance–covariance matrix relative to the observed integrated intensities (Kockelmann et al., 1985).

In spite of the above difficulties, the ab initio phasing techniques so far summarized may often succeed, provided that the peak overlapping is not too severe. To compensate for the low (with respect to single crystal data) quality of powder diffraction data, some low-level prior information may be used in the phasing process. If such information is not specific for the molecule under study, but nevertheless is valid for a wide range of materials, such methods may be considered ab initio. The two techniques summarized below, say *POLPO* (Altomare et al., 2000b; Giacovazzo et al., 2002) and *COVMAP* (Altomare et al., 2012a; 2012b), may be better considered as subsidiary to the ab initio methods, designed to improve their convergence.

POLPO is based on the fact that direct methods applied to powder diffraction data often provide well-located heavy atoms and unreliable light-atom positions; completion of the crystal structure is then not always straightforward. Prior knowledge of the heavy-atom polyhedral coordination is used in a Monte Carlo procedure aimed at locating the light atoms under the restraints of the experimental heavy-atom connectivity model. The correctness of the final model is assessed by criteria based on the agreement between the whole experimental diffraction pattern and the calculated one. The procedure requires little cpu computing time and has been implemented as a routine in *EXPO2009* (see, for example, Altomare et al. (2009a) and literature quoted therein). The method is sufficiently robust against the distortion of the coordination polyhedra and has been successfully applied to some test structures.

POLPO may also succeed if only some of the cations are correctly located; it is also able to locate missing cations and surrounding anions when the cation coordination is tetrahedral or octahedral.

As an example, in Fig. 12.18, the case of *chromium-chromate-tetrachromate*, Cr_8O_{21} is given. Three chromium positions (Fig. 12.18a) are well located by direct methods, and are used in *POLPO* as the starting model for recovering the missing cation and all of the anions. The distance Cr_1–Cr_2 was compatible only with a fourth Cr (the missed one) with octahedral coordination. Therefore, some directives were given to *POLPO* to find the anions of the three located Crs in tetrahedral coordination, with cation–anion distances equal to 1.75 Å, and the missed Cr (and corresponding anions) with octahedral coordination, with distances cation–anion equal to 1.90 Å (see Fig. 12.18a). *POLPO* provided the complete and correct structure model depicted in Fig. 12.18b.

COVMAP is based on the concept of *covariance* between two points of an electron density map (see Section 7.4); i.e. the density at one point depends on the density at another point of the map if their covariance is not vanishing. Let us suppose that the available structural model is not interpretable, because it contains a large percentage of misplaced peaks. Around each correctly positioned peak (i.e. the *pivot peak*), one or more atoms should be present at bond distance, unless the pivot atom is isolated. Since the current model is so poor that the corresponding electron density does not achieve sufficiently high values at the pixels located at bond distance from the pivot peak, *COVMAP* modifies the corresponding current density. In practice the largest peaks of the

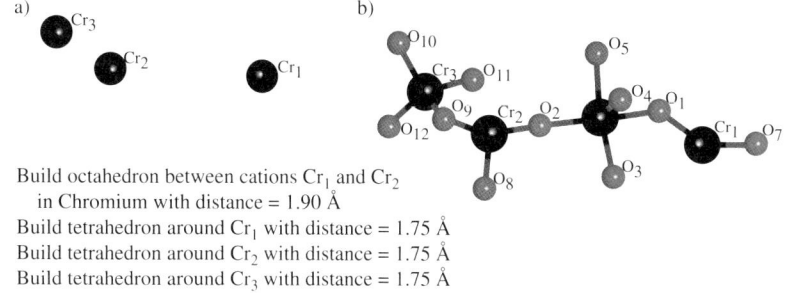

Fig. 12.18
Chromium-chromate-tetrachromate, Cr_8O_{21} (Norby, P. et al. (1991)). (a) the three chromium positions located by direct methods and the directives given to *POLPO* for completing the structure; (b) correct solution provided by *POLPO*.

Fig. 12.19

S-Bupivacaine hydrocloride, $C_{18}H_{28}N_2O \cdot HCl$ (Niederwanger, V. et al. (2009)). (a) the structure model obtained at the end of the ab initio solution process of *EXPO2009* standard version; (b) correct model obtained at the end of the *RBM–COVMAP–wLSQ* procedure.

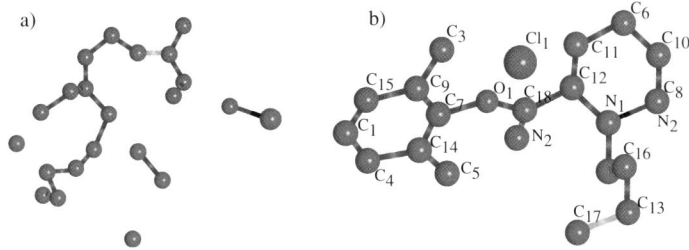

map are, in turn, considered pivots, and for each pair of these, the reasonable expectation that some other peak would be present at bond distance from them is transformed into electron density modifications. In this way, new correct atomic positions can be generated which, submitted to two other important refining tools such as *RBM* (resolution bias minimization algorithm, mentioned in point (c) above) and *wLSQ* (weighted least squares, mentioned in point (d)), often lead to the correct structure, even if the starting model is very poor. The *RBM–COVMAP–wLSQ* procedure has been introduced into *EXPO2009*. In Fig. 12.19, the case of S-bupivacaine hydrochloride, $C_{18}H_{28}N_2O \cdot HCl$ (22 non-H atoms in the asymmetric unit) is considered. In (a) the structure model, obtained at the end of the ab initio solution process of *EXPO2009*, standard version, appears chemically non-interpretable. Figure 12.19b shows the correct model derived through application of the *RBM–COVMAP–wLSQ* approach. It is very close to the true one; only amendable errors in the chemical labels are present.

The current version of *COVMAP* may be considered to be a tool for completing an incomplete and disturbed model provided by direct methods. However, a new version of *COVMAP* has been set up (Altomare et al., 2013) which is applied to random starting models; the method may then be considered to be a self-consistent ab initio approach.

12.9 Non-ab initio phasing methods

Let us suppose that:

(a) for an organic compound, substantial prior information on the molecular geometry is available (the same hypothesis also holds for inorganic structures with well-defined building units);
(b) a chemically feasible model, fixing the connectivity of the various atoms in the molecule, has been constructed by some model-building program (such as *Cerius²*, *Chem3Dultra*, or *Sybyl*, with the support of the *Cambridge Structural Database* (*CSD*));
(c) the model is described in terms of internal coordinates; i.e. bond lengths, bond angles, and torsion angles. Bond lengths and angles may be considered to be approximately fixed, while torsion angles are allowed to vary freely. An example is shown in Fig. 12.20, where the torsion angles of an organic molecule are emphasized.

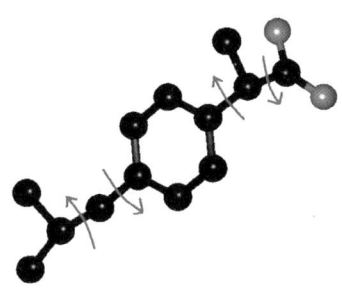

Fig. 12.20

The very simple molecule of ibuprofen (Shankland, K. et al. (1998)). The arrows show the torsion angles.

To solve the structure, one needs to determine the torsion angles (internal degrees of freedom, or internal DOFs) and the global parameters defining position and orientation of the molecular fragment(s) (external DOFs). This may be accomplished by using *direct-space techniques* (*DST*), which employ *global optimization algorithms* to find the best DOF values. These algorithms are able, starting from any random point, to escape from local minima up to when the global minimum is found. An appropriate figure of merit, called *cost function* (*CF*), measures the fitting agreement between experimental and calculated (from the current model) powder diffraction patterns; pattern decomposition is therefore not needed. Grid search, Monte Carlo, simulated annealing, and genetic algorithm are the search methods most commonly used for solving the structure.

Grid search methods. Systematic translations and rotations are performed by a suitably fine grid in the asymmetric region (Chernishev and Schenk, 1998). The method is intuitive, but rather time consuming; indeed it is often applied to rigid models.

Monte Carlo methods. Let us describe these methods in steps:

1. The parameter space is randomly sampled, that is, random values are associated with the various parameters defining the starting configuration. This is the initial model.
2. Any new trial configuration is generated from the preceding one (*Markov chain*) through small random displacements in the parameter space (Harris et al., 1994; Andreev et al., 1997; Tremaine et al., 1997). For example, the generic parameter p_i is changed into

$$(p_i)_{new} = (p_i)_{old} + r_i s_i \Delta p_i,$$

where r_i is a random number in the interval (0,1), s_i may be +1 or −1, it is chosen in a random way, and defines the sense of the displacement, Δp_i is the pre-definite maximum step allowed for the *i*th parameter.
3. The cost function is calculated; the new trial model is accepted (Metropolis et al., 1953) if $CF < CF_{old}$. It is also accepted with probability

$$\exp\left[-(CF - CF_{old})/T\right] \qquad (12.5)$$

if $\exp[-(CF-CF_{old})/T] > r$, where r is a random number between 0 and 1. T is an appropriate scaling factor.
4. When the new model is accepted, it is considered to be a new configuration in the Markov chain, and the algorithm returns to step 2. If the new model is rejected, the algorithm returns to step 2, but the search then starts from the old model.

   The efficiency of Monte Carlo methods is increased if the potential energy of the structural models is calculated; this avoids the acceptance of unrealistic structures.

Simulated annealing techniques. Sampling Monte Carlo techniques are used but the scaling factor T is varied in accordance with an annealing schedule. According to Kirkpatrick et al. (1983), the cost function should be interpreted as the energy of the physical system, the global minimum configuration assumed as the ground state, and the Metropolis criterion considered as the Boltzmann factor $\exp(-E/k_B T)$, for the energy level E: k_B is then the Boltzmann constant and T is the temperature. The ground state may be reached by slowly lowering its temperature, and avoiding the system being trapped in local minima (David et al., 2001; Putz et al., 1999; Engel et al., 1999; Coelho, 2003; Le Bail, 2001). A powerful implementation of the simulated annealing approach has been proposed by Favre-Nicolin and Černy (2002), who used *parallel tempering* techniques in performing a small number of parallel optimizations.

Genetic algorithm techniques. The evolution from a random to the correct model is interpreted on the basis of the Darwinian theory of evolution; selection rules push the population to evolve to the state of best fitness (see Goldberg, 1989). The crystallographic problem may be expressed in terms of biological evolution terminology if the following equivalences are used: a DOF corresponds to a gene (both vary under selection rules), the sequence of DOFs characterizing the current model corresponds to a chromosome (the properties of the current model and of living beings are fully characterized by the DOF sequence and by the chromosome), the CF corresponds with the fitness associated with each model (a favourable CF value corresponds with a good fitness). If we consider a single evolution step (and, equivalently, a step in the model modification), three different operations may be performed: *selection* (drives the population towards the best fit by the selection rules), *mating* (mixes the genetic information of the two parents), *mutation* (this may occur and prevent stagnation of the characters).

An important feature of the genetic algorithm is parallelism, which allows us to treat different members of the population simultaneously (Kariuki et al., 1997; Shankland et al., 1997a,b)

Hybrid approaches. These global optimization methods have the merit of combining the best features of two different algorithms. Of particular interest is the combination direct methods–simulated annealing. In the case of organic compounds (Altomare et al., 2003), direct methods may be unable to provide a fully interpretable electron density map; it is, however, probable that some atomic positions are correct. Such positions may be used as a pivot for model searching via simulated annealing; then, the three DOF global translation parameters are no longer necessary, and the total number of DOFs is correspondingly reduced.

APPENDIX 12.A MINIMIZING TEXTURE EFFECTS

In Fig. 12.A.1, we show the different diffraction patterns of AGPZ, obtained by Masciocchi et al. (1994) by varying the sample preparation technique. In the first of the three patterns, owing to a strong preferred orientation, only $(0k0)$ lines are present; a nearly perfect powder, devoid of texture effects, is obtained only after the third sample preparation. Sometimes, in spite of great efforts, texture effects cannot be avoided; then one has to derive, from the observed intensities, those corresponding to a randomly oriented specimen.

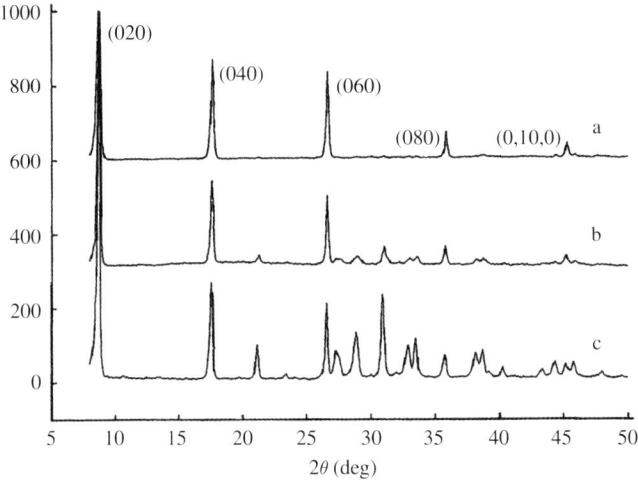

Fig. 12.A.1
AGPZ: [Ag(C3H3N2)]$_n$ (Masciocchi et al., 1994). Different diffraction patterns obtained by varying the sample preparation technique. (a) Impressive texture effects.

This may be done by measuring pole density distributions for a number of reflections, followed by inversion of data to produce pole density for all of the other reflections (see Bunge et al., 1989 and references therein). Such a procedure requires a multi-axis goniometer and additional experimental work. A variant to this approach is the symmetrized harmonic method (see Järvinen, 1993 and references therein), which represents pole figures by spherical harmonic functions and may be applied to specimens with cylindrical symmetry.

An alternative method for texture correction (Altomare et al., 1994a, 1996; Peschar et al., 1995) may be applied to usual powder diffractometer data, provided that the sample has cylindrical symmetry (frequently flat powder samples or capillary tube samples composed of effectively rod- or disc-shaped crystallites tend naturally towards axial symmetry). The method may be described as follows:

(a) The diffraction pattern is decomposed into single intensities. In the absence of any information on the texture, computer programs will provide biased values, $|F'_k|^2$, which are related to the true $|F_k|^2$ by the relation

$$|F'_k|^2 = |F_k|^2 O_k. \qquad (12.A.1)$$

where O_k is the preferred orientation correction factor.
(b) A normalization program derives the normalized structure factor moduli, $|E'|$, from the $|F'|$s.
(c) n reciprocal lattice rows $[hkl]$ are selected, whose directions are expected to be uniformly distributed. In practice, the n reflections (n between 10 and 30) with the smallest $\sin\theta/\lambda$ are considered as a sufficiently exhaustive set of planes candidate for being preferred orientation planes. For each reflection h in the set, steps (d) and (e) are executed in sequence.
(d) Reciprocal space is divided into cones around the direction $[h] = [hkl]$ (see Fig. 12.A.2), each cone contained in the successive one. Each shell (i.e. the region of reciprocal space between two neighbouring cones) will have approximately the same volume so as to include, on average, an equal

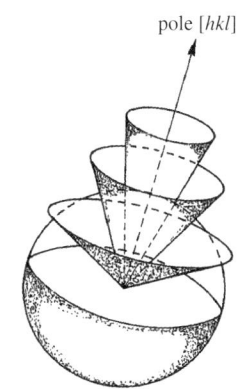

Fig. 12.A.2
The reciprocal space is divided into cones around a direction $[hkl]$.

number of reflections. For all reflections k in the shell, the angle α between the ks and h will be approximately constant.

(e) For each shell, $\langle |E'_k|^2 \rangle$ is calculated; the various $\langle |E'_k|^2 \rangle$ values are then plotted against α.

(f) The von Mises distribution,

$$O = \exp(G \cos 2\alpha)$$

best fitting the experimental data is calculated by a least squares procedure. O is the preferred orientation correction factor.

(g) The plane h (among the candidates) with the final highest value of G is selected as the preferred orientation plane.

(h) The normalized intensities, $|E'_k|^2$, are renormalized to $|E_k|^2$ according to

$$|E_k|^2 = |E'_k|^2 O_k^{-1}.$$

The $|E_k|^2$ so obtained will constitute the new 'observed' normalized moduli for the direct phasing process. $|E_k|$ are expected to be closer to the true values than the $|E'_k|$s; as a consequence, the phasing process should be more straightforward giving more accurate atomic parameters.

Molecular replacement

13.1 Introduction

Modern phasing methods may be subdivided into:

(a) *ab initio* approaches, which include *direct methods*, *Patterson techniques*, *charge flipping*, and *VLD* (*vive la différence*). These approaches do not use (but, suitably modified, some of them can) any prior information on the molecular geometry.

(b) *non-ab initio* methods. In this category, we include *molecular replacement* (*MR*), *isomorphous derivatives* (*SIR-MIR*) and *anomalous dispersion* (*SAD-MAD*) approaches. MR exploits information on the molecular geometry (i.e. the target molecule is known to be similar to that present in another previously solved structure), SIR-MIR uses the supplementary information contained in the experimental data from one or more isomorphous structures, and SAD-MAD exploits anomalous dispersion effects (we will see that such effects simulate isomorphism).

It is immediately clear that classification into ab initio and non-ab initio categories may be questionable, because it hides substantial diversities in the prior information. For example, SAD-MAD, unlike SIR-MIR, may use the native protein data only, and no prior information on the molecular geometry is necessary; apparently, this may be considered to belong to the ab initio category. MR does not use supplementary experimental data, and therefore seems not to be similar to SAD-MAD and SIR-MIR. The latter two techniques are often referred to as *experimental phasing approaches*, but also this appellation is questionable; indeed, the experiment does not provide phases, these are derived by treating the experimental data, as in any other phasing approach.

The above considerations suggest that a more precise, even if conventional, definition for ab initio methods is necessary; in this book, they are identified as *those techniques which do not use the molecular geometry as prior information and exploit only native data, without anomalous dispersion effects*. We have seen in Section 12.8 that some approaches use low-level prior information, not specific to the current structure, but valid for a large range of compounds (e.g. the coordination of some heavy atoms and corresponding bond angles and distances). Also such procedures may be considered as ab initio approaches; to this category we add *ARCIMBOLDO*, which combines the 'trivial' information that a protein consists of smaller molecular fragments of known geometry (among which are α-helices) with MR. *ARCIMBOLDO* is summarized in Section 13.9.

This chapter is devoted to MR (SIR-MIR and SAD-MAD techniques will be described in Chapters 14 and 15, respectively). Basically, MR may be defined as *a method for phasing a target structure when a model molecule, structurally similar to the target, is available*. It implies the use of mathematical techniques for correctly orienting and locating the model molecule in the unit cell of the target structure, and may be applied to small, medium, and macro molecules (see Section 12.9 for its application to powders).

A basic difference exists between the MR approaches used for small molecules (e.g. see Section 12.9 for non-ab initio methods in powder diffraction crystallography) and the typical MR used in protein phasing. When small molecules are treated, the model is usually non-rigid; indeed the solution is found when the correct values of the internal (e.g. torsion angles) and external (e.g. orientation and location of the full model molecule) free parameters are determined. In this case, the small size of the structural problem allows the use of internal degrees of freedom. Conversely, when proteins are treated, the large size of the problem requires the use of rigid models.

MR is particularly important for proteins to which ab initio methods cannot be applied; indeed MR is by far the most popular phasing technique in macromolecular crystallography and we will essentially focus our attention on its applications to proteins. The reasons for its success are twofold:

(a) The cost and effort necessary for preparation of one or more isomorphous derivatives, or, in the case of anomalous scattering, for setting up *Se-Met* labelled proteins or for collecting data based on *S* anomalous scattering, are not needed.
(b) Models similar to the target protein molecule (the concept of similarity is essentially based on sequence identity) are becoming increasingly available. Recent statistics on the *Protein Data Bank* (*PDB*) is depicted in Fig. 13.1; the new fold topologies are becoming increasingly rare. It is not surprising, therefore, that more than 75% of protein structures are today solved by MR.

MR may also be applied to defining non-crystallographic symmetry (NCS); if more than one protein molecule is packed in the same asymmetric unit, different molecules will be related by local symmetry axes. The orientation and position of the NCS axes may be found using MR techniques; this information may be used to improve the efficiency of phase refinement procedures (see Section 8.2).

MR for proteins is essentially a six-dimensional problem; at least three rotation angles and three components of the translation vector must be found. Performing a six-dimensional search directly is too demanding in terms of computer resources and is still not popular. In Section 13.3, a short description of six-dimensional searching is given.

Far fewer computational resources are required if the search is broken into a three-dimensional rotation search, followed by a three-dimensional

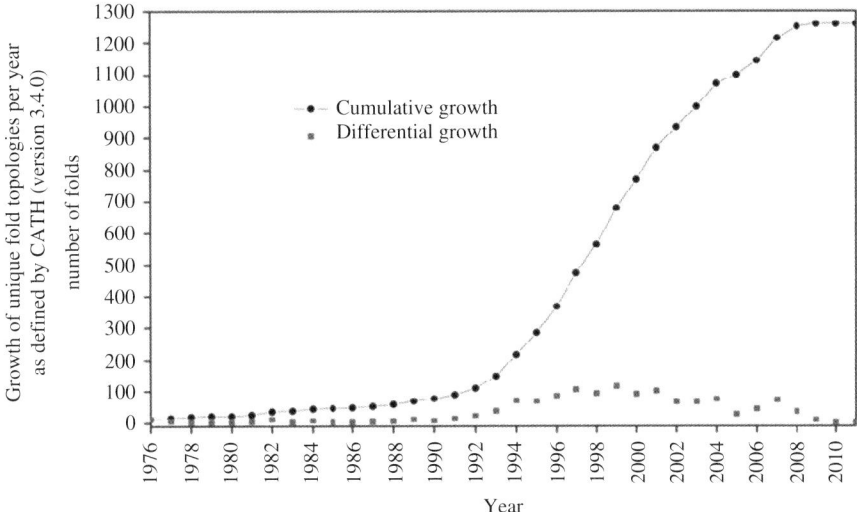

Fig. 13.1
Growth of unique fold topologies per year as defined by *CATH* (version 3.4.0). The two curves represent the cumulative and the differential growths. CATH categorizes structures according to type of structure and assumed evolutionary relations.

translation search (Rossmann and Blow, 1962). Various efficient programs have been written using this strategy [*MERLOT* (Fitzgerald, 1988), *GLRF-REPLACE* (Tong and Rossmann, 1990; Tong, 1993), *X-PLOR/CNS* (Brünger et al., 1998; Grosse-Kunstleve and Adams, 2001), *AMoRe* (Navaza, 1994), *MOLREP* (Vagin and Teplyakov, 1997); *Beast* (Read, 2001), *ACORN_MR* (Yao, 2002), *REMO* (Caliandro et al., 2006), *Phaser* (McCoy et al., 2007), *REMO09* (Caliandro et al., 2009)].

Our description will substantially follow the basic Rossmann and Blow approach because this provides the simple mathematical basis of the method. This point of view will be integrated through illustration of more modern algorithms.

13.2 About the search model

MR is very efficient when the search model is sufficiently similar to the target structure. Its application is, therefore, straightforward for mutants or ligand-bound proteins where the conformational changes are local. The sequence identity (say *ID*) may be used as a criterion for estimating such a similarity; in particular, one can estimate the *root mean square deviation* (in short, *RMS*) between pairwise $C\alpha$ backbone positions according to the Chothia and Lesk (1986) relation,

$$RMS = 0.4 \cdot \exp[1.87(1 - ID)]. \quad (13.1)$$

Some programs also use

$$RMS = \max\{0.8, 0.4 \cdot \exp[1.87(1 - ID)]\}$$

to avoid the underestimation of *RMS*. The following two lines may be used as a practical guide for the estimation of *RMS*:

ID (%)	100	64	63	50	40	30	20	→ 0
RMS (Å)	0.80	0.80	0.80	1.02	1.23	1.48	1.78	2.60

High values of *ID*, and therefore small values of *RMS*, will make MR straightforward; indeed, proteins with similar sequences very often fold in similar structures. For example, as frequently occurs when the same protein, under different crystallization conditions (e.g. change of pH, change of solvent, etc.), crystallizes in different space groups. In this case, the model molecule found in a given space group may be used for phasing in the other space groups. However, the extraordinary flexibility of some molecules can make the use of (13.1) critical; indeed, two molecules may have a high sequence identity but an unusually high value of *RMS*. Conversely, two molecules with low sequence identity may show a small value of *RMS*; this usually occurs when they have a large functional identity.

As a rule of thumb, if $ID \leq 0.30$ (or $RMS \geq 1.5$ Å) MR techniques rarely produce correct solutions; quite crucial, therefore, is the correct identification of a good search model, a context in which the alignment of the protein and of the model sequences (the set of residue-by-residue equivalences between the two sequences) may play a central role. Various software packages may be applied for this purpose, among which we quote the dynamic alignment algorithm by Needleman and Wunsch (1970). When there is a low identity between the target structure sequence and those of possible homologues, identification of the best search model is time consuming. To increase the MR productivity some automated pipelines have been proposed such as *NORMA* (Delarue, 2008), *MrBUMP* (Keegan and Winn, 2008), *BALBES* (Long et al., 2008), part of the *JSCS* software (Scwarzenbacher et al., 2008), or automated servers like *OCA* (Boutselakis et al., 2003) and *PSI-BLAST* (Altschul et al., 1997); they automate identification of the best model by using the amino acid sequence and a customized version of the *PDB* database.

To reduce the risk of failure, large flexible parts of the model may be pruned, as well as those parts for which there is a lack of sequence alignment; indeed, it is not unusual for polyalanine models to be submitted to MR with more chances of solving the structure than the models including side chains. Various pruning procedures may be adopted. For example, in *CHAINSAW* (Schwarzenbacher et al., 2004), one of the programs in *CCP4* (see Winn et al., 2011), given a sequence alignment between template and target, the template structure is modified on a residue by residue basis by pruning non-conserved residues, while conserved residues are left unchanged. It is usually assumed that gaps and insertions are impossible within sequence segments corresponding to helices and strands. Sheet structures have higher percentages of flexible loop residues and therefore they are more sensitive to low sequence identities than largely helical structures.

In *MOLREP*, the search model is modified as follows: (i) residues that align with gaps in the target sequence are deleted; (ii) in case of a pair aligned residues, atoms in the search model without correspondence in the target are deleted.

NMR-based search models are rarely useful for MR (Chen et al., 2000), while search models based on ab initio *modelling* start to play an important role. Prediction of protein structures from their amino acid sequences is still a challenge in computational structural biology: the main obstacles are the large number of degrees of freedom in a protein chain and the complicated energy landscape defined by the strong atomic repulsion at short distances. Here, we mainly refer to *ROSETTA* (Qian et al., 2007), a very popular program for ab initio prediction, which is able to produce a large number of models by combining fragments of known structures having sequences locally compatible with that of the target. The fragments are clustered, and the final conformations are submitted to a process, minimizing a physically realistic energy function. A seminal attempt for exploiting *ROSETTA* models for MR has been made by Ridgen et al. (2008); they selected 16 test cases for which a maximum of 30% sequence identity to previously determined structures was allowed. For ten of them, sufficient reliable trial models were produced, to which MR was applied. In two cases, a complete crystal structure was obtained, via *Phaser,* from the ab initio models. The number increased to three when a special *EDM* procedure was applied to the electron density provided by the MR step (Caliandro et al., 2009). The conclusion is that while ab initio modelling is, today, unable to solve structures on its own, it may be useful when it is coupled with another source of independent information like the crystallographic experimental data.

A last parameter, say n, the number of monomers in the asymmetric unit, deserves to be discussed. Indeed, less straightforward is the application of MR when more molecules are in the asymmetric unit of the target structure; if there are n monomers, but the model consists of one monomer, then n suitable orientation and translation movements are necessary to find the target structure. Since the scattering power of the model structure is $1/n$ (at best) of the target scattering power, recognition of the correct rotations and translations will be difficult, and the difficulty will increase with n. If the n monomers have strong intermolecular contacts, and a complete model constituted by n monomers is available, then a structure solution may be more easily attempted (provided that the monomers maintain the same configuration in both the target and model structures). For example, tetramers often exhibit a 222 point symmetry because they are dimers of dimers. If the dimer also maintains the same configuration in the model, it may be used as a search molecule.

13.3 About the six-dimensional search

Suppose that we have found a good model protein molecule and that we have collected the target experimental data. How do we exploit such information to solve the target crystal structure directly via a six-dimensional search? The most naive idea is to divide the asymmetric unit into grid points. After having located the molecule in a grid point, orient the molecule in all possible orientations and calculate the corresponding structure factor. If the crystallographic residual is never sufficiently small, move the molecule to another grid point and repeat the calculations. The correct structure will be found when the crystallographic residual is sufficiently small. In practice, the algorithm may be too time consuming even for modern computers because:

(i) the grid must be sufficiently fine otherwise possible solutions will be lost. Thus, if we subdivide the asymmetric unit into $100 \times 100 \times 100$ grid points, the total number of grid points to explore would be 10^6;

(ii) for each grid point we should calculate 360^3 structure factor calculations (one structure factor calculation for each molecule orientation, if we rotate in independent 1° intervals), for a total of $10^6 \times 360^3 = 4.7 \times 10^{13}$ structure factor calculations, a huge number of calculations. The reader should consider that each structure factor calculation involves thousands of reflections and for each reflection, thousands of atoms.

There are, however, algorithms for making the six-dimensional search practical (among the various computing programs we quote: Chang and Lewis, 1997; Sheriff et al., 1999; the program *ULTIMA* by Rabinovich and Shakked, 1984; *Queen of Spades*, by Glykos and Kokkinidis, 2000a,b; *SOMoRe*, by Jamrog et al., 2003; *EPMR* by Kissinger et al., 1999). Two problems must be solved:

(a) find a very fast structure factor computation algorithm and replace the naive search approach by an efficient search algorithm. A first attempt to speed up structure factor calculation may be made by using the FFT (fast Fourier transform) algorithm; first, the shape of the electron density is approximated by a Gaussian function, and then the Fourier back transform provides the required structure factors. The method is much faster than calculation from an atomic model, and the difference increases with the size of the structure. This algorithm is, however, not able to substantially reduce the computing time involved in a six-dimensional search. A superfast algorithm may be used, based on a Fourier transform interpolation algorithm (Lattman and Love, 1970; Huber and Schneider, 1985; Kissinger et al., 1999). Since the search molecule is rigid, and the relative coordinates do not change, if the electron density is sampled in a sufficiently fine grid, then the structure factors at any orientation may be obtained by interpolation, while the translation is quantified by the corresponding phase shifts.

(b) replace the exhaustive search algorithms by superfast search approaches, which may be selected from among those used in stochastic optimization methods (see Section 12.9 for their applications to powders), like genetic and evolutionary algorithms (e.g. in *EPMR*, by Kissinger et al., 1999).

However, in spite of the great work done in making six-dimensional searching viable, most of the MR applications are performed via programs in which a six-dimensional search is broken up into a three-dimensional rotation search, followed by a three-dimensional translation search. This is the method we will describe in the following sections.

13.4 The algebraic bases of vector search techniques

Let us consider a P1 structure with two similar or identical molecules in arbitrary orientations: **X** and **X'** are the coordinates of a pair of corresponding atoms in the crystallographic reference system. According to Fig. 13.2, a linear transformation of type

$$\mathbf{X}' = \mathbf{MX} + \mathbf{N}, \tag{13.2}$$

will (approximately, if the molecules are not perfectly identical) relate the corresponding atomic coordinates of the two molecules. **M** and **N** represent a rotation and a translation operator respectively.

The Patterson map of such a structure may be subdivided into three different parts (see Fig. 13.3):

1. the *self-Patterson vectors* of molecule 1, corresponding to the interatomic vectors within molecule 1;
2. the self-Patterson vectors of molecule 2;
3. the *cross-Patterson vectors* corresponding to interatomic vectors relating atoms in molecule 1 with atoms in molecule 2.

Self-Patterson vectors depend on the orientation of the molecule, not on its location; cross-Patterson are quite sensitive to the relative location of the two molecules.

The self-Patterson vectors would lie in a volume U, centred at the origin, whose dimensions are defined by the dimension of the molecule. Some or most of the cross-vectors should lie outside U (indeed, intermolecular vectors are expected to be longer than intramolecular ones). For example, a good choice of the domain U is shown in Fig. 13.3: the circle U contains all the intramolecular vectors while the intermolecular vectors are outside. Of course the choice of U is not always straightforward, mostly when the molecule is rod-like.

Let us now superimpose a rotated version of the Patterson map on the Patterson map itself; there will be no special agreement except when the set of intramolecular vectors of one molecule superimpose with the set of the other molecule. This suggests that one could use the overlapping criterion for assessing the relative orientation of the two molecules.

We will consider two practical cases in which the overlapping criterion may be applied successfully:

(a) Find the reciprocal orientation of two or more molecules lying in the asymmetric unit of the target structure and referred by NCS.
(b) Find the relative orientation of a model molecule, similar or identical to the target molecule, with respect to that of the target molecule.

Once the correct orientation has been found, the set of intermolecular vectors, which depend on the relative location of the two molecules, could be used to locate the molecule. The correct shift would occur when the intermolecular vectors of the model structure superimpose with those of the target structure.

In accordance with the above observations, we have to analyse in the following the so-called *rotation functions* (Sections 13.5 to 13.6) and the *translation functions* (Section 13.7); the first used to orient the search molecules, the second to locate them.

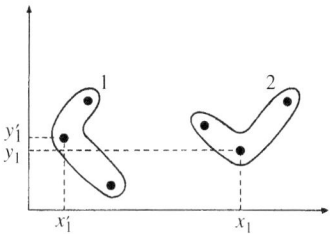

Fig. 13.2
A linear non-crystallographic symmetry operator transforms the set of coordinates $X'_i = (x'_i, y'_i)$ of molecule 1 into the set of coordinates $X_i = (x_i, y_i)$ of molecule 2.

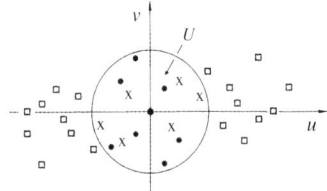

Fig. 13.3
Patterson peaks corresponding to the example in Fig. 13.2. ●, self-Patterson end vectors for molecule 1; X, self-Patterson end vectors for molecule 2. □, cross-Patterson end vectors between molecules 1 and 2.

13.5 Rotation functions

Let us consider the following very simple case: in the asymmetric unit of the protein to solve (*the target*) there is a molecule very similar to another molecule with known geometry (the *model molecule*). How do we solve the target structure by exploiting such prior information?

One might calculate the self-convolution of the isolated model molecule, set in a random orientation (say, $P_{mol}(\mathbf{u}')$), which provides the intramolecular vectors for the chosen random orientation), and try to superimpose $P_{mol}(\mathbf{u}')$, through continuous rotations, to the Patterson map of the target structure (say $P_{targ}(\mathbf{u})$). The rotation function may be written as

$$RF = \int_U P_{mol}(\mathbf{u}')P_{targ}(\mathbf{u})d\mathbf{u}', \qquad (13.3a)$$

where U is the integration domain defined in Section 13.4. Any point $\mathbf{u}'$ in $P_{mol}(\mathbf{u}')$ is related to a corresponding point $\mathbf{u}$ in $P_{targ}(\mathbf{u})$ through a rotation matrix, say

$$\mathbf{u}' = \mathbf{Mu}.$$

Equation (13.3a) becomes

$$RF = \int_U P_{mol}(\mathbf{Mu})P_{targ}(\mathbf{u})d\mathbf{u}. \qquad (13.3b)$$

RF is expected to be a maximum when the peaks of the two Patterson functions overlap, that is when the orientation of the model molecule coincides with the orientation of the target molecule. We will use the following notation:

$$P_{mol}(\mathbf{u}') = \frac{1}{V}\sum_{\mathbf{p}}|F^2_{\mathbf{p}mol}|\exp(-2\pi i\mathbf{p}\cdot\mathbf{u}')$$

and

$$P_{targ}(\mathbf{u}) = \frac{1}{V}\sum_{\mathbf{h}}|F^2_{\mathbf{h}targ}|\exp(-2\pi i\mathbf{h}\cdot\mathbf{u}).$$

In accordance with (13.3b), $P_{mol}(\mathbf{u}')$ may be rewritten as

$$P_{mol}(\mathbf{u}) = \frac{1}{V}\sum_{\mathbf{p}}|F^2_{\mathbf{p}mol}|\exp(-2\pi i\mathbf{p}\mathbf{M}\cdot\mathbf{u}),$$

where $\mathbf{pM}$ is in general a non-integral reciprocal vector. Then

$$RF = \frac{1}{V^2}\int_U \sum_{\mathbf{p}}|F^2_{\mathbf{p}mol}|\exp(-2\pi i\mathbf{pM}\cdot\mathbf{u})\cdot\sum_{\mathbf{h}}|F^2_{\mathbf{h}targ}|\exp(-2\pi i\mathbf{h}\cdot\mathbf{u})d\mathbf{u}$$

$$= \frac{1}{V^2}\sum_{\mathbf{p}}|F^2_{\mathbf{p}mol}|\left(\sum_{\mathbf{h}}|F^2_{\mathbf{h}targ}|G_{\mathbf{hp}}\right),$$

(13.4)

where

$$G_{\mathbf{hp}} = \int_U \exp[-2\pi i(\mathbf{h}+\mathbf{pM})\cdot\mathbf{u}]\,d\mathbf{u}. \qquad (13.5)$$

Expression (13.5) is an interference factor which depends on the integration volume U around the Patterson origin.

Equation (13.4) involves a double summation, over integral indices **h** and non-integral indices **pM**: therefore its calculation may be time-consuming. However, if U is a sphere with radius d_0, $G_{\mathbf{hp}}$ reduces to

$$G_{\mathbf{hp}} = \frac{3\sin(2\pi K d_0) - (2\pi K d_0)\cos(2\pi K d_0)}{(2\pi K d_0)^3},$$

where

$$K = |\mathbf{h} + \mathbf{pM}|.$$

The slope of the interference function is shown in Fig. 13.4, against $K d_0$. Its maximum value is 1 and it is never larger than 0.086 for $|K d_0| > 0.72$. This feature allows (Tollin and Rossmann, 1966) the computing time of the rotation function to be reduced if one eliminates the contribution of the reflections **h** having too small $G_{\mathbf{hp}}$ values. For example, let us suppose that the protein diameter is 50 Å and the cell is 100 Å in each direction; then $d_0 = 0.5$. If we want to neglect, for a given **pM** reflection, the contribution of the reflections for which $G_{\mathbf{hp}} < 0.086$, we should neglect the reflections **h** for which

$$|K d_0| = |\mathbf{h} + \mathbf{pM}|\, 0.5 \leq 0.72,$$

or equivalently,

$$|\mathbf{h} + \mathbf{pM}| < 1.5.$$

In this way, the inner summation in (13.4) may be performed over a limited set of points **h**, which are sufficiently close to the non-integral point $-\mathbf{pM}$.

Two ways have been found to make calculation of the rotation function faster (by more than 100 times):

(i) The so-called *fast rotation function* (Crowther, 1972). Each Patterson function may be expanded in terms of spherical harmonics and spherical Bessel functions. Navaza (1987, 1993) suggested that the above expansions are slowly convergent and may cause non-negligible errors. It was shown that errors can be drastically reduced if the expansion in radial functions is replaced by suitable numerical integration rules. Such modifications and greater automation have made the Navaza package *AMoRe* one of the most popular phasing tools.

(ii) The MFT method (Rabinovitch et al., 1998). So far we have assumed that the model structure is rotated in small steps and that, at each step,

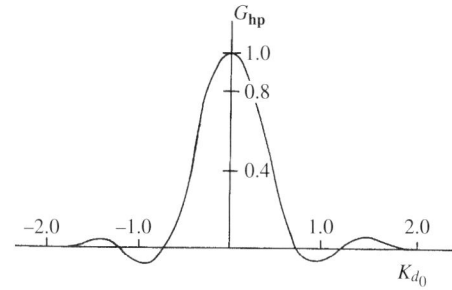

Fig. 13.4
Shape of G_{hp} against $K d_0$.

the corresponding structure factor is calculated. According to the MFT approach, the structure factor of the molecular model is calculated only once, and fitting is achieved by rotating the observed reciprocal lattice with respect to the model lattice. As specified in point 3 of Section 13.6, the model is accommodated in an enlarged cubic cell. The indices **h** of the protein structure, once transformed from the original protein unit cell to the enlarged cubic cell defined for the model structure, are systematically rotated via the matrix **M**, corresponding to each sampling point of the orientation space. Since the algorithm involves only the indices, it is very fast; it has been adopted in the programs *REMO* and *REMO09*.

Equation (13.4) represents the classical rotation function calculated in reciprocal space. But the Patterson map superposition may also be performed in direct space (Hoppe, 1957; Nordman and Nakatsu, 1963; Huber, 1965, 1969; Braun et al., 1969); the fit criterion in this case is the product of the map and of its rotated version at corresponding grid points, as suggested by equation (13.3b). Since the grid points of the search map may not coincide with the grid points of the target map during rotation, it is necessary to interpolate the values; this limits the accuracy of the method. To keep computing time low, only the strongest Patterson peaks of the model map (generally a few thousand) are used in the search. During rotation the origin peaks in the two Patterson maps are always overlapping, which reduces fluctuations in the figure of merit. This effect can be removed by removing the origin peaks from both of the Patterson maps. Sharpening of the Patterson may also be useful.

13.6 Practical aspects of the rotation function

The rotation of a solid object is more easily governed by using a Cartesian frame, rather than a crystallographic reference system. The mathematics necessary to rotate a molecule in a crystallographic ambient is described in Appendix 13.A; here, we illustrate some practical recipes for the use of rotation functions.

When using a MR program the crystallographer has to take (or the program automatically takes) several decisions to make the rotation search effective. Among others we quote:

1. *The choice of an optimal model to rotate*. These aspects have been summarized in Section 13.2.
2. *Limits of data resolution*. Two resolution cut-offs are applied in MR applications. The first concerns the low-resolution reflections; these are omitted from MR procedures because their intensity depends strongly on the solvent (see Section 8.A.3). The high resolution cut-off depends on the similarity between model and target molecules; the smaller the similarity, the more severe the cut-off should be. Appropriate limits can be found by trial (the common range is between 2.5 and 5 Å resolution). Once a good MR solution has been found, EDM techniques (see Section 8.2) are generally applied for phase extension and refinement, eventually followed by the application of *automatic model building procedures* (see Section 8.3).

3. *The choice of the volume U.* This may have an arbitrary shape, but it is mostly chosen as a spherical domain in order to take advantage of the properties of the rotation group. Its radius should exclude intermolecular vectors; in practice, it has to be chosen so as to maximize the ratio number of intramolecular/number of intermolecular vectors in U. An appropriate choice for most of the problems should be using a value of 75–80% of the molecular diameter (Blow, 1985). When several intermolecular vectors lie in U, the efficiency of the method decreases.

   In most modern MR computing programs no use is made of U. In order to reduce interpolation problems, the rotation is performed in a very dense orthogonal reciprocal lattice. Its direct dimensions are chosen to be four times the maximum molecular dimension.

4. *The use of the symmetry.* A rotation function possesses a symmetry which is a direct consequence of the symmetries of the Patterson functions. Calculations are greatly reduced if such symmetry is taken into account (see Appendix 13.A.2), because the search may be limited to the asymmetric unit of the rotation space. The computing time is also considerably reduced, particularly in the case of high symmetries (Tollin et al., 1966; Burdina, 1973; Rao et al., 1980; Moss, 1985).

5. *The significance of the possible solution.* The correct rotation is usually found via the correlation coefficient,

$$FOM_R = \frac{\sum_h \left(X_{targ} - <X_{targ}>\right) \cdot (X_{mol} - <X_{mol}>)}{\left[\sum_h \left(X_{targ} - <X_{targ}>\right)^2\right]^{1/2} \left[\sum_h (X_{mol} - <X_{mol}>)^2\right]^{1/2}}, \quad (13.6)$$

where $|X_{targ}| = |F_{targ}|^2$ is the squared structure factor modulus of the target structure, and X_{mol} is a quantity varying with the orientation of the intramolecular vectors. X_{mol} is defined, and its use justified, in Appendix 13.C.

The solution landscape in MR is rather flat and the FOM average value varies from case to case. Quite often, rather than the absolute FOM value, it is the high contrast between a FOM and the others which indicates a good solution. Thus, the orientations corresponding to the highest values of FOM_R may be selected according to the normalized variable,

$$\rho_R = \frac{FOM_R - FOM_{R\,min}}{FOM_{R\,max} - FOM_{R\,min}}. \quad (13.7)$$

6. A selection threshold is applied in order to take into account the expected difficulty of the rotation case; e.g. if the sequence identity is low and/or if *NCS* is present. Then, usually, a larger number of possible solutions are selected. The orientations corresponding to the highest values of ρ_R are refined by performing a finer rotational search to constitute a better starting point for application of the translational search. The translation search itself would provide, through suitable figures of merit, quite effective score functions able to identify the correct orientations.

7. Let us now consider the case in which the angular relations between the copies are known, e.g. the orientations of the NCS axes of a proper point group have been predetermined via a study of the Patterson self-rotation function (see Appendix 13.B). Then, a strong constraint on the rotations

arises, which may be exploited by the use of a *locked rotation function*, RF_L (Tong and Rossmann, 1990; Tong, 2001).

Let S_n be one of the τ NCS operators, including the identity. If **M** is a correct orientation of the cross-rotation function, then $S_n\mathbf{M}$ also corresponds to a correct orientation (if the rotational NCS does not form a proper group, either $S_n\mathbf{M}$ or $S_n^{-1}\mathbf{M}$, but not both, correspond to a correct orientation). Then the average of the cross-rotation values at orientations related by NCS may be used in equation (13.3b), leading to the locked rotation function,

$$RF_L = \int_U P_{mol}(\mathbf{Mu}) \sum_{n=1}^{\tau} [P_{targ}(S_n\mathbf{u})] d\mathbf{u}.$$

RF_L is expected to increase the signal-to-noise ratio.

13.7 The translation functions

We will consider the case in which the model molecule has been correctly oriented, but its absolute position is unknown. The problem of finding its position does not exist in the space group P1, at least for the first monomer, because the origin may be arbitrarily fixed; for the other oriented monomers the correct location with respect to the first has to be found.

In space groups with symmetry higher than P1 the problem may be solved by observing that, when a molecule is translated in the unit cell, symmetry-related molecules move accordingly. As a consequence, all intermolecular vectors change, while intramolecular vectors remain unmodified. Thus the absolute position of the molecule would correspond to a maximum of the overlapping between the cross-vectors calculated for the molecule and the cross-vectors of the target Patterson. The problem of locating the molecule is not simple, because the function monitoring the overlapping can show many maxima. Sources of noise are the imperfect orientation of the molecule, the presence of intramolecular vectors mixed with intermolecular ones, and a limited similarity between model and target.

In analogy with the rotation function (13.3), the translation function may be expressed as the convolution of the target Patterson with the Patterson map of the shifted model:

$$TF = \int_V P_{targ}(\mathbf{r}) P_{mol}(\mathbf{r} - \mathbf{N}_{mol}) dV, \quad (13.8)$$

where $\mathbf{N}_{mol}$ is the shift vector between the model molecule and the target molecule, and $P_{mol}(\mathbf{r} - \mathbf{N}_{mol})$ is the shifted Patterson map of the model structure. Application of the convolution theorem allows us to rewrite (13.8) in the form of a discrete summation:

$$TF(\mathbf{N}_{mol}) = \sum_{\mathbf{h}} |F_{targ}|^2 |F_{mol}(\mathbf{h}, \mathbf{N}_{mol})|^2. \quad (13.9)$$

In equation (13.9) we emphasize the fact that TF is a function of the shift vector $\mathbf{N}_{mol}$. The explicit form of the $TF(\mathbf{N}_{mol})$ map is given in Appendix 13.C.

More translation functions have been suggested (a review has been made by Beurskens et al., 1987). Function (13.9) has a basic advantage in that it may be calculated rapidly via FFT. The largest peaks in the TF map are expected to correspond to possible translation vectors.

Once a reasonable number of peaks in the TF map have been selected, they may be submitted to a score function, say FOM_{TF}, to select the most probable translation vector $\mathbf{N}_{mol}$. Typical FOM_{TF} are the crystallographic residue or the correlation between $|F_{targ}|$ and $|F_{mol}(\mathbf{h}, \mathbf{N}_{mol})|$ (the latter is the structure factor modulus calculated at the current position of the model molecule). In practice, the feasible translation vectors are ranked by the normalized criterion,

$$\rho_{TF} = \frac{FOM_{TF} - (FOM_{TF})_{\min}}{(FOM_{TF})_{\max} - (FOM_{TF})_{\min}}.$$

The correct solution is expected to be among the trials with the largest values of ρ_{TF}.

The probabilistic approach in *REMO09* suggests the criterion,

$$FOM_{TF} = \sum_{\mathbf{h}} M_{u\mathbf{h}} X_{\mathbf{h}} m_1(X_{\mathbf{h}}) = \max, \qquad (13.10)$$

where X and m are defined by equations (7.7) and (7.8) (R_{targ} and R_{mol} replace R and R_p, respectively), and $M_{u\mathbf{h}}$ is the multiplicity of the reflection $\mathbf{h}$. The probabilistic nature of the criterion (13.10) suggests to consider only reflections for which $X > 1$ (these give the largest contribution to the sum). A relevant point to stress is that using $\sum_{\mathbf{h}} X_{\mathbf{h}} m_1(X_{\mathbf{h}})$ is not equivalent to using $< Xm_1(X) >$. Indeed, for the correct translation it is expected that the number of reflections for which $|F_{mod}|$ and $|F_{obs}|$ are both large or small is bigger than for a trial translation. Dividing $\sum_{\mathbf{h}} X_{\mathbf{h}} m_1(X_{\mathbf{h}})$ by the number of terms in the summation would deplete the score of the correct translation.

To better explain the relation between the rotation and the translation step in MR procedures we notice the following:

(a) As specified above, the orientations corresponding to the highest values of ρ_R (see equation (13.7)) are submitted to translational search (multisolution approach). The top translation solutions usually undergo a final refinement process. For example, they may be submitted to rigid body refinement in which the overall orientation and translation parameters are reconsidered. Rigid refinement may also be applied to segments of the model molecule which are thought to be independent, or to different units of a multimeric assembly related by NCS. Alternatively, optimization of the overall orientation and translation parameters may be more finely assigned using a subspace-searching simplex method for unconstrained optimization (Rowan, 1990), which is a generalization of the downhill simplex method (Nelder and Mead, 1965). This has been the choice for *REMO* and *REMO09*.
(b) The bulk solvent contribution to the structure factors (see Section 8.A.3) may be taken into consideration to increase the efficiency of the MR translation search. Indeed, such a correction improves the correlation between observed and calculated structure factors and therefore increases the efficiency of the search.
(c) For the first oriented molecule, not all of the unit cell needs to be explored in the translation search. For example, in P2$_1$ the origin may be freely

chosen along the dyad axis. Therefore, the vector $\mathbf{N}_{mol}$ may be restricted to the family of vectors [x0z]. In Pm, the origin is free along the direction x and z, therefore $\mathbf{N}_{mol}$ may be restricted to the family of vectors [0y0]. If the symmetry of the space group is taken fully into account, it may be seen that only the Cheshire cell (Hirshfeld, 1968), a limited region of the unit cell, must be covered to obtain the solution; it depends on the spatial arrangement of the positions of the allowed origins described in Sections 3.4 and 3.6. In the space group P1, the Cheshire cell is a point; the molecule may be arbitrarily located anywhere.

(d) A steric clash between symmetry-related molecules may occur after the translation step; when a search for multiple molecules in the asymmetric unit of the target structure is performed, a clash may also be found between molecules related by NCS. The clash may be numerically estimated by calculating the molecular envelopes (Hendrickson and Ward, 1976) or via an analytical function (Harada et al., 1981), or by calculating the number of contacts closer than 3 Å between Cα atoms in the two clashing molecules. Severe clashes cannot be admitted; translations corresponding to less severe clashes (e.g. due to some inadequacy of the model) are usually downloaded, so contributing to an increase in the signal for the correct solution.

(e) If the target structure contains $n > 1$ monomers in the asymmetric unit, a cyclic procedure is started. The first monomer is located as illustrated in point (a) above. Such a candidate, locally optimized, is combined with those selected in the rotational search to form candidate pairs. The first (that is the located) monomer of the pair is kept fixed and the translation TF function for two independent models is applied to provide the position of the second monomer. To save computing time, for each pair only the peaks in the TF map with the highest figure of merit are considered. Once the second monomer has been located, the procedure may be iterated.

A different approach has been proposed by Vagin and Teplyakov (2000) for simultaneously locating two properly oriented monomers. If the nrf largest peaks of the rotation function have been stored, each pair combination may fit the orientations of a monomer dyad (in all, we have $nrf^2/2$ pair combinations). How do we determine, for every putative dyad, the intermolecular vector relating the two monomers? The first step aims at locating the intermolecular vectors between the two monomers by exploiting the relation,

$$|F_\mathbf{h}|^2 = F_\mathbf{h} F_{-\mathbf{h}} = F_{1\mathbf{h}} F_{1\mathbf{h}}^* + F_{2\mathbf{h}} F_{2\mathbf{h}}^* + F_{1\mathbf{h}} F_{2\mathbf{h}}^* \exp[-2\pi i \mathbf{h}(\mathbf{s}_1 - \mathbf{s}_2)] + F_{1\mathbf{h}}^* F_{2\mathbf{h}} \exp[-2\pi i \mathbf{h}(\mathbf{s}_2 - \mathbf{s}_1)],$$

where $\mathbf{s}_1$ and $\mathbf{s}_2$ define the centres of mass of the two monomers, and $F_{1\mathbf{h}}$ and $F_{2\mathbf{h}}$ are the structure factors of the two monomers centred at the origin. For each putative dyad, a translation function is applied in the space group P1 to find the intermolecular vectors that relate the monomers in the dyad. When that has been done, a positional search for each dyad is made by means of a conventional translation function.

(f) If an electron density map has been obtained by another phasing method, the phased translation function may be applied (Bentley and Houdusse,

1992; Read and Schierbeek, 1988; Tong, 1993). Such a function may be very useful when there are several molecules in the asymmetric unit and the last few must be located. In this case, the phase information from the located molecules can be applied to the observed structure factor amplitudes, so improving the contrast signal to noise.

13.8 About stochastic approaches to MR

So far we have emphasized the algebraic aspects of MR. The incompleteness of the model, the (unknown) differences between search model and target molecule, and the errors in the experimental data suggest that the stochastic treatment of MR can make the algorithm more robust. The task may be accomplished via a study of the joint probability distribution function, $P(E_{targ}, E_{mol})$. Study should reflect the specific conditions under which the particular MR problem has to be solved. A general condition to be considered in the formulation of a joint probability distribution is the degree of similarity between search and target molecules. In orientation problems this condition is represented by the similarity of the corresponding interatomic distances, in the translation problem, the positional vectors $\mathbf{r}_{jmol}$ of the jth atom of the model molecule may not exactly coincide with the corresponding positional vector in the target molecule, after application of the correct translation $\mathbf{N}_{mol}$ (say, $\mathbf{r}_{jmol}$ may be close but not identical with $\mathbf{r}_{jtarg} + \mathbf{N}_{mol}$).

Furthermore, the study should take into account information that is available at each step of the MR procedure. In particular, the study should be adapted for:

(i) Orienting a monomer when one or more other monomers have already been oriented;
(ii) Orienting a monomer when one or more other monomers have already been oriented and located;
(iii) Translating a well-oriented monomer when one or more monomers have been correctly oriented and located.

Such an approach has been followed in *PHASER* (McCoy et al., 2007) and in *REMO09* (Caliandro et al., 2009). The first package uses the *maximum likelihood approach*, the second, the *method of joint probability distribution functions*.

13.9 Combining MR with 'trivial' prior information: the *ARCIMBOLDO* approach

We have often stressed in this book the important role of the positivity and atomicity of the electron density in ab initio phasing. Actually, these properties may guarantee success when data resolution is sufficiently high and/or when structural complexity is sufficiently small. The presence of heavy atoms makes phasing less difficult; even complex proteins at 2 Å resolution can be solved by exploiting atomicity and positivity (see Chapter 10). When data resolution is poor and the structure does not include heavy atoms, positivity and atomicity are no longer sufficient for successful phasing; some additional prior information is necessary, like that exploited by MR, SIR-MIR and SAD-MAD techniques.

There is, however, some low-level prior information which is widely available, not specific to the current structure, but valid for a wide range of compounds. For example (see Section 12.8), in the small molecule field, *POLPO* (Altomare et al., 2000b) exploits the known coordination of some heavy atoms and *COVMAP* (Altomare et al., 2012) profits from the known average bond distance and angles of carbon. Both require an electron density map, usually disturbed and not interpretable, to guide the location of the atoms.

In the macromolecular field a number of approaches have been developed to automatically accommodate molecular fragments, even of small size, in noisy electron density maps. Optimization techniques are used to make less computationally demanding the six-dimensional search necessary for orienting and translating the fragments (Kleywegt and Jones, 1997a; Jones, 2004; Cowtan, 1998, 2008).

If no phase information is available, ab initio approaches may be attempted by exploiting some low-level prior information based on well-conserved domains (like, for example, α-helix polyalanine fragments) the overall geometry of which is nearly the same, no matter which protein. Apparently, the trivial application of MR techniques to fragments which are a very small portion of the full structure is unlikely to succeed. The result should be a very long list of possible solutions among which it is very difficult to find the correct solution by suitable figures of merit.

A general approach to protein ab initio crystal structure solution which exploits the 'trivial' information that a protein consists of smaller molecular fragments of known geometry (among which are α-helices) is that of *ARCIMBOLDO* (Rodriguez et al., 2009, 2012). *ARCIMBOLDO* exploits in a more efficient way an important previous result obtained using *ACORN* (Yao, 2002); orienting and locating a perfect fragment representing 13% of the structure can be enough for a successful protein phasing (followed by application of EDM techniques). *ARCIMBOLDO* made the approach much more efficient; indeed, it can use different types of prior information to locate the small fragments and has extended data resolution limits up to 2 Å. Its procedure may be described as follows:

1. The orientation and positions of α-helical polyalanine fragments of about 14 residues are searched via the program *PHASER*, after having truncated the experimental diffraction data at 2.5 Å resolution. They represent a very low fraction of the scattering mass and their positions cannot be fixed without ambiguity when the target helices are fragments with more than 14 residues (e.g. an α-helix of 20 amino acids may accommodate a helix of 14 amino acids in seven displaced positions). The larger the number of fragments to correctly locate, the larger will be the number of allowed positions.
2. The application of *PHASER* may return a huge number of partial solutions (i.e. hundreds or thousands) with very similar figures of merit. At this level, good partial solutions cannot be discriminated from the false ones.

3. *PHASER* is restarted and all the solutions are used for searching additional new fragments. Then the EDM procedures start again; each cycle of density modification includes structure factor extrapolation beyond the experimental resolution (Caliandro et al., 2007b) to improve the phases of the observed reflections and to make the electron density map more interpretable. At this stage, figures of merit may discriminate the correct solutions. Main chain autotracing may allow us to interpret the map, assemble the fragments, and control the solutions; better figures can then be applied which take into account the number of residues the program has been able to trace and the correlation coefficient of the partial structure against the experimental data.

The present version of *ARCIMBOLDO* is very demanding in terms of computational power. The calculations (such as rotation search, translation search refinement, density modification, etc.) are distributed on a computer grid and executed in a parallel way.

13.10 Applications

The continuous improvements in *MR* methods allowed that 2/3 of the structures deposited in the *PDB* (> 84000 entries) are solved by *MR*. As anticipated in the § 13.2 several pipelines are today available to automatize the entire structure solution process, that is, from the identification of the best search model up to the automated model building. In some of these pipelines efforts are also dedicated to distort the template in such a way that it becomes, locally, more similar to the target: this operation may be made before submitting the template to the *MR* step (as described in the §13.2), and later on, during the phase refinement step, to improve the electron density map (Terwilliger et al., 2012). An example may be the following: residues in a β-sheet and in an adjacent α-helix have a similar relation in the model and in the target structure, but the orientation and the location of the sheet with respect to the helix could be different in the two structures. Model and target structures may be made locally closer by searching for a translation leading in overlapping corresponding fragments of the target and of the model. This is made by selecting a group of atoms in a 12 Å diameter sphere and applying the criterion according to which the shift should maximize the correlation between the electron density map and the density calculated from the shifted atoms. The deformed model is then refined to improve the geometry, and then the process is iterated until convergence.

Of particular interest for future *MR* developments is the sequence of programs connected to *ROSETTA* and to *Phaser*, as described by DiMaio et al. (2011), where algorithms for protein structure modeling are combined with those developed for the crystal structure solution. In other words, let us suppose that a search model has been identified and possibly modified as described in the §13.2, but:

(i) the resulting template is sufficiently similar to the target (e.g., with 0.20 < SI < 0.30) to allow a successful *MR* run, but the correct *MR* solution (i.e., that with the template correctly positioned in the target unit cell) is not recognizable among the other trials.

(ii) the correctly positioned model is different enough to hinder the generation of an electron density map of enough quality to rebuild successfully.

The a posteriori analysis of these cases show that difficulties are mainly due to the fact that large fragments of the target main chain differ by 2-3 Å from the corresponding fragments in the template. For such cases *ROSETTA* is able to distinguish the correct *MR* solution from a large set of candidates (see point i)) and also to improve the selected model so that it may generate an automatically interpretable electron density map (see point ii)).

It should be too long to describe the applications of the various methods and to compare their results. It is probably more useful for the reader to be informed, by a few examples, on some practical details of a typical *MR* approach. In the following we describe them by using the automatic pipeline available in *SIR2011*, as settled by Carrozzini et al. (2013). The pipeline runs in sequence:

(a) the program *REMO09*, the main characteristic of which have been described in previous paragraphs of this chapter.

(b) *REFMAC* (Murshudov et al., 2011), available from *CCP4* (*Collaborative Computational Project*, Number 4, 1994). The program automatically reads the output of *REMO09* and submits positions and temperature factors of the model atoms to five cycles of a maximum likelihood refinement procedure. The final phases are submitted to the modulus VLD.

(c) the VLD-EDM approach, for extending and refining the phases provided by *REMO09*. This modulus is a combination of the *VLD* method described in Section 9.3 and of the EDM techniques. After the EDM cycles the phases are submitted to *VLD* and then resubmitted to EDM. A few cycles usually allows a good phase extension and a reduction in the phase error.

(d) the *FREE LUNCH* procedure, described in Section 8.2, for final phase refinement of the observed reflections via structure factor extrapolation.

(e) *ARP/wARP*, for automatic model building (see Section 8.3 and Appendix 8.C).

Some numerical data concerning four examples are listed below. *ID* is the sequence identity, *RMS* is the root mean square deviation between pairwise $C\alpha$ backbone positions (see Section 13.2). Average phase errors follow the strings *MR, REFMAC, VLD-EDM* and *FREE-LUNCH*, as obtained after their applications. The percentage of docked residues follow the string *ARP/wARP* (in the case of more numerical values, more automatic cycles of *ARP/wARP* have been run).

Ex. n.1. Target structure: *1bxo*, 1 chain with 323 residues. $RES = 0.9$ Å

Model structure: *1er8*, 2 chains, the first with 330 residues, the second with 8 residues; $ID = 0.55$, $RMS = 1.15$ Å.

$MR = 74°$; $VLD\text{-}EDM = 21°$; *FREE LUNCH* not run because of high experimental data resolution; $ARP/wARP = 0.99$.

Ex. n.2 Target structure: *2hyu*, 1 monomer in the a.u. with 308 residues. $RES = 1.86$ Å.

Model structure *1xjl*: one monomer with 319 residues; $ID = 0.99$, $RMS = 0.50$ Å.
$MR = 50°$; $VLD\text{-}EDM = 40°$; $FREE\ LUNCH = 37°$; $ARP/wARP = 0.99$.
Ex. n.3 Target structure: *2b5o*, 2 monomers in the a.u., each with 292 residues. $RES = 2.5$ Å.
Model structure: *1b2r*, one monomer with 295 residues; $ID = 0.63$, $RMS = 1.16$ Å.
$MR = 50°$; $VLD\text{-}EDM = 44°$; $FREE\ LUNCH = 43°$; $ARP/wARP = 0.72, 0.78, 0.88$.
Ex. n.4 Target structure: *2qu5*, 1 monomer with 292 residues. $RES = 2.86$ Å.
Model structure: *2p2i*, one monomer with 289 residues; $ID = 1$, $RMS = 0.81$ Å.
$MR = 44°$; $VLD\text{-}EDM = 34°$; $FREE\ LUNCH = 35°$; $ARP/wARP = 0$.

The above examples show that:

(a) if *ID* is large and/or *RMS* is small the phase error at the end of the *MR* step is usually small. The application of *VLD* and *FREE LUNCH* is not essential for the success of *ARP/wARP*, they only make it more easy.
(b) If *ID* is small and/or *RMS* is large the *MR* step ends with a large phase error. If the data resolution is good, the error may be easily minimized by *VLD-EDM*, which makes successful the application of *ARP/wARP*.
(c) If *ID* is small and/or *RMS* is large, and if the data resolution is bad, then reduction of the large *MR* phase error is more difficult via *VLD-EDM* and *FREE LUNCH*. Then, the automatic model building by *ARP-wARP* may succeed or may fail, according to circumstances. In example n.3, *ARP/wARP* succeeds (in three cycles, automatically run by *SIR2011*). In example n.4, *ARP/wARP* fails in spite of the small average phase error, probably because of the very low data resolution. If, in example n.4, the *REMO* model is submitted to restrained least squares cycles of *REFMAC*, and then the phases are submitted to *VLD-EDM* and *FREE LUNCH*, then the final average phase error is only 25°, but still *ARP/wARP* is unable to obtain the structure coverage. The reason should be identified in the low data resolution, close to the *ARP/wARP* limit.

The above scheme is just one of many possible EDM schemes; any program has its own preferred recipe for driving the model towards the target structure. It may be worthwhile mentioning that some efforts are today being directed towards EDM procedures which locally deform the model located by MR, in order to make the model closer to the target atoms (Terwilliger et al., 2012). As stated before, a typical example may be the following: residues in a β-sheet and in an adjacent α-helix have similar relations in the model and in the target structure, but the orientation and location of the sheet with respect to the helix could be different in the two structures. One can correctly orient and locate the sheet but not, simultaneously, the helix; the inverse may also be true. In order to make model and target structures closer, a shift in the coordinates of each residue is calculated, smoothed, and applied, so leading to local deformations of the model which may improve the match between model and map. The deformed model is then refined to improve the geometry, and the process is then iterated until convergence.

APPENDIX 13.A CALCULATION OF THE ROTATION FUNCTION IN ORTHOGONALIZED CRYSTAL AXES

Let us suppose that, in a given crystallographic system, we want to rotate a molecule in such a way that, after rotation, it overlaps with another identical molecule. According to equation (13.2), a linear transformation will relate the point with coordinate $\mathbf{X}$ with the point with coordinate $\mathbf{X}'$, corresponding to $\mathbf{X}$ after the rotation:

$$\mathbf{X}' = \mathbf{M}\mathbf{X} \quad (13.A.1)$$

The calculation of $\mathbf{M}$ may be performed in three steps: orthogonalization of the crystallographic reference system, rotation in Cartesian space, and return to the crystallographic reference system. In the case where we explore the rotation space in steps, the crystallographic symmetry has to be taken into account to reduce the computing time.

In the following we will describe the various steps by using the following notation: $\mathbf{a}, \mathbf{b}, \mathbf{c}$ are the crystallographic axes, α, β, γ their interaxial angles, $\mathbf{a}^*, \mathbf{b}^*, \mathbf{c}^*$ the reciprocal axes, $\alpha^*, \beta^*, \gamma^*$ the reciprocal angles, and $\mathbf{e}_1, \mathbf{e}_2, \mathbf{e}_3$ the Cartesian axes, respectively.

13.A.1 The orthogonalization matrix

Transformation from fractional crystallographic coordinates $\mathbf{X}$ (dimensionless) to orthogonal Cartesian coordinates $\mathbf{X}_{\text{ort}}$ (in Å), may be performed via the orthogonalization matrix $\boldsymbol{\beta}$ under different conventions (see *Fundamentals of Crytallography*, Chapter 2). For example, if we assume (see Fig. 13.A.1)

$$\mathbf{e}_1 \parallel \mathbf{a}, \mathbf{e}_2 \parallel (\mathbf{a} \wedge \mathbf{b}) \wedge \mathbf{e}_1, \mathbf{e}_3 \parallel (\mathbf{a} \wedge \mathbf{b}) \text{ (or equivalently, } \mathbf{e}_3 \parallel \mathbf{c}^*) \quad (13.A.2)$$

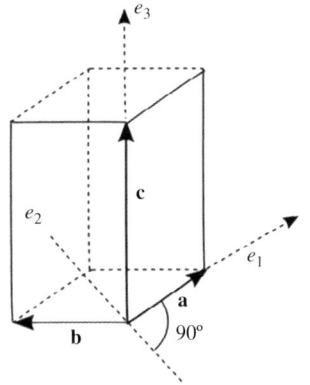

Fig. 13.A.1
For a hexagonal unit cell, the convention $\mathbf{e}_1 \parallel \mathbf{a}, \mathbf{e}_2 \parallel (\mathbf{a} \wedge \mathbf{b}) \wedge \mathbf{e}_1, \mathbf{e}_3 \parallel (\mathbf{a} \wedge \mathbf{b})$ is used.

then

$$\boldsymbol{\beta} = \begin{vmatrix} a & b\cos\gamma & c\cos\beta \\ 0 & b\sin\gamma & \dfrac{c(\cos\alpha - \cos\beta\cos\gamma)}{\sin\gamma} \\ 0 & 0 & \dfrac{V}{ab\sin\gamma} \end{vmatrix},$$

where

$$V = \det(\boldsymbol{\beta}) = abc(1 - \cos^2\alpha - \cos^2\beta - \cos^2\gamma + 2\cos\alpha\cos\beta\cos\gamma)^{1/2}.$$

If the convention (see Fig. 13.A.2)

$$\mathbf{e}_1 \parallel \mathbf{a}^*, \mathbf{e}_2 \parallel \mathbf{b}, \mathbf{e}_3 \parallel (\mathbf{b}, \mathbf{c}) \text{ plane} \quad (13.A.3)$$

is used, then

$$\boldsymbol{\beta} = \begin{vmatrix} a\sin\gamma\sin\beta^* & 0 & 0 \\ a\cos\gamma & b & c\cos\alpha \\ a\sin\gamma\cos\beta^* & 0 & c\sin\alpha \end{vmatrix}$$

is obtained.

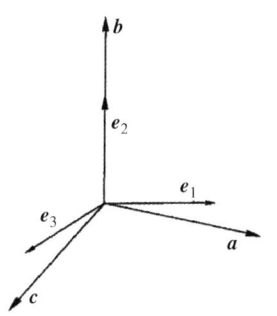

Fig. 13.A.2
Orthonormal axes, $\mathbf{e}_1, \mathbf{e}_2, \mathbf{e}_3$ and crystallographic axes, $\mathbf{a}, \mathbf{b}, \mathbf{c}$, according to the convention (13.A.3).

Calculation of the rotation function in orthogonalized crystal axes

13.A.2 Rotation in Cartesian space

Any rotation in three-dimensional space is defined by three parameters. Two methods are usually employed to perform a rotation.

(i) The *method of spherical polar angles*. Define first the direction of the rotation axis $\boldsymbol{E}$ (called the *principal Euler axis*) relative to the reference system and then fix the rotation, χ (called the *principal Euler angle*), about this axis. Often χ is called k.

(ii) The *method of Eulerian angles*. Rotate the object three times in succession about any three non-planar directions.

In both cases the rotation is performed via the so called *direction cosine matrix* $\boldsymbol{\rho}$, from which it is possible to define the direction of $\boldsymbol{E}$ via its three direction cosines l, m, and n (these are the cosines of the angles that $\boldsymbol{E}$ makes with the positive axes of the orthogonal system) and the value of χ.

In terms of Eulerian angles, the rotation matrix $\boldsymbol{\rho}_{Eu}$ may be represented as a product of three successive rotation matrices around three independent axes, which are applied to the generic point X_{ort} to obtain X'_{ort}, according to:

$$X'_{ort} = \boldsymbol{\rho}_{Eu} X_{ort} = \mathbf{R}_3 \left(\mathbf{R}_2 \left(\mathbf{R}_1 X_{ort} \right) \right).$$

$\mathbf{R}_3$, $\mathbf{R}_2$, and $\mathbf{R}_1$ may be rotations around the Cartesian axes. Different conventions may be used, among which we quote:

ZYZ convention : $\boldsymbol{\rho}_{Eu} = \mathbf{R}(\theta_1, \theta_2, \theta_3) = \mathbf{R}_z(\theta_3)\mathbf{R}_y(\theta_2)\mathbf{R}_z(\theta_1)$

ZXZ convention : $\boldsymbol{\rho}_{Eu} = \mathbf{R}(\theta_1, \theta_2, \theta_3) = \mathbf{R}_z(\theta_3)\mathbf{R}_x(\theta_2)\mathbf{R}_z(\theta_1)$.

In the following, we will describe the mathematics connected to the ZXZ convention (see Fig. 13.A.3). The point defined by the Cartesian coordinates X_{ort} is rotated to the point X'_{ort} by the rotation matrix $\boldsymbol{\rho}_{Eu}$, defined by:

$$\boldsymbol{\rho}_{Eu} = R_z(\theta_3)R_x(\theta_2)R_z(\theta_1)$$

$$= \begin{vmatrix} c\theta_3 & s\theta_3 & 0 \\ -s\theta_3 & c\theta_3 & 0 \\ 0 & 0 & 1 \end{vmatrix} \begin{vmatrix} 1 & 0 & 0 \\ 0 & c\theta_2 & s\theta_2 \\ 0 & -s\theta_2 & c\theta_2 \end{vmatrix} \begin{vmatrix} c\theta_1 & s\theta_1 & 0 \\ -s\theta_1 & c\theta_1 & 0 \\ 0 & 0 & 1 \end{vmatrix} \quad (13.A.4)$$

$$= \begin{vmatrix} c\theta_1 c\theta_3 - s\theta_1 c\theta_2 s\theta_3 & s\theta_1 c\theta_3 + c\theta_1 c\theta_2 s\theta_3 & s\theta_2 s\theta_3 \\ -s\theta_1 c\theta_2 c\theta_3 - c\theta_1 s\theta_3 & -s\theta_1 s\theta_3 + c\theta_1 c\theta_2 c\theta_3 & s\theta_2 c\theta_3 \\ s\theta_1 s\theta_2 & -c\theta_1 s\theta_2 & c\theta_2 \end{vmatrix}$$

where $c\theta_i$ and $s\theta_i$ stand for $\cos\theta_i$ and $\sin\theta_i$.

Let us now define a rotation in terms of spherical polar coordinates. Any rotation may be described in terms of the azimuthal angle ϕ (horizontal rotation), the lateral angle ψ (up/down rotation), and the rotation angle χ around the new axis defined by the ϕ and ψ rotations. Several conventions may be used; two of these are shown in Figs. 13.A.4a and 13.A.4b. In the first case, the Cartesian coordinates (x, y, z) are referred to polar coordinates by

$$x = r \sin\psi \cos\phi, \quad y = r \cos\psi, \quad z = -r \sin\psi \sin\phi,$$

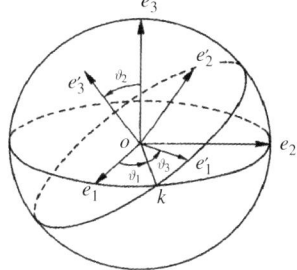

Fig. 13.A.3
Eulerian axes. Two orthonormal frameworks, $A = [0, \boldsymbol{e}_1, \boldsymbol{e}_2, \boldsymbol{e}_3]$ and $A' = [0, \boldsymbol{e}'_1, \boldsymbol{e}'_2, \boldsymbol{e}'_3]$, are shown. The axis ok, called the *line of nodes*, is the intersection of the $(\boldsymbol{e}_1, \boldsymbol{e}_2)$ and $(\boldsymbol{e}'_1, \boldsymbol{e}'_2)$ planes, and is perpendicular both to $\boldsymbol{e}_3$ and to $\boldsymbol{e}'_3$. A and A' may be superimposed by three anticlockwise rotations in the following order: (1) rotate about $\boldsymbol{e}_3$ by an angle θ_1 (ok and $\boldsymbol{e}_1$ are now identical); (2) rotate through θ_2 about ok, which will bring $\boldsymbol{e}_3$ into coincidence with $\boldsymbol{e}'_3$; (3) rotate about $\boldsymbol{e}'_3$ by θ_3, which brings $\boldsymbol{e}_1$ to $\boldsymbol{e}'_1$ and $\boldsymbol{e}_2$ to $\boldsymbol{e}'_2$.

the direction cosines are defined by

$$\begin{vmatrix} l \\ m \\ n \end{vmatrix} = \begin{vmatrix} \sin\psi\cos\phi \\ \cos\psi \\ -\sin\psi\sin\phi \end{vmatrix},$$

and the direction cosine matrix becomes

$$\boldsymbol{\rho}_{sp} = \begin{vmatrix} c\chi + (1-c\chi)s^2\psi c^2\phi & -s\psi s\phi s\chi + (1-c\chi)c\psi s\psi c\phi & -c\psi s\chi - (1-c\chi)s^2\psi c\phi s\phi \\ s\psi s\phi s\chi + (1-c\chi)c\psi s\psi c\phi & c\chi + (1-c\chi)c^2\psi & s\psi c\phi s\chi - (1-c\chi)c\psi s\psi s\phi \\ c\psi s\chi - (1-c\chi)s^2\psi c\phi s\phi & -s\psi c\phi s\chi - (1-c\chi)c\psi s\psi s\phi & c\chi + (1-c\chi)s^2\psi s^2\phi \end{vmatrix}$$

(13.A.5a)

$\boldsymbol{\rho}_{sp}$ may be expressed in terms of direction cosines l, m, n:

$$\boldsymbol{\rho}_{sp} = \begin{vmatrix} c\chi + l^2(1-c\chi) & ns\chi + lm(1-c\chi) & -ms\chi + ln(1-c\chi) \\ -ns\chi + lm(1-c\chi) & c\chi + m^2(1-c\chi) & ls\chi + mn(1-c\chi) \\ ms\chi + ln(1-c\chi) & -ls\chi + mn(1-c\chi) & c\chi + n^2(1-c\chi) \end{vmatrix}$$

(13.A.5b)

If the convention depicted in Fig. 13.A.4b is adopted, then

$$x = r\sin\psi\cos\phi, \quad y = r\sin\psi\sin\phi, \quad z = r\cos\psi,$$

$$\begin{vmatrix} l \\ m \\ n \end{vmatrix} = \begin{vmatrix} \sin\psi\cos\phi \\ \sin\psi\sin\phi \\ \cos\psi \end{vmatrix},$$

and

$$\boldsymbol{\rho}_{sp} = \begin{vmatrix} s^2\psi c^2\phi + (s^2\psi s^2\phi + c^2\psi)c\chi & s^2\psi s\phi c\phi(1-c\chi) - c\psi s\chi & s\psi c\psi c\phi(1-c\chi) + s\psi s\varphi s\chi \\ s^2\psi s\phi c\phi(1-c\chi) + c\psi s\chi & s^2\psi s^2\phi + (s^2\psi c^2\phi + c^2\psi)c\chi & s\psi c\psi s\phi(1-c\chi) - s\psi c\varphi s\chi \\ s\psi c\psi c\phi(1-c\chi) - s\psi s\phi s\chi & s\psi c\psi s\phi(1-c\chi) + s\psi c\phi s\chi & c^2\psi + s^2\psi c\chi \end{vmatrix}$$

(13.A.5c)

In terms of direction cosines, $\boldsymbol{\rho}_{sp}$ becomes

$$\boldsymbol{\rho}_{sp} = \begin{vmatrix} l^2 + (m^2 + n^2)\cos\chi & lm(1-\cos\chi) - n\sin\chi & nl(1-\cos\chi) + m\sin\chi \\ lm(1-\cos\chi) + n\sin\chi & m^2 + (n^2 + l^2)\cos\chi & mn(1-\cos\chi) - l\sin\chi \\ nl(1-\cos\chi) - m\sin\chi & mn(1-\cos\chi) + l\sin\chi & n^2 + (m^2 + l^2)\cos\chi \end{vmatrix}$$

(13.A.5d)

The matrices $\boldsymbol{\rho}_{Eu}$ and $\boldsymbol{\rho}_{sp}$ have a number of interesting properties:

(i) they are real square matrices for which $\bar{\boldsymbol{\rho}} = \boldsymbol{\rho}^{-1}$ and $\det(\boldsymbol{\rho}) = 1$;
(ii) for a 180° rotation the matrix is symmetric; indeed, for such a rotation, $\boldsymbol{\rho} = \boldsymbol{\rho}^{-1}$ and therefore $\boldsymbol{\rho} = \bar{\boldsymbol{\rho}}$. Furthermore, $trace(\boldsymbol{\rho}) = -1$, a very useful indication for recognizing twofold axes directly from $\boldsymbol{\rho}$;
(iii) $\boldsymbol{\rho}$ has three eigenvalues, say $\{1, \exp(i\chi), \exp(-i\chi)\}$: the eigenvector (l, m, n) corresponding with the real eigenvalue 1 is the Euler axis $\mathbf{E} = (l, m, n)$, where

$$\begin{vmatrix} l \\ m \\ n \end{vmatrix} = \frac{1}{2\sin\chi} \begin{vmatrix} \rho_{32} - \rho_{23} \\ \rho_{13} - \rho_{31} \\ \rho_{21} - \rho_{12} \end{vmatrix}. \quad (13.A.6)$$

Calculation of the rotation function in orthogonalized crystal axes

ρ_{ij} are elements of the matrix $\boldsymbol{\rho}$. According to point (ii), the elements of the matrix on the right-hand side of (13.A.6) vanish when $\boldsymbol{\rho}$ represents a twofold axis.

(iv) χ, the principal Euler angle, may be derived from

$$\text{trace}(\boldsymbol{\rho}) = 1 + 2\cos\chi. \tag{13.A.7}$$

The reader may immediately verify the above properties in the case of spherical polar angles; then the axis $\boldsymbol{E}$ is defined by the rotations about ϕ and ψ, and χ is just the principal Euler angle. To verify (13.A.7), one has to calculate the trace of the matrices (13.A.5b) and (13.A.5d) by using the well-known property $l^2 + m^2 + n^2 = 1$; to verify the property (13.A.6), one must introduce the elements of the same matrices.

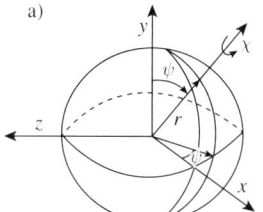

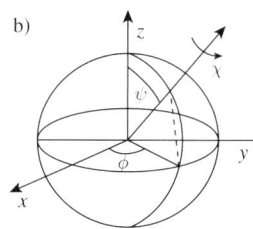

Fig. 13.A.4
Spherical polar coordinates: the variables ψ and ϕ specify a direction about which the coordinate system may be rotated by an angle χ. The axes are first rotated about the lateral angle ϕ, then rotated up/down by ψ, and finally the χ rotation is performed. (a) $x = r\sin\psi\cos\phi$, $y = r\cos\psi$, $z = -r\sin\psi\sin\phi$. (b) $x = r\sin\psi\cos\phi$, $y = r\sin\psi\sin\phi$, $z = r\cos\psi$.

13.A.3 Conversion to fractional coordinates

If the convention (13.A.2) is used, the rotated coordinates $\boldsymbol{X}'_{ort}$ are converted to fractional coordinates $\boldsymbol{X}'$ by

$$\boldsymbol{X}' = \boldsymbol{\alpha}\boldsymbol{X}'_{ort},$$

where

$$\boldsymbol{\alpha} = \begin{vmatrix} \dfrac{1}{a} & -\dfrac{\cos\gamma}{a\sin\gamma} & \left(\dfrac{bc\cos\gamma(\cos\alpha - \cos\beta\cos\gamma)}{\sin\gamma} - bc\cos\beta\sin\gamma\right)\dfrac{1}{V} \\ 0 & \dfrac{1}{b\sin\gamma} & \dfrac{-ac(\cos\alpha - \cos\beta\cos\gamma)}{V\sin\gamma} \\ 0 & 0 & \dfrac{ab\sin\gamma}{V} \end{vmatrix},$$

and by

$$\boldsymbol{\alpha} = \begin{vmatrix} \dfrac{1}{as\gamma s\beta^*} & 0 & 0 \\ \dfrac{1}{bt\alpha t\beta^*} - \dfrac{1}{bt\gamma s\beta^*} & \dfrac{1}{b} & \dfrac{-1}{bt\alpha} \\ \dfrac{-1}{cs\alpha t\beta^*} & 0 & \dfrac{1}{cs\alpha} \end{vmatrix}$$

if the convention (13.A.3) is used.

$t\alpha, t\beta^*, \ldots,$ stands for $\tan\alpha, \tan\beta^*, \ldots$

At the end of the three-step procedure (orthogonalization, rotation, deorthogonalization), the rotation $\mathbf{M}$ in the crystallographic reference system may be represented by

$$\boldsymbol{X}' = \mathbf{M}\boldsymbol{X} = \boldsymbol{\alpha}\boldsymbol{\rho}\boldsymbol{\beta}\boldsymbol{X},$$

where $\mathbf{M}$ is given by

$$\mathbf{M} = \boldsymbol{\alpha}\boldsymbol{\rho}\boldsymbol{\beta}. \tag{13.A.8}$$

If the deorthogonalization procedure implies a return to the original crystallographic frame then, $\alpha = \beta^{-1}$ and

$$\mathbf{M} = \beta^{-1} \rho \beta. \tag{13.A.9}$$

If Eulerian angles are used the following identity holds:

$$\mathbf{R}(\theta_1, \theta_2, \theta_3) = \mathbf{R}(\theta_1 + 2n_1\pi, \theta_2 + 2n_2\pi, \theta_3 + 2n_3\pi). \tag{13.A.10}$$

Furthermore, redundancy in definitions leads to

$$\mathbf{R}(\theta_1, \theta_2, \theta_3) = \mathbf{R}(\theta_1 + \pi, -\theta_2, \theta_3 + \pi), \tag{13.A.11}$$

which is an n glide perpendicular to θ_2. Therefore, the full range of rotation operations is

$$0 \le \theta_1 < \pi, \quad 0 \le \theta_2 < 2\pi, \quad 0 \le \theta_3 < 2\pi.$$

If polar coordinates are used (in the convention defined by Fig. 13.A.4a),

$$\mathbf{R}(\chi, \psi, \phi) = \mathbf{R}(\chi + 2n_1\pi, \psi + 2n_2\pi, \phi + 2n_3\pi) \tag{13.A.12}$$

and

$$\mathbf{R}(\chi, \psi, \phi) = \mathbf{R}(\chi, 2\pi - \psi, \phi + \pi) \tag{13.A.13}$$

which is a ϕ glide perpendicular to ψ in polar space. (13.A.13) relies on the fact that, if ψ is greater than π, the rotation is the same as that in which ϕ is increased by π, and ψ becomes $2\pi - \psi$. Since a rotation by ψ about any axis is equivalent to a rotation $-\chi$ about an opposite directed axis,

$$\mathbf{R}(\chi, \psi, \phi) = \mathbf{R}(-\chi, \pi - \psi, \phi + \pi). \tag{13.A.14}$$

All rotation operations are therefore included in

$$0 \le \chi < 2\pi, \quad 0 \le \psi < \pi, \quad 0 \le \phi < \pi.$$

If the convention defined in Fig. 13.A.4b is used, the angular ranges that must be covered are

$$0 \le \chi < \pi, \quad 0 \le \psi < \pi, \quad 0 \le \phi < 2\pi.$$

How do we use the above mathematical formalism in a MR procedure? Once an orthogonal frame has been defined (e.g. in accordance with convention (13.A.2) or (13.A.3)), the corresponding orthogonal lattice has to be constructed. At each rotation step, defined by equation (13.A.9), the model structure factors may be computed and associated with each grid point of the lattice. If the MFT method is used (see Section 13.5), the observed structure factor is associated with each grid point; the grid points are solidly moved during the rotation without being recalculated. In modern MR programs, to avoid calculation of the factor G, defined by equation (13.5), an orthogonal lattice grid is generated, the direct-space dimensions of which are chosen to be four times the maximum molecular dimension. The high-resolution limit of the lattice may be chosen by the user or automatically determined by the program. The same resolution limit is applied to select the observed reflections to be used for the MR search.

13.A.4 Symmetry and the rotation function

In Section 13.5, we defined the rotation function via the integral (13.3b), which estimates the degree of coincidence between target and model Patterson maps. In Section 13.A.3, we obtained the general expression for the rotation matrix valid for any crystallographic reference system, and we defined the limits of the rotation search. In all of the above mathematical formalism, the Laue symmetries of the two Patterson functions have not been taken into account. This point is of paramount importance (Rossmann and Blow, 1962; Tollin et al., 1966; Burdina, 1973; Rao et al., 1980); indeed, point group symmetry in the reciprocal lattice will cause the same value of the rotation function to be found for distinct rotations. This reduces the range of angles to be explored before all independent rotation operations have been considered. It may be shown that the symmetry of the two Patterson functions, P_{mol} and P_{targ} in (13.3b), allows identification of a minimum range of rotations which is a multiple of the product of the orders n_1 and n_2 of the groups of rotation of the original Patterson functions. The reader is addressed to *International Tables for Crystallography* (1993), vol. B (ed. U. Shmueli), Tables 2.3.6.3 and 2.3.6.4, for definitions of ranges for the asymmetric unit in rotation space (cubic groups excluded). The entries in such tables are justified below via some algebraic calculations and specific examples; rotations by Eulerian angles will be considered.

Let us consider the equation (13.3b) in orthogonal coordinates; it may be written as

$$\mathcal{R} = \int_{U_{ort}} P_2(\rho \mathbf{u}_{ort}) P_1(\mathbf{u}_{ort}) d\mathbf{u}_{ort}, \tag{13.A.15}$$

where, for simplicity, we have replaced P_{mol} by P_2 and P_{targ} by P_1.

Let $\mathbf{R}_1$ and $\mathbf{R}_2$ be symmetry point transformations of the distributions P_1 and P_2 for orthogonal coordinates. Then, (13.A.15) may be rewritten as

$$\mathcal{R}(\boldsymbol{\rho}) = \int_{U_{ort}} P_2(\mathbf{R}_2 \rho \mathbf{u}_{ort}) P_1(\mathbf{R}_1 \mathbf{u}_{ort}) d\mathbf{u}_{ort}, \tag{13.A.16}$$

which will have the same value for any $\mathbf{R}_2$ and $\mathbf{R}_1$. By replacing $\mathbf{u}'_{ort} = \mathbf{R}_1 \mathbf{u}_{ort}$, equation (13.A.16) becomes,

$$\mathcal{R}(\boldsymbol{\rho}) = \int_{U'_{ort}} P_2(\mathbf{R}_2 \rho \mathbf{R}_1^{-1} \mathbf{u}'_{ort}) P'(\mathbf{u}'_{ort}) d\mathbf{u}'_{ort} = \mathcal{R}(\mathbf{R}_2 \rho \mathbf{R}_1^{-1}). \tag{13.A.17}$$

Relationship (13.A.17) suggests that rotations $\boldsymbol{\rho}$ and $\boldsymbol{\rho}' = \mathbf{R}_2 \rho \mathbf{R}_1^{-1}$ are equivalent positions of the rotation function $\mathcal{R}$. It may also be noted that $\boldsymbol{\rho}'$ is a pure rotation (i.e. without inversion or reflection), only in the case in which $\mathbf{R}_2$ and $\mathbf{R}_1$ are simultaneously proper or simultaneously improper rotations. In the latter case, we could multiply $\boldsymbol{\rho}'$ on the left and on the right side by the inversion operation $\bar{1}$, without changing the rotation; indeed,

$$\boldsymbol{\rho}' = \bar{1} \boldsymbol{\rho}' \bar{1} = \bar{1} \mathbf{R}_2 \rho \mathbf{R}_1^{-1} \bar{1} = \mathbf{R}'_2 \rho \mathbf{R}'^{-1}_1,$$

where $\mathbf{R}'_2 = \bar{1} \mathbf{R}_2$ and $\mathbf{R}'^{-1}_1 = \mathbf{R}_1^{-1} \bar{1}$ are now pure rotations. Such a result allows us to restrict the analysis to the subgroups of the Patterson symmetry groups, which consist of pure rotation symmetry elements (see Table 13.A.1).

Let us denote by $\mathbf{R}_2 \times \mathbf{R}_1$, the symmetry operation transforming $\boldsymbol{\rho}$ into $\boldsymbol{\rho}'$. If $\mathbf{R}_4 \times \mathbf{R}_3$ is a further symmetry operator, the successive application of

Table 13.A.1 The 11 Laue groups, the corresponding subgroups of pure rotation, distinguished by the arrangement of the principal axes

Laue group	Proper rotation group	Laue group	Proper rotation group
$\bar{1}$	1	$\bar{3}$	3
$2/m$ (b-axis unique)	$2_{[010]}$	$\bar{3}m$	321
$2/m$ (c-axis unique)	$2_{[010]}$	$6/m$	6
mmm	222	$6/mmm$	622
$4/m$	4	$m3$	23
$4/mmm$	422	$m3m$	432

$(\mathbf{R}_2 \times \mathbf{R}_1)$ and $(\mathbf{R}_4 \times \mathbf{R}_3)$ is equivalent to the operator $(\mathbf{R}_2\mathbf{R}_4) \times (\mathbf{R}_1\mathbf{R}_3)$, which may be considered to be the product of the first two operators; indeed,

$$\boldsymbol{\rho}' = \mathbf{R}_2\boldsymbol{\rho}\mathbf{R}_1^{-1}, \boldsymbol{\rho}'' = \mathbf{R}_4\boldsymbol{\rho}'\mathbf{R}_3^{-1} = \mathbf{R}_4\mathbf{R}_2\boldsymbol{\rho}\mathbf{R}_1^{-1}\mathbf{R}_3^{-1}.$$

This suggests that the symmetry operators form a group according to the rule,

$$(\mathbf{R}_4 \times \mathbf{R}_3) \cdot (\mathbf{R}_2 \times \mathbf{R}_1) = (\mathbf{R}_4\mathbf{R}_2) \times (\mathbf{R}_3\mathbf{R}_1); \qquad (13.A.18)$$

such a group is a direct product of the groups $\{P_2\}$ and $\{P_1\}$. Since Patterson groups show 11 groups of pure rotation (see Table 13.A.1) $11 \cdot 11 = 121$ groups of symmetry operators for the rotation function may be constructed.

It may be noted from Table 13.A.1 that only six different elements of symmetry are necessary to describe the rotation groups, i.e.

$$2_{[010]}, 2_{[001]}, 4_{[001]}, 3_{[001]}, 6_{[001]}, 3_{[111]}.$$

In Table 13.A.2, we give the corresponding elementary rotation matrices.

The order in which the Patterson functions are arranged in (13.A.15) must be taken into account; different angular relationships are generated if such an order is reversed (Eulerian relation matrices are not Hermitian). Let us denote by

$$\mathscr{R}_1 = \int_{U_{\text{ort}}} P_2(\boldsymbol{u}_{2\text{ort}})P_1(\boldsymbol{u}_{1\text{ort}})\mathrm{d}\boldsymbol{u}_{1\text{ort}},$$

the rotation function for which $\boldsymbol{u}_{2\text{ort}} = \boldsymbol{\rho}\boldsymbol{u}_{1\text{ort}}$, and by

$$\mathscr{R}_2 = \int_{U_{\text{ort}}} P_2(\boldsymbol{u}_{2\text{ort}})P_1(\boldsymbol{u}_{1\text{ort}})\mathrm{d}\boldsymbol{u}_{2\text{ort}},$$

the rotation function for which $\boldsymbol{u}_{1\text{ort}} = \boldsymbol{\rho}'\boldsymbol{u}_{2\text{ort}}$. Since

$$\boldsymbol{u}_{2\text{ort}} = \boldsymbol{\rho}'^{-1}\boldsymbol{u}_{1\text{ort}},$$

it follows that $\mathscr{R}_1 = \mathscr{R}_2$ when

$$\boldsymbol{\rho}'^{-1} = \boldsymbol{\rho}. \qquad (13.A.19)$$

Thus, reversal of the Patterson functions generates different angular relationships. Relation (13.A.19) may be written in a more explicit form,

$$\mathscr{R}_1(\theta_1, \theta_2, \theta_3) = \mathscr{R}_2(-\theta_3, -\theta_2, -\theta_1). \qquad (13.A.20)$$

In conclusion, reversal of the Pattersons in (13.A.17) will give rise to different, though related, rotation groups. Accordingly, of the 121 rotation groups, 11 are

Table 13.A.2 Symmetry operators and corresponding rotation matrices for orthogonal coordinates

Symmetry element	Rotation matrix
$2_{[010]}$	$\begin{vmatrix} \bar{1} & 0 & 0 \\ 0 & 1 & 0 \\ 0 & 0 & \bar{1} \end{vmatrix}$
$2_{[001]}$	$\begin{vmatrix} \bar{1} & 0 & 0 \\ 0 & \bar{1} & 0 \\ 0 & 0 & 1 \end{vmatrix}$
$4_{[001]}$	$\begin{vmatrix} 0 & \bar{1} & 0 \\ 1 & 0 & 0 \\ 0 & 0 & 1 \end{vmatrix}$
$3_{[001]}$	$\begin{vmatrix} -1/2 & -\sqrt{3}/2 & 0 \\ \sqrt{3}/2 & -1/2 & 0 \\ 0 & 0 & 1 \end{vmatrix}$
$6_{[001]}$	$\begin{vmatrix} 1/2 & -\sqrt{3}/2 & 0 \\ \sqrt{3}/2 & 1/2 & 0 \\ 0 & 0 & 1 \end{vmatrix}$
$3_{[111]}$	$\begin{vmatrix} 0 & 1 & 0 \\ 0 & 0 & 1 \\ 1 & 0 & 0 \end{vmatrix}$

Table 13.A.3 Symmetry operators S_i and $_jS$ for all proper rotation groups except the cubic one. Each group includes the operation $\pi + \theta_1, -\theta_2, \pi + \theta_3$, which is an identity operation (see text) in Eulerian systems

Laue group	Proper rotation group	Symmetry elements $\mathbf{S}_i$	Symmetry elements $_j\mathbf{S}$
$\bar{1}$	1	$\pi + \theta_1, -\theta_2, \pi + \theta_3$	$\pi + \theta_1, -\theta_2, \pi + \theta_3$
$2/m$ (b-axis unique)	2	$\pi + \theta_1, -\theta_2, \pi + \theta_3$ $\pi - \theta_1, \pi + \theta_2, \theta_3$	$\pi + \theta_1, -\theta_2, \pi + \theta_3$ $\theta_1, \pi + \theta_2, \pi - \theta_3$
$2/m$ (c-axis unique)	2	$\pi + \theta_1, -\theta_2, \pi + \theta_3$ $\pi + \theta_1, \theta_2, \theta_3$	$\pi + \theta_1, -\theta_2, \pi + \theta_3$ $\theta_1, \theta_2, \pi + \theta_3$
mmm	222	$\pi + \theta_1, -\theta_2, \pi + \theta_3$ $\pi - \theta_1, \pi + \theta_2, \theta_3$ $\pi + \theta_1, \theta_2, \theta_3$	$\pi + \theta_1, -\theta_2, \pi + \theta_3$ $\theta_1, \pi + \theta_2, \pi - \theta_3$ $\theta_1, \theta_2, \pi + \theta_3$
$4/m$	4	$\pi + \theta_1, -\theta_2, \pi + \theta_3$ $-\pi/2 + \theta_1, \theta_2, \theta_3$	$\pi + \theta_1, -\theta_2, \pi + \theta_3$ $\theta_1, \pi + \theta_2, \pi/2 + \theta_3$
$4/mmm$	422	$\pi + \theta_1, -\theta_2, \pi + \theta_3$ $\pi - \theta_1, \pi + \theta_2, \theta_3$ $-\pi/2 + \theta_1, \theta_2, \theta_3$	$\pi + \theta_1, -\theta_2, \pi + \theta_3$ $\theta_1, \pi + \theta_2, \pi - \theta_3$ $\theta_1, \theta_2, \pi/2 + \theta_3$
$\bar{3}$	3	$\pi + \theta_1, -\theta_2, \pi + \theta_3$ $-2\pi/3 + \theta_1, \theta_2, \theta_3$	$\pi + \theta_1, -\theta_2, \pi + \theta_3$ $\theta_1, \theta_2, 2\pi/3 + \theta_3$
$\bar{3}m$	321	$\pi + \theta_1, -\theta_2, \pi + \theta_3$ $\pi - \theta_1, \pi + \theta_2, \theta_3$ $-2\pi/3 + \theta_1, \theta_2, \theta_3$	$\pi + \theta_1, -\theta_2, \pi + \theta_3$ $\theta_1, \pi + \theta_2, \pi - \theta_3$ $\theta_1, \theta_2, 2\pi/3 + \theta_3$
$6/m$	6	$\pi + \theta_1, -\theta_2, \pi + \theta_3$ $-\pi/3 + \theta_1, \theta_2, \theta_3$	$\pi + \theta_1, -\theta_2, \pi + \theta_3$ $\theta_1, \theta_2, \pi/3 + \theta_3$
$6/mmm$	622	$\pi + \theta_1, -\theta_2, \pi + \theta_3$ $\pi - \theta_1, \pi + \theta_2, \theta_3$ $-\pi/3 + \theta_1, \theta_2, \theta_3$	$\pi + \theta_1, -\theta_2, \pi + \theta_3$ $\theta_1, \pi + \theta_2, \pi - \theta_3$ $\theta_1, \theta_2, \pi/3 + \theta_3$

squares of rotation groups of the Laue groups, 55 groups are related to the other 55 through isomorphic groups differing in the arrangement of the terms of the direct product. In Table 13.A.3 we show equivalent points in the rotation Eulerian space generated by the various symmetry elements (non-cubic space groups only). According to Tollin et al. (1966), S_i and $_jS$ are symmetry operators of the rotation groups operating in P_1 (the map which is rotated) and in P_2, respectively, $_jS_i$ is the 'product' of S_i and $_jS$ (the symmetry operations which satisfy (13.A.17) coincide with the set $_jS_i$).

Owing to the fact that symmetry operations constitute a group and angles show a 2π translational symmetry along each axis (see relation (13.A.12)), the symmetry of the rotation function can be described by a rotation space group. Some examples will be given to clarify this statement (cubic space groups lead to different types of phase relationships, which are not linear).

Example 1. The Laue group of P_1 (the map which is rotated) is $\bar{1}$, the Laue group of P_2 is $2/m$. Then (see Table 13.A.3)

$$S_1 : \pi + \theta_1, -\theta_2, \pi + \theta_3$$
$$_1S \equiv S_1 : \pi + \theta_1, -\theta_2, \pi + \theta_3$$
$$_2S : \theta_1, \pi + \theta_2, \pi - \theta_3$$
$$_1S_2 = {_1S} \cdot {_2S} : \pi + \theta_1, \pi - \theta_2, -\theta_3.$$

We observe that S_1 is an n glide plane perpendicular to $\boldsymbol{b}$, $_2S$ is a b glide plane perpendicular to $\boldsymbol{c}$, and $_1S \cdot _2S$ is a screw axis parallel to $\boldsymbol{a}$. The resulting space group is $P2_1nb$ (the reader may usefully compare the angular relations with the equivalent positions in $P2_1nb$, say $[(x,y,z),(x+\pi,-y,z+\pi),(x,y+\pi,\pi-z),(x+\pi,\pi-y,-z)]$). The range of the asymmetric unit for $P2_1nb$, corresponding to the minimum range of θ_2, is

$$0 \leq \theta_1 < 2\pi$$
$$0 \leq \theta_2 \leq \pi/2$$
$$0 \leq \theta_3 < 2\pi.$$

Example 2. The Laue group of P_1 in $2/m$, the Laue group of P_2 is $\bar{1}$. Then (see Table 13.A.3),

$$S_1 : \pi + \theta_1, -\theta_2, \pi + \theta_3$$
$$S_2 : \pi - \theta_1, \pi + \theta_2, \theta_3$$
$$_1S = S_1 : \pi + \theta_1, -\theta_2, \pi + \theta_3$$
$$_1S_2 = S_2 \cdot _1S : -\theta_1, \pi - \theta_2, \pi + \theta_3.$$

We note that S_1 is an n glide plane perpendicular to $\boldsymbol{b}$, S_2 is a b glide plane perpendicular to $\boldsymbol{a}$, and $_1S_2$ is a screw axis parallel to $\boldsymbol{c}$. Therefore, the space group is $Pbn2_1$ (the equivalent positions in this space group are $[(x,y,z),(x+\pi,-y,z+\pi),(\pi-x,y+\pi,z),(-x,\pi-y,\pi+z)]$). The range of the asymmetric unit for $Pbn2_1$, corresponding to the minimum range of θ_2, is

$$0 \leq \theta_1 < 2\pi$$
$$0 \leq \theta_2 \leq \pi/2$$
$$0 \leq \theta_3 < 2\pi.$$

Example 3. Let us consider the case in which the proper rotation group of P_1 (the map which is rotated) is 2, and 2 is also the proper rotation group of P_2. Then, from Table 13.A.3, we will have,

$$S_1 : \pi + \theta_1, -\theta_2, \pi + \theta_3 \qquad _1S : \pi + \theta_1, -\theta_2, \pi + \theta_3$$
$$S_2 : \pi - \theta_2, \pi + \theta_2, \theta_3 \qquad _2S : \theta_1, \pi + \theta_2, \pi - \theta_3$$
$$_2S_1 : \pi + \theta_1, \pi - \theta_2, -\theta_3 \qquad _2S_2 : \pi - \theta_1, \theta_2, \pi - \theta_3$$
$$_1S_2 := -\theta_1, \pi - \theta_2, \pi + \theta_3 \qquad S_1 \cdot _2S \cdot S_2 : -\theta_1, -\theta_2, -\theta_3.$$

It may be observed that S_2 is a b glide plane perpendicular to $\boldsymbol{a}$, S_1 is an n glide plane perpendicular to $\boldsymbol{b}$, and $_2S$ is a b glide plane perpendicular to $\boldsymbol{c}$. Thus, the rotation space group $Pbnb$ arises, which is plotted in Fig. 13.A.5. The size of the asymmetric unit, corresponding to the minimum range of θ_2, is

$$0 \leq \theta_1 \leq \pi/2$$
$$0 \leq \theta_2 < \pi$$
$$0 \leq \theta_3 < 2\pi.$$

Calculation of the rotation function in orthogonalized crystal axes

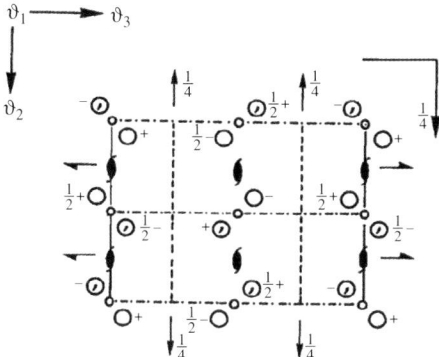

Fig. 13.A.5
Diagram of the rotation group *Pbnb* (retains the order $\theta_1, \theta_2, \theta_3$). Equivalent values of the Eulerian angles are also shown.

Example 4. Let *mmm* be the Laue symmetry of P_1 and $2/m$ that of P_2. Then, from Table 13.A.3, we have

$$S_1 : \pi + \theta_1, -\theta_2, \pi + \theta_3$$
$$S_2 : \pi - \theta_1, \pi + \theta_2, \theta_3$$
$$S_3 : \pi + \theta_1, \theta_2, \theta_3$$
$$_1S = S_1$$
$$_2S : \theta_1, \pi + \theta_2, \pi - \theta_3$$
$$_2S_1 = S_1 \cdot {_2S}: \pi + \theta_1, \pi - \theta_2, -\theta_3$$
$$_2S_2 = S_2 \cdot {_2S}: \pi - \theta_1, \theta_2, \pi - \theta_3$$
$$_2S_3 = S_3 \cdot {_2S}: \pi + \theta_1, \pi + \theta_2, \pi - \theta_3$$
$$S_1 \cdot S_2 = -\theta_1, \pi - \theta_2, \pi + \theta_3$$
$$S_1 \cdot S_3 = \theta_1, -\theta_2, \pi + \theta_3$$
$$S_2 \cdot S_3 = -\theta_1, \pi + \theta_2, \theta_3$$

When all the products are exhaustively made, we will obtain 16 equivalent angular positions in which it is easy to recognize a glide plane of type *b*, perpendicular to **a** (say S_2), a *c* glide plane perpendicular to **b** (say $S_1 \cdot S_3$), and a *b* glide plane perpendicular to **c** (say $_2S$). The space group is therefore *Pbcb*. Since S_3 repeats the equivalent positions after a π translation along θ_1, the space group diagram will be that shown in Fig. 13.A.6, containing 16 instead of the standard 8 equivalent positions. The size of the asymmetric unit will then be

$$0 \leq \theta_1 \leq \pi/2$$
$$0 \leq \theta_2 \leq \pi/2$$
$$0 \leq \theta_3 < 2\pi.$$

Example 5. Let $6/m$ be the Laue symmetry of P_1 and $2/m$ that of P_2. From Table 13.A.3, we derive,

$$S_1 : \pi + \theta_1, -\theta_2, \pi + \theta_3$$
$$S_2 : -\pi/3 + \theta_1, \theta_2, \theta_3$$
$$_1S \equiv S_1$$
$$_2S : \theta_1, \pi + \theta_2, \pi - \theta_3.$$

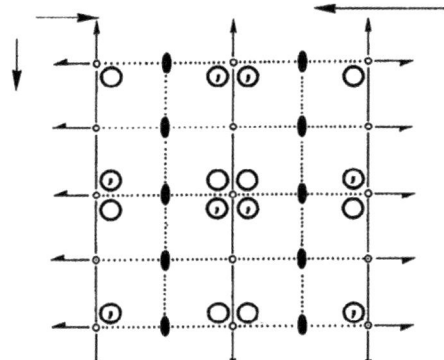

Fig. 13.A.6
Rotation space group diagram for the rotation function of a Patterson function P_1 with Pmmm symmetry, against a P_2 Patterson function with P2/m symmetry.

Multiple application of the sixfold axis S_2 and an exhaustive combination of the operators S_i and $_jS$ will give rise to 24 equivalent angular positions. The rotation space group is P2cb, but there are consecutive translations by $\pi/3$ along the θ_1 axis; this brings the number of equivalent positions from the standard (for P2cb) number of 4 to 24. For example, the reader will find that, besides $\theta_1, \theta_2, \theta_3$, the additional positions

$$(\pi/3 + \theta_1, \theta_2, \theta_3); \quad (2\pi/3 + \theta_1, \theta_2, \theta_3);$$
$$(\pi + \theta_1, \theta_2, \theta_3);$$
$$(4\pi/3 + \theta_1, \theta_2, \theta_3); \quad (5\pi/3 + \theta_1, \theta_2, \theta_3)$$

exist. The asymmetric unit to explore is then

$$0 \leq \theta_1 \leq \pi/3$$
$$0 \leq \theta_2 \leq \pi/2$$
$$0 \leq \theta_3 < 2\pi.$$

APPENDIX 13.B NON-CRYSTALLOGRAPHIC SYMMETRY

13.B.1 NCS symmetry operators

Crystallographic symmetry operators, when considered in the crystallographic reference system, hold for the full crystal; the corresponding rotation matrices **R** only contain −1,0,1 elements, and the translation components are simply related to the unit cell axes. When reconsidered in a Cartesian reference system, two situations may occur:

(i) If the original crystallographic frame is orthogonal, the elements of the rotation matrix **ρ** (calculated in Cartesian space) are still −1, 0, +1; indeed, orthogonalization and de orthogonalization matrices are both diagonal, with diagonal elements one the reciprocal of the other.
(ii) If the original crystallographic frame is oblique, the elements of the matrix **ρ** do not maintain −1 ,0, 1 entries.

In general, the above rules are no more valid when NCS operators are considered (they do not bring lattice points in overlapping).

So, why is there such a high interest in NCS? We observe that:

(a) it is present both in small molecule and in macromolecule areas;
(b) half of all structures in the PDB are dimeric or in a higher oligomerization state;
(c) the diffraction pattern contains redundant information, since the atomic coordinates of one molecule define the atomic coordinates of the molecules referred by NCS. This feature implies strong restraints on the structure factors;
(d) if NCS symmetry operators are well located, the electron density map may be improved by averaging the density over equivalent points; indeed, the noise may be averaged out and consistent structural features may be enhanced. The signal-to-noise improvement (at maximum) is proportional to $n^{1/2}$, where n is the number of molecular copies in the asymmetric unit.

NCS operators may be divided into two categories, marked by *proper* or *improper rotations*. When proper NCS is present, equivalent regions may be related by a local point group symmetry which is not limited to 2-, 3-,4-, and 6-fold rotations. In this case, the symmetry is independent of the sense of rotation. When improper rotations are present, both the rotational and translational components may contribute to the NCS; the rotation angle and the translation shift can be arbitrary. Figure 13.B.1 represents the asymmetric unit of *1mi1*; the two molecules in the asymmetric unit are related by an improper NCS, with an angle of rotation of 163°. In this case, the *NCS* operator brings the molecule one in overlapping with the molecule two by a 163° but not by a −163° rotation. In Fig. 13.B.2, five subunits are related by a proper local fivefold axis. Such axes are present in icosahedral viruses; no matter which sense of rotation, one fifth of a revolution moves one region onto a symmetry-equivalent region. Near-perfect rotations relating dimers, trimers, and higher assemblies are frequent in protein crystallography. If more proper axes meet at the origin of the unit cell, as in viruses, no translation component is found.

NCS usually shows translational components associated with the rotation. In accordance with Euclidean geometry, two equivalent objects may be related by different combinations of rotation and translation; the orientation of the rotation axis is fixed, but not its location. It is always convenient to leave the translation perpendicular to the axis at zero after rotation (see Fig. 13.B.3).

A special type of NCS is the so-called (in the small molecule field) *pseudotranslational symmetry*, or *translational NCS symmetry* (in the macromolecular area). This will be described in Section 13.B.3.

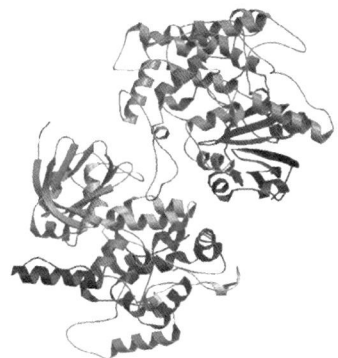

Fig. 13.B.1
Asymmetric unit of *1mi1*: two molecules in the asymmetric unit are related by improper NCS, with an angle of rotation of 163°.

13.B.2 Finding NCS operators

In Section 13.5 we defined the rotation function using equation (13.3b), repeated here for convenience:

$$RF = \int_U P_{mol}(\mathbf{Mu})P_{targ}(\mathbf{u})d\mathbf{u}, \qquad (13.B.1)$$

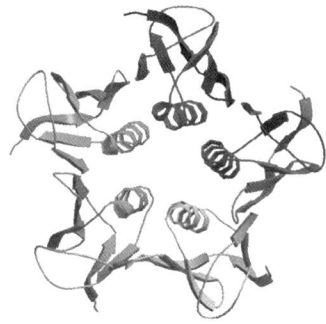

Fig. 13.B.2
1c48: a fivefold NCS axis relates five subunits.

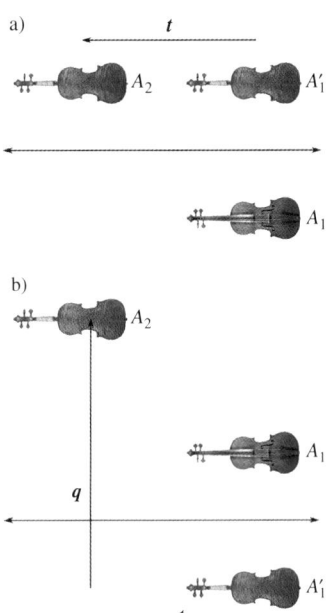

Fig. 13.B.3
Two objects A_1 and A_2 are related by a two-old rotation axis followed by a translation. (a) The intermediate object A'_1 is obtained by applying only the rotation operator to A_1: the translation t, parallel to the twofold rotation axis, brings A'_1 on A_2. (b) The two objects A_1 and A_2 are in the same positions occupied in case (a), but the rotation axis has been moved to a new position. The intermediate object A'_1 is obtained by applying only the rotation operator to A_1, but this time, to bring A'_1 on A_2, we need to apply the translation t, plus a supplementary translation q perpendicular to the rotation axis. It is concluded that the orientation of the rotation axis is fixed, but its location may be freely varied.

where P_{mol} and P_{targ} are the Patterson maps of the model molecule and of the target structure respectively, $\mathbf{M}$ is a matrix performing the rotation of the model map with respect to the target map. The *self-rotation function*, RF_S, is obtained from (13.B.1) by replacing P_{mol} by P_{targ}:

$$RF_S = \int_U P_{targ}[\mathbf{M}_{\phi,\psi,\chi}\mathbf{u}]P_{targ}(\mathbf{u})d\mathbf{u}, \quad (13.B.2)$$

where $\mathbf{M}_{\phi,\psi,\chi}$ emphasizes that the rotation is performed via the spherical polar angles described in the Section 13.A.2.

We will see that RF_S provides information about the content and organization of the asymmetric unit and suggests which oligomer model may be used as a search model in MR. It is worthwhile noticing that (i) a Patterson self-rotation function is intrinsically more reliable than a *cross-rotation function*, because it only depends on the observed target diffraction amplitudes; (ii) most of the considerations discussed in Section 13.6 hold for the self-Patterson.

Let us now suppose that RF_S is plotted as a function of the rotation angle. It will always show a huge value (the so-called *origin peak*) for zero rotation; indeed, the two Patterson maps will perfectly overlap if no rotation has been applied. This peak does not say anything about NCS and may therefore be ignored.

RF_S will also show huge peaks for any rotation that represents a crystallographic symmetry operation. For example, if a crystal has threefold rotational symmetry, a rotation corresponding to this symmetry axis will create a copy of the Patterson, completely identical to the original. Also, these peaks do not say anything about NCS and may therefore be ignored.

Information on NCS is contained in smaller peaks that are substantially larger than any other peak and several standard deviations larger than the noise.

Computer programs devoted to identification of the NCS usually store RF_S in *stereographic projections*, the basic geometry of which is illustrated in Fig. 13.B.4, where the convention

$$\begin{vmatrix} l \\ m \\ n \end{vmatrix} = \begin{vmatrix} \sin\psi\cos\phi \\ \sin\psi\sin\phi \\ \cos\psi \end{vmatrix}$$

has been chosen (see Section 13.A.2). As Fig. 13.B.4b suggests, to reveal proper NCS operators, RF_S is plotted in separate projections, each corresponding to a fixed value of χ (i.e. for $\chi = 360°/n$, say for $180°, 120°, 90°, 72°, 60°$, etc.); indeed, the order of a NCS operator is not limited to crystallographic values. The projections corresponding to $\chi = 360°/n$ will show peaks corresponding to n-fold local axes. Thus fivefold local axes will show peaks at $\chi = 72°$ and twofold axes, at $\chi = 180°$ (crystallographic axes of order two, if present, will produce the largest peaks in this projection).

As an example, we show in Fig. 13.B.5, the two monomers related by a twofold NCS axis in the protein structure identified by the PDB code *3hpe* (Sisinni et al., 2010), space group P2$_1$. In Fig. 13.B.6, the stereographic projections obtained by *MOLREP* at $\chi = 180°, 120°, 90°$, and $60°$ are shown; the only one with appreciable peaks corresponds to $\chi = 180°$. The results indicate that only NCS binary axes are allowed (the projection at $\chi = 72°$, not shown for brevity, is devoid of meaningful peaks). In the projection corresponding to $\chi = 180°$ we notice:

(a) The two strongest peaks are located at

$$\phi = 90°, \psi = 90°, \chi = 180°,$$
$$\phi = -90°, \psi = -90°, \chi = 180°,$$

both corresponding to the crystallographic binary axis along **b**.

(b) Two strong peaks are located at

$$\phi = 0°, \psi = 68°, \chi = 180°,$$
$$\phi = 180°, \psi = 202°, \chi = 180°,$$

suggesting the presence of a non-crystallographic binary axis with direction close (i.e. tilted by 22°) to that of **a**.

We have already seen in Section 13.A.3 that the angular asymmetric unit to cover is

$$0 \leq \chi < \pi, \quad 0 \leq \psi < \pi, \quad 0 \leq \phi < 2\pi.$$

However, as for the cross-Patterson function (see Section 13.A.4), additional symmetry in the stereographic projections would arise (and therefore a smaller asymmetric unit should be considered) as the effect of the native Patterson symmetry. For example, if the space group of the target is P2$_1$ and there is a dimer in the asymmetric unit with a NCS axis perpendicular to **b**, the self-Patterson will create a third twofold, orthogonal to the other two.

NCS symmetry operators must be accurately defined; indeed, the error increases with increasing distance from the symmetry element.

Let us now suppose that proper NCS axes are parallel to crystallographic symmetry axes of the same or of high order; then, self-rotation peaks due to NCS operators will be obscured by the corresponding crystallographic peaks. The ambiguity may be solved by checking the target Patterson map at the Harker sections. As an example, let us suppose that in P2$_1$ a twofold NCS axis is parallel to the 2$_1$ crystallographic axis. If (x_0, y, z_0), with y a free value, represents the generic point of the twofold axis, then (x, y, z) and ($-x + 2x_0$, y, $-z + 2z_0$) are equivalent positions. The full set of equivalent positions (including crystallographic and NCS symmetry) is

$$(x, y, z), \ (-x + 2x_0, y, -z + 2z_0), \ (-x, y + 0.5, -z),$$
$$(x - 2x_0, y + 0.5, z - 2z_0),$$

and the corresponding Patterson vectors are

$$(2x + 2x_0, 0, 2z + 2z_0), \ (2x, 0.5, 2z), \ (2x_0, 0.5, 2z_0),$$
$$(+ 2x_0, 0.5, + 2z_0), \ (-2x + 4x_0, 0.5, -2z + 4z_0),$$
$$(-2x + 2x_0, 0, -2z + 2z_0) + \text{cs. positions.}$$

The vector ($2x_0$, 0.5, $2z_0$) lies in the crystallographic Harker section, and does not depend on the positions (x,y,z); thus all of the atoms of an oligomer contribute to it. Accordingly, the position of the NCS twofold axis is immediately derivable by inspection of the Harker section. The reader should,

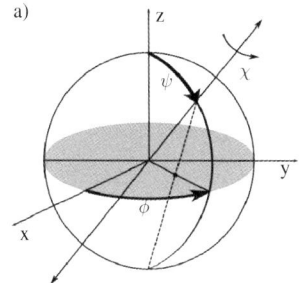
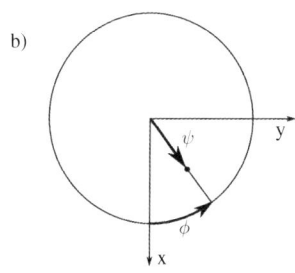

Fig. 13.B.4
Spherical polar angles according to the convention, $x = r \sin \psi \cos \phi, y = r \sin \psi \sin \phi, z = r \cos \psi$. (a) geometry of the stereographic projection; (b) spherical polar coordinates on the two-dimensional map.

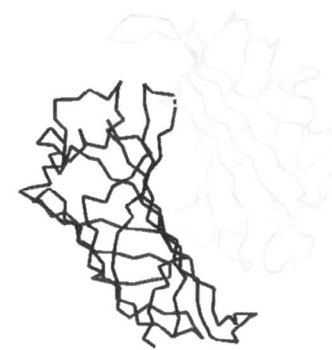

Fig. 13.B.5
3hpe. The two monomer backbones in the asymmetric unit.

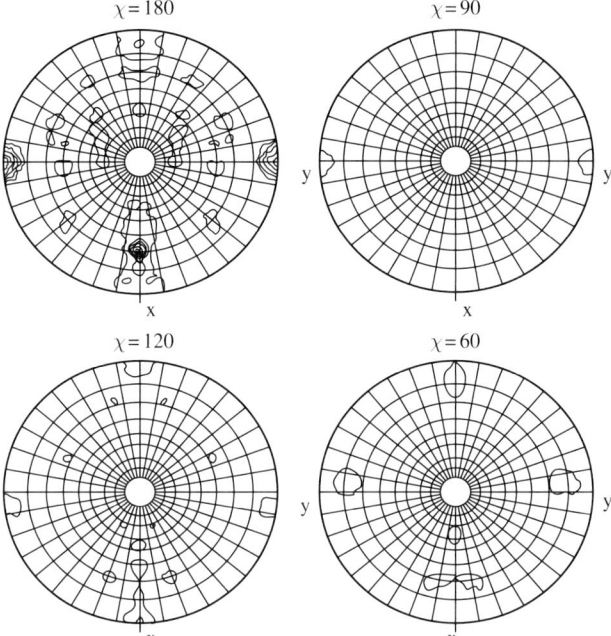

Fig. 13.B.6
3hpe. Self-rotation function: stereographic projections obtained by *MOLREP* at various χ angles. The convention is the following: ϕ is the angle in the plane (x,y) between x and the projected rotation axis; ψ is the angle between z and the rotation axis; χ is the rotation angle around the rotation axis (by courtesy of Giuseppe Zanotti).

however, remember that if the NCS axis deviates by a few degrees from the proper alignment with the crystallographic axis, the peak in the Harker section disappears.

13.B.3 The translational NCS

Let us suppose that a non-negligible amount of electron density $\rho(\mathbf{r})$ satisfies the condition

$$\rho(\mathbf{r}) \approx \rho(\mathbf{r} + \mathbf{q}),$$

where $\mathbf{q}$ is a rational fraction of the lattice periods,

$$\mathbf{q} = \frac{\nu_1}{\mu_1}\mathbf{a} + \frac{\nu_2}{\mu_2}\mathbf{b} + \frac{\nu_3}{\mu_3}\mathbf{c}, \quad \text{with} \quad 0 \le \nu_i \le \mu_i, \ i = 1, 2, 3.$$

Then, a pseudotranslational symmetry occurs and $\mathbf{q}$ is its pseudotranslational vector.

The index n of the pseudotranslation is the smallest integer for which $n\mathbf{q}$ is a lattice vector; it coincides with the least common multiple of μ_1, μ_2, μ_3. For example, $\mathbf{q} = \frac{1}{2}\mathbf{a} + \frac{1}{2}\mathbf{b}$ and $\mathbf{q} = \frac{1}{2}\mathbf{a} + \frac{1}{3}\mathbf{b}$ are of order two and six, respectively.

If the fraction of electron density $\rho(\mathbf{r})$ satisfying the pseudotranslational symmetry is not negligible, a strong Patterson peak will be visible in positions $n\mathbf{q}$; the intensity of the peak will be proportional to the fraction in pseudosymmetry. Here, we prefer to characterize the pseudotranslational symmetry by reciprocal space calculations.

It is usual to say (Buerger, 1956, 1959) that in these conditions the crystal structure is divided into two parts: the *substructure*, comprising that part of the electron density which conforms to the periodicity of the subcell, and the *complement structure* or *superstructure*, comprising the rest of the electron density. Accordingly, the reciprocal lattice $\{H\}$ is divided into two subsets: subset $\{H_A\}$ referred to substructure reflections and set $\{H_B\}$, associated with superstructure reflections.

The Fourier transform of the electron density distribution of the substructure is zero everywhere, except at set $\{H_A\}$, while the Fourier transform of the complement structure contributes to the entire set $\{H\}$.

A simple example is shown in Fig. 13.B.7. For a one-dimensional cell with period **a** and density ρ, the substructure ρ_{sub} with $\mathbf{u} = \mathbf{a}/3$ and the complement structure $\rho_{sup} = \rho - \rho_{sub}$ are emphasized. For this example, subset $\{H_A\}$ contains reflections for which $h \equiv 0 \mod(3)$, while set $\{H_B\}$ is consists of those for which $h \neq 0 \mod(3)$.

More pseudotranslational vectors may coexist simultaneously; for example, $\mathbf{q}_1 = \mathbf{a}/3$ and $\mathbf{q}_2 = \mathbf{c}/4$. In this case, because of the pseudosymmetry, a pseudocell ρ_{sub} with 1/12 of the volume of the original cell may be identified (the simultaneous presence of $\mathbf{q}_1$ and $\mathbf{q}_2$ is equivalent to the presence of the pseudotranslational vector $\mathbf{q} = \mathbf{a}/3 + \mathbf{c}/4$).

Sometimes pseudotranslational symmetry combines with crystallographic symmetry with multiplicative effects. For example, if $\mathbf{q}_1 = \mathbf{a}/2$ in the P4 space group, $\mathbf{q}_2 = \mathbf{b}/2$ also exists.

Real structures do not comply exactly with the mathematical models shown in Fig. 13.B.7. Indeed, atoms related by pseudotranslational symmetry may not be exactly referred (because of *displacive deviations* from ideal pseudosymmetry), or are of a different chemical nature (*replacive deviations* from ideal pseudosymmetry). Such situations produce special effects in the reciprocal space which deserve to be mentioned. A simple example of replacive deviation is shown in Fig. 13.B.8, where only three atoms satisfy the pseudotranslational vector $\mathbf{u} = \mathbf{a}/4$, the fourth being missed. If substructure reflections only are used, the average or *Takeuchi substructure* $\hat{\rho}_{sub}$ (Takeuchi, 1972) is seen instead of ρ_{sub}, so dividing ρ into two components, $\hat{\rho}_{sub}$ and $\rho - \hat{\rho}_{sub}$ (the latter is negative in a region of the unit cell), instead of into ρ_{sub} and $\rho - \rho_{sub}$.

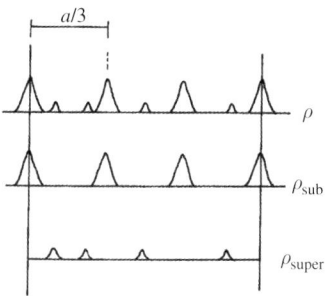

Fig. 13.B.7
A one-dimensional density ρ with pseudotranslational vector, $\mathbf{u} = \mathbf{a}/3$.

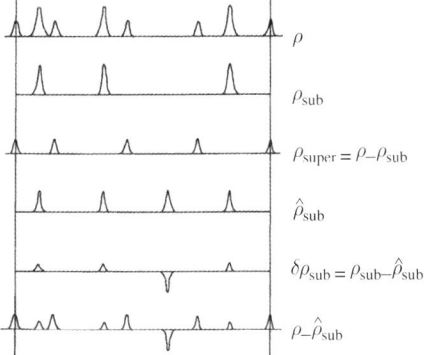

Fig. 13.B.8
Gaussian 'atoms' are located in a one-dimensional unit cell. Three heavy atoms satisfy the pseudotranslation vector, $\mathbf{u} = \mathbf{a}/4$, the fourth being missed. Four light atoms belong to ρ_{super}, but if substructure reflections only are used, $\hat{\rho}_{sub}$ is seen.

Fig. 13.B.9
Gaussian 'atoms' are located in a one-dimensional unit cell with pseudotranslational vector, **u** = **a**/4. Atoms related by **u** are of the same type, but slightly shifted from ideal positions.

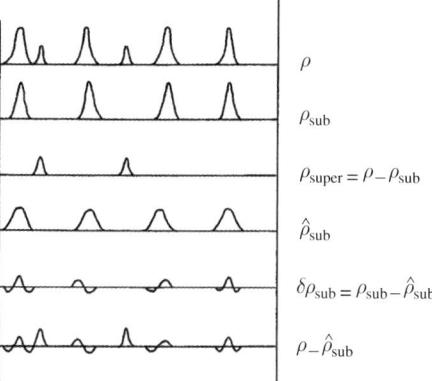

An example of displacive deviation from ideal pseudosymmetry is shown in Fig. 13.B.9, where a pseudotranslation vector, $\mathbf{q} = \mathbf{a}/4$, is shown, referring atoms of the same type but slightly shifted from ideal positions. Again, the Takeuchi substructure will be seen by X-ray diffraction, and ρ will be interpreted as having the two components $\hat{\rho}_{sub}$ and $\rho - \hat{\rho}_{sub}$.

It may be shown (Mackay, 1953; Cascarano et al., 1988a,b) that suitable statistical analysis of diffraction data can reveal the presence of displacive deviation from ideal pseudosymmetry (while replacive deviations cannot be revealed at all). We will not describe the mathematical modelling of pseudo-translational symmetry settled by Cascarano et al. (1985a,b; 1987b) (see also Böhme, 1982, 1983 and Gramlich, 1984). We will only mention that it allows us, via a statistical analysis of the diffraction amplitudes, to:

(i) detect the pseudotranslational symmetry;
(ii) define the pseudotranslational vectors, **q**;

Table 13.B.1 *FREIES. SIR2011* statistical analysis for revealing the presence and the nature of pseudotranslational symmetry. The program searched for pseudotranslational symmetry classes of reflections affected by pseudotranslational effects. Classes of reflections probably affected by pseudotranslational effects:

Condition	NREF	$<|E|^2>$	FOM	$\sum_p / \sum_N$
h = 2n & k = 3n	349	3.30	6.92	0.52
3h+2k = 6n	349	3.30	6.92	0.52
h = 2n	1028	1.55	4.85	0.67
3h+4k = 12n	173	3.38	4.45	0.24
h+2k+2l = 4n	516	2.23	4.14	0.46
k = 3n	577	2.05	3.83	0.46
3h+2k+3l = 6n	273	2.27	2.90	0.21
2k+3l = 6n	301	2.08	2.62	0.20
h = 2n & l = 2n	514	1.65	2.18	0.24

Remarkable deviations (of displacive type) from ideal pseudotranslational symmetry:

$$\text{at } (\sin\theta/\lambda)^2 = 0, \sum_p / \sum_N = 0.78, \text{ at } (\sin\theta/\lambda)^2 = \max, \sum_p / \sum_N = 0.32$$

Table 13.B.2 *2hyw. SIR2011* statistical analysis for revealing the presence and the nature of pseudotranslational symmetry. The program searched for pseudotranslational symmetry classes of reflections affected by pseudotranslational effects. 5130 reflections selected using resolution range (4.40–22.02) Å. Classes of reflections affected by pseudotranslational effects:

Condition	NREF	$<|E|^2>$	FOM	$\sum_p / \sum_N$
k + l = 2n	4903	1.90	18.67	0.89
k + l = 4n	3671	1.93	4.36	0.55
2h + k + l = 4n	3672	1.92	4.24	0.54
h = 2n & k + l = 2n	2444	1.92	2.78	0.30
h + k = 2n & h + l = 2n	2434	1.87	2.63	0.28
k = 2n & l = 2n	2449	1.84	2.56	0.28

Modest deviations (of displacive type) from ideal pseudotranslational symmetry:

at $(\sin\theta/\lambda)^2 = 0$, $\sum_p / \sum_N = 0.97$, at $(\sin\theta/\lambda)^2 = \max$, $\sum_p / \sum_N = 0.84$.

(iii) estimate the percentage of the electron density satisfying the pseudosymmetry;
(iv) reveal the possible displacive nature of the pseudotranslational symmetry.

Two examples are shown in Tables 13.B.1 and 13.B.2, the first refers to a small molecule (*FREIES*), the second to a protein (*2hyw*). The program *SIR2011* explores a large number of classes of reflections; for each class the average value of $<|E|^2>$ is calculated, and this calculation is also performed for each resolution shell. The percentage of the electron density satisfying the pseudosymmetry is computed ($\sum_p / \sum_N$ in the tables), a figure of merit (FOM) identifies the most important pseudotranslational symmetry (the figure takes into account the order of the pseudosymmetry and the percentage above defined). By comparing the $<|E|^2>$ values at high and low values of $\sin\theta/\lambda$, the displacive nature of the pseudosymmetry is estimated. In *2hyw*, the pseudotranslational symmetry is nearly perfect; in *FREIES* it has a strong displacive character.

APPENDIX 13.C ALGEBRAIC FORMS FOR THE ROTATION AND TRANSLATION FUNCTIONS

The canonical structure factor expression for the target structure is

$$F_{\mathbf{h}targ} = \sum_{j=1}^{t} f_j \sum_{s=1}^{m} \exp[2\pi i\mathbf{h}(\mathbf{R}_s\mathbf{r}_j + \mathbf{T}_s)], \quad (13.C.1)$$

where t is the number of atoms in the asymmetric unit and m is the number of symmetry operators, $\mathbf{C}_s \equiv (\mathbf{R}_s, \mathbf{T}_s)$.

Let us now suppose that a model molecule is randomly oriented and wrongly positioned: its *j*th atomic position, $\mathbf{r}'_j$, is referred to the corresponding target atomic position $\mathbf{r}_j$ by the relation, $\mathbf{r}'_j = \mathbf{M}\mathbf{r}_j + \mathbf{N}$. The corresponding structure factor is,

$$F_{\mathbf{h}mol} = \sum_{j=1}^{t} f_j \sum_{s=1}^{m} \exp\{2\pi i \mathbf{h}(\mathbf{R}_s \mathbf{r}'_j + \mathbf{T}_s)\} = \sum_{s=1}^{m} a_s \gamma(\mathbf{h}\mathbf{R}_s), \quad (13.\text{C}.2)$$

where

$$a_s = \exp[2\pi i \mathbf{h} \mathbf{T}_s]$$

$$\gamma(\mathbf{h}\mathbf{R}_s) = \sum_{j=1}^{t} f_j \exp(2\pi i \mathbf{h}\mathbf{R}_s \mathbf{r}'_j).$$

From (13.C.2), the following relation arises:

$$|F_{\mathbf{h}mol}|^2 = \sum_{s=1}^{m} |\gamma(\mathbf{h}\mathbf{R}_s)|^2 + \sum_{s_1 \neq s_2 = 1}^{m} \gamma(\mathbf{h}\mathbf{R}_{s_1}) \gamma^*(\mathbf{h}\mathbf{R}_{s_2}) a_{s_1} a_{s_2}^*.$$

The explicit form of $\sum_{s=1}^{m} |\gamma(\mathbf{h}\mathbf{R}_s)|^2$ is

$$\sum_{s=1}^{m} |\gamma(\mathbf{h}\mathbf{R}_s)|^2 = \sum_{j_1, j_2 = 1}^{t} f_{j_1} f_{j_2} \exp\left[2\pi i \mathbf{h} \mathbf{R}_s \mathbf{M} (\mathbf{r}_{j_1} - \mathbf{r}_{j_2})\right].$$

The corresponding interatomic vectors are of self-Patterson type (see Section 13.4), do not depend on the translation vector **N**, but only on the rotation matrix **M**. Therefore, $\sum_{s=1}^{m} |\gamma(\mathbf{h}\mathbf{R}_s)|^2$ may provide information on the orientation of the model molecule, and is the frequent expression of the parameter X_{mol} in equation (13.6).

Let us now suppose that the model molecule has been well oriented but is still wrongly positioned. We can write the corresponding structure factor as,

$$F_{mol}(\mathbf{h}, \mathbf{N}) = \sum_{j=1}^{t} f_j \sum_{s=1}^{m} \exp\{2\pi i \mathbf{h} \left[\mathbf{R}_s (\mathbf{r}_j + \mathbf{N}) + \mathbf{T}_s\right]\}, \quad (13.\text{C}.3)$$

where $\mathbf{r}_j + \mathbf{N}$ is the trial position of the jth atom in the model, shifted with respect to the correct location by the vector **N**. We can rewrite (13.C.3) as,

$$F_{mol}(\mathbf{h}, \mathbf{N}) = \sum_{s=1}^{m} a_s \gamma(\mathbf{h}\mathbf{R}_s),$$

with

$$a_s = \exp\left[2\pi i \mathbf{h} (\mathbf{R}_s \mathbf{N} + \mathbf{T}_s)\right]$$

$$\gamma(\mathbf{h}\mathbf{R}_s) = \sum_{j=1}^{t} f_j \exp(2\pi i \mathbf{h}\mathbf{R}_s \mathbf{r}_j).$$

The translation function (13.9) may then be rewritten as,

$$TF(\mathbf{N}) = \sum_{\mathbf{h}} |F_{\mathbf{h}targ}|^2 \sum_{s_1, s_2 = 1}^{m} \gamma(\mathbf{h}\mathbf{R}_{s_1}) \gamma^*(\mathbf{h}\mathbf{R}_{s_2}) a_{s_1} a_{s_2}^*, \quad (13.\text{C}.4)$$

which may be represented as a Fourier series, as reported by Vagin and Teplyakov (1997):

$$TF(\mathbf{N}) = V^{-1} \sum_{\mathbf{H}} A_{\mathbf{H}} \exp[-2\pi i \mathbf{H} \mathbf{N}], \quad (13.\text{C}.5)$$

with coefficients,

$$A_{\mathbf{H}} = \sum_{\mathbf{h}} |F_{\mathbf{h}targ}|^2 \sum_{s_1, s_2 = 1}^{m} \gamma(\mathbf{h}\mathbf{R}_{s_1}) \gamma^*(\mathbf{h}\mathbf{R}_{s_2}) \exp\left[2\pi i \mathbf{H} (\mathbf{T}_{s_1} - \mathbf{T}_{s_2})\right], \quad (13.\text{C}.6)$$

where summation is over those **h** for which $\mathbf{H} = \mathbf{h}(\mathbf{R}_{s_2} - \mathbf{R}_{s_1})$. The maxima of the TF map should correspond to the most probable translation vectors.

The TF expression, (13.C.6), may be modified by replacing the term $|F_{\mathbf{h}targ}|^2$ by $\Delta|F_{\mathbf{h}targ}|^2 = |F_{\mathbf{h}targ}|^2 - <|F_{\mathbf{h}targ}|^2>$, as suggested by Navaza (1994); then the term corresponding to the condition $s_1 = s_2$ is excluded from summation (13.C.6). The effect of this is that the origin peak is removed and the contrast between the correct and the false solutions increases.

14 Isomorphous replacement techniques

14.1 Introduction

The isomorphous replacement method is a very old technique, used incidentally by Bragg to solve *NaCl* and *KCl* structures: it was later formulated in a more general way by Robertson (1935, 1936) and by Robertson and Woodward (1937). Its modern formulation is essentially due to Green et al. (1954) and to Bragg and Perutz (1954), who applied the method to haemoglobin. The technique has made possible the determination of the first three macromolecular structures, myoglobin, haemoglobin, and lysozyme.

The approach may be summarized as follows. Suppose that the target structure is difficult to solve (e.g. it is a medium-sized structure, resistant to any phasing attempt, or it is a protein with bad data resolution) and we want to adopt isomorphous replacement techniques. Then we should perform the following steps:

(a) Collect the diffraction data of the *target structure*; in the following we will suppose that it is the *native protein*.
(b) Crystallize a new compound in which one or more heavy atoms are incorporated into the target structure. This new compound is called *derivative*.
(c) Check if the operations in (b) heavily disturb the target structure. If not, the derivative is called *isomorphous*; then, only local (in the near vicinity of the binding site) structural modifications are induced by the heavy atom addition. Non-isomorphous derivative data are useless.
(d) Use the two sets of diffraction data, say set $\{|F_P|\}$ of the target structure and set $\{|F_d|\}$ of the isomorphous derivative, to solve the target structure.

The above case is referred to as *SIR* (*single isomorphous replacement*). The reader should notice that redundant experimental information is available; indeed, two experimental sets of diffraction data relative to two isomorphous structures may be simultaneously used for solving the native protein. The redundancy of the experimental information allows crystal structure solution even if data resolution is far from being atomic (e.g. also when RES is about 3 or 4 Å, and even more in lucky cases).

Imperfect isomorphism may hinder crystal structure solution. Then, more derivatives may be prepared; their diffraction data may be used in a combined way and may more easily lead the phasing process to success. This case is referred as *MIR* (*multiple isomorphous replacement*), characterized by the fact that more sets of diffraction data are available:

$$\{|F_P|\}, \{|F_{d1}|\}, \{|F_{d2}|\}, \ldots \ldots$$

Roughly speaking, the more redundant information that is available, the easier structure solution will be.

In Section 14.2, we briefly recall the laboratory techniques for preparing derivatives, with special regard to proteins; in Sections 14.3 and 14.4, the algebraic SIR-MIR bases are stated. In Section 14.5, the most popular approaches for putting on absolute scale native and derivative data are described. Sections 14.6 and 14.7 illustrate the classical two-step procedure (i.e. first solve the heavy atom substructure, and then the native structure), as seen from a modern probabilistic point of view. Some schematic applications are described in Section 14.8. A one-step procedure (i.e. the intermediate step of solving the heavy atom substructure is no longer necessary) is also possible and is summarized in Appendix 14.B; even if appealing and interesting theoretically, it never became popular.

In this chapter, the capital subscript P and the lower-case subscript d will indicate the native protein and the derivative, respectively.

14.2 Protein soaking and co-crystallization

Ligands or heavy atoms can be soaked into, or co-crystallized with, the native protein (the *target structure*). Co-crystallization is used when ligands induce strong conformational changes into the protein.

Soaking is not allowed for small molecules, because the solvent channels do not permit diffusion of large molecules (their diameter varies between 20 and 100 Å). Usually, crystals are soaked with various heavy-atom salt solutions; the mechanism requires the existence of pores in the crystal through which the reagent can diffuse to reach the protein surface. The soaking time ranges from several minutes to months, and depends on the protein under study, on the pH, on the temperature, and on the precipitating agents.

Commonly used heavy-atom derivatives are *Pt*, *Hg*, and *Au* compounds, uranyl salts, and rare earth elements. Very recently, quick soak derivatization by heavy halide atoms (iodine and bromine anions), or heavy alkali (cesium and rubidium cations) have been proposed (Dauter et al., 2001; Nagem et al., 2003); these ions are used in strong concentrations and bind to the surface of the protein, frequently with partial occupancy. Xenon–protein complexes are obtained by pressurizing native proteins with xenon gas.

We notice:

1. The site of attachment of heavy atoms is not due to chance; for example, histidine residues frequently act as ligands, sulphur atoms in methionine often bind platinum compound, etc.

2. The number of binding sites per protein molecule is small, generally from 1 to 10.
3. Their chemical occupancy may be inferior to unity. The most useful derivatives are those for which heavy-atom binding is not disordered; unfortunately, it usually occurs that, at a given position, the heavy atoms are not present in all of the unit cells, and this fact simulates chemical occupancies smaller than unity. Full occupancy has usually good effects on the success of SIR-MIR techniques.
4. The isomorphism is never perfect, because of the chemical reaction between reagent and protein. When heavy atoms are incorporated into the native protein, two effects may arise. A change in the cell constants, which may be easily recognized by the diffraction experiment, and a change in the orientation of molecular fragments (which does not necessarily produce cell changes). A preliminary control of isomorphism may be made according to the following criteria: the unit cell of the target and of the derivative structure are expected to be different for less than 1.5%, the corresponding diffraction intensities are not significantly different. For the latter parameter, the crystallographic residue R_{cryst} may be used as a criterion; if it lies between 0.15 and 0.25, the isomorphism condition may be accepted. A smaller value of R_{cryst} indicates little incorporation of the heavy atoms. Crick and Magdoff (1956) first showed that incorporation of heavy atoms produces detectable differences in the diffraction intensities (see Appendix 14.C).
5. The derivatization of large biological molecules (or of aggregates of them, like *ribosomes*) requires special care. Indeed, numerous atoms would be necessary in this case to produce measurable signals; the location of such multiple derivatives would then be extremely difficult, and therefore, this approach is not suitable. Alternatively, one can take advantage of compact and dense compounds containing, in a small volume, a large number of heavy atoms directly linked to each other (see, e.g. Thygesen et al., 1996; Schlünzen et al., 1995). The large size of these clusters has some practical consequences: multiple-site binding is not frequent, but diffusion of the clusters into a crystal may be difficult. The cluster may bind in several modes (e.g. more than one atom is capable of coordination with the macromolecule). The cluster may be treated as a unique scatterer, and heavy atoms cannot be resolved; isomorphism is then maintained up to 6–8 Å resolution. If the cluster has perfect symmetry, the individual positions of the cluster atoms can be resolved and the phasing power of the derivative is higher.

From the above statements, the following conclusion arises: the purpose of soaking is not to obtain derivatives, but to obtain isomorphous derivatives. The method is a trial and error technique, and often requires patience and determination. For the designing of derivatization experiments it may be useful to consult the heavy-atom protein data bank (available at <http: //www.bmm.icnet.uk/had/index.html>), which contains information on heavy-atom derivatives for more than 1000 protein crystals.

14.3 The algebraic bases of SIR techniques

Let us suppose that diffraction data of the native protein and of one isomorphous derivative have been collected. F_P and F_d are the corresponding structure factors and ϕ_P and ϕ_d are the related phases. If F_H and ϕ_H are the structure factor and the phase of the heavy atom substructure, and if we assume a perfect, ideal isomorphism between native protein and derivative, then the relation

$$F_d = F_P + F_H \qquad (14.1)$$

holds. At this initial step of the phasing process, $|F_d|$ and $|F_P|$ are known from the diffraction experiment; ϕ_d and ϕ_P as well as $|F_H|$ and ϕ_H are unknown. The number of heavy-atom sites and their chemical occupancy are also unknown at this stage.

Equation (14.1) is represented in Fig. (14.1); no estimate can be made at this stage on the value of ϕ_P without some additional information.

Let us now suppose that $|F_H|$ and ϕ_H are known; we will show that estimating ϕ_P is now possible. In Fig. 14.2, we draw, in the Argand plane, the *Harker diagram*, consisting of two circles, the first centred at the origin O with radius $|F_P|$, the second centred at the tip of the vector $-F_H$ with radius $|F_d|$. The two circles meet at two points, the only ones compatible with equation (14.1); accordingly, ϕ_P may only assume two values. We are here explicitly excluding the case in which $|F_H| = 0$, where the circles are essentially concentric and do not have well-defined points of intersection.

The Harker diagram may easily be transferred into simple trigonometric calculations. Indeed, equation (14.1) may be rewritten as,

$$|F_d|^2 = |F_P|^2 + |F_H|^2 + 2\,|F_P F_H| \cos(\phi_P - \phi_H). \qquad (14.2)$$

Then,

$$\cos(\phi_P - \phi_H) = \left(|F_d|^2 - |F_P|^2 - |F_H|^2\right) / (2|F_P F_H|)$$

and

$$\phi_P = \phi_H + \cos^{-1}\left[\left(|F_d|^2 - |F_P|^2 - |F_H|^2\right) / (2|F_P F_H|)\right] = \phi_H \pm \alpha. \quad (14.3)$$

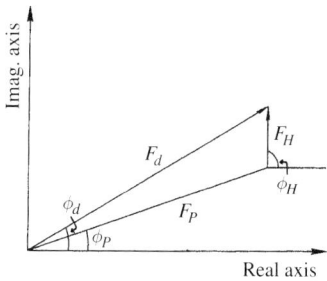

Fig. 14.1
Vector triangle showing the relation between F_p, F_d, and F_H.

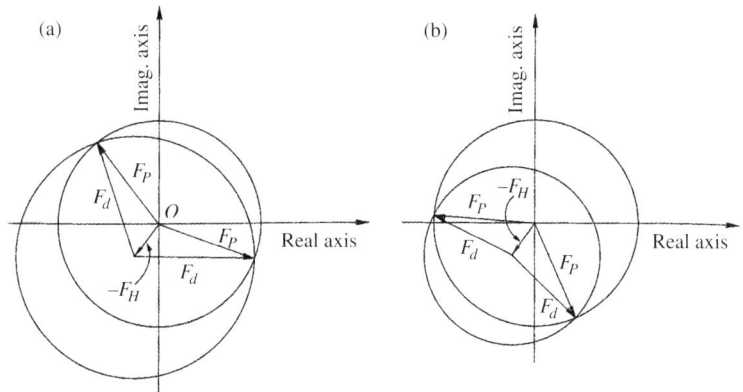

Fig. 14.2
Harker construction illustrating the SIR method for the cases: (a) $|F_d| > |F_p|$; the two allowed solutions assign to ϕ_P a value close to ϕ_H. (b) $|F_d| < |F_p|$; the two allowed solutions assign to ϕ_P a value close to $\phi_H + \pi$.

Because of the two possible values of the cosine term, two ϕ_P solutions arise, which are symmetric with respect to ϕ_H. If $|F_d| > |F_p|$, the two allowed solutions for ϕ_P are close to ϕ_H. If $|F_d| < |F_p|$, the two allowed solutions for ϕ_P are closer to $\phi_H + \pi$ (see Fig. 14.2).

The above conclusion suggests the following two-step phasing procedure:

Step 1: *Solve the heavy atom substructure, in order to estimate $|F_H|$ and ϕ_H;*
Step 2: *Use this information to phase the protein, by solving the phase ambiguity fixed by equation (14.3).*

But how do we solve the heavy-atom substructure? This is a very small structure, and, on the basis of experience accumulated in the small molecule field, a fairly good estimate of $|F_H|$ is needed. If this information is available, to accomplish Step 1, two approaches may be used: calculate a $|F_H|^2$ Patterson synthesis (to be deconvoluted in accordance with the techniques described in Chapter 10), or, to submit the estimated $|F_H|$ to traditional direct methods, charge flipping, or *VLD* techniques, according to personal preference (see Chapters 6, 9, and 10). Figure 14.3 suggests the following approximation:

$$\Delta_{iso} \approx |F_H| \cos(\phi_d - \phi_H), \tag{14.4}$$

where
$\Delta_{iso} = |F_d| - |F_P|$ is the so-called *isomorphous difference*. From (14.4),

$$|F_H| \approx \Delta_{iso}/\cos(\phi_d - \phi_H)$$

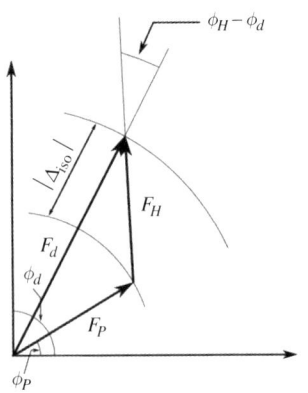

Fig. 14.3
The relation between Δ_{iso}, F_H, and $\cos(\phi_d - \phi_H)$.

arises. Unfortunately, at this step ϕ_d and ϕ_H are unknown, and (14.4) cannot be applied.

Perutz (1956) suggested use of the relation $|F_H|^2 \approx ||F_d|^2 - |F_P|^2|$ and the corresponding Patterson is called a *difference Patterson*, but is not very convenient. Indeed, if we expand $|F_d|^2$ in its components we have,

$$|F_d|^2 - |F_P|^2 = (F_P + F_H)(F_P^* + F_H^*) - F_P F_P^* = |F_H|^2 + F_H F_P^* + F_P F_H^*.$$

The result is that $|F_d|^2 - |F_P|^2$ contains not only the correct squared heavy-atom structure factor, but also substantial noise arising from native–heavy atomic distances. Blow (1958), Blow and Crick (1959), and Rossmann (1960) showed that a better approximation is:

$$|F_H|^2 \approx (|F_d| - |F_P|)^2 \equiv \Delta_{iso}^2. \tag{14.5}$$

Since (see Fig. 14.1),

$$|F_H|^2 = |F_P|^2 + |F_d|^2 - 2|F_d F_P| \cos(\phi_P - \phi_d), \tag{14.6}$$

if $|F_d|$ and $|F_P|$ are large with respect to $|F_H|$ (which is frequent in practice), then F_d and F_P will have similar phase values and (14.6) may be approximated by

$$|F_H|^2 = |F_P|^2 + |F_d|^2 - 2|F_d F_P| \equiv |\Delta_{iso}|^2.$$

The resulting *isomorphous difference Patterson* is therefore closer to the heavy-atom Patterson synthesis than the difference Patterson synthesis (Terwilliger and Eisenberg, 1987). However, considerable noise is present in the $|\Delta_{iso}|^2$ synthesis also; the reader can easily verify from Fig. 14.1, that if F_H is perpendicular to F_d, $\Delta_{iso} \approx 0$, irrespective of the value of $|F_H|$.

Deeper algebraic analyses were performed by Phillips (1966), Dodson and Vijayan (1971), Blessing and Smith (1999), and Grosse-Kunstleve and Brunger (1999), who suggested useful weighting schemes.

Once the heavy-atom substructure is solved, Step 2 may be started. But, how do we solve the phase ambiguity fixed by equation (14.3)? Indeed, this equation does not seem to be very useful, because the two allowed ϕ_P values may be far away from each other and it is not known which of the two values should be preferred. The solution may be obtained by looking at Fig. 14.3; if $|\Delta_{iso}|$ is sufficiently large (then $|F_H|$ must be sufficiently large, because $|F_H| > |\Delta_{iso}|$), then $\cos(\phi_d - \phi_P)$ must be close to unity, at least for ideal isomorphism and perfect experimental data; otherwise, $|F_H|$ is obliged to take values too large for the small scattering power of the heavy-atom substructure. In this case (that is, when $|\Delta_{iso}|$ is sufficiently large), the two allowed values of ϕ_P are close to each other, and the following assumptions would not (at least in favourable cases) imply a large error:

$$\phi_P \approx \phi_H \quad \text{if} \quad \Delta_{iso} > 0, \quad \text{(see Fig. 14.2a)}$$

and

$$\phi_P \approx \phi_H + \pi \quad \text{if} \quad \Delta_{iso} < 0. \quad \text{(see Fig. 14.2b)}$$

The centrosymmetric case is treated in Appendix 14.A.

Some practical details should be discussed, to better explain the main obstacles to a straightforward phasing:

(i) The $|\Delta_{iso}|$ represent always lower bound estimates of $|F_H|$, and are therefore affected by systematic errors which, inevitably, make the phasing process more difficult, no matter whether direct methods or Patterson deconvolution techniques are used. In the first case, the errors affect the reliability of phase relationships, in the second, they lower the quality of the *isomorphous difference Patterson*. The situation is not different if a third approach is used: considering the task of locating heavy atoms as a special case of molecular replacement, for which the trial model consists of only one atom (Vagin and Teplyakov, 1998; Grosse-Kunstleve and Brunger, 1999). The rotation function is not necessary because of the spherical symmetry of the model, only the translation function (similar to that defined in Section 13.7) has to be applied.

(ii) As stated above, SIR phasing is based on amplitude differences; therefore, small errors on the values of the amplitudes may also generate large errors in their differences and in the related phases. For a numerical insight into the problem, let $|F_d|/\sigma_d$ and $|F_P|/\sigma_P$ be the ratio *amplitude/standard deviation* for the derivative and for the protein, respectively. Frequently, both may be larger than five; conversely, $(|F_d| - |F_P|)/\sigma_\Delta$, where $\sigma_\Delta = (\sigma_d^2 + \sigma_P^2)^{1/2}$ is the standard deviation which may be associated with the difference in amplitudes, and is not frequently larger than five. The errors on $|F_H|$ estimates will increase even more if scaling of $|F_d|$ on $|F_P|$ is imperfect; thus, special procedures are needed to put derivative data onto their absolute scales (see Section 14.5).

(iii) Usually the experimental data resolution of the derivative is worse than that of the native protein. Therefore, the above approach can only be applied to reflections which are in common with both native and derivative data; as a consequence, phase expansion and refinement via EDM techniques is a necessary supplemental step for improving the model obtained via isomorphous data.

(iv) The heavy-atom substructure determined at the end of Step 1 does not necessarily have the correct handedness; there is only a 50% chance of having determined the correct enantiomorph.

(v) There is a second source of ambiguity: two values are allowed for the protein phases when the heavy-atom substructure is known. Taking their average probably leads to an electron density map where two structural images are present. This ambiguity may be overcome by improving and extending the phases via EDM techniques. Let us now suppose that the procedure converges to a final map where one of the two images has been cancelled. We will still have a 50% chance of obtaining the correct enantiomorph. A decision may be taken by looking at the helices; all of their densities should appear left-handed in the correct map. If the handedness of the substructure is recognized to be wrong, the correct structure may be obtained by changing the fractional coordinates from **r** to –**r**. If the space group is one of the enantiomorphic pairs (e.g. $P3_1$ or $P3_2$, $P3_121$ or $P3_2 21$), then the space group also has to be inverted (e.g. from $P3_1$ to $P3_2$, or vice versa; from $P3_1 21$ to $P3_2 21$, or vice versa). In the chiral space groups $I4_1$, $I4_122$, and $I4_132$, the crystallographic symmetry, applied to the 4_1 axis, generates a 4_3 operator (and vice versa). Thus, the chiral partners of these space groups (say $I4_3$, $I4_322$, and $I4_332$) are not distinct from the originals, and the *International Tables for Crysallography* do not need to report them. In these space groups, the origin is not located on an enantiomorph axis, and the centre of inversion cannot coincide with the origin. The handedness inversion therefore requires inversion operators like $(-x, -y + 1/2, -z)$, $(-x, -y + 1/2, -z + 1/4)$, and $(-x + 1/4, -y + 1/4, -z + 1/4)$ (Sheldrick, 2007).

14.4 The algebraic bases of MIR techniques

Using more derivatives (*MIR*) is the traditional way for reducing the noise and emphasizing the signal. To better understand the MIR advantages, let us return to Fig. 14.2a and modify it in order to include a new derivative (see Fig. 14.4). Besides the heavy atom and the derivative structure factors considered in Fig. 14.2a (say F_{H1} and F_{d1}), the heavy atom and the second derivative structure factors (say F_{H2} and F_{d2}) have also been included into the figure. Since the heavy-atom binding sites in the second derivative usually do not coincide with those of the first, $F_{H1} \neq F_{H2}$, $F_{d1} \neq F_{d2}$. In the ideal case (i.e. no error in the data and no lack of isomorphism), only one of the two points on the $|F_P|$ circle, compatible with F_{H1} and F_{d1}, is also compatible with F_{H2} and F_{d2}. The SIR ambiguity is now overcome. Figure 14.4 is equivalent to solving the system of equations

$$\phi_P = \phi_{H_1} + \cos^{-1}\left[\left(|F_{d_1}|^2 - |F_P|^2 - |F_{H_1}|^2\right)/2|F_P F_{H_1}|\right]$$

$$\phi_P = \phi_{H_2} + \cos^{-1}\left[\left(|F_{d_2}|^2 - |F_P|^2 - |F_{H_2}|^2\right)/2|F_P F_{H_2}|\right].$$

The algebraic bases of MIR techniques

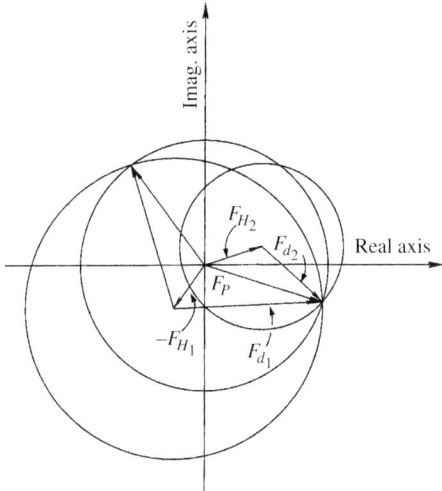

Fig. 14.4
Harker diagram for two heavy-atom derivatives.

However, the two equations may only be used if the heavy-atom positions are referred to the same origin. An easy and widely used method of finding a common origin is to calculate a difference Fourier synthesis with coefficients

$$(|F_{d_2}| - |F_P|)\exp(i\phi_P),$$

where the phase angles ϕ_P come from the first derivative. This synthesis directly provides the heavy-atom positions of the second derivative, referred to the same origin as the first.

A generalization of the above techniques to three or more derivatives is straightforward.

In early times, when experimental errors were larger (on average) and algebraic methods were used, frequently, more than two derivatives were used. Today, thanks to the advent of probabilistic approaches, one derivative is usually sufficient to phase the protein, provided that it is sufficiently isomorph.

Let us now check the phase ambiguities in the MIR case. When the first heavy-atom substructure has been determined, its handedness has a 50% chance of being correct. The other substructures, determined via the above described difference Fourier syntheses, are congruent with the first; therefore, if the first has a wrong handedness, all the others will show the same handedness.

The SIR phase ambiguity is not present in MIR and only one phase value is suggested for the protein. Accordingly, the application of EDM techniques to MIR data will more easily converge to a good final structure, which again, however, has a 50% chance of being the correct enantiomorph. The choice of enantiomorph may be undertaken by looking at the helices (all their densities should appear left-handed in the correct map), and inverting the coordinates if the final structure has wrong handedness; exactly as described for the SIR case.

14.5 Scaling of experimental data

Data scaling is a preliminary step for any probabilistic or algebraic approach to SIR-MIR. It is critical to the success of the phasing procedure, because differences between isomorphous data are small, and therefore very sensitive to scaling; even small errors in scaling may lead phasing to fail.

The aim of a scaling process is to set native and derivative diffraction amplitudes on their absolute scales before any further numerical treatment. For the probabilistic approach to SIR, a second preliminary step is necessary: structure factor normalization, which implies the supplementary estimate of the overall temperature factors for both native and derivative. Wilson scaling procedures, described in Chapter 2, may be used for both scaling and normalization, but their success is hampered by the strong Debye effects on the Wilson plot, and by the fact that derivative data resolution may be significantly smaller than that of native. We showed in Section 2.9 (see Fig. 2.10), that if scaling and temperature factors are calculated for the same experimental data but at different resolutions (e.g. one time using all the data up to RES and one time, cutting them at a poorer resolution), the resulting two least squares lines may be quite different in slope and intercept. It may therefore be expected that Debye effects can heavily disturb the scaling of two isomorphous structures, if the data resolution of one is significantly smaller than the other. To reduce Debye effects, the procedure suggested by Blundell and Johnson (1976) may be followed:

(a) Firstly, K_P and B_P (scale and overall thermal factor of the protein at derivative resolution) are calculated by the standard Wilson method, and then K_d/K_P and $B_d - B_P$ are estimated through the equation,

$$\ln \left(\sum_d < |F_P|^2 > / \sum_P < |F_P|^2 > \right) = \ln(K_P/K_d) + 2(B_d - B_P) \sin^2 \theta / \lambda^2 \quad (14.7)$$

In accordance with Chapter 2, the following definitions hold:

$$\sum_d = \sum_{j=1}^{N_d} f_j^2, \sum_P = \sum_{j=1}^{N_P} f_j^2, \sum_H = \sum_{j=1}^{N_H} f_j^2,$$

where N_d, N_P, N_H are the number of non-hydrogen atoms in the unit cell for the derivative, the native, and the heavy-atom structure, respectively.

Equation (14.7) is easily obtained by dividing each by the following two equations (these are the basis of the Wilson plot):

$$< |F_P|^2 > = K_P \sum_{0P} \exp(-2B_P \sin^2 \theta / \lambda^2),$$

$$< |F_d|^2 > = K_d \sum_{0d} \exp(-2B_d \sin^2 \theta / \lambda^2).$$

The Debye effects are minimized if (14.7) is used.

(b) While $\sum_{0P}$ is usually known, the number and the occupancy factors of the heavy-atom sites are unknown at this stage; thus, the approximation $\sum_{0d} \approx \sum_{0P}$ is necessarily introduced into (14.7), which implies that derivative structure factors are put on the same scale as that of the native structure factors.

(c) Two probabilistic procedures, outlined in Appendix 14.C, may be used to estimate the scattering power of the substructure. Once this information is obtained, the estimated $\sum_d = \sum_P + \sum_H$ values may be used in equation (14.7) to put the derivative amplitudes on their absolute scale.

(d) In the MIR case, each derivative may be submitted to the above described procedure to set the corresponding amplitude on the absolute scale.

It is worthwhile mentioning some practical aspects related to the scaling problem:

(i) Sometimes, local scaling is used (Matthews and Czerwinski, 1975) to reduce the effects of significant non-random errors in the observed amplitudes. The technique was reconsidered by Blessing (1997), who introduced a local variable scale factor, $q = q(\mathbf{h})$. The variable,

$$\Delta |F| = |F_P| - q|F_d|$$

was introduced, where q is estimated by a least squares fit minimizing

$$\chi^2 = \sum_{\mathbf{h}-\Delta \mathbf{h}}^{\mathbf{h}+\Delta \mathbf{h}} w[|F_P|/|F_d| - q]^2. \quad (14.8)$$

In relation (14.8), the summation over the indices runs in a local block of reciprocal lattice points surrounding, but not including, the point $\mathbf{h}$, for which we want to estimate $\Delta|F|$. A typical block consist of $(3 \times 3 \times 3) - 1$ points, but criteria may be used to optimize their size.

(ii) If the crystal diffracts anisotropically, an anisotropic scaling is attempted.

(iii) Most scaling programs try to detect outliers, marked by an extremely high $|\Delta_{iso}|$: these may dominate direct methods or Patterson phasing, hindering heavy-atom substructure solution (Read, 1999). For example, differences which exceed four times the root mean square deviation in the corresponding resolution shell are rescaled or eliminated.

(iv) High-resolution data usually contain larger noise, and are removed from the calculations; truncation of data usually increases the chance of solving the structure.

14.6 The probabilistic approach for the SIR case

The number of symmetry-independent heavy-atom sites in the isomorphous derivative structure exceeds 10 in exceptional cases. Structural problems of this size are trivial for Patterson or direct phasing techniques if $|\Delta_{iso}|$ is an accurate estimate of the true modulus R_H. Unfortunately, in protein crystallography this is not the case, as the crystallographic residue R_{cryst} between $|\Delta_{iso}|$ and R_H may also exceed 0.50.

Luckily, substructure solution is facilitated by the fact that the few heavy atoms are dispersed in a large volume (the native protein unit cell); thus, their peaks are far from each other and do not overlap, no matter what is the data resolution.

The algebraic techniques described in Section 14.3 are not the best way to face the challenges connected with errors in experimental data or to minimize the effects of the lack of isomorphism. Indeed,

(a) They do not take into account experimental errors in the observed (of native and of derivative) amplitudes.
(b) The heavy-atom substructure model is often incomplete (e.g. when sites with minor occupancy should still be discovered). Then errors in the modulus and in the phase of F_H arise.
(c) The hypothesis of perfect isomorphism does not hold in practice, e.g. equation (14.1) should be replaced by the more realistic relation,

$$F_d = F_P + F_H + |\mu_d| \exp(i\theta), \quad (14.9)$$

where μ_d is a complex vector arising from the imperfect isomorphism. It is generated by displacement of protein atoms and of solvent molecules close to the heavy atom-binding sites, and remains unknown up to when the native protein has been completely solved; it varies with the reflection index **h**.

As a consequence of the points (a) to (c), Fig. 14.1 is no longer realistic; the triangle defined by the vectors F_H, F_P, and F_d presents a lack of closure, indicated by ε in Fig. 14.5.

Clearly, the probabilistic approaches are more suitable than algebraic techniques to face such challenges; they were pioneered by Blow and Crick (1959), whose work strongly influenced subsequent scientific contributions. Any modern approach should today consider suitable joint probability distribution functions, in which careful treatment is made of various sources of error and of the correlations among them. We will below describe the logical steps involved in such approaches with a minimum of mathematical treatment. References to maximum likelihood procedures may be found in two contributions by Pannu et al. (2003) and Read (2003a).

In order to consider the correlations between F_d, F_P, and F_H, the joint probability distribution,

$$P(R_d, R_P, R_H, \phi_d, \phi_P, \phi_H) \quad (14.10)$$

should be studied. We notice that this distribution is very similar to that described in Section 9.A.1, where the structure factors of a target structure, of a model, and of a difference structure were studied; such a distribution turns into equation (9.A.1) if the role of the target structure is played by the derivative, the model by the heavy-atom substructure, and the difference structure by the native protein.

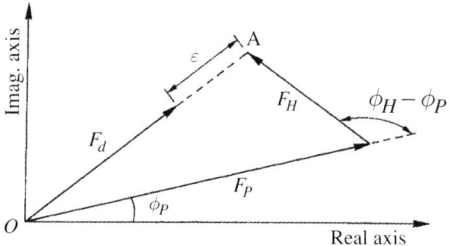

Fig. 14.5
Vector triangle emphasizing the lack of closure when lack of isomorphism occurs.

From the distribution (14.10), all the parameters necessary for phasing the native protein may be derived. We will analyse separately the logical operations necessary for treating Step 1 and Step 2, as defined in Section 14.3, without providing the mathematical detail.

Step 1 We are interested in estimating R_H, given the observations; say, given R_P and R_d (phases are unknown at this stage). Thus:

(i) We integrate equation (14.10) over the phases ϕ_d, ϕ_P, ϕ_H, so as to obtain the marginal distribution,

$$P(R_d, R_P, R_H). \tag{14.11}$$

(ii) From (14.11), the conditional distribution,

$$P(R_H | R_d, R_P) \tag{14.12}$$

is obtained by standard techniques.

(iii) From (14.12), the conditional average value

$$<R_H | R_d, R_P> \tag{14.13}$$

is derived (from now on denoted $<R_H>$ for simplicity).

$<R_H>$ is the value to submit to direct phasing procedures (traditional direct methods, charge flipping, *VLD*) for solving the heavy-atom substructure, and $<R_H^2>$ is the best coefficient for calculating the *heavy-atom substructure Patterson map*.

According to Giacovazzo et al. (2004),

$$<|R_H|> = \frac{1}{(1+<|\sigma_d|^2>)^{1/2}} \left[\frac{\pi}{4} <|\sigma_d|^2> + \frac{\Delta_{iso}^2}{\sum_H + <|\mu_d|^2>} \right]^{1/2} \tag{14.14}$$

and

$$<|R_H|^2> = \frac{1}{(1+<|\sigma_d|^2>)} \left[<|\sigma_d|^2> + \frac{\Delta_{iso}^2}{\sum_H + <|\mu_d|^2>} \right], \tag{14.15}$$

with

$$\sigma_{R_H} = \left[\left(1 - \frac{\pi}{4}\right) \frac{<|\sigma_d|^2>}{(1+<|\sigma_d|^2>)} \right]^{1/2}.$$

Thus, $<|R_H|^2>$ is estimated by equation (14.15) and σ_{R_H} is the standard deviation of the estimate. σ_{R_H} depends on $<|\sigma_d|^2> = <|\mu_d|^2> / \sum_H$ which is the pseudonormalized standard deviation connected to the global error on the derivative, including the lack of isomorphism.

Equation (14.15) suggests that if $<|\mu_d|^2> = 0$, the probabilistic approach confirms the Blow and Rossmann approximation; if $<|\mu_d|^2> \neq 0$, the Blow and Rossmann approximation should be affected by a systematic error, increasing with $<|\mu_d|^2>$.

Step 2 We assume in this step that R_H and ϕ_H are known (in other words, that Step 1 has been successfully accomplished). Then:

(i) From (14.10), the conditional distribution,

$$P(R_d, R_P, \phi_P, \phi_d | R_H, \phi_H) \qquad (14.16)$$

is calculated.

(ii) The assumption $\phi_P \approx \phi_d$ may be made; indeed, in the most interesting cases, $|F_H|$ is much smaller than $|F_d|$ and $|F_P|$, and therefore cannot cause a big difference between ϕ_d and ϕ_P. The distribution,

$$P(R_d, R_P, \phi_P | R_H, \phi_H) \qquad (14.17)$$

is then obtained.

(iii) The conditional distribution,

$$P(\phi_P | R_d, R_P, R_H, \phi_H) \qquad (14.18)$$

is derived.

(iv) From (14.18), the best ϕ_P value may be deduced.

According to Giacovazzo and Siliqi (2002),

$$P(\phi_P | R_d, R_P, R_H, \phi_H) = [2\pi I_0(G)]^{-1} \exp[G \cos(\phi_P - \phi_H)] \qquad (14.19)$$

is obtained, where G is the phase reliability factor given by,

$$G = \frac{2(|F_d| - |F_P|)|F_H|}{<|\mu_d|^2>} = \frac{2\Delta_{iso}|F_H|}{<|\mu_d|^2>} \qquad (14.20a)$$

In accordance with (14.19) and (14.20), the expected value of ϕ_P is ϕ_H if $\Delta_{iso} > 0$, and is $\phi_P + \pi$ if $\Delta_{iso} < 0$; the larger the product $\frac{|\Delta_{iso} F_H|}{<|\mu_d|^2>}$, the more accurate the phase estimate will be. It may be useful to rewrite equation (14.20a) in terms of normalized structure factors (with respect to the protein scattering power),

$$G = \frac{2(R_d - R_P)R_H}{\sigma^2}, \qquad (14.20b)$$

where

$$\sigma^2 = <|\mu_d|^2> / \sum_P.$$

Let us now suppose that the phases have been assigned. An electron density map with observed R_P coefficients and best ϕ_P values cannot provide an accurate model of the native protein, unless some *phase refinement* is made. Rossmann (1960) suggested refinement of the substructure parameters by minimizing

$$S = \sum w[|F_{Hobs}| - |F_{Hcalc}|]^2.$$

While $|F_{Hcalc}|$ may be computed from the substructure parameters, $|F_{Hobs}|$ is only available experimentally for centric reflections; this detail makes the procedure inefficient. A better method (Dickerson et al., 1961; Muirhead et al., 1967; Dickerson et al., 1968; Terwilliger and Eisenberg, 1983) involves minimization of the function,

$$S = \sum w\varepsilon^2, \qquad (14.21)$$

where

w is a weight associated with each reflection;
$\varepsilon = (|F_d|_{obs} - |F_d|_{calc})$, this is called the *lack of closure error*;
$|F_d|_{obs}$ is the measured amplitude of the derivative structure factor;
$|F_d|_{calc} = |F_P + F_H|$ is the calculated derivative structure factor, a function of the moduli and of the phases of the native and the derivative structures.

Once a set of approximated protein phases are estimated via the distribution (14.19), the moduli $|F_{d_{calc}}|$ may be calculated and then (14.21) may be minimized, so refining positions, occupancies, and temperature factors of the heavy atoms. The refined parameters are then used to calculate a new set of protein phases through equation (14.19). The process goes on via alternate cycles of parameter refinement and native phase calculation, until convergence is reached.

The progress of phase refinement is usually monitored using three factors:

$$R_{Cullis} = \sum ||F_{dobs} \pm F_{Pobs}| - |F_{Hcalc}|| / \sum |F_{dcalc} \pm |F_{Pobs}||,$$

which was originally defined for centric reflections, but it is also a useful statistical figure for acentric reflections.

$$R_{Kraut} = \sum ||F_{dobs}| - |F_{dcalc}|| / \sum |F_{dobs}|,$$

which is equal to the sum of closure residuals divided by the sum of derivative amplitudes.

$$PhP = \sum |F_H| / \sum ||F_{dobs}| - |F_{dcalc}||,$$

defined as the sum of heavy-atom contributions divided by the sum of closure residuals.

Acceptable values of R_{Cullis} are between 0.4 and 0.6, while PhP, also called the *phasing power*, is expected to be larger than unity.

The same considerations on handedness hold for the probabilistic approach to SIR as are described in Section 14.3 for the algebraic approach.

14.7 The probabilistic approach for the MIR case

In Section 14.6 we have described the two steps necessary for phasing the native protein in the SIR case via a probabilistic approach. The procedure must certainly be modified if more derivatives are available, but there are some steps in common with the SIR case. For example, the diffraction amplitudes of the native and of each derivative are settled on their absolute scales, the corresponding structure factors are normalized, and the heavy-atom substructures are found and refined. These steps were described in the preceding sections, and we will assume that the corresponding procedures are applied to each derivative.

At this stage, application of equations (14.19) and (14.20) to each of the n derivatives provides the set $\{\phi_P\}_i$, $i = 1, \ldots, n$, each of which is an estimate of the native protein phases. The ith and jth estimates may be quite unlike each other, because the two derivatives may have different isomorphism, data resolution, and measurement errors. The most general way for handling the MIR case when the substructures are known is to calculate the joint probability distribution function (Giacovazzo and Siliqi, 2002),

$$P(E_P, E_{d_1}, \ldots, E_{d_n} | E_{H_1}, \ldots, E_{H_n}), \qquad (14.22)$$

from which the marginal distribution,

$$P(\phi_P | \ldots) = P(\phi_P | R_P, R_{d_1}, \ldots, R_{d_n}, E_{H_1}, \ldots, E_{H_n}) \qquad (14.23)$$

may be obtained. Distribution (14.23) is of a von Mises type:

$$P(\phi_P | \ldots) = [2\pi I_0(\alpha_P)]^{-1} \exp[\alpha_P \cos(\phi_P - \theta_P)], \qquad (14.24)$$

where

$$\tan \theta_P = \frac{\sum_{j=1}^{n} G_j \sin \phi_{H_j}}{\sum_{j=1}^{n} G_j \cos \phi_{H_j}} = \frac{T}{B}, \qquad (14.25)$$

$$G_j = \frac{2(R_{d_j} - R_{P_j}) R_{H_j}}{\sigma_j^2}, \qquad (14.26)$$

$$\alpha_P = (T^2 + B^2)^{1/2} \qquad (14.27)$$

θ_P is the most probable value of ϕ_P and α_P is its reliability parameter.

In the distributions (14.22) and (14.23) the substructures are assumed to be known; in practice, the substructure models available at the end of Step 1 may be poor, and consequently sets $\{\phi_P\}_i$ obtained by application of equation (14.25) may also be a poor approximation of the true native phases. It is then rewarding to spend some additional computing time to improve the substructure models and then, as a consequence, the native phase estimates. This task may be performed according to the following procedure:

(a) The '*best*' derivative is selected, which is the one for which the SIR phasing process is expected to provide the minimum phase error. Appropriate figures of merit should be used for its recognition; Giacovazzo et al. (2004) proposed the figure,

$$PON = (PhP/R_{Cullis}) \cdot Nr,$$

where $Nr = NREFL/NREFL_{max}$ is the ratio between the number of observed amplitudes for the current derivative and the maximum number of observed amplitudes among the various derivatives.

(b) The set of phases $\{\phi_{Pbest}\}$, obtained from the best derivative at the end of the refinement step, is employed to calculate, for each derivative, the differential Fourier synthesis with coefficient $(|F_{d_i}| - |F_P|) \exp(i\phi_{Pbest})$, where subscript *i* refers to the current derivative. The rationale is that, even if the *n* substructures are uncorrelated, the sets $\{\phi_P\}_i, i = 1, \ldots, n$ are expected to be correlated with each other, because each $\{\phi_P\}_i$ is an estimate of the same set of phases, the native protein phases. Thus, if we associate $\{\phi_{Pbest}\}$ with the *i*th Δ_{iso} set, we should obtain the map which better approximates the *i*th true substructure. As observed in Section 14.4 for the algebraic approach to the MIR case, the various substructures so obtained refer to the same origin.

The new model substructure may be refined by minimizing (14.21).

(c) The new sets of phases $\{\phi_H\}_i$, $i = 1, \ldots, n$ are combined via relationships (14.24)–(14.27).
(d) Steps (a) to (c) may be cyclically repeated.
(e) EDM procedures are applied to improve and to extend the phase information up to native resolution.

The same considerations on the handedness hold for the probabilistic approach to MIR as are described in Section 14.4 for the algebraic approach.

14.8 Applications

Many fully automated programs exist for protein crystal structure solution via SIR-MIR techniques. We quote *SHELXD* (Schneider and Sheldrick, 2002), *SOLVE* (Terwilliger and Berendzen, 1999), *CNS* (Brünger et al., 1998). Programs explicitly based on a probabilistic background include *IL MILIONE* (Burla et al., 2007c) and *PHENIX* (Adams et al., 2002). The heavy-atom substructures may be found by Patterson techniques or by direct methods, according to personal preference. Maximum likelihood approaches for the refinement of substructures (and more generally for structure refinement) are very popular. The reader is referred to the above mentioned papers by Pannu et al. (2003) and Read (2003b) for theoretical aspects, and to the programs *SHARP* (Bricogne et al., 2003), *MLPHARE* (Otwinowski, 1991a), *SOLVE* (Terwilliger and Berendsen, 2001), and *PHENIX* (Adams et al., 2002).

We will now schematize some simple SIR-MIR examples for an easier understanding of the machinery. In Table 14.1, some relevant data for one SIR and three MIR cases are reported. The reader should notice that at this initial stage, the nature of the heavy atoms is known while their number per unit cell

Table 14.1 Essential data for one SIR case (*APP*) and three MIR cases (*BPO*, *M-FABP*, *NOX*). For the native protein, the table provides information on the space group (*S.G.*), the number of non-hydrogen atoms in the asymmetric unit (*NASYM*), the protein data resolution (*RES$_P$*), and the number of measured reflections. For each derivative, the information concerns the nature of the heavy atoms (*H.A.*), the data resolution (*RES$_d$*), and the number of measured reflections (*NREFL$_d$*)

CODE	S.G.	native protein			derivatives		
		NASYM	RES$_P$ (Å)	NREFL$_P$	H.A.	RES$_d$ (Å)	NREFL$_d$
APP	C2	302	0.99	17058	Hg	2.00	2108
BPO	P2$_1$3	4529	2.35	23956	Au	2.80	15741
					Pt	2.76	17433
M-FABP	P2$_1$2$_1$2$_1$	1101	2.14	7595	Hg	2.18	7125
					Pt	2.15	6586
NOX	P4$_1$2$_1$2	1689	2.26	9400	Pt$_1$	2.26	9068
					Hg	2.59	5425
					Au	2.38	7299
					Pt$_2$	2.37	6752

Table 14.2 For each SIR-MIR case quoted in Table 14.1, this table gives the initially estimated scattering power of the heavy-atom substructure ($\sum_H / \sum_P$), the figure of merit (PON) for estimating which is the best derivative, the final correlation between the electron density map obtained at the end of the SIR-MIR procedure, and the published map

PDB code	H. A.	$\sum_H / \sum_P$	PON	CORR
BPO	Au	0.11	1.22	0.78
	Pt	0.06	0.61	
MFABP	Hg	0.06	1.54	0.74
	Pt	0.03	1.35	
NOX	Pt_1	0.03	1.34	0.74
	Hg	0.22	0.94	
	Au	0.18	1.19	
	Pt_2	0.03	0.90	

and the corresponding site occupancies are unknown. The derivative resolution is often worse than that of the native protein.

In Table 14.2, we report for each case:

(a) The first estimate of the scattering power of the heavy-atom substructure via the probabilistic procedure described in Appendix 14.C. This suggests the capacity of the heavy atoms to bind protein residues, but still does not state their occupancy (which may be established only at the refinement stage).
(b) The PON value (see Section 14.7) for each derivative. It may be assumed that the largest PON characterizes the best derivative, that is, the one with the minimum lack of isomorphism. As described in Section 14.7, the best derivative is used for deriving the substructures of the other derivatives via the difference Fourier syntheses, $(|F_{d_i}| - |F_P|) \exp(i\phi_{Pbest})$.
(c) CORR is the correlation factor between the electron density of the refined protein model (that obtained at the end of the cyclic procedure described in Section 14.7) and the published structure. In spite of full automation of the process, the final protein model is suitable for easy interpretation.

APPENDIX 14.A THE SIR CASE FOR CENTRIC REFLECTIONS

Let us consider equation (14.1), say

$$F_d = F_P + F_H.$$

We want to guess the modulus and phase of F_H for centric reflections, given $|F_d|$ and $|F_P|$. Let us suppose:

(a) $|F_H| > 0$;
(b) $|F_H| << |F_P|$ and $|F_H| << |F_d|$.

Condition (b) is justified by the small scattering power of the heavy-atom substructure with respect to scattering power of the native or of the derivative.

For centric reflections, ϕ_P, ϕ_d, ϕ_H must be collinear because all three phases are submitted to the same symmetry restrictions. This result, together with conditions (a) and (b), imply that

$$\phi_P = \phi_d \text{ and } |F_H| = ||F_d| - |F_P||;$$

the extraordinary case, $\phi_P = \phi_d + \pi$, is excluded if condition (b) holds.

In Fig. 14.A.1a,b, two cases are shown, $|F_d| > |F_P|$ and $|F_d| < |F_P|$. In the first case, $\phi_H = \phi_P$, in the second, $\phi_H = \phi_P + \pi$.

The extraordinary case in which $\phi_P = \phi_d + \pi$ is shown in Fig. 14.A.1c; in this case, called *overcross*, $|F_H| = ||F_d| + |F_P||$ and $\phi_H = \phi_P + \pi$. There are very few cases in which the overcross occurs; necessarily, $|F_d|$ and $|F_P|$ should be very small, and this condition subtracts interest in the case.

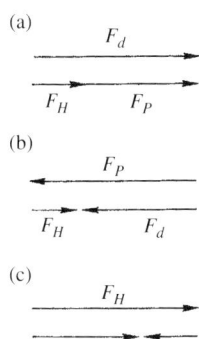

Fig. 14.A.1
Centric reflections: (a) if $|F_p| < |F_d|$, then $\phi_p = \phi_H$; (b) if $|F_p| > |F_d|$, then $\phi_p = \phi_H + \pi$; (c) 'cross-over case'. This occurs when $|F_H|$ is larger than $|F_d|$ and $|F_P|$.

APPENDIX 14.B THE SIR CASE: THE ONE-STEP PROCEDURE

In Chapters 6 and 9 we showed that traditional direct methods, as well as charge flipping and *VLD*, are unable to solve ab initio protein structures unless of limited size and with atomic or quasi-atomic resolution. Size limits may be extended to some thousands of non-hydrogen atoms in the asymmetric unit, and resolution limits may be expanded to about 2 Å, if the structure contains heavy atoms and Patterson techniques are used (see Section 10.4).

If one isomorphous derivative is available, its diffraction data constitute a source of supplementary information which may extend the limits of traditional direct methods. The integration of DM with SIR techniques was introduced by Hauptman (1982a) and revisited by Giacovazzo (1988). Both authors studied the joint probability distribution,

$$P(E_{P1}, E_{d1}, E_{P2}, E_{d2}, E_{P3}, E_{d3}), \quad (14.\text{B}.1)$$

where the subscripts 1, 2, and 3 replace **h**, **k**, and **-h-k**, respectively. From (14.B.1), the conditional distribution,

$$P(\Phi_P | R_{P1}, R_{d1}, R_{P2}, R_{d2}, R_{P3}, R_{d3}) \quad (14.\text{B}.2)$$

was derived, where $\Phi_P = \phi_{p1} + \phi_{p2} + \phi_{p3}$. According to Giacovazzo,

$$P(\Phi_P | \ldots) \approx [2\pi I_0(G)]^{-1} \exp(G \cos \Phi_P), \quad (14.\text{B}.3)$$

where

$$G = \frac{2 R_{P1} R_{P2} R_{P3}}{N_{Peq}^{-1/2}} + \frac{2 \Delta_1 \Delta_2 \Delta_3}{N_{Heq}^{-1/2}} \quad (14.\text{B}.4)$$

$$\Delta = \Delta_{iso} / {\sum}_H^{1/2}.$$

N_{Peq} and N_{Heq} are defined by equation (5.5) for the native and derivative structures, respectively. According to (14.B.4), the concentration parameter G is the sum of two contributions: the Cochran term, calculated for the native, plus an additional term which depends on the pseudonormalized difference Δ. Since the number of heavy atoms in the unit cell is much smaller than the number of protein atoms, $N_{Peq} \gg N_{Heq}$. The consequence is that for usual proteins, the Cochran contribution to G is negligible with respect to the second. This is quite a relevant result; it is shown that supplementary prior knowledge of the derivative diffraction data reduces the complexity of the phasing problem from order N_{Peq} to order N_{Heq}.

The expected sign of the triplet cosine mainly depends on the sign of the product $\Delta_1 \Delta_2 \Delta_3$; a positive value of this product indicates positive cosine values. Negative values of $\Delta_1 \Delta_2 \Delta_3$ are also frequent and suggest Φ_P values close to π.

According to equation (14.B.3), the native structure phases may be found as in traditional DM (see Chapter 6); i.e. the reflections with the largest values of $|\Delta|$ are selected, triplet invariants are found among them, random phases are generated and submitted to the tangent formula, and the most promising solutions are selected via suitable figures of merit.

The above approach allows us to solve the native protein structure in one step; preliminary solution of the heavy-atom substructure is no longer necessary. Even if the one-step approach is highly appealing, the two-step approach seems more effective, because it confronts the phasing difficulties gradually.

An interesting exercise is to establish a relation between the two-step SIR approach and the above one-step procedure. Let us suppose that $\Delta_1 > 0$, $\Delta_2 > 0$ and $\Delta_3 > 0$; in this case, the distribution (14.B.3) suggests the relation,

$$\Phi_P \approx 0. \qquad (14.B.5)$$

Classical two-step SIR techniques suggest the following relationships:

$$\phi_{P_1} \approx \phi_{H_1}, \ \phi_{P_2} \approx \phi_{H_2}, \ \phi_{P_3} \approx \phi_{H_3},$$

which, summed, give

$$\Phi_P \approx \Phi_H, \qquad (14.B.6)$$

where $\Phi_H = \phi_{H_1} + \phi_{H_2} + \phi_{H_3}$.

Expressions (14.B.6), provided by the two-step SIR relations, and (14.B.5), provided by traditional direct methods, coincide if $\Phi_H \approx 0$. This will occur if $|\Delta_1|, |\Delta_2|, |\Delta_3|$ are sufficiently large and if the number of heavy atoms in the unit cell is small (as is usual in SIR cases).

Suppose now that $\Delta_1 > 0$, $\Delta_2 > 0$ and $\Delta_3 < 0$. Then, distribution (14.B.3) suggests

$$\Phi_P \approx \pi. \qquad (14.B.7)$$

Two-step SIR relations provide the following indications:

$$\phi_{P_1} \approx \phi_{H_1}, \phi_{P_2} \approx \phi_{H_2}, \phi_{P_3} \approx \phi_{H_3} + \pi$$

which, summed, give

$$\Phi_P \approx \Phi_H + \pi. \qquad (14.B.8)$$

Since $\Phi_H \approx 0$, again (14.B.7) and (14.B.8) provide similar phase indications.

APPENDIX 14.C ABOUT METHODS FOR ESTIMATING THE SCATTERING POWER OF THE HEAVY-ATOM SUBSTRUCTURE

We will describe two simple procedures for estimating the scattering power of a heavy-atom substructure. The first has an algebraic basis and is due to Crick and Magdoff (1956); the second has a probabilistic background and was suggested by Giacovazzo et al. (2002).

The algebraic relation. Let us suppose that one or more heavy atoms have been added to the native protein, so giving rise to a perfect isomorph derivative. Let $<I_P>$ be the average intensity for the protein at a given $\sin\theta/\lambda$, $<I_H>$ the corresponding average intensity for the heavy atom structure, and $<I_d>$ the value for the derivative. An estimate of the average relative change in intensity is given by the ratio,

$$<\Delta> = \frac{[<(I_d - I_P)^2>]^{1/2}}{<I_P>}. \qquad (14.C.1)$$

Firstly, let us consider the acentric reflections. In accordance with equation (14.C.1),

$$I_d = |F_P + F_H|^2 = |F_P|^2 + |F_H|^2 - 2|F_P F_H|\cos(\phi_P - \phi_H),$$

from which

$$I_d - I_P = |F_H|^2 - 2|F_P F_H|\cos(\phi_P - \phi_H). \qquad (14.C.2)$$

Introducing (14.C.2) into (14.C.1) gives,

$$<\Delta> = \frac{\{<|F_H|^4> + <4|F_P F_H|^2 \cos^2(\phi_P - \phi_H)> - <4|F_P||F_H|^3 \cos(\phi_P - \phi_H)>\}^{1/2}}{<I_P>}$$

Since F_P and F_H may be considered to be uncorrelated, $<\cos(\phi_P - \phi_H)>$ is expected to be close to zero and $<\cos^2(\phi_P - \phi_H)>$ close to 1/2. Then,

$$<\Delta> \approx \frac{(<|F_H|^4> + 2<|F_P|^2><|F_H|^2>)^{1/2}}{<|F_P|^4>}. \qquad (14.C.3)$$

Since $<|F_H|^4>$ is very much smaller than $2<|F_P|^2><|F_H|^2>$, we can approximate (14.C.3) by,

$$<\Delta> \approx \sqrt{2}\sqrt{\frac{<I_H>}{<I_P>}}, \qquad (14.C.4)$$

which is the Crick and Magdoff relation. Let us apply (14.C.4) to two proteins of quite different size: the first with 1000 atoms per molecule, the second 20 000 atoms per molecule. On assuming $Z_j = 7$ for the average protein atom, one mercury atom per molecule, we have at $\sin\theta/\lambda = 0$:

1. For the first protein,

$$<I_P> = 1000 \times 49 = 49000$$
$$<I_H> = 1 \times 6400 = 6400$$
$$<\Delta> = 0.51.$$

2. For the second protein,

$$<I_P> = 20000 \times 49 = 980000$$
$$<I_H> = 1 \times 6400 = 6400$$
$$<\Delta> = 0.11.$$

We see that the average change in intensity is quite detectable, even for large proteins. It is also clear that a partial occupancy for the mercury atom gives less favourable values of $<\Delta>$.

It is easy to show that for centric reflections,

$$<\Delta> = 2\sqrt{\frac{<I_H>}{<I_P>}} \qquad (14.C.5)$$

If we compare (14.C.4) with (14.C.5), we see that $<\Delta>$ is larger for restricted phases than for general reflections.

Since $<I_H>/<I_P> \approx \sum_H / \sum_P$, Equation (14.C.4) (and obviously equation (14.C.5)), may also be applied to estimate the scattering power of the heavy-atom substructure: i.e.

$$\sum_H / \sum_P \approx \frac{<\Delta>^2}{4}.$$

The probabilistic relation. In Section 7.2, we illustrated the joint probability distribution $P(E, E_p)$, where E was the normalized structure factor of the target structure, and E_p was the normalized structure factor of a model structure. The same distribution (say equation (7.3)) is valid if E and E_p are replaced by structure factors E_P and E_d of the native protein and of the derivative, respectively.

From $P(E_P, E_d)$, the conditional distribution

$$P(|\Delta|) = c^{-1} \exp(-|\Delta|/c) \qquad (14.C.6)$$

may be derived, where

$$\Delta = R_d'^2 - R_P^2, \quad c = \sum_H / \sum_d, \quad R_d' = |F_d|/\left(\sum_P\right)^{1/2}.$$

Distribution (14.C.6) only depends on the scattering power of the unknown heavy-atom substructure ($\sum_d$ is a known parameter). Therefore, the best c value is that for which the theoretical distribution (14.C.6) fits the experimental $|\Delta|$ distribution. Finding c is equivalent to estimating the scattering power of the heavy-atom substructure.

Anomalous dispersion techniques

15

15.1 Introduction

The term anomalous scattering originates from the first research on light dispersion in transparent materials. It was found that, in general, the index of refraction increases when the wavelength decreases (this was considered to be *normal*). It was also found that, close to the absorption edges, the refractive index shows a negative slope, and this effect was called *anomalous*.

Today, it is clear that anomalous dispersion is a resonance effect. Indeed, atomic electrons may be considered to be oscillators with natural frequencies; they are bound to the nucleus by forces which depend on the atomic field strength and on the quantum state of the electron. If the frequency of the primary beam is near to some of these natural frequencies, resonance will take place (the concept of dispersion involves a change of property with frequency). The scattering is then called *anomalous*, it occurs in correspondence with the so-called *absorption edges* of a chemical element, and is expressed analytically via the complex quantity,

$$f = f_0 + \Delta f' + if'', \qquad (15.1)$$

where f_0 is the scattering factor of the atom in the absence of anomalous scattering. $\Delta f'$ and f'' (with $f'' > 0$) are called the *real and imaginary dispersion corrections*.

f'' is proportional to the absorption coefficient of the atom, μ_λ, at the given X-ray energy E_λ:

$$f''_\lambda = \left(mc/4\pi e^2 \hbar\right) E_\lambda \mu_\lambda,$$

where m and e are the electronic mass and charge respectively, λ is the wavelength, c is the speed of light, and $h = 2\pi\hbar$, the Planck constant.

$\Delta f'$ may be obtained from an absorption scan via the Kramers–Kronig transform relating the real to the imaginary component:

$$\Delta f'(\lambda) = \frac{2}{\pi} \int_0^\infty E' \frac{f''(E')}{E^2 - E'^2} dE'.$$

An important question is whether $\Delta f'$ and f'' vary with diffraction angle. Some theoretical treatments suggest changes of a few percent with $\sin\theta/\lambda$, but no rigorous experimental check has been described; therefore, in most of the applications (and also in this book), $\Delta f'$ and f'' are considered to be constant. As a consequence, the relative modification of the scattering process due to anomalous dispersion is stronger at high $\sin\theta/\lambda$ values, where f_0 is smaller, while $\Delta f'$ and f'' remain constant.

For most substances, at most X-ray wavelengths from conventional sources, dispersion corrections are rather small. Calculated values for $CrK_\alpha (\lambda = 2.229\,\text{Å})$, $CuK_\alpha (\lambda = 1.542\,\text{Å})$, and $MoK_\alpha (\lambda = 0.7107\,\text{Å})$ are listed in the *International Tables for Crytallography*. Only in some special cases can ordinary X-ray sources generate relevant dispersion effects. For example, the following dispersion corrections are calculated for *holmium*, which has the L_3 absorption edge, (~ 1.5368 Å) close to CuK_α radiation:

$$Cu\,K_{\alpha_1}(\lambda = 1.5406\,\text{Å}); \quad \Delta f' = -15.41, \quad f'' = 3.70$$

$$Cu\,K_{\alpha_2}(\lambda = 1.5444\,\text{Å}); \quad \Delta f' = -14.09, \quad f'' = 3.72.$$

Larger anomalous effects are obtained by using *synchrotron radiation*; from its intense continuous spectrum, specific wavelengths may be selected with high precision, in order to provoke, in several cases, exceptionally large anomalous scattering.

Why do anomalous dispersion effects help to solve protein structures (Peederman and Bijvoet, 1956)? We will see in the following that they simulate isomorphism and more derivatives are simulated when multiwave techniques are used. A pioneering study by Hoppe and Jakubowski (1975) on erythrocruorin, an iron-containing protein, showed the feasibility of this method. These authors used NiK_α (1.66 Å) and CoK_α (1.79 Å) radiation to vary $\Delta f'$ and f'' sizes of Fe around its K edge (1.74 Å) and obtained phase values with a mean phase error of 50°.

What are the limits and the advantages of anomalous dispersion with respect to isomorphous derivative techniques? We notice two disadvantages, in (a) and (b) in the following, and a big advantage in (c):

(a) The signal provoked by anomalous scattering is inferior to that usually obtained by isomorphous replacement. The case of praseodynium (Templeton et al., 1980), for which $\Delta f' = -26, 2f'' = 55$ and of gadolinium (Templeton et al., 1982), for which $\Delta f' = 31.9, 2f'' = 62.4$ are exceptional. Therefore, the use of anomalous effects requires high measurement accuracy.
(b) Since the modulus of f'' is proportional to the absorption coefficient, the corresponding correction for absorption should be carefully calculated. Luckily, heavy atoms usually have absorption edges in the short wavelength range (say $0.6\,\text{Å} < \lambda < 1.1\,\text{Å}$), for which absorption is greatly reduced.
(c) The isomorphism of the derivatives with respect to the protein is never ideal. We saw in Chapter 14 how challenging bad derivatives may be for the success of the phasing process. If anomalous dispersion techniques are used, the scattering structure coincides with the target structure, and therefore no lack of isomorphism occurs.

Furthermore, in case of MAD, the anomalous scattering substructure does not change with the wavelength (in the MIR approach, the heavy-atom substructure changes by changing the heavy atom). As a result, the substructure is overdetermined by MAD data.

The above considerations suggest that we should deal with different cases:

1. The *SAD* (single wavelength anomalous scattering) case;
2. The *SIRAS* (single isomorphous replacement combined with anomalous scattering) case. Typically, protein and heavy-atom derivative data are simultaneously available, with heavy atoms as anomalous scatterers;
3. The MAD (multiple wavelengths technique) case;
4. The MIRAS (multiple isomorphous replacement combined with anomalous scattering) case.

15.2 Violation of the Friedel law as basis of the phasing method

Suppose that an n.cs. crystal contains N_P non-H atoms in the unit cell and that all of them are anomalous scatterers. Usually, the number of efficient anomalous scatterers is a very small fraction of the scatterers in the unit cell; to include this hypothesis in our mathematical treatment, the reader may set to zero the anomalous scattering of the atoms the contribution of which they want to neglect. In the following, for shortness, we will indicate $F_\mathbf{h}$ and $F_{-\mathbf{h}}$ by $F^+ = |F^+|\exp(i\phi^+)$ and $F^- = |F^-|\exp(i\phi^-)$, respectively. Then,

$$F^+ = \sum_{j=1}^{N_P} \left(f_{0j} + \Delta f'_j + if''_j\right) \exp(2\pi i \mathbf{h} \cdot \mathbf{r}_j) \qquad (15.2)$$
$$= F_0^+ + \Delta F'^+ + F''^+ = F'^+ + F''^+,$$

where

$F_0^+ = \sum_{j=1}^{N_P} f_{0j} \exp(2\pi i \mathbf{h} \cdot \mathbf{r}_j) = |F_0^+| \exp(i\phi_0^+),$

$\Delta F'^+ = \sum_{j=1}^{N_P} \Delta f'_j \exp(2\pi i \mathbf{h} \cdot \mathbf{r}_j),$

$F''^+ = \sum_{j=1}^{N_P} if''_j \exp(2\pi i \mathbf{h} \cdot \mathbf{r}_j) = \sum_{j=1}^{N_P} f''_j \exp[i(2\pi \mathbf{h} \cdot \mathbf{r}_j + \pi/2)] = |F''^+| \exp(i\phi''^+),$

$F'^+ = F_0^+ + \Delta F'^+ = \sum_{j=1}^{N_P} (f_{0j} + \Delta f'_j) \exp(2\pi i \mathbf{h} \cdot \mathbf{r}_j) = |F'^+| \exp(i\phi'^+).$

Analogously:

$$F^- = \sum_{j=1}^{N_P} \left(f_{0j} + \Delta f'_j + if''_j\right) \exp(-2\pi i \mathbf{h} \cdot \mathbf{r}_j) \qquad (15.3)$$
$$= F_0^- + \Delta F'^- + F''^- = F'^- + F''^-,$$

where

$F_0^- = \sum_{j=1}^{N_P} f_{0j} \exp(-2\pi i \mathbf{h} \cdot \mathbf{r}_j) = |F_0^-| \exp(i\phi_0^-),$

$\Delta F'^- = \sum_{j=1}^{N_P} \Delta f'_j \exp(-2\pi i \mathbf{h} \cdot \mathbf{r}_j),$

$F''^- = \sum_{j=1}^{N_P} if''_j \exp(-2\pi i \mathbf{h} \cdot \mathbf{r}_j) = \sum_{j=1}^{N_P} f''_j \exp[i(-2\pi \mathbf{h} \cdot \mathbf{r}_j + \pi/2)] = |F''^-| \exp(i\phi''^-),$

$F'^- = F_0^- + \Delta F'^- = \sum_{j=1}^{N_P} (f_{0j} + \Delta f'_j) \exp(-2\pi i \mathbf{h} \cdot \mathbf{r}_j) = |F'^-| \exp(i\phi'^-).$

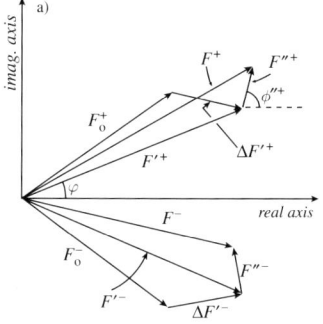

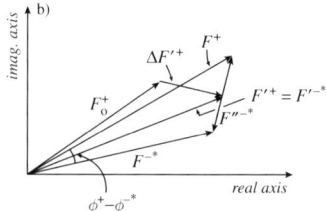

Fig. 15.1
(a) Relation between F^+ and F^- when anomalous dispersion is present. (b) Relation between F^+ and F^{-*} when anomalous dispersion is present.

F^+ and its component vectors F_0^+, $\Delta F'^+$, F'^+, F''^+, as well as F^- and its component vectors F_0^-, $\Delta F'^-$, F'^- and F''^-, are shown in Fig. 15.1a.

The reader will notice that F_0^+ and F_0^- components also contain the normal scattering contribution of the anomalous scatterers and that they are related by the Friedel law. The Friedel law also holds for the pairs $(\Delta F'^+, \Delta F'^-)$, (F'^+, F'^-). In detail,

$$|F_0^+| = |F_0^-|, \quad \phi_0^- = -\phi_0^+,$$
$$|F'^+| = |F'^-|, \quad \phi'^- = -\phi'.$$

The Friedel law does not hold for the pair F''^+, F''^-; from definitions, the relations

$$|F''^+| = |F''^-|, \quad \phi''^- = \pi - \phi''^+$$

are easily obtained (see Fig. 15.2). Therefore, while ϕ'^+ and ϕ'^- are symmetrical with respect to the zero angle, ϕ''^+ and ϕ''^- are symmetrical with respect to $\pi/2$.

It may be useful to remember another vectorial relationship. In general conditions, $\Delta F'^+$ and F''^+ are not perpendicular. They are perpendicular only if the anomalous scattering arises from the same atomic species. Then,

$$\Delta F'^+ = \Delta f' \sum_{j=1}^{N_P} \exp(2\pi i \mathbf{h} \cdot \mathbf{r}_j),$$

$$F''^+ = if'' \sum_{j=1}^{N_P} \exp(2\pi i \mathbf{h} \cdot \mathbf{r}_j) = f'' \sum_{j=1}^{N_P} \exp[i(2\pi \mathbf{h} \cdot \mathbf{r}_j + \pi/2)] \perp \Delta F'^+.$$

The relation between F^+ and F^- is more clearly understood if we compare F^+ with F^{-*} (the star indicates the complex conjugate), as in Fig. 15.1b. From definitions, the following phase relations hold:

$$\phi'^{-*} = \phi'^+, \quad \phi''^{-*} = -\phi''^- = \phi''^+ + \pi. \tag{15.4}$$

Equivalently, in vectorial form,

$$F''^{-*} = -F''^+ \text{ and } F''^{+*} = -F''^-. \tag{15.5}$$

Because of (15.5), from now on we will denote the moduli $|F'^+|, |F'^-|$ by $|F'|$ and the moduli $|F''^+|, |F''^-|$ by $|F''|$.

From Fig. 15.1a, it is very easy to derive the following relations:

$$|F^+|^2 = |F'|^2 + |F''|^2 + 2|F'F''|\cos(\phi''^+ - \phi'^+) \tag{15.6a}$$

and

$$|F^-|^2 = |F'|^2 + |F''|^2 + 2|F'F''|\cos(\phi''^- - \phi'^-).$$

In accordance with equation (15.4) and Fig. 15.2b, the above equation may be modified to,

$$|F^-|^2 = |F'|^2 + |F''|^2 - 2|F'F''|\cos(\phi''^+ - \phi'^+). \tag{15.6b}$$

From (15.6a) and (15.6b),

$$\Delta I = |F^+|^2 - |F^-|^2 = 4|F'F''|\cos(\phi''^+ - \phi'^+). \tag{15.7}$$

Violation of the Friedel law as basis of the phasing method

and

$$\frac{|F^+|^2 + |F^-|^2}{2} = |F'|^2 + |F''|^2 \quad (15.8)$$

are easily obtained.

Relation (15.7) suggests that $|F_\mathbf{h}| = |F_{-\mathbf{h}}|$ is no longer valid; in other words,

in n.cs. space groups the Friedel law is not satisfied in the presence of anomalous dispersion.

Figure 15.3 suggests that, if F'^+ is perpendicular to F''^+, the Friedel law concerning the moduli $|F_\mathbf{h}|$ and $|F_{-\mathbf{h}}|$ is satisfied, even if the structure is n.cs.; but this happens only by chance. On the contrary, the Friedel law:

(i) is satisfied in an n.cs. crystal if it is composed entirely of the same anomalous scatterer;
(ii) is systematically satisfied in cs. structures. Indeed, if the structure is cs. (see Fig. 15.4), then $F'^+ = F'^- = F'^{-*}$ is a real value, while $F''^+ \equiv F''^-$ is an imaginary value; indeed, $\phi''^+ = \phi''^- = \pm\pi/2$.

Accordingly, $F^+ \equiv F^-$, and the following statement arises:

in cs. space groups the Friedel law is satisfied even in the presence of anomalous scattering.

The Friedel law is also satisfied for phase restricted reflections in n.cs. space groups.

Let us now consider how anomalous dispersion modifies the reciprocal space symmetry. In the absence of anomalous dispersion, the number of reflections with equal (because of the symmetry) amplitude is fixed by the Laue group. If an anomalous signal is present in an n.cs. space group of order m, the following relations hold:

$$|F_\mathbf{h}| = |F_{\mathbf{hR}_2}| = \ldots\ldots = |F_{\mathbf{hR}_m}|$$

and

$$|F_{-\mathbf{h}}| = |F_{-\mathbf{hR}_2}| = \ldots\ldots = |F_{-\mathbf{hR}_m}| \quad \text{with} \quad |F_{-\mathbf{h}}| \neq |F_{-\mathbf{h}}|.$$

For example, for C2,

$$|F_{hkl}| = |F_{\bar{h}k\bar{l}}| \neq |F_{h\bar{k}l}| = |F_{\bar{h}\bar{k}\bar{l}}|.$$

It may be concluded that:

for normal scattering, the equivalence of reflections is fixed by the Laue group symmetry, while, when anomalous scattering occurs, the equivalence agrees with the point group symmetry (Ramaseshan, 1963).

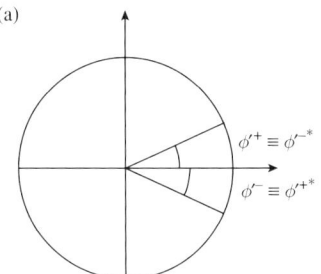

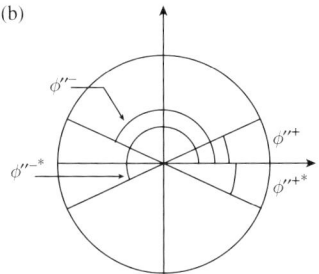

Fig. 15.2

Phase relationships between: (a) ϕ'^+ and ϕ'^-; (b) ϕ''^+ and ϕ''^-. ϕ'^+ and ϕ'^- are the phases of the reflections F'^+ and F'^-; ϕ'^{+*} and ϕ'^{-*} are the phases of the reflections F'^{+*} and F'^{-*}, complex conjugates of F'^+ and F'^-, respectively. Since F'^+ and F'^- obey the Friedel relationship, then $\phi'^+ = \phi'^{-*}$ and $\phi'^- = \phi'^{+*}$. F''^+ and F''^- do not obey the Friedel relationship: we have $\phi''^+ = \phi''^{-*} + \pi$ and $\phi''^- = \phi''^{+*} + \pi$.

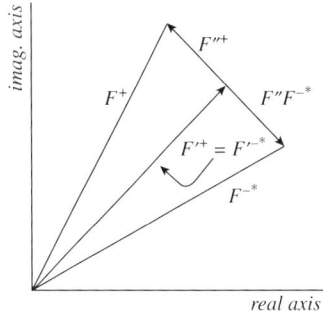

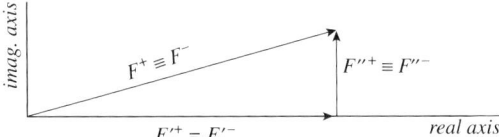

Fig. 15.3

$\Delta I = 0$ if F'^+ is perpendicular to F''^+.

Fig. 15.4

The cs. case: F^+ and F^- when anomalous dispersion is present.

The reader will certainly have understood that a violation of the Friedel law is just the source of information necessary to solve a structure. Indeed, the *anomalous difference*,

$$\Delta_{ano} = |F^+| - |F^-|,$$

depends on the nature of the anomalous scatterers and on their positions; conversely, the anomalous substructure may, in principle, be derived from the anomalous differences.

If the anomalous scattering from light atoms like C, O, N, H, (these are nearly all the atoms in proteins) may be neglected (see Section 15.3), the anomalous substructure mainly consists of some (relatively) few anomalous scatterers. Thus, as for the isomorphous derivative technique, the phasing approach may be subdivided into two steps: defining the anomalous substructure first, and then phasing the protein. If we compare this just outlined approach with the isomorphous derivative method, we see that the role of the isomorphous difference, Δ_{iso}, is now played by Δ_{ano}. This is the reason why it is usual to state the following (Pepinsky and Okaya, 1956; Ramaseshan and Venkatesan, 1957; Mitchell, 1957):

anomalous dispersion effects simulate isomorphism.

15.3 Selection of dispersive atoms and wavelengths

Anomalous dispersion may be used in small- as well as in macro-molecule crystallography. Since the phase problem is practically solved for small-sized structures, in this area anomalous effects are mostly used for other purposes (see Helliwell, 2000). For example:

(i) by tuning the wavelength close to the absorption edge of specific elements, it is possible to distinguish atoms which have close atomic numbers, even when they occupy the same site.
(ii) in microporous materials, for determining the concentrations of transition metals incorporated into the frameworks.
(iii) in powder crystallography, as an additional tool for phasing (see Section 15.9).

In this chapter, we are mainly interested in macromolecular crystallography. For such structures we will give some practical recipes for designing good SAD or MAD experiments.

MAD experiments are very demanding in terms of beamline properties. The question is: which strategy should be chosen in order to collect the minimum amount of data allowing a straightforward crystal structure solution (Gonzalez, 2003a,b)? To decide which strategy, the characteristics of the crystal sample (e.g. resistance to radiation damage, diffracting power, chemical composition) and the properties of the X-ray source (e.g. intensity of the beamline at the wavelengths of interest, ease of tunability, stability, and reproducibility) should be considered. If the beamline does not fulfil one or more of the above requirements, a SAD experiment is advisable.

In the case of a weak anomalous signal (e.g. on the S anomalous dispersion), some additional parameters like exposure time and data redundancy may become more critical. According to Cianci et al. (2008) $<|\Delta_{ano}|/\sigma(\Delta_{ano})>$ greater than 1.5 for all resolution shells is a necessary requirement for a successful phasing attempt.

The above considerations suggest that the phasing process would become more straightforward if the anomalous signal is maximized; that may be done by introducing (if necessary) stronger anomalous scatterers into the crystal, and/or by proper selection of the wavelengths. Let us examine the various aspects.

The usable energy range for most synchrotrons is in the range 5–15 keV; to have good anomalous differences, the absorption edges (K, or L, or M) to exploit should lie in this range. Luckily, most of the elements of the periodic table show edges in this interval (see Fig. 15.5). In particular, many metalloenzymes, oxidase, reductase, etc. naturally contain transition metals like Fe, Zn, Cu, etc. with absorption edges in the range 5–15 keV, and also very heavy atoms (like Hg, Pt, Au, Br) show a strong L-edge in that range. Only elements belonging to the 5th period between Rb and Te lie outside the range; we will see below, however, that anomalous data collection may also be made at wavelengths above the absorption edge, even if this choice does not maximize the anomalous difference (Leonard et al., 2005).

The S absorption edge falls outside the K-range, which is the reason why special *selenoproteins* are grown, i.e. proteins where the S atom in *methionines* is replaced by Se. There are two main reasons for preferring *seleno-methionine* to methionine. The first is that Se is a more efficient anomalous scatterer; the second is that, at a given wavelength λ, the best achievable resolution is $RES = \lambda/2$. Since the K-edge for sulphur falls at 5.018 Å, the best resolution experimentally attainable is about 2 Å; the K-edge for Se falls at 0.98 Å, so allowing more complete experimental datasets.

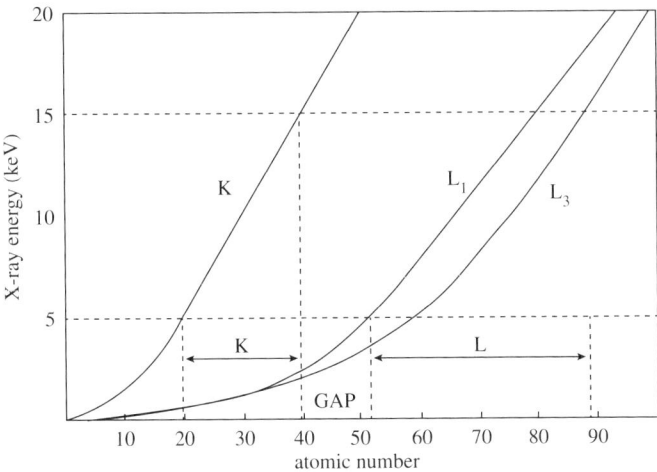

Fig. 15.5

Absorption edges as a function of the atomic number. The interval 5–15 keV approximately corresponds to the interval 2.48–0.83 Å.

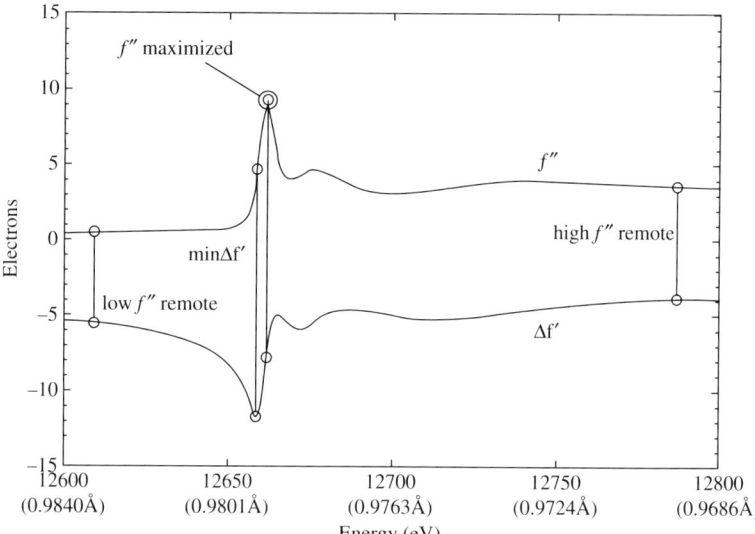

Fig. 15.6
Se dispersion pattern close to the K absorption edge.

Let us now consider how the wavelengths may be chosen in a typical MAD experiment. We will see in the following paragraphs that, besides Δ_{ano}, also the *dispersive differences*, Δ_{disp} (say $|F^+_{\lambda_i}| - |F^+_{\lambda_j}|$ or $|F^-_{\lambda_i}| - |F^-_{\lambda_j}|$) and the *mixed differences*, $|F^+_{\lambda_i}| - |F^-_{\lambda_j}|$ may contribute to the phase problem solution.

The first data are usually measured by using a wavelength which maximizes the f'' (*peak dataset*; see Fig. 15.6).

Accordingly, the second wavelength is usually chosen at the negative peak of the $\Delta f'$ curve, to maximize the dispersive differences between wavelengths. This wavelength coincides with the inflection point of the f'' scan, and the corresponding data are called the *inflection point dataset*. It may be useful to notice that at such an inflection point, the f'' value is about half the value at the peak, while the dispersive difference with any other wavelength is very high. If the radiation damage is low, a third wavelength may be chosen; the corresponding data are denoted as a *high energy remote dataset*. This shows a non-negligible f'' signal and a good dispersive contrast against the inflection dataset.

If radiation decay is slowly progressing, a fourth low energy wavelength may be chosen (then, *low energy remote data* are collected). Even if there is a high f'' contrast with the peak and a still good $\Delta f'$ contrast with the inflection point, this wavelength is the last choice because of the large absorption of light atoms, which may rapidly destroy the crystal.

X-ray absorption edges are very sharp in many cases; thus the energies of the peak and of the inflection points are seperated by a few eV. In addition, the exact position of the edge depends on the chemical environment of the anomalous scatterers. It is therefore mandatory to record the absorption edge at the time of a MAD experiment.

Sometimes, the anisotropy of anomalous scattering should be taken into account. Templeton and Templeton (1988) showed that a remarkable anisotropy can be found for the Sc K-edge in selenomethionine proteins, caused by the non-spherical symmetry of the antibonding valence orbitals. In such cases, the anomalous dispersion effects depend on the orientation of the incident and diffracted beams relative to those molecular orbitals. Templeton and Templeton developed a tensorial formalism for the anomalous dispersion, which can be successfully applied in practical cases (see Hendrickson et al., 1989). Schiltz and Bricogne (2008) observed that, as an effect of the anisotropy of anomalous scattering, intensity differences between symmetry-related reflections rise. They considered this effect as a supplementary source of extra experimental information and described a new formalism for the solution of the phase problem.

Anomalous dispersion effects may, in magnitude, be of the same size as the measurement errors. To minimize the noise, particularly that caused by crystal decay, for each unique reflection all measurements at different wavelengths are taken from the same asymmetric region of the reciprocal space. Furthermore, the anomalous differences of Friedel related pairs are collected in short temporal intervals. To reduce the speed of crystal decay, modern cryopreservation techniques are used. Thus, with the synchrotron, exposure of the crystals to X-ray doses up to 1000 times larger than with laboratory sources is possible. By the laws of statistics, this makes it possible to measure signals up to 30 times smaller, with, in general, a corresponding increase in the precision of all measurements. However, long irradiation of protein crystals must be avoided; this may lead to disruption of disulfide bridges, decarboxylation of acidic groups, and changes in the unit cell parameters (Ravelli et al., 2002).

Not all of the collected reflections are equally useful for phasing (see Section 15.7). For example, if one calculates, for a given wavelength, the internal residual factor, R_{int}, over symmetry-related reflections (see Section 2.6), where

$$R_{int} = \frac{\sum ||F_{obs\mathbf{h}}| - \langle |F_{obs\mathbf{h}}| \rangle |}{\sum |F_{obs\mathbf{h}}|},$$

and

$$\langle |F_{obs}| \rangle = \frac{\sum_{\mathbf{h}} w_{\mathbf{h}} |F_{obs\mathbf{h}}|}{\sum w_{\mathbf{h}}},$$

R_{int} usually increases with the diffraction angle because of the rapid f_0 fall-off, even if the anomalous signal is constant with it. This is the reason why MAD data, at the resolution limit of the crystal, are rarely useful.

It is now apparent that SAD techniques are gaining wider applicability, in particular sulphur single wavelength anomalous dispersion (S-SAD). Its appeal derives from the following observation: the average frequency of sulphur-containing amino acids is about 3.3, for 100 amino acids, and a technique exploiting this natural source of phasing information may reduce experimental effort. The method has been pioneered by Hendrickson and Teeter (1981), who demonstrated that even the small anomalous S signal could lead to solution of a small protein (crambin), with 46 residues and six sulphurs in disulfide bridges. Some later papers (e.g. by Wang, 1985; Weiss et al., 2001; Dauter et al., 2002; Ramagopal et al., 2003) demonstrated that SAD may also

344 Anomalous dispersion techniques

Table 15.1 Anomalous scattering of P, S, and Cl at different wavelengths (W)

W (Å)	P $\|\Delta f'\|; f''$	S $\|\Delta f'\|; f''$	Cl $\|\Delta f'\|; f''$
2.29 (Cr K_α)	0.377; 0.899	0.375; 1.141	0.333; 1.423
1.74	0.318; 0.544	0.349; 0.697	0.368; 0.876
1.54 (Cu K_α)	0.282; 0.433	0.317; 0.556	0.345; 0.701
1.28 (Au L_α)	0.228; 0.304	0.261; 0.393	0.294; 0.498
0.71 (Mo K_α)	0.082; 0.094	0.102; 0.124	0.122; 0.159

succeed for P and Cl (naturally present in several macromolecular families), by choosing wavelengths far above their K-edge energies.

To understand limits and advantages of the technique, we show in Table 15.1, the values of $|\Delta f'|$ and f'' for P, S, and Cl at specific wavelengths far away from their absorption edges. The anomalous signal is less than 1 electron unit at wavelengths shorter than 2 Å, with ratio $<|\Delta_{ano}|>/<|F|>$ of a few per cent; therefore, the errors in estimation of intensities should not exceed 2%. This implies very accurate diffraction data, not available until recent decades, and only obtainable via well-conducted experiments and large data redundancy (i.e. multiple measurements of the same reflection intensity, averaged to reduce the error). Therefore a compromise has to be made between data redundancy, resolution limit, and crystal exposure.

The reader should not believe that successful anomalous effects are only possible when synchrotron data are used. Indeed, SAD data with S atoms as anomalous scatterers, collected in-house using current data collection methods, may also succeed; highly redundant data are often necessary, as well as the use of techniques to minimize the effects of radiation damage and absorption. The anomalous signal may be collected at the Cu K_α wavelength (Yang and Pflugrath, 2001; Nagem et al., 2005), or also at a chromium radiation source ($\lambda = 2.29$ Å; Yang et al., 2003; Nan et al., 2009). This last choice is suggested by the fact that $\Delta f'' = 1.14$ e$^-$ when Cr K_α radiation is used, about double as compared with that for Cu K_α radiation. It may also be mentioned that an increased $\Delta f''$ value is obtained for many other elements intrinsic to macromolecules, such as calcium, zinc, and phosphorus.

15.4 Phasing via SAD techniques: the algebraic approach

Let us suppose that the diffraction data of a protein have been collected, and that some efficient anomalous scatterers are present naturally; their number is usually a small fraction of N_P. At the initial step of the phasing process, $|F^+|$ and $|F^-|$ are known from the diffraction experiment, while $|\Delta F'|$ and $|F''|$ are unknown; no estimate can be made on the values of ϕ^+ and ϕ^- without some additional information.

We will show that if the anomalous scatterer substructure has been solved (then $|F''|$ and ϕ''^+ are known), estimating ϕ^+ and ϕ^- is possible. In Fig. 15.7, two circles are drawn in the Argand plane, one centred at the tip of the

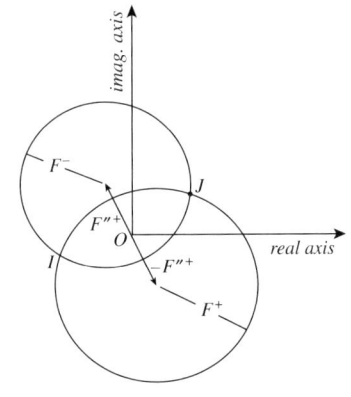

Fig. 15.7
Harker diagram illustrating phase determination via SAD techniques. Circles of radii $|F^+|$ and $|F^-|$ are drawn, with centres at F''^+ and F''^-. The two points of intersection, I and J, indicate two possible vectors for F', say OI and OJ.

vector $-F''^+$ with radius $|F^+|$, and the second centred on the tip of F''^+, with radius $|F^{-*}| \equiv |F^-|$ (in the figure the amplitudes of $|F''^+|$ and $|F''^-|$ have been magnified to make the drawing more clear). Because of equation (15.2),

$$F^+ - F''^+ = F'^+$$

and, because of equation (15.5),

$$F^{-*} + F''^+ = (F_0^{-*} + \Delta F'^{-*} + F''^{-*}) + F''^+ = (F_0^{-*} + \Delta F'^{-*}) = F'^{-*} = F'^+,$$

Consequently, the two circles meet at two points, I and J in the figure, symmetrically disposed about F', which are the only ones compatible with the modulus $|F'|$: accordingly, ϕ'^+ may only take two values. Since we have assumed that the number of anomalous scatterers is usually a small fraction of N_P, F' may be considered to be a good approximation, in modulus and phase, of the protein structure factor. Therefore, we can also state that the protein phase ϕ_0^+ may only assume two values.

The above result suggests the following two-step phasing procedure:

Step 1 *Solve the anomalous scatterer substructure;*
Step 2 *Use the information gained at* Step 1 *to solve the phase ambiguity present in Fig. 15.7 and obtain the target phases.*

But how do we solve the anomalous scatterer substructure? What is needed is a sufficiently good estimate of $|F''|$. Having achieved that, one of two approaches may then be used to accomplish Step 1: calculate a Patterson synthesis (to be deconvoluted in accordance with the techniques described in Chapter 10), or submit the estimated $|F''|$ values to traditional direct methods, charge flipping, or *VLD* techniques, according to personal preference (see Chapters 6, 9, and 10).

Blow (1957) and Rossmann (1961) suggested the so-called *anomalous-difference Patterson synthesis,*

$$P_{ano}(\mathbf{u}) = \frac{1}{V} \sum |\Delta_{ano}|^2 \exp(-2\pi i \mathbf{h} \cdot \mathbf{u}), \quad (15.9)$$

where

$$|\Delta_{ano}| = ||F^+| - |F^-||, \quad (15.10)$$

is the required estimate of $|F''|$. To check the conditions under which the approximation $|\Delta_{ano}| \approx |F''|$ holds, we return to Fig. 15.1b, which suggests the relation,

$$4|F''|^2 = |F^+|^2 + |F^-|^2 - 2|F^+ F^-|\cos(\phi^+ - \phi^{-*}). \quad (15.11)$$

$(\phi^+ - \phi^{-*})$ represents the angle between F^+ and F^{-*}. Since $(\phi^+ - \phi^{-*})$ is expected to be small, equation (15.11) reduces to

$$4|F''|^2 \approx (|F^+| - |F^-|)^2 = |\Delta_{ano}|^2, \quad (15.12)$$

from which $|\Delta_{ano}| \approx 2|F''|$.

Equation (15.9) is therefore a Patterson synthesis, approximately equivalent to that having coefficients $|F''|^2$ and would reveal the positions of anomalous scatterers via peak height, proportional to f'' (Strahs and Kraut, 1968; Hendrickson and Teeter, 1981). Mukherjee et al. (1989) proved that the anomalous scatter positions could really be found by submitting $|\Delta_{ano}|$ to traditional direct methods, in spite of the lack of anomalous differences in the centric data.

Once the anomalous scattering substructure is solved, Step 2 may be started. Since the anomalous contribution to the structure factor of a protein is small compared with the non-anomalous part, the following approximation holds (see Fig. 15.1b):

$$|F^+| + |F^-| \approx 2|F'|. \tag{15.13}$$

Introducing (15.13) into (15.7) provides,

$$\cos(\phi''^+ - \phi'^+) = \frac{(|F^+| + |F^-|)(|F^+| - |F^-|)}{4|F'F''|} \approx \frac{\Delta_{ano}}{2|F''|}, \tag{15.14}$$

from which,

$$\phi'^+ \approx \phi''^+ \pm \cos^{-1}\left(\frac{\Delta_{ano}}{2|F''|}\right). \tag{15.15}$$

Since the anomalous scattering is usually much smaller than normal scattering, relationship (15.15) reduces to,

$$\phi_0 \approx \phi'' \pm \cos^{-1}\left(\frac{\Delta_{ano}}{2|F''|}\right). \tag{15.16}$$

Let us rewrite (15.16) in a different form. Since $\Delta F'^+$ and F'' are approximately perpendicular (they are rigorously perpendicular if the anomalous scatterers are all of the same atomic species), then,

$$\phi''^+ = \phi'^+_\Delta + \pi/2,$$

where ϕ'^+_Δ is the phase of $\Delta F'^+$. Accordingly, (15.16) may be rewritten as

$$\phi_0 = \phi'_\Delta + \pi/2 \pm \cos^{-1}\left(\frac{\Delta_{ano}}{2|F''|}\right), \tag{15.17}$$

which allows ϕ_0 to be estimated from the normal scattering part of the anomalous scattering atoms.

Neither (15.16) nor (15.17) fix ϕ_0 unambiguously; two values are allowed which, according to (15.16), are symmetrically disposed around ϕ'', and often far away from each other. So far it is unclear which of the two values should be preferred. However, if $|\Delta_{ano}|$ is sufficiently large (see Fig. 15.8), then it is expected to be close to $2|F''|$, and the allowed ϕ'^+ values defined by equation (15.15) are expected to be very close to each other; taking the average of the two values may not incur a huge error. The reason for the above expectation is quite simple: the scattering power of the substructure is very small with respect to that of the full structure. Thus, if $|\Delta_{ano}|$ is large, $\cos(\phi'' - \phi'^+)$ must be close to unity (see equation (15.14)). Accordingly,

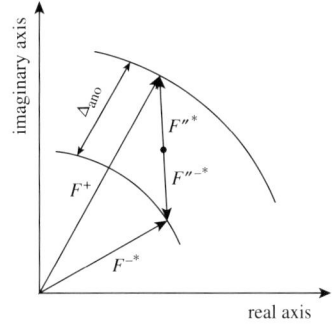

Fig. 15.8
Relation between Δ_{ano} and F''. If $|\Delta_{ano}|$ is sufficiently large, it is expected to be close to $2|F''|$.

$\phi \approx \phi''$ if $\Delta_{ano} > 0$ and sufficiently large.
$\phi \approx \phi'' + \pi$ if $\Delta_{ano} < 0$ and sufficiently large in modulus.

Some observations now about phase ambiguities in the SAD case.

A source of ambiguity arises from the fact that two values are allowed for the protein phases when the anomalous scatterer substructure has been determined. Taking their average certainly introduces noise into the corresponding electron density map.

A next source of ambiguity is that the anomalous scatterer substructure determined at the end of Step 1 does not necessarily have the correct handedness: there is only a 50% chance of having determined the correct enantiomorph.

Let us now imagine that the anomalous scatterer substructure has the wrong handedness. Assigning the protein phases produces an electron density map with poor contrast and low connectivity. No interpretation of the map is possible, and any improvement of the map via EDM techniques is without hope. Conversely, the map obtained by using the correct handedness of the substructure is much better: solvent channels are distinguishable, connectivity is larger, and contrast between solvent and protein is high. EDM techniques may lead easily to a final interpretable electron density map.

In conclusion, if it is not possible to define the correct handedness of the anomalous scatterer substructure by some figure of merit, the two hands should be used one after the other; which is the correct hand should be decided on the basis of the quality of the resulting final map.

A specific phasing difficulty is met when the anomalous scatterer distribution is centrosymmetric (in P1, two identical anomalous scatterers always simulate a centrosymmetric substructure). In this case, the EDM procedure will succeed only if the centrosymmetric nature of the phases is broken. This process may be monitored by a special Fourier synthesis (called the FF synthesis) described by Burla et al. (2006b).

15.5 The SIRAS algebraic bases

Let us suppose that the native protein does not contain sufficiently good anomalous scatterers: a SAD experiment would then be discouraged. Since the anomalous scattering phenomenon is relevant for most of the heavy atoms, a heavy-atom derivative may be used to produce detectable anomalous effects (SIRAS case). Two sets of data are then available: the native protein data and the derivative data, the latter with detectable anomalous diffraction effects.

We will subdivide this section into two parts: the first dedicated to establish the SIRAS algebraic bases, and the second to describe a typical algebraic phasing procedure.

The SIRAS algebraic bases. Derivative and native protein data are related by the following equations (see Fig. 15.9a):

$$F_d^+ = F_P^+ + F_H^+, \quad F_d^- = F_P^- + F_H^-, \tag{15.18}$$

where F_d^+ and F_d^- are the derivative structure factors for **h** and **−h**, respectively,

$$F_P = \sum_{j=1}^{N_P} f_{0j} \exp(2\pi i \mathbf{h} \cdot \mathbf{r}_j) = |F_P| \exp(i\phi_P),$$

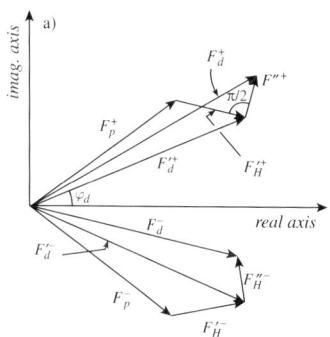

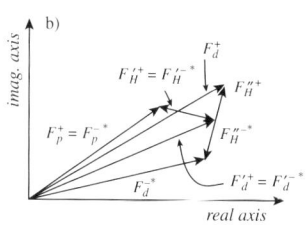

Fig. 15.9
SIRAS case: (a) Relations among the vectors, $F_d^+, F_d^-, F_P^+, F_P^-, F_H'^+, F_H'^-, F_H''^+$, $F_H''^-, F_H^+, F_H^-$; (b) Relations between F_d^+, $F_d^{-*}, F_P^+, F_P^{-*}, F_H'^+, F_H'^{-*}, F_H''^+, F_H''^{-*}$, F_H^+, F_H^{-*}.

and

$$F_H^+ = F_H'^+ + F_H''^+ = |F_H^+|\exp(i\phi_H^+),$$
$$F_H^- = F_H'^- + F_H''^- = |F_H^-|\exp(i\phi_H^-),$$

are the structure factors of the heavy-atom substructure. Let N_P be the number of non-H atoms in the native protein unit cell; $N_d = N_P + N_H$ will be the number of non-H atoms in the derivative unit cell, and N_H the corresponding number of heavy atoms. $F_H'^+$ and $F_H''^+$ are the heavy-atom structure factor components, defined by

$$F_H'^+ = \sum_{j=1}^{N_H} f_j' \exp(2\pi i \mathbf{h}\cdot\mathbf{r}_j) = |F_H'^+|\exp(i\phi_H'^+),$$

$$F_H'^- = \sum_{j=1}^{N_H} f_j' \exp(-2\pi i \mathbf{h}\cdot\mathbf{r}_j) = |F_H'^-|\exp(i\phi_H'^-)$$

$$F_H''^+ = i\sum_{j=1}^{N_H} f_j'' \exp(2\pi i \mathbf{h}\cdot\mathbf{r}_j) = \sum_{j=1}^{N_H} f_j'' \exp[i(2\pi\mathbf{h}\cdot\mathbf{r}_j + \pi/2)]$$
$$= |F_H''^+|\exp(i\phi_H''^+),$$

$$F_H''^- = i\sum_{j=1}^{N_H} f_j'' \exp(-2\pi i \mathbf{h}\cdot\mathbf{r}_j) = \sum_{j=1}^{N_H} f_j'' \exp[i(-2\pi\mathbf{h}\cdot\mathbf{r}_j + \pi/2)]$$
$$= |F_H''^-|\exp(i\phi_H''^-).$$

We may factorize equations (15.18) in a different way,

$$F_d^+ = F_d'^+ + F_H''^+, \quad F_d^- = F_d'^- + F_H''^-,$$

where

$$F_d'^+ = F_P^+ + F_H'^+, \quad F_d'^- = F_P^- + F_H'^- \qquad (15.19)$$

$F_d'^+$ and $F_d'^-$ are not affected by anomalous dispersion effects, and therefore are related by the Friedel law (see Fig. 15.10a and Fig. 15.9b); say,

$$|F_d'^+| = |F_d'^-|, \quad \phi_d'^- = -\phi_d'^+.$$

In Fig. 15.10, the stars indicate the phases of the complex conjugate structure factors.

The Friedel law also holds for $F_H'^+$ and $F_H'^-$; indeed, $|F_H'^+| = |F_H'^-|$ and (see Fig. 15.9b and Fig. 15.10b):

$$\phi_H'^- = -\phi_H'^+, \quad \phi_H'^{-*} = \phi_H'^+.$$

The Friedel law is violated for the pair $F_H''^+$ and $F_H''^-$. Indeed, $|F_H''^+| = |F_H''^-|$, but (see Fig. 15.9b and Fig. 15.10c),

$$\phi_H''^- = \pi - \phi_H''^+ = -\phi_H''^{-*}$$
$$\phi_H''^{+*} = -\phi_H''^+, \quad \phi_H''^{-*} = -\phi_H''^-.$$

The reader may notice that, while $\phi_d'^+$ and $\phi_d'^-$, as well as $\phi_H'^+$ and $\phi_H'^-$, are symmetrical with respect to the zero angle, $\phi_H''^+$ and $\phi_H''^-$ are symmetrical with respect to $\pi/2$.

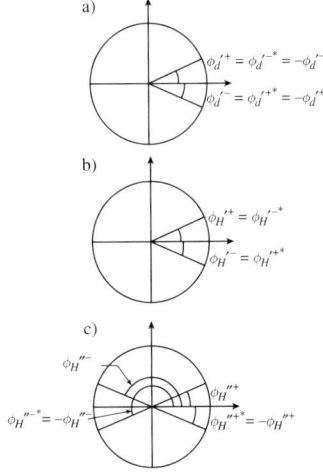

Fig. 15.10
SIRAS case: (a) Phase relationships among $\phi_d'^+, \phi_d'^-, \phi_d'^{+*}, \phi_d'^{-*}$; (b) phase relationships between $\phi_H'^+, \phi_H'^-, \phi_H'^{+*}, \phi_H'^{-*}$; (c) phase relationships among $\phi_H''^+, \phi_H''^-, \phi_H''^{+*}, \phi_H''^{-*}$.

From (15.19), the following relations hold:

$$|F_d^+|^2 = |F_d'^+|^2 + |F_H''^+|^2 + 2|F_d'^+ F_H''^+| \cos\left(\phi_d'^+ - \phi_H''^+\right). \quad (15.20a)$$

$$|F_d^-|^2 = |F_d'^-|^2 + |F_H''^-|^2 + 2|F_d'^- F_H''^-| \cos\left(\phi_d'^- - \phi_H''^-\right). \quad (15.20b)$$

Equations (15.20) may be rewritten as

$$|F_d^+|^2 = |F_d'^+|^2 + |F_H''^+|^2 + 2|F_d'^+ F_H''^+| \cos\left(\phi_d'^+ - \phi_H''^+\right). \quad (15.21a)$$

$$|F_d^-|^2 = |F_d'^-|^2 + |F_H''^-|^2 - 2|F_d'^- F_H''^-| \cos\left(\phi_d'^+ - \phi_H''^+\right). \quad (15.21b)$$

For simplicity, from now on, we will denote the moduli $|F_d'^+|$ and $|F_d'^-|$ by $|F_d'|$ and the moduli $|F_H''^+|$ and $|F_H''^-|$ by $|F_H''|$. From equations (15.21), the relations

$$\Delta I = |F_d^+|^2 - |F_d^-|^2 = (|F_d^+| + |F_d^-|)\Delta_{ano} = 4|F_d' F_H''| \cos(\phi_H''^+ - \phi_d'^+). \quad (15.22)$$

and

$$\frac{|F_d^+|^2 + |F_d^-|^2}{2} = |F_d'|^2 + |F_H''|^2. \quad (15.23)$$

arise. Equations (15.22) and (15.23) may be usefully compared with equations (15.7) and (15.8), obtained for the SAD case.

The reader will certainly have understood that violation of the Friedel law is just the source of information necessary to solve a structure. Indeed, the *anomalous difference*, $|F_d^+|^2 - |F_d^-|^2$ depends on the heavy-atom substructure and conversely, the anomalous substructure may, in principle, be derived from the anomalous differences.

The typical algebraic SIRAS phasing procedure. At the initial phasing step, $|F_P^+|, |F_d^+|, |F_d^-|$ are known from the diffraction experiment, while F_H^+ and F_H^- are unknown; no estimate can be made on the values of ϕ_P. We will show that, if the heavy-atom substructure has been identified (then F_H^+ and its components $F_H'^+$ and $F_H''^+$ are known), estimating ϕ_P is possible. In Fig. 15.11, two circles are drawn in the Argand plane, one centred at the tip of the vector, $-F_H'^+ - F_H''^+$ and the second centred at the tip of the vector, $-F_H'^+ - F_H''^+$; the first circle has radius $|F_d^+|$ and the second, $|F_d^{-*}| = |F_d^-|$.

The generic point of the first circle is the tip of the vector,

$$-F_H'^+ - F_H''^+ + F_d^+ = -F_H^+ + F_d^+ = F_P^+.$$

The generic point of the second circle is the tip of the vector,

$$-F_H'^+ + F_H''^+ + F_d^{-*} = -F_H'^{-*} - F_H''^{-*} + F_d^{-*} = (F_d^- - F_H^-)^* = F_P^{-*}.$$

Consequently, the two circles meet at two points, R and S in the figure, which are the only ones compatible with the modulus $|F_P|$; accordingly, ϕ_P^+ may only take two values.

In the SIRAS case, however, the protein amplitude $|F_p|$ is also available experimentally. Thus, a third circle, centred at O with radius $|F_p|$, may be added to the two drawn in Fig. 15.11. As a result (see Fig. 15.12), only one point is compatible with the experimental data, in our case, point R; the phase ambiguity noticed for the SAD case is cancelled if SIRAS data are available.

Fig. 15.11
Harker diagram illustrating phase determination in the SIRAS case. Circles of radii $|F_d^+|$ and $|F_d^-|$ are drawn, with their centres at the tips of the vectors $-F'^{+}_H - F''^{+}_H$ and $-F'^{+}_H + F''^{+}_H$, respectively. The two intersection points, R and S, indicate two possible vectors for $|F_p|$, say OR and OS.

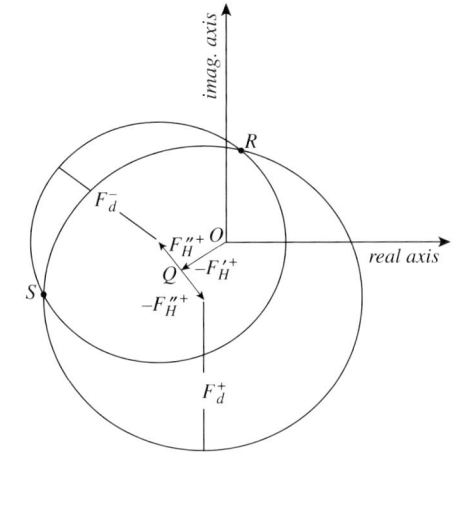

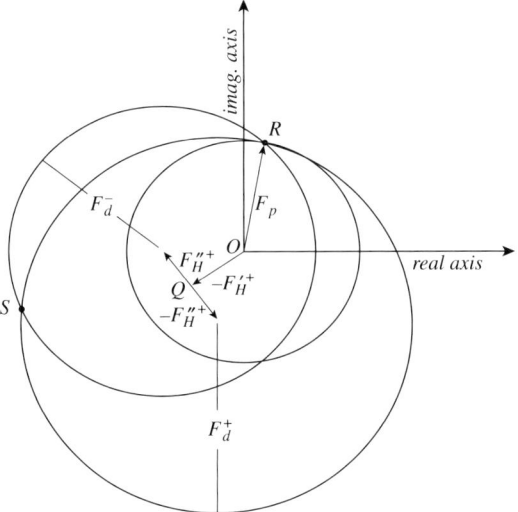

Fig. 15.12
Harker diagram illustrating phase determination in the SIRAS case. To the circles shown in Fig. 15.11, a third circle may be added, centred on O and radius $|F_p|$. R is the unique solution.

The first important conclusion of this section may now be stated: knowing the anomalous scatterer substructure, and therefore knowing F''_H, allows us to define the protein phases. This suggests the following two-step phasing procedure:

Step 1 *Solve the anomalous scatterer substructure;*
Step 2 *Use the information gained at Step 1 to estimate the protein phases.*

But how do we solve the anomalous scatterer substructure? The ab initio methods described in Chapters 6, 9, and 10 indicate that the only necessary prior information is knowledge of the modulus $|F''|$. Unfortunately, $|F''|$ is not available experimentally, but can only be estimated statistically. A relevant difference between SIR (or SAD) and SIRAS is that, in the case of SIRAS, simultaneous knowledge of anomalous differences and isomorphous differences is available experimentally. Let us see how these may be combined to provide a better estimate of $|F''_H|$.

Since (see Fig. 15.9b),

$$|F_d^+| + |F_d^-| \approx 2|F'_d|, \phi_H''^+ \approx \phi_H'^+ + \pi/2, \quad (15.24)$$

replacing (15.24) into (15.22) gives,

$$\Delta_{ano}^+ \approx 2|F'_H|\sin(\phi_d'^+ - \phi_H'^+). \quad (15.25)$$

Let us now find a useful expression for Δ_{iso} in the SIRAS case, where two different values of Δ_{iso} exist, say $\Delta_{iso}^+ = |F_d^+| - |F_P|$ and $\Delta_{iso}^- = |F_d^-| - |F_P|$. In the SIR case (in the absence of anomalous dispersion), the following relation was found (see equation (14.4)):

$$\Delta_{iso} \approx |F_H|\cos(\phi_d - \phi_H).$$

The above equation may be transformed, so as to be valid for the SIRAS case, into

$$\Delta_{iso}^+ \approx |F'_H|\cos(\phi_d'^+ - \phi_H'^+). \quad (15.26)$$

Taken in isolation, equations (15.25) and (15.26) are of no use for Step 1 (because $\phi_d'^+$ and $\phi_H'^+$, are still unknown) but, if combined, they provide a useful estimate of $|F'_H|$. Indeed,

$$|\Delta_{iso}^+|^2 + \frac{1}{4}|\Delta_{ano}|^2 \approx |F'_H|^2 \left[\cos^2\left(\phi_d'^+ - \phi_H'^+\right) + \sin^2\left(\phi_d'^+ - \phi_H'^+\right) \right] = |F'_H|^2. \quad (15.27)$$

It is then clear that, within the framework of the approximations made, the coefficients (15.27), no matter whether used for calculating a Patterson map or being processed by direct methods, will provide heavy-atom positions with less noise than maps obtained from separated coefficients $|\Delta_{iso}^+|$ or $|\Delta_{ano}|$ (Kartha and Parthasarathy, 1965).

It is worthwhile stressing that $|\Delta_{iso}^+|$ will in general be larger than Δ_{ano}, and that Δ_{iso}^+ is usually affected by lack of isomorphism. A suitable weighting function should then be introduced into (15.27).

Once the anomalous scattering substructure is solved, Step 2 may be started. In the SIRAS case, we can combine the phase indication suggested by isomorphous difference with that provided by the anomalous difference. For example, we can rewrite equation (15.22) as

$$\cos\left(\phi_H''^+ - \phi_d'^+\right) \approx \frac{\Delta_{ano}}{2|F''|}. \quad (15.28)$$

or also as

$$\phi_d'^+ \approx \phi_H''^+ \pm \cos^{-1}\left(\frac{\Delta_{ano}}{2|F''|}\right) \approx \phi_H'^+ + \pi/2 \pm \cos^{-1}\left(\frac{\Delta_{ano}}{2|F''|}\right). \quad (15.29)$$

Equation (15.29) exploits the information contained in the anomalous differences. From the isomorphous derivative method, the following indication may be obtained (see equation (14.3)):

$$\phi_d \approx \phi_H \pm \cos^{-1}\left(\frac{|F_d|^2 - |F_P|^2 + |F_H|^2}{2|F_d F_H|}\right). \quad (15.30)$$

Since ϕ_d and ϕ'_d may be considered to be very close, equations (15.29) and (15.30) may provide a unique value of ϕ_d.

Once ϕ_d has been estimated, the protein phase may be obtained by subtracting F_H from F_d. Then,

$$\tan \phi_P = \frac{|F_d|\sin\phi_{d-}|F_H|\sin\phi_H}{|F_d|\cos\phi_{d-}|F_H|\cos\phi_H}. \qquad (15.31)$$

The above mathematical approach does not take into account the lack of isomorphism, errors in measurement of $|F_d^+|, |F_d^-|$, and $|F_P|$, and errors arising from the model heavy-atom substructure. In practice, (15.31) is not strictly valid. To understand, graphically, the effects of the above factors, it may easily be argued that, in practical applications, the radii and the centres of the three circles in Fig. 15.11 do not usually coincide with the correct values and the three circles do not necessarily intersect at one point. Suitable weighting schemes are usually applied to reduce the effect of errors.

15.6 The MAD algebraic bases

Let us suppose that a n-wavelength experiment has been carried out providing the measured amplitudes of n pairs $|F_j^+|, |F_j^-|$ for $j = 1, \ldots, n$. Once the anomalous scatterer substructure has been identified, algebraic MAD techniques may be applied to solve the phase problem. A successful approach was suggested by Ramakrishnan and Biou (1997) and by Terwilliger (1997): the MAD case is treated as a MIR case.

We will see in this section that, by using a mathematical formalism which separates the normal from the anomalous scattering, the phase problem may be reduced to the solution of a linear system of equations. The idea was first introduced by Mitchell (1957), and then used by Corby and Black (1973) and by Black and Corby (1975) to solve two small structures. Karle (1980; 1983a,b; 1984a,b; 1989a,b; see also Chapuis et al., 1985) gave a more general mathematical treatment, valid even for the case in which different types of anomalous scatterers are present. Let us follow the Karle approach for the *simplest case in which there is only one kind of anomalous scatterer in the unit cell*.

For a generic wavelength, in accordance with equation (15.2),

$$F^+ = F_0^+ + \Delta F'^+ + F''^+ = F_0^+ + F_a^+, \quad F^- = F_0^- + F_a^-, \qquad (15.32)$$

where, $F_a^+ = \Delta F'^+ + F''^+$ and $F_a^- = \Delta F'^- + F''^-$ are the structure factors of the anomalous scatterer substructure, calculated from anomalous scattering only. The reader should notice that, in accordance with equation (15.2), F_0^+ and F_0^- also contain the normal scattering contribution (say F_{0a}^+ and F_{0a}^-, respectively) of the anomalous scatterers, where

$$F_{0a}^+ = f_0 \sum_{j=1}^{N_a} \exp(2\pi i \mathbf{h} \cdot \mathbf{r}_j) = |F_{0a}^+|\exp(i\phi_{0a}^+),$$

$$F_{0a}^- = f_0 \sum_{j=1}^{N_a} \exp(-2\pi i \mathbf{h} \cdot \mathbf{r}_j) = |F_{0a}^-|\exp(i\phi_{0a}^-).$$

Then,

$$F_a^+ = (\Delta f' + if'') \sum_{j=1}^{N_a} \exp(2\pi i \mathbf{h} \cdot \mathbf{r}_j) \qquad (15.33a)$$

$$= \frac{\Delta f' + if''}{f_{0a}} F_{0a}^+ = \gamma(\exp i\delta) F_{0a}^+ = \gamma |F_{0a}^+| \exp\left[i(\phi_{0a}^+ + \delta)\right],$$

where

$$\gamma = \frac{\left(\Delta f'^2 + f''^2\right)^{1/2}}{f_{0a}}, \quad \delta = \tan^{-1}\left(f''/\Delta f'\right).$$

Analogously, the following equation may be derived:

$$F_a^- = \gamma |F_{0a}^-| \exp\left[i(\phi_{0a}^- + \delta)\right], \qquad (15.33b)$$

Let us now introduce (15.33a) into the first of the equations (15.32), and (15.33b) into the second equation (15.32), and then multiply the resulting formulas by their complex conjugates. Two equations are obtained,

$$|F^+|^2 = |F_0^+|^2 + \gamma^2 |F_{0a}^+|^2 + 2\gamma |F_0^+ F_{0a}^+| \cos\left(\phi_0^+ - \phi_{0a}^+ - \delta\right) \qquad (15.34a)$$

and

$$|F^-|^2 = |F_0^-|^2 + \gamma^2 |F_{0a}^-|^2 + 2\gamma |F_0^- F_{0a}^-| \cos\left(\phi_0^- - \phi_{0a}^- - \delta\right). \qquad (15.34b)$$

Since

$$\frac{\Delta f'}{f_{0a}} = \gamma \cos\delta, \quad \frac{f''}{f_{0a}} = \gamma \sin\delta,$$

equations (15.34) may be rewritten as

$$|F^+|^2 = |F_0^+|^2 + \gamma^2 |F_{0a}^+|^2 + 2|F_0^+ F_{0a}^+| \left[\frac{\Delta f'}{f_{0a}} \cos\left(\phi_0^+ - \phi_{0a}^+\right) + \frac{f''}{f_{0a}} \sin\left(\phi_0^+ - \phi_{0a}^+\right)\right] \qquad (15.35a)$$

and

$$|F^-|^2 = |F_0^-|^2 + \gamma^2 |F_{0a}^-|^2 + 2|F_0^- F_{0a}^-| \left[\frac{\Delta f'}{f_{0a}} \cos\left(\phi_0^- - \phi_{0a}^-\right) + \frac{f''}{f_{0a}} \sin\left(\phi_0^- - \phi_{0a}^-\right)\right] \qquad (15.35b)$$

In accordance with the definitions,

$$|F_0^+| = |F_0^-|, \quad |F_{0a}^+| = |F_{0a}^-|, \quad (\phi_0^- = -\phi_0^+, \phi_{0a}^- = -\phi_{0a}^+).$$

If we denote the first two moduli by $|F_0|$, the second two by $|F_{0a}|$, and the phases ϕ_0^+ and ϕ_{0a}^+ by ϕ_0 and ϕ_{0a}, respectively, the two equations (15.35) may be rewritten in a more useful and simple form:

$$|F^+|^2 = |F_0|^2 + \gamma^2 |F_{0a}|^2 + 2|F_0 F_{0a}| \left[\frac{\Delta f'}{f_{0a}} \cos\left(\phi_0 - \phi_{0a}\right) + \frac{f''}{f_{0a}} \sin\left(\phi_0 - \phi_{0a}\right)\right] \qquad (15.36a)$$

$$|F^-|^2 = |F_0|^2 + \gamma^2 |F_{0a}|^2 + 2|F_0 F_{0a}| \left[\frac{\Delta f'}{f_{0a}} \cos(\phi_0 - \phi_{0a}) - \frac{f''}{f_{0a}} \sin(\phi_0 - \phi_{0a}) \right]$$
(15.36b)

Let us examine the main feature of the above equation for each wavelength:

(i) $|F^+|^2$ and $|F^-|^2$ are known from the diffraction experiment for each wavelength;
(ii) $\frac{\Delta f'}{f_{0a}}$ and $\frac{f''}{f_{0a}}$ are functions of λ, but may be derived experimentally through a study of the absorption coefficient.
(iii) The unknown quantities are $|F_0|$, $|F_{0a}|$ and $(\phi_0 - \phi_{0a})$; all of them are independent of λ.

It may be concluded that a SAD experiment provides two equations of type (15.36), with three unknowns; a two-wavelength experiment provides four equations with three unknowns, and is therefore sufficient for estimating the values of $|F_0|$, $|F_{0a}|$, and $(\phi_0 - \phi_{0a})$. The three unknowns are overdetermined for a three or more wavelength experiment. The above approach has been used in the *MADSYS* package (Wu and Hendrickson, 1996).

The SAD phase ambiguity (see Section 15.4) is not present in MAD, but the handedness ambiguity of the anomalous scatter substructure remains. Both of the substructure enantiomorphs should be submitted to phase assignment and refinement via EDM techniques if, via suitable figures of merit, the correct handedness of the anomalous scatterer substructure has not been clearly determined.

15.7 The probabilistic approach for the SAD-MAD case

The probabilistic approach is more suitable than algebraic techniques to face the challenges connected with errors in experimental data and in the model. For simplicity, we will not enter into mathematical details, but we will only describe the logical steps necessary to find conclusive formulas, and we will discuss their main characteristics. As in the preceding paragraphs, the two-step phasing procedure (i.e. first find the anomalous substructure (Step 1), and then phase the protein given the substructure (Step 2)) will be described.

A one-step procedure is also possible for the SAD case. This was first suggested by Kroon et al. (1977), Heinerman et al. (1978), and Karle (1984a,b). The most general approach has been described by Hauptman (1982b) and independently by Giacovazzo (1983b), via the study of the joint probability distribution function,

$$P(E_\mathbf{h}, E_\mathbf{k}, E_{\mathbf{h+k}}, E_{-\mathbf{h}}, E_{-\mathbf{k}}, E_{-\mathbf{h-k}}),$$

but is not competitive with the two-step approach (Giacovazzo et al., 2003), and will not be reported.

Before describing the two-step approach, some preliminary considerations are necessary in order to answer a basic question: are the experimental data collected at the various wavelengths equally useful for solving the phase problem, or the data quality changes with the wavelength and with the diffraction angle?

About the experimental data selection. In a MAD experiment, each wavelength is characterized by specific values of $\Delta f'$ and f'', by specific absorption values, and by proper measurement errors; therefore, the various wavelengths are not expected to be equally informative. The phasing efficiency should improve if each wavelength is suitably weighted or if the worst ones are eliminated from the calculations. A criterion is therefore necessary to predict the most informative wavelength combinations, otherwise all of the combinations should be explored in order to identify the correct solution. Thus, if $n = 4$, one should explore four one-wavelength, six two-wavelength, four three-wavelength, and one four-wavelength combination, and, at the end of the calculations, rely on a suitable figure of merit in order to identify the correct solution.

Schneider and Sheldrick (2002) made a basic assumption: good experimental multi-wavelength data should show high correlation values between the various Δ_{ano} values. Thus, if the data collected at a given wavelength are badly correlated with the other experimental data, that wavelength should be eliminated or underweighted.

The quality of the information provided by a multi-wavelength experiment varies with the resolution; usually, high resolution diffraction data should be eliminated from the calculation (this is also true for a SAD experiment). Thus, a good algorithm should not only be able to identify the best wavelength combination, but also to fix a threshold for data resolution, in order to eliminate the experimental data with weakly correlated Δ_{ano}. A very efficient automatic algorithm for limiting the data resolution and for predicting the most informative wavelength combination has been proposed by Burla et al. (2004).

Step 1 In any probabilistic approach, the first item is to define the primitive random variables; we choose the atomic positions as the primitive random variables, while the indices are kept fixed. The structure factors are defined according to

$$F_j^+ = F_{aj}^+ + F_{naj}^+ + |\mu_j^+| \exp(i\theta_j^+)$$
$$F_j^- = F_{aj}^- + F_{naj}^- + |\mu_j^-| \exp(i\theta_j^-),$$

where F_j^+ and F_j^- are the values of $F_\mathbf{h}$ and $F_{-\mathbf{h}}$ for the jth wavelength, F_{aj}^+ and F_{aj}^-, the corresponding structure factors for the anomalous scatterer substructure, F_{naj}^+ and F_{naj}^-, the corresponding structure factors for the non-anomalous scatterer substructure. μ_j^+ and μ_j^- are parameters which take into account measurement and model errors; they are unknown and are treated as additional primitive random variables.

To accomplish Step 1, we need to previously estimate $|F_{0a}|$, say the amplitude for normal scattering of the anomalous scatterer substructure, given the amplitudes measured in the n-wavelength MAD experiment. If the joint probability distribution approach (see Chapter 4) is used, the steps are the following:

1. Calculate the $4n+2$-dimensional joint probability distribution,

$$P_n = P(E_{0a}, E_1^+, E_2^+, \ldots, E_n^+, E_1^-, E_2^-, \ldots, E_n^-)$$
$$= P(A_{0a}, A_1^+, A_2^+, \ldots, A_n^+, A_1^-, A_2^-, \ldots, A_n^-, B_{0a}, B_1^+, B_2^+, \ldots, B_n^+, B_1^-, B_2^-, \ldots, B_n^-), \quad (15.37)$$

where $E = A + iB$ is the generic normalized structure factor, normalized with respect to the scattering power of the non-anomalous substructure (i.e. $E = F / \left[\sum_{i=1}^{N_{na}} f_i^2\right]^{1/2}$, where N_{na} is the number of non-H atoms in the non-anomalous scatterer substructure).

2. Express (15.37) in terms of polar coordinates:

$$P_n = P(R_{0a}, R_1^+, R_2^+, \ldots, R_n^+, R_1^-, R_2^-, \ldots, R_n^-, \phi_{0a}, \phi_1^+, \phi_2^+, \ldots, \phi_n^+, \phi_1^-, \phi_2^-, \ldots, \phi_n^-) \quad (15.38)$$

via the change of variables,

$$A_{0a} = R_{0a} \cos \phi_{0a}, \quad B_{0a} = R_{0a} \sin \phi_{0a}$$
$$A_j^+ = R_j^+ \cos \phi_j^+, \quad B_j^+ = R_j^+ \sin \phi_j^+$$
$$A_j^- = R_j^- \cos \phi_j^-, \quad B_j^- = R_j^- \sin \phi_j^-.$$

3. Integrate (15.38) over the phase variables to obtain the marginal distribution,

$$P(R_{0a}, R_1^+, R_2^+, \ldots, R_n^+, R_1^-, R_2^-, \ldots, R_n^-).$$

4. Calculate the conditional distribution,

$$P(R_{0a} | R_1^+, R_2^+, \ldots, R_n^+, R_1^-, R_2^-, \ldots, R_n^-). \quad (15.39)$$

5. Calculate the expected conditional value,

$$< R_{0a} | R_1^+, R_2^+, \ldots, R_n^+, R_1^-, R_2^-, \ldots, R_n^- > . \quad (15.40)$$

The results are as follows (Burla et al., 2002, 2003): $P(R_{0a} | \ldots\ldots)$ is a Gaussian probability distribution which may be written in the form,

$$P(R_{0a} | \ldots\ldots) = \pi^{-(2n+1)} (\det \mathbf{K})^{1/2} \exp\left(-\frac{1}{2} \tilde{\mathbf{T}} \mathbf{K}^{-1} \mathbf{T}\right),$$

where $\mathbf{K} = \{k_{ij}\}$ is a symmetric square matrix of order $(4n+2)$ taking into account the variances of and the covariances among the $(4n+2)$ variables, $\mathbf{K}^{-1} = \{\lambda_{ij}\}$ is its inverse, and $\mathbf{T}$ is a suitable vector with components defined in terms of the $(4n+2)$ variables.

$$< R_{0a} | R_1^+, \ldots, R_n^- > = \frac{1}{2} (\pi/\lambda_{11})^{1/2} \left[1 + 4X^2/(\pi \lambda_{11})\right]^{1/2}, \quad (15.41)$$

where

$$X^2 = Q_1^2 + Q_2^2$$
$$Q_1 = \lambda_{12} R_1^+ + \lambda_{13} R_2^+ + \cdots + \lambda_{1,n+1} R_n^+ + \lambda_{1,n+2} R_1^- + \cdots + \lambda_{1,2n+1} R_n^-$$
$$Q_2 = \lambda_{1,2n+3} R_1^+ + \lambda_{1,2n+4} R_2^+ + \cdots + \lambda_{1,3n+2} R_n^+ + \cdots - \lambda_{1,3n+3} R_1^- \cdots - \lambda_{1,4n+2} R_n^-.$$

The standard deviation of the estimate is

$$\sigma_{R_{0a}} = \left(<R_{0a}^2|\ldots> - <R_{0a}|\ldots>^2\right)^{1/2} = \left[\left(1 - \frac{\pi}{4}\right)\lambda_{11}^{-1}\right]^{1/2}, \quad (15.42)$$

from which,

$$\frac{<R_{0a}|\ldots>}{\sigma_{R_{0a}}} = \left[\frac{[(\pi/4) + (X^2)/\lambda_{11}]}{1 - (\pi/4)}\right]^{1/2}. \quad (15.43)$$

We will not provide mathematical expressions for all of the various parameters in equation (15.41), but the reader is referred to the original papers. We are more interested to describe why the probabilistic approach makes the procedure more robust.

The success of Step 1 depends on how accurately the $<R_{0a}|\ldots>$ estimates approximate the true values; the advantage of probabilistic methods is that they provide the standard deviation of the estimate and therefore they may suggest which estimates are more reliable. To clarify these aspects we will formulate some questions and then answer them by applying the theory to experimental data from three test proteins: *AEPT* (PDB code, *1m32*), *CYANASE* (PDB code, *1dw9*), and *TGEV* (PDB code, *1lvo*), with three-wavelength data for all of the three structures. Given below are the questions and the possible answers:

(a) Is the standard deviation (15.42) strictly correlated or anti-correlated with the $<R_{0a}|\ldots>$ value? To answer this question we show, in Fig. 15.13, distribution (15.39) for selected pairs (X, λ_{11}). The different location of the maxima and the different sharpness of the curves suggest that highest values of $<R_{0a}|\ldots>$ are not necessarily correlated with the sharpest distributions. In other words, the largest $<R_{0a}|\ldots>$ values do not necessarily correspond to the best estimates.

(b) Is the reliability of the estimate provided by (15.40) dependent on the value $<R_{0a}|\ldots>$? We have calculated the parameter,

$$R_{cryst} = \frac{\sum |(R_{0a})_t - S <R_{0a}|\ldots>|}{\sum (R_{0a})_t},$$

where $(R_{0a})_t$ is the true value of R_{0a}, $<R_{0a}|\ldots>$ is its estimate (15.41), and S is a suitable scale factor. In Fig. 15.14, we show, for the three test structures, the trend of R_{cryst} as a function of $<R_{0a}|\ldots>$. Quite wrong estimates are frequent for small values of $<R_{0a}|\ldots>$; the best estimates

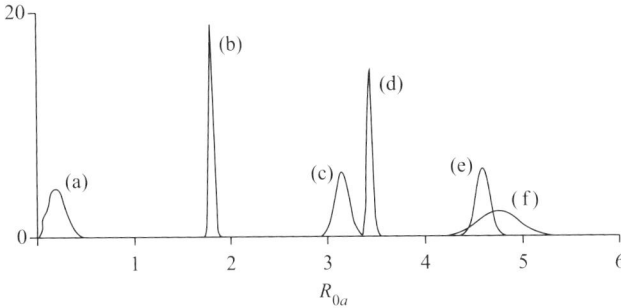

Fig. 15.13
The probability distribution $P(R_{0a}|R_1^+, R_2^+, \ldots, R_n^+, R_1^-, R_2^-, \ldots, R_n^-)$, defined by equation (15.39) is plotted for some selected pairs of (X, λ_{11}). (a) $(X, \lambda_{11}) = (7.5, 49.8)$; (b) $(X, \lambda_{11}) = (2043, 1136)$; (c) $(X, \lambda_{11}) = (318.8, 101.4)$; (d) $(X, \lambda_{11}) = (2488.9, 724.5)$ (e) $(X, \lambda_{11}) = (505.7, 110.3)$; (f) $(X, \lambda_{11}) = (63, 13.3)$.

Fig. 15.14
The R_{cryst} value versus $<R_{0a}|\ldots>$ for the three test structures when, for each structure, the experimental data from three wavelengths are used simultaneously.

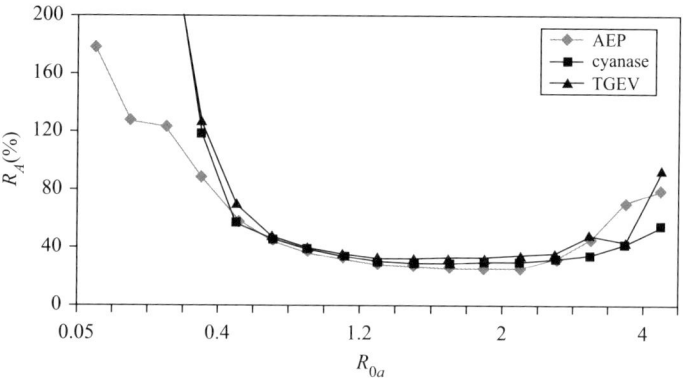

Fig. 15.15
The R_{cryst} value versus $<R_{0a}|\ldots>/\sigma_{R_{0a}}$ for the three test structures when, for each structure, the experimental data from three wavelengths are used simultaneously.

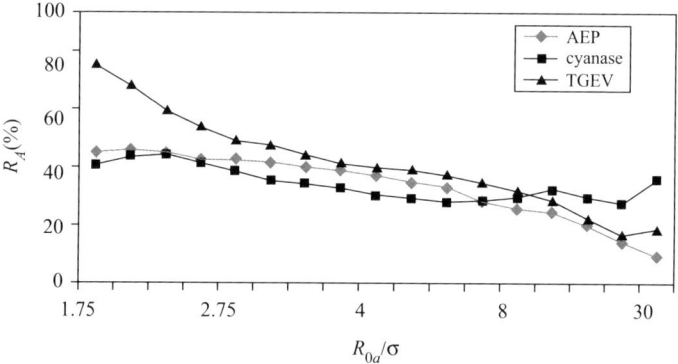

are attained for medium values of $<R_{0a}|\ldots>$, while a loss of accuracy may be noted for the largest $<R_{0a}|\ldots>$ values. This last behaviour, even if it concerns a limited number of estimates, is not ideal for the application of Patterson or direct methods, the success of which is based on the accuracy of the largest structure factor moduli.

(c) May the ratio (15.43) be considered as a ratio signal to noise? In this case smaller values of R_{cryst} should be obtained for subsets of reflections characterized by large values of $<R_{0a}|\ldots>/\sigma_{R_{0a}}$. This expectation is confirmed by Fig. 15.15, where R_{cryst} is plotted against $<R_{0a}|\ldots>/\sigma_{R_{0a}}$. The reader may usefully compare the relatively small values of R_{cryst} for the reflections with the largest values of $<R_{0a}|\ldots>/\sigma_{R_{0a}}$ in Fig. 15.14, with the relatively high values of R_{cryst} for the reflections with the largest values of $<R_{0a}|\ldots>$ in Fig. 15.13.

In conclusion, the probabilistic approach allows us to estimate the accuracy of each conditional average $<R_{0a}|\ldots>$. The reflections with the highest values of $<R_{0a}|\ldots>/\sigma_{R_{0a}}$ are those which should be submitted to direct methods or to Patterson deconvolution procedures to find the anomalous scatterer substructure.

Step 2 Let us now suppose that the anomalous scatterer substructure has been found. We want to phase the protein structure by using the substructure as prior information. Any approach based on the theory of the joint probability distribution (e.g. McCoy et al., 2004; Pannu and Read, 2004; Giacovazzo and Siliqi, 2004) should perform the following steps in sequence:

1. Calculate the conditional probability,

$$P_n = P(E_1^+, E_2^+, \ldots, E_n^+, E_1^-, E_2^-, \ldots, E_n^- | E_{a1}^+, E_{a2}^+, \ldots, E_{an}^+, E_{a1}^-, E_{a2}^-, \ldots, E_{an}^-),$$

where E_{aj} is the generic structure factor of the anomalous scatterer substructure, normalized with respect to the scattering power of the non-anomalous substructure. In formulas, $E_{aj} = F_{aj}/(\sum_{na})^{1/2}$, where $\sum_{na} = \sum_{i=1}^{N_{na}} f_i^2$, and N_{na} is the number of non-H atoms in the non-anomalous scatterer substructure. In Step 2 the E_{aj}s are known in modulus and phase.

2. Calculate the marginal conditional distribution,

$$P(\phi_1^+, \ldots, \phi_n^- | R_1^+, \ldots, R_n^+, R_1^-, \ldots, R_n^-, E_{a1}^+, \ldots, E_{an}^+, E_{a1}^-, \ldots, E_{an}^-). \tag{15.44}$$

3. Simplify the calculations by introducing the following approximations (justified by the fact that the scattering power of the anomalous scatterer substructure is small compared with the total protein scattering power):

$$\phi_1^+ \approx \phi_2^+ \approx \phi_n^+ \approx -\phi_1^- \approx -\phi_2^- \approx \ldots \approx -\phi_n^-,$$

Then, (15.44) reduces to

$$P(\phi_1^+ | \ldots \ldots) \approx [2\pi I_0(L)]^{-1} \exp[L\cos(\phi_1^+ - \theta_1^+)]. \tag{15.45}$$

The most probable ϕ_1^+ value is the phase of the vector (Giacovazzo and Siliqi, 2004),

$$V = \sum_{j=1}^{n} \left(w_j^+ E_{aj}^+ + w_j^- E_{aj}^{-*} \right)$$

$$+ \sum_{j,p=1, p>j}^{n} w_{jp} \left(E_{aj}^+ - E_{ap}^+ \right) + w_{n+j,n+p} \left(E_{aj}^{-*} - E_{ap}^{-*} \right) \tag{15.46}$$

$$+ \sum_{j,p=1}^{n} w_{j,n+p} \left(E_{aj}^+ - E_{ap}^{-*} \right)$$

and the reliability parameter L of the phase estimate is nothing else but its modulus.

We do not provide the mathematical details defining the various parameters involved in distribution (15.45). We want to characterize the main features of the distribution to make its advantages clear. The following may be noticed:

(a) The larger the number of wavelengths, the larger the number of terms in the summations (15.46), and therefore the better (on average) will be the reliability of the phase estimate. In conclusion, MAD is better than SAD (provided that the data are sufficiently good for each wavelength!).
(b) The first of the three terms on the right-hand side of equation (15.46) relates the protein phase to the model phases for each wavelength. This Sim (1959a,b) type term should always be present when a model is available (in our case, the model is the anomalous scatterer substructure).

(c) The other two terms on the right-hand side of (15.46) depend on the anomalous differences and the dispersive differences in all of the possible combinations. While SAD exploits only one anomalous difference, in MAD techniques dispersive differences provide a non-negligible contribution.

(d) w are suitable weights which are not defined here for brevity; the reader is referred to the original paper.

To clarify the main features of the vector V, we describe its expression for the SAD case in Appendix 15.A.

Let us now suppose that the protein phases have been assigned to each reflection; then a least squares procedure usually starts to improve the phase estimates. It is described in Appendix 15.B.

The same considerations made for the algebraic approaches described in Sections 15.4 and 15.6 hold for phase ambiguity related to the probabilistic SAD-MAD approach.

15.8 The probabilistic approach for the SIRAS-MIRAS case

Let us suppose that the diffraction data for the native protein and the n derivatives, obtained by addition of anomalously scattering heavy atoms, have been collected; we are dealing with the SIRAS or the MIRAS case, according to whether $n = 1$ or $n > 1$. The following experimental amplitudes are then available:

$$|F_P|, |F_{d_1}^+|, \ldots, |F_{d_n}^+|, |F_{d_1}^-|, \ldots, |F_{d_n}^-|.$$

which will be settled on their absolute scales via statistical methods (as described for other phasing techniques).

While in the MAD case there is a unique anomalous scatterer substructure, in the MIRAS case there is one substructure for each derivative. The two-step approach, described in preceding sections for SIR-MIR and SAD-MAD cases, will also be the method of choice for the SIRAS-MIRAS case. Again, we have first to determine the anomalous scatterer substructures, and then use this information to phase the protein.

Step 1 Each derivative is characterized by a proper heavy-atom substructure; it may be supposed that, for the jth derivative, the following relation holds:

$$F_{d_j}^+ = F_P^+ + F_{H_j}^+ + |\mu_{d_j}^+| \exp\left(i\theta_j^+\right),$$

$$F_{d_j}^- = F_P^- + F_{H_j}^- + |\mu_{d_j}^-| \exp\left(i\theta_j^-\right),$$

where $\mu_{d_j}^\pm$ are complex vectors arising from experimental measurements and imperfect isomorphism. The jth substructure will be determined if sufficiently good estimates of the corresponding $|F_{H_j}|$ s are obtained. Such result may be obtained by studying, for each jth derivative, the joint probability distribution,

$$P(R_{H_j}, R_P, R_{d_j}^+, R_{d_j}^-, \phi_{H_j}, \phi_P, \phi_{d_j}^+, \phi_{d_j}^-), \tag{15.47}$$

from which the conditional,

$$P(R_{H_j}, R_P, R_{d_j}^+, R_{d_j}^-) \quad (15.48)$$

is obtained, by integrating (15.47) over $\phi_{H_j}, \phi_P, \phi_{d_j}^+, \phi_{d_j}^-$.

From (15.48), the conditional distribution

$$P(R_{H_j}|R_P, R_{d_j}^+, R_{d_j}^-)$$

is obtained by standard techniques, from which,

$$< R_{H_j}|R_P, R_{d_j}^+, R_{d_j}^- >$$

may be derived.

A simple satisfactory approximation to R_{H_j} is obtained by the relation,

$$<R_{H_j}^2|\ldots> \approx w_1|\Delta_{isoj}|^2 + w_2|\Delta_{anoj}|^2, \quad (15.49)$$

suggested by the algebraic approach described in Section 15.5 (see equation (15.27)).

The set of amplitudes (15.46) are then submitted to direct or Patterson methods to find the heavy-atom substructure of the *j*th derivative.

Step 2 At the end of Step 1, n substructure models become available. From each model, the protein phases may be obtained via the following probabilistic approach. The joint probability distribution function,

$$P\left(A_P, A_{d_j}^+, A_{d_j}^-, B_P, B_{d_j}^+, B_{d_j}^-|A_{H_j}^+, B_{H_j}^+, A_{H_j}^-, B_{H_j}^-\right)$$

is calculated and transformed into polar variables,

$$P\left(R_P, R_{d_j}^+, R_{d_j}^-, \phi_P, \phi_{d_j}^+, \phi_{d_{j_1}}^-|R_{H_j}^+, \phi_{H_j}^+, R_{H_j}^-, \phi_{H_j}^-\right). \quad (15.50)$$

The following approximation is introduced:

$$\phi_{d_j}^+ \approx \phi_P, \quad \phi_{d_j}^- \approx -\phi_P,$$

which modifies (15.50) to,

$$P\left(R_P, R_{d_j}^+, R_{d_j}^-, \phi_P|R_{H_j}^+, \phi_{H_j}^+, \phi_{H_j}^-\right),$$

from which the conditional phase distribution,

$$P(\phi_P|\ldots\ldots) \approx [2\pi I_0(X)]^{-1} \exp[X \cos(\phi_P - \theta_P)] \quad (15.51)$$

is derived, where

$$\tan \theta_P = \frac{(G_j^+ \sin \phi_{Hj}^+ - G_j^- \sin \phi_{Hj}^-)}{G_j^+ \cos \phi_{Hj}^+ - G_j^- \cos \phi_{Hj}^-} = \frac{T}{B}, \quad (15.52)$$

$$G_j^+ = 2|F_{H_j}^+|\Delta_{isoj}^+/|\mu_j^+|^2, \quad G_j^- = 2|F_{H_j}^-|\Delta_{isoj}^-/|\mu_j^-|^2$$

$$X = (T^2 + B^2)^{1/2}.$$

θ_P is the most probable value of ϕ_P, given the prior information on $R_P, R_{d_j}^+, R_{d_j}^-, R_{H_j}^+, \phi_{H_j}^+$.

Equation (15.52) is very simple in practical use, but some details are necessary to disclose its internal mechanism. We provide this information in Appendix 15.C.

If we apply equation (15.52) to each substructure, n protein models are obtained. The quality of such models depends on the quality of the diffraction data collected at the various wavelengths (indeed, the derivatives may have different degrees of isomorphism, different data resolution, and different measurement errors). Combining the various protein models in these conditions is also not rewarding, because they are probably referred to different origins and have different handedness (according to the occasional handedness of the substructure models).

To avoid such difficulties, a procedure similar to that used for the MIR case may be undertaken:

(a) The 'best' derivative is selected: this is the one for which the SIRAS phasing process is expected to provide the minimum phase error; appropriate figures of merit should be used towards recognizing it.

(b) The set of phases $\{\phi_{Pbest}\}$, obtained from the best derivative at the end of the refinement step, is employed to calculate, for each jth derivative, the differential Fourier synthesis with coefficient $(|\bar{F}_{d_j}| - |F_p|)\exp(i\phi_{Pbest})$, where $|\bar{F}_{d_j}| = (|F_{d_j}^+| + |F_{d_j}^-|)/2$. The rationale is as follows: even if the n substructures are uncorrelated, the protein models provided by the different derivatives are expected to be correlated with each other. Thus, if we associate $\{\phi_{Pbest}\}$ to the jth $(|\bar{F}_{d_j}| - |F_p|)$ coefficient, we should obtain a substructure map which better approximates the jth true substructure. As explained for the MIR case, the procedure provides an additional advantage: the same handedness and the same origin are secured for all of the substructures.

(c) The new model substructures may be refined by special least squares procedures, similar to that used for the MIR case. Then, from the improved substructures, new phase estimates ϕ_P are obtained via equation (15.51). The new sets of phases $\{\phi_P\}_i, i = 1, \ldots, n$ are combined according to point (d).

(d) The most general way for handling the MIRAS case when the n substructures are known is to calculate the joint probability distribution function,

$$P(A_P, A_{d_1}^+, \ldots, A_{d_n}^+, B_P, A_{d_1}^-, \ldots, B_{d_n}^- | A_{H_1}^+, \ldots, A_{H_n}^-, B_{H_1}^+, \ldots, B_{H_n}^-),$$

transform it into polar variables,

$$P(R_P, R_{d_1}^+, \ldots, R_{d_n}^-, \phi_P, \phi_{d_1}^+, \ldots, \phi_{d_n}^- | R_{H_1}^+, \ldots, R_{H_n}^-, \phi_{H_1}^+, \ldots, \phi_{H_n}^-),$$
(15.53)

and, for each derivative, introduce the following approximation:

$$\phi_{d_j}^+ \approx \phi_P, \ \phi_{d_j}^- \approx -\phi_P, \quad \text{for } j = 1, \ldots, n.$$

Then, the conditional probability (15.53) reduces to

$$P(R_P, R_{d_1}^+, \ldots, R_{d_n}^-, \phi_P | R_{H_1}^+, \ldots R_{H_n}^-, \phi_{H_1}^+, \ldots, \phi_{H_n}^-),$$

from which the conditional probability,

$$P(\phi_P | \ldots \ldots) \approx [2\pi I_0(X)]^{-1} \exp[X \cos(\phi_P - \theta_P)]$$

may be derived, where

$$\tan\theta_P = \frac{\sum_{j=1}^{n}\left(G_j^+ \sin\phi_{Hj}^+ - G_j^- \sin\phi_{Hj}^-\right)}{\sum_{j=1}^{n}\left(G_j^+ \cos\phi_{Hj}^+ - G_j^- \cos\phi_{Hj}^-\right)} = \frac{T}{B}, \quad (15.54)$$

$$G_j^+ = 2|F_{H_j}^+|\Delta_{isoj}^+/|\mu_j^+|^2, \quad G_j^- = 2|F_{H_j}^-|\Delta_{isoj}^-/|\mu_j^-|^2$$

$$X = \left(T^2 + B^2\right)^{1/2}.$$

The summation varies over the n different derivatives, θ_P is the most probable value of ϕ_P and X is its reliability parameter.

(e) Steps (a) to (c) may be cyclically repeated.
(f) EDM procedures are applied to improve and to extend the phase information up to native resolution.

15.9 Anomalous dispersion and powder crystallography

MAD techniques are not very beneficial for powder crystallography, owing to an unavoidable loss of experimental information. In general:

(a) The reflections F^+ and F^- systematically overlap in the diffraction pattern; consequently, anomalous differences $|F^+|^2 - |F^-|^2$ cannot be measured, only the intensities $I_\mathbf{h} = |F^+|^2 + |F^-|^2$ are available experimentally. This excludes the usefulness of a SAD experiment and it also excludes the possibility of identifying the absolute structure from powder data affected by anomalous dispersion effects.

(b) Dispersive differences between $(|F_i^+|^2 + |F_i^-|^2)$ and $(|F_j^+|^2 + |F_j^-|^2)$ (subscripts denote the wavelengths) may be estimated from the experiment, but their accuracy is questionable in most cases, because of the systematic and casual overlapping present in the two diffraction patterns.

The apparently minor experimental information provided by powder diffraction experiments discouraged the use of MAD and its applications; pioneering contributions were made by Prandl (1990, 1994), Gu et al. (2000), and Helliwell et al. (2005). In a more recent paper (Altomare et al., 2009d,e), the probabilistic bases of the method were established. Since Bijvoet pairs cannot be estimated separately, the distribution $P\left(E_{oa}, E_1^+, E_1^-, \ldots, E_n^+, E_n^-\right)$, so basic for applications to single crystal data, cannot be used for powder data. The joint probability distribution function, $P\left(E_{oa}, \bar{E}_1, \bar{E}_2\right)$, was then considered, where $\bar{E}_p = \frac{1}{2}\left(E_p^+ + E_p^{-*}\right)$, from which the conditional distribution, $P\left(R_{oa}|\bar{R}_1, \bar{R}_2\right)$, and therefore the value of $<R_{oa}|\bar{R}_1, \bar{R}_2>$ could be obtained. This value may then be used as input for Patterson and direct methods, to find the anomalous scatterer substructure.

In accordance with the two-step procedure, a probabilistic approach may then used for finding the full structure given the anomalous substructure. Details on the theory and applications may be found in the original papers.

15.10 Applications

Several computing packages are available for phasing proteins via anomalous dispersion effects.

Anomalous scatterer substructures (Step 1) may be identified by applying any of the third generation direct methods programs (*SHELXD* (Schneider and Sheldrick, 2002), *ACORN* (Foadi et al., 2000), *IL MILIONE* (Burla et al., 2007c), *SHAKE & BAKE* (Weeks and Miller, 1999), *SUPERFLIP* (Dumas and van der Lee, 2008)) to Bijvoet differences or to the more accurate $|F_H|$ estimates described in Sections 15.7 and 15.8. *IL MILIONE* (as well as *PHENIX* (Terwilliger et al., 2009)), may also use Patterson techniques, which have proved to be particularly efficient; see Burla et al. (2007b).

The success of the above phasing procedures for data at non-atomic resolution may appear to be unexpected; it mainly comes about because the anomalous scatterer substructures consist of a relatively small number of atoms, which are therefore dispersed in an enormous (with respect to the number of anomalous scatterers) empty unit cell. As a result, peak overlapping in the Patterson map is strongly reduced, which makes the deconvolution process easier. If we consider the quality of the electron density maps, it may be seen that the electron density peaks of the substructure do not overlap even for low resolution data, so they may be clearly distinguished from one other. In these conditions, the diffraction data overdetermines the anomalous scatterer substructure (see the discussion in Section 1.6 on the ratio number of observations/number of structural parameters to define).

The protein structure (Step 2) may be found via programs like *SHELXE* (Sheldrick, 2002), *MLPHARE* (Otwinowski, 1991a,b), *SHARP* (La Fortelle and Bricogne, 1997), *OASIS* (Hao et al., 2000), *IL MILIONE*, and *PHENIX*. Each program uses its own phasing recipe, but each recipe may be reconducted to the theoretical considerations described in the preceding paragraphs. To give a small account of the potential of the methods we describe some applications of the program, *IL MILIONE*.

In Tables 15.2 and 15.3, we show lists of MAD and SAD test structures, respectively, with the necessary information on crystal chemical data and on the diffraction experiment. In order to provide an a posteriori estimate of the efficiency of the phasing technique, in the last columns of each table are given the average phase errors (*ERRP* and *ERRF*, in degrees) corresponding to some phasing steps of the procedure.

The probabilistic approach described in the Section 15.7 (see equation (15.40)) was applied to find the estimates of R_{0a}, say

$$< R_{0a} | R_1^+, R_2^+, \ldots, R_n^+, R_1^-, R_2^-, \ldots, R_n^- > .$$

The reflections with the largest values of $< R_{0a} | R_1^+, R_2^+, \ldots, R_n^+, R_1^-, R_2^-, \ldots, R_n^- > /\sigma_{R_{0a}}$ (see equation (15.43)) were used for calculating the Patterson map, which was then submitted to Patterson deconvolution techniques (see Chapter 10) to find the positions of the anomalous scatterers. In both Tables 15.2 and 15.3, *Found* is the number of symmetry-independent anomalous scatterers found by the default run, to be compared with the number of scatterers really present in the asymmetric unit (*An. Scatt.*). In most of the cases, all anomalous scatterer positions were found, in some cases a high percentage of

Table 15.2 Set of test structures for MAD experiments. *PDB* is the Protein Data Bank code, *SG* the space group, *NRES* is the number of residues, *nw* is the number of wavelengths used in the experiment, *An. Scatt.* is the atomic species of the anomalous scatterers (in parentheses the number of anomalous scatterers per asymmetric unit), *Res* is the limiting resolution to which the data are measured. When native data are available, the resolution is quoted in parentheses. For the meanings of the other headings, see text

PDB	SG	NRES	nW	RES(Å)	An. Scatt.	Found	RANK	ERRP(CC)	ERRF(CC)
1srv	C222$_1$	145	4	2.27 (1.7)	Se(3)	3	1	62(0.62)	57(0.73)
1c8u	C222$_1$	570	4	1.90	Se(8)	8	1	57(0.54)	40(0.88)
1ga1	P6$_2$	372	3	1.40	Br(13)	12	1	69(0.45)	44(0.83)
1m3u	P2$_1$	2640	3	3.0 (1.8)	Se(160)	158	5	60(0.61)	54(0.78)
1lvo	P2$_1$	1812	4	2.70 (1.95)	Se(60)	60	1	57(0.58)	67(0.72)
1ks9	P4$_2$2$_1$2	291	3	1.70	Se(8)	8	1	57(0.62)	60(0.75)
1m32	P2$_1$	2196	3	2.60 (2.2)	Se(66)	66	1	56(0.56)	50(0.73)
1j6n	P2$_1$	1212	3	2.60 (1.8)	Se(45)	45	1	47(0.74)	44(0.82)
1fi4	P2$_2$2$_1$2	832	3	2.28	Se(9)	9	1	64(0.53)	60(0.69)
1i94	P4$_1$2$_1$2	364	2	2.40	Se(8)	8	4	63(0.55)	50(0.82)

Table 15.3 Set of test structures for SAD experiments. *PDB* is the Protein Data Bank code, *SG* the space group, *NRES* is the number of residues, *An. Scatt.* is the atomic species of the anomalous scatterers (in parentheses the number of anomalous scatterers per asymmetric unit), *Res* is the limiting resolution to which the data are measured

STRUCT.	SG	NRES	RES(Å)	An. Scatt.	Found	Rank	ERRP(CC)	ERRF(CC)
2fdn	P4$_3$2$_1$2	55	0.94	Fe(8)	8	2	42(0.60)	36(0.86)
8xia	I222	388	1.50	Mn(1);Mg(1)	1	1	65(0.35)	32(0.90)
1fj2	P2$_1$	464	1.80	Br(22)	22	1	65(0.49)	54(0.68)
1svn	P2$_1$2$_1$2$_1$	269	1.74	Ca(4);S(3);Cl(2)	8	1	62(0.42)	51(0.68)
1l78	P4$_3$2$_1$2	258	1.53	S(10)Cl(8);	18	18	56(0.57)	46(0.78)
1ick	P2$_1$2$_1$2$_1$	12	0.95	P(10)	10	1	48(0.53)	44(0.82)
1w92	R32	148	1.90	Fe(2)	2	1	57(0.51)	39(0.88)
P4	C2	6432	2.5	Se(144)	120	9	---	---

them were located. *RANK* is the order of the correct substructure, among the trial solutions provided by the Patterson deconvolution techniques, as fixed by a suitable figure of merit. In most cases *RANK* is one, however, in very difficult cases *RANK* may be in a lower position.

Protein phasing was undertaken according to equations (15.45) and (15.46), in agreement with the procedure described in Section 15.7. The average phase error, calculated just after location of the anomalous scattererers, is reported in the two tables as *ERRP*, and the final average error after the application of EDM procedures over all the protein measured reflections is reported as *ERRF*. *CC* is the correlation between the final electron density map and the published map. In all cases, the densities are immediately interpretable.

APPENDIX 15.A   A PROBABILISTIC FORMULA FOR THE SAD CASE

The distribution (15.45) and equation (15.46) were obtained under the following assumptions: an *n*-wavelength experiment has been carried out and the anomalous scatterer substructure has been identified. Here, we provide some details on the specific form of V for the *SAD* case. If $n = 1$, V reduces to

$$V = w^+ E_a^+ + w^- E_a^{-*} + w\left(E_a^+ - E_a^{-*}\right),$$

where

$$w^+ = \frac{2}{1+S}\frac{R^+}{(\sigma^+)^2}, \quad w^- = \frac{2}{1+S}\frac{R^-}{(\sigma^-)^2}, \quad w = \frac{2}{1+S}\frac{(R^+ - R^-)}{(\sigma^+\sigma^-)^2}$$

$$S = (\sigma^+)^2 + (\sigma^-)^2.$$

σ^+ and σ^- are the normalized (with respect to the non-anomalous scatterer substructure) values of μ^+ and μ^-, and $S^{1/2}$ may be considered to be the standard deviation associated with $|E_a^+| - |E_a^-|$. Accordingly,

$$V = \frac{2}{1+S}\left[\frac{R^+ E_a^+}{(\sigma^+)^2} + \frac{R^- E_a^{-*}}{(\sigma^-)^2} + \frac{(R^+ - R^-)}{(\sigma^+\sigma^-)^2}\left(E_a^+ - E_a^{-*}\right)\right] \quad (15.A.1)$$

Taking into account the E definitions, the expression (15.A.1) may be rewritten in terms of structure factors as,

$$V \approx 2\left(\frac{|F^+||F_a^+|}{(\mu^+)^2} + \frac{|F^-||F_a^{-*}|}{(\mu^-)^2}\right) + 2\sum_{na}\frac{\Delta_{ano}}{(\mu^+\mu^-)^2}\left(F_a^+ - F_a^{-*}\right)$$

$$\approx 2\left(\frac{|F^+||F_a^+|}{(\mu^+)^2} + \frac{|F^-||F_a^{-*}|}{(\mu^-)^2}\right) + 4\sum_{na}\frac{\Delta_{ano}}{(\mu^+\mu^-)^2}F_a''^+ \quad (15.A.2)$$

We notice:

(i) The first term on the right-hand side of (15.A.2) is the Sim-like contribution for the **h** and **−h** Miller indices. The smaller the error in the model and in the measurements, the larger its contribution will be. Since F_a^+ and F_a^{-*} are vectors in the Argand plane, the direction of the first term in (15.A.2) is approximately the direction of $F_a'^+$ (see Fig. 15.A.1).

(ii) The second term is the product of two differences, say Δ_{ano} and $(F_a^+ - F_a^{-*})$. The amplitude of the product is generally very small, but its contribution to V is magnified by the factor $\sum_{na}$. Thus, it is the most important term in (15.A.2). The resulting vector in the Argand plane (see again Fig. 15.A.1) has the direction of $F_a''^+$ if Δ_{ano} is positive; the opposite direction if Δ_{ano} is negative.

(iii) The first term is important only when the scattering power of the anomalous scatterer substructure is not very small.

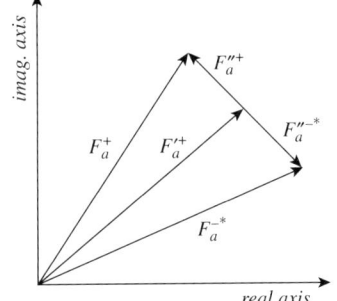

Fig. 15.A.1
SAD case. Vectorial representation of the V components in equation (15.A.2).

APPENDIX 15.B STRUCTURE REFINEMENT FOR MAD DATA

We have seen in Section 15.7 that the most probable value of the protein phase in a n-wavelength MAD experiment depends on the anomalous differences and the dispersive differences $(E_{aj}^+ - E_{ap}^+)$, $(E_{aj}^{-*} - E_{ap}^{-*})$, $(E_{aj}^+ - E_{ap}^{-*})$, for $j,p = 1, \ldots, n$. The refinement process (Otwinowski, 1991a,b) may be performed by refining not only the parameters of the anomalous scatterer substructure but also the anomalous scattering components $\Delta f'$ and f'' of the anomalously scattering atoms; indeed they are not perfectly known at the time of the experiment. We will shortly describe the procedure followed by Giacovazzo and Siliqi (2004).

The quantities,

$$\sum_{\mathbf{h}} \sum_{j} \left[\Delta_{anoj|} - K_j \left(|F_{aj}^+ - F_{aj}^{-*}| \right) \right]^2 \qquad (15.B.1)$$

and

$$\sum_{\mathbf{h}} \sum_{j,p} \left[|\bar{\Delta}_{dispj,p}| - K_{j,p} \left(|\bar{F}_{aj} - \bar{F}_{ap}| \right) \right]^2, \qquad (15.B.2)$$

are minimized, where j and p denote the wavelengths, K_j and $K_{j,p}$ are suitable scale factors, and

$$\bar{\Delta}_{dispj,p} = \bar{F}_j - \bar{F}_p, \quad \bar{F}_j = \frac{|F_j^+| + |F_j^-|}{2}, \quad \bar{F}_{aj} = \frac{F_{aj}^+ + F_{aj}^{-*}}{2}.$$

Minimization of (15.B.1) involves the anomalous differences, the dispersive differences are taken into account in the minimization of (15.B.2). The procedure may be described as follows:

(i) The experimental values of $|F_j^+|, \sigma(|F_j^+|), |F_j^-|, \sigma(|F_j^-|)$ are read together with the expected $\Delta f'_j, f''_j$ values for each jth wavelength. If the reflection is centric, we set

$$|F_j^+| = |F_j^-| = \frac{|F_j^+| + |F_j^-|}{2},$$

$$\sigma(|F_j^+|) = \sigma(|F_j^-|) = \frac{1}{2} \left[\sigma^2 \left(|F_j^+| \right) + \sigma^2 \left(|F_j^-| \right) \right]^{1/2}.$$

(ii) All of the diffraction intensities are normalized with respect to the scattering power of the non-anomalous scatterers.
(iii) The least squares procedure is applied; the refinement is controlled by suitable weights. The atomic positional parameters of the anomalous scatterers, their occupancy, and thermal factors are considered to be *global parameters* (a unique structural model is refined via all of the measured intensities); the f'' and $\Delta f'$ values are treated as *local parameters* (they are refined via the intensities collected at specific wavelengths).
(iv) The global parameters and the f'' values are refined by minimizing the quantity (15.B.1). The summation over $\mathbf{h}$ includes 70% of the measured reflections (those with the largest values of $<|\Delta_{ano}|>$, where the average is taken over all the wavelengths).
(v) The model obtained at step (iv), the occupancies excluded, is kept fixed when the quantity (15.B.2) is minimized for defining the differences $\Delta f'_j - \Delta f'_p$. In this case, the summation over $\mathbf{h}$ uses only centric reflections, if their number is sufficiently large.
(vi) After least squares convergence, formula (15.46) is calculated.

APPENDIX 15.C ABOUT PROTEIN PHASE ESTIMATION IN THE SIRAS CASE

For convenience, we rewrite equation (15.52), providing the best value θ_P for the protein phase given $R_P, R_{d_j}^+, R_{d_j}^-, R_{H_j}^+, \phi_{H_j}^+$:

$$\tan \theta_P = \frac{(G_j^+ \sin \phi_{Hj}^+ - G_j^- \sin \phi_{Hj}^-)}{G_j^+ \cos \phi_{Hj}^+ - G_j^- \cos \phi_{Hj}^-} = \frac{T}{B}, \quad (15.C.1)$$

where

$$G_j^+ = 2|F_{H_j}^+|\Delta_{isoj}^+/|\mu_j^+|^2, \, G_j^- = 2|F_{H_j}^-|\Delta_{isoj}^-/|\mu_j^-|^2$$

$$X = (T^2 + B^2)^{1/2}.$$

We will give some examples in order to clarify how the phase indications obtained from equation (15.C.1) may be interpreted in terms of vectors in the Argand plane.

If $|F_{H_j}^+|$ is not very different from $|F_{H_j}^-|$, we can approximate equation (15.C.1) by

$$\tan \theta_P \approx \frac{(\Delta_{isoj}^+ \sin \phi_{Hj}^+ + \Delta_{isoj}^- \sin \phi_{Hj}^{-*})}{(\Delta_{isoj}^+ \cos \phi_{Hj}^+ + \Delta_{isoj}^- \cos \phi_{Hj}^{-*})} = \frac{T}{B}, \quad (15.C.2)$$

where $\phi_{Hj}^{-*} = -\phi_{Hj}^-$.

The tangent formula (15.C.2) will be useful to illustrate some SIRAS didactical cases:

1. If f_H'' is very small compared with f_H', then

$$|F_H'| \gg |F_H''|, \Delta_{iso}^+ \approx \Delta_{iso}^-, \phi_H^+ \approx \phi_H^{-*} \approx \phi_H^{'+},$$

and (15.C.2) reduces to

$$\tan \theta_P = \frac{\Delta_{iso}^+ \sin \phi_H^{'+}}{\Delta_{iso}^+ \cos \phi_H^{'+}}.$$

Accordingly,

$$\theta_P \approx \phi_H^{'+} \text{ if } \Delta_{iso}^+ \approx \Delta_{iso}^- > 0,$$

$$\theta_P \approx \phi_H^{'+} + \pi \text{ if } \Delta_{iso}^+ \approx \Delta_{iso}^- < 0.$$

This is the classical SIR case, discussed in Section 14.6; indeed, if f_H'' is very small, the anomalous scattering does not add any valuable information to the phase indication provided by the SIR method (see Fig. 15.C.1a).

2. If f_H'' is non-negligible with respect to f_H' (then $|F_H''|$ is also comparable with $|F_H'|$), and if

$$\Delta_{iso}^+ > 0, \Delta_{iso}^- > 0 \text{ with}, |\Delta_{iso}^-| < |\Delta_{iso}^+|,$$

then (15.C.2) may be written as

$$\tan \theta_P = \frac{(|\Delta_{iso}^+| \sin \phi_H^+ + |\Delta_{iso}^-| \sin \phi_H^{-*})}{(|\Delta_{iso}^+| \cos \phi_H^+ + |\Delta_{iso}^-| \cos \phi_H^{-*})},$$

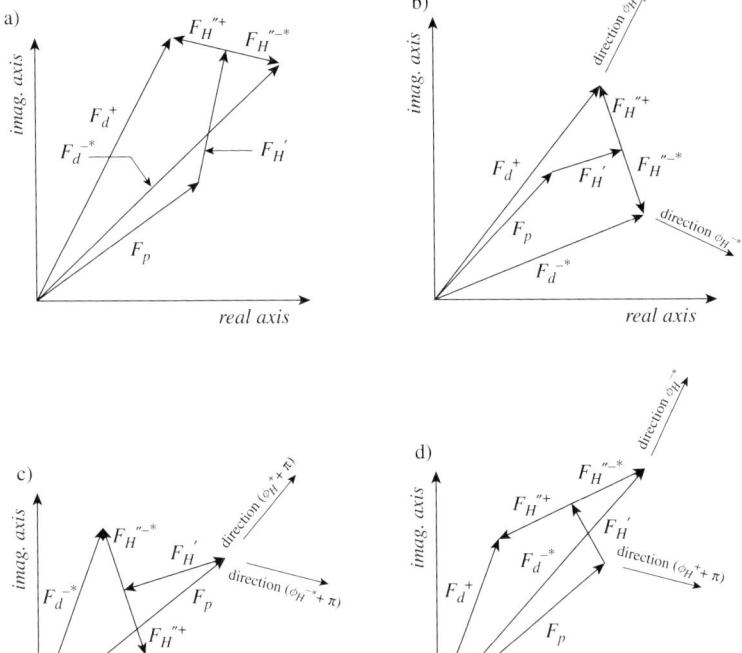

Fig. 15.C.1
SIRAS case. Geometrical details for clarifying the internal mechanism of equation (15.52), which estimates the protein phase, ϕ_P, given the experimental data and the heavy-atom substructure. Four typical cases are illustrated: (a) $|F'_H| \gg |F''_H|$, $\Delta^+_{iso} \approx \Delta^-_{iso}$. Then $\phi^+_H \approx \phi^{-*}_H \approx \phi'^+_H$ and the phase indication is $\theta_P \approx \phi'^+_H$, if $\Delta^+_{iso} \approx \Delta^-_{iso} > 0$, $\theta_P \approx \phi'^+_H + \pi$ if $\Delta^+_{iso} \approx \Delta^-_{iso} < 0$. (b) $\Delta^+_{iso} > 0$, $\Delta^-_{iso} > 0$ with $|\Delta^-_{iso}| < |\Delta^+_{iso}|$. θ_P is estimated between ϕ^+_H and ϕ^{-*}_H, closer to ϕ^{-*}_H. (c) $\Delta^+_{iso} < 0$, $\Delta^-_{iso} < 0$ with $|\Delta^-_{iso}| < |\Delta^+_{iso}|$. Then ϕ_P is expected between $(\phi^+_H + \pi)$ and $(\phi^{-*}_H + \pi)$, closer to $(\phi^+_H + \pi)$. (d) $\Delta^+_{iso} < 0$, $\Delta^-_{iso} > 0$ with $|\Delta^-_{iso}| > |\Delta^+_{iso}|$. Then ϕ_P is expected between $(\phi^+_H + \pi)$ and ϕ^{-*}_H, closer to ϕ^{-*}_H.

according to which, ϕ_P is estimated to be between ϕ^+_H and ϕ^{-*}_H, closer to ϕ^+_H. This situation is illustrated in Fig.15.C.1b. If $|\Delta^-_{iso}| > |\Delta^+_{iso}|$, then the expected value of ϕ_P will be closer to ϕ^{-*}_H.

3. f''_H is comparable with f'_H, $\Delta^+_{iso} < 0$, $\Delta^-_{iso} < 0$ with $|\Delta^-_{iso}| < |\Delta^+_{iso}|$. Then, (15.C.2) reduces to

$$\tan \theta_P = \frac{[|\Delta^+_{iso}| \sin(\phi^+_H + \pi) + |\Delta^-_{iso}| \sin(\phi^{-*}_H + \pi)]}{[|\Delta^+_{iso}| \cos(\phi^+_H + \pi) + |\Delta^-_{iso}| \cos(\phi^{-*}_H + \pi)]},$$

which estimates ϕ_P between $(\phi^+_H + \pi)$ and $(\phi^{-*}_H + \pi)$, closer to $(\phi^+_H + \pi)$. This situation is illustrated in Fig. 15.C.1c. If $|\Delta^-_{iso}| > |\Delta^+_{iso}|$, then the expected value of ϕ_P will be closer to $\phi^{-*}_H + \pi$.

4. f''_H is comparable with f'_H, $\Delta^+_{iso} < 0$, $\Delta^-_{iso} > 0$ with $|\Delta^-_{iso}| > |\Delta^+_{iso}|$. In this case, (15.C.2) reduces to

$$\tan \theta_P = \frac{[|\Delta^+_{iso}| \sin(\phi^+_H + \pi) + |\Delta^-_{iso}| \sin \phi^{-*}_H]}{[|\Delta^+_{iso}| \cos(\phi^+_H + \pi) + |\Delta^-_{iso}| \cos \phi^{-*}_H]},$$

which estimates ϕ_P between $(\phi^+_H + \pi)$ and ϕ^{-*}_H, closer to ϕ^{-*}_H (see Fig. 15.C.1d). If $|\Delta^-_{iso}| < |\Delta^+_{iso}|$, then the expected value of ϕ_P will be closer to $(\phi^+_H + \pi)$.

Appendices

Some readers of this book may be interested in crystallographic phasing methods, not only for their practical use in crystal structure solution procedures, but also for a deeper insight into their theoretical background, and possibly, for their improvement. To simplify their learning, some mathematical appendices, of a quite general type and frequent application in Chapters 1 to 15, are collected together here. The first appendix is devoted to the basics of probability theory and to the concepts of moments and cumulants of a multivariate distribution. The other appendices are dedicated to some basic results in *gamma* function theory, to *Hermite* and *Laguerre* polynomials, and to *Bessel* functions. The most frequent integration formulas are also reported.

APPENDIX M.A
SOME BASIC RESULTS IN PROBABILITY THEORY
M.A.1 Probability distribution functions

In textbooks on probability theory, the term *probability distribution function* of the random variable ξ is used to denote the probability that ξ is less than or equal to x:

$$F(x) = P(\xi \leq x).$$

To be consistent with the crystallographic literature, we shall use the term *cumulative distribution function* (or simply *cumulative function*) for $F(x)$. The following properties hold:

1. $F(-\infty) = 0$; $F(+\infty) = 1$.
2. $F(x)$ shall not decrease at any point:

$$F(x_1) \leq F(x_2), \quad x_1 < x_2.$$

3. $F(x_2) - F(x_1) = P(x_1 < x \leq x_2)$.

Let $F(x)$ be defined at every point in a continuous range and be continuous, except perhaps at specific points. If it possesses a derivative, we have $dF = f(x) \, dx$. $f(x)$ is called the *probability density function*. The following properties hold:

1. $f(x) \geq 0$.
2. $\int_{-\infty}^{+\infty} f(x) dx = F(\infty) - F(-\infty) = 1$.
3. $F(x) = \int_{-\infty}^{x} f(\xi) \, d\xi$, from which

$$F(x_2) - F(x_1) = \int_{x_1}^{x_2} f(x) \, dx,$$

or
$$P(x_1 \leq x \leq x_2) = \int_{x_1}^{x_2} f(x)\, dx.$$

M.A.2 Moments of a distribution

The mathematical expectation of any function $g(x)$ (x is supposed to range from $-\infty$ to $+\infty$) is given by,
$$\langle g(x) \rangle = \int_{-\infty}^{+\infty} g(x) f(x)\, dx. \tag{M.A.1}$$

provided that this integral exists. The mathematical expectation (or *expectation value*) of the power x^ν with positive integer exponent is called the moment of order ν or the νth moment. We shall denote it by m_ν so that
$$m_\nu = \int_{-\infty}^{+\infty} x^\nu f(x)\, dx = \langle \xi^\nu \rangle.$$

The νth moment of the *variable x about the point c* is then
$$\langle (x-c)^\nu \rangle = \int_{-\infty}^{+\infty} (x-c)^\nu f(x)\, dx.$$

If m_1 is the *mean value* defined by
$$m_1 = \langle x \rangle = \int_{-\infty}^{+\infty} x f(x)\, dx,$$

we can now define the *central moments* (they will be denoted by μ_ν):
$$\mu_\nu = \int_{-\infty}^{+\infty} (x - m_1)^\nu f(x)\, dx = \langle (x - m_1)^\nu \rangle. \tag{M.A.2}$$

The reader can easily verify that
$$\mu_0 = 1,$$
$$\mu_1 = 0,$$
$$\mu_2 = m_2 - m_1^2,$$
$$\mu_3 = m_3 - 3 m_1 m_2 + 2 m_1^3,$$
$$\mu_4 = m_4 - 4 m_1 m_3 + 6 m_1^2 m_2 - 3 m_1^4,$$
etc.

M.A.3 The characteristic function

The *characteristic function* $C(u)$ of a random variable x is defined to be the expectation value of the complex function $\exp(ixt)$, where t is a real variable:
$$C(u) = \int_{-\infty}^{+\infty} \exp(iux) f(x)\, dx = \langle \exp(iux) \rangle. \tag{M.A.3}$$

$C(u)$ is nothing but the Fourier transform of the probability density function $f(x)$ and has the following properties:

$$C(0) = 1, \quad |C(u)| \leq \int_{-\infty}^{+\infty} f(x)dx = 1.$$

If the moments up to the tth order for the random variable x exist, (M.A.3) may be differentiated v times ($0 < v \leq t$), giving rise to

$$C^{(v)}(u) = i^v \int_{-\infty}^{+\infty} x^v \exp(iux) f(x) dx. \qquad (\text{M.A.4})$$

When $u = 0$, (M.A.4) becomes

$$C^{(v)}(0) = i^v \int_{-\infty}^{+\infty} x^v f(x) dx = i^v m_v.$$

By expanding $C(u)$ in a Taylor series, we obtain

$$C(u) = 1 + \sum_{v=1}^{t} \frac{m_v}{v!} (iu)^v + R_t, \qquad (\text{M.A.5})$$

R_t is the truncation error ($\lim_{t \to \infty} R_t = 0$). Thus the characteristic function is also a *moment generating function*.

The probability density function of ξ is given by the inverse Fourier transform of $C(u)$:

$$P(\xi) = \frac{1}{2\pi} \int_{-\infty}^{+\infty} C(u) \exp(-iu\xi) du.$$

Knowledge of $C(u)$ is therefore sufficient for calculating $P(\xi)$.

By replacing $C(u)$ by (M.A.5), we obtain a further basic result: the probability density of ξ may be calculated if the moments m_v are known.

One of the qualities of the characteristic function is that it aids derivation of the probability density function of a sum of random variables. For example, if we have two independent random variables, ξ and η, the characteristic distribution $C(u)$ of their sum, $\xi + \eta$, is given by

$$C(u) = \langle \exp[iu(\xi + \eta)] \rangle = \langle \exp(iu\xi) \exp(iu\eta) \rangle$$
$$= \langle \exp(iu\xi) \rangle \langle \exp(iu\eta) \rangle = C_1(u) C_2(u).$$

The result is easily generalized to the case of the sum of n independent random variables. The characteristic function, $C(u)$, of $\xi = \xi_1 + \xi_2 + \cdots + \xi_n$, is then

$$C(u) = C_1(u) C_2(u) \cdots C_n(u). \qquad (\text{M.A.6})$$

The probability density function of ξ is given by the inverse Fourier transform of (M.A.6):

$$P(\xi_1 + \xi_2 + \cdots + \xi_n) = \frac{1}{2\pi} \int_{-\infty}^{+\infty} C(u) \exp[-iu(\xi_1 + \xi_2 + \cdots + \xi_n)] du.$$

Again, the probability density of ξ may be calculated by means of the characteristic functions C_j or, because of (M.A.5), by means of the moments in which $C(u)$ may expressed. In this last case, the concept of *cumulant* will be useful.

M.A.4 Cumulants of a distribution

It is often convenient to use the logarithm of the characteristic function, which is called the *cumulant generating function*: $K(iu) = \log C(u)$. Expanding them in a Taylor series, we obtain

$$\log C(u) = \sum_v \frac{k_v (iu)^v}{v!},$$

where k_v is the *cumulant* of the vth order of the distribution. An important property of the cumulants is that they are additive when a random variable is a sum of a set of independent random variables; i.e. if $\xi = \xi_1 + \xi_2$, then $C(u) = C_1(u)C_2(u)$; therefore

$$K(iu) = \log C(u) = \log C_1(u) + \log C_2(u)$$
$$= K_1(iu) + K_2(iu).$$

When $\xi = \xi_1 + \xi_2 + \cdots + \xi_n$, we can also write

$$C(u) = \exp\left(\sum_{s=1}^n K_s(iu)\right).$$

Then, according to Section M.A.3, we can write

$$P(\xi) = \frac{1}{2\pi} \int_{-\infty}^{+\infty} \exp\left(\sum_{s=1}^n K_s(iu)\right) \exp(-iu\xi)\, du$$

$$= \frac{1}{2\pi} \int_{-\infty}^{+\infty} \exp\left(-iu\xi + \sum_{s=1}^n K_s(iu)\right) du.$$

Knowledge of cumulants is therefore a condition for the calculation of $P(\xi)$. The cumulants are related to the moments of the distribution and vice versa. The relations may be found by observing that

$$\log(1+z) = z - \frac{z^2}{2} + \cdots + (-1)^{r-1}\frac{z^r}{r} + \cdots.$$

We obtain

$$\begin{aligned}
k_1 &= m_1, \\
k_2 &= m_2 - m_1^2, \\
k_3 &= m_3 - 3m_1 m_2 + 2m_1^3, \\
k_4 &= m_4 - 3m_2^2 - 4m_1 m_3 + 12 m_1^2 m_2 - 6 m_1^4, \\
&\text{etc.}
\end{aligned} \quad \text{(M.A.7)}$$

If $m_1 = 0$, equations (M.A.7) become,

$$\begin{aligned}
k_1 &= 0, \\
k_2 &= m_2, \\
k_3 &= m_3, \\
k_4 &= m_4 - 3m_2^2, \\
k_5 &= m_5 - 10 m_2 m_3, \\
k_6 &= m_6 - 10 m_3^2 - 15 m_2 m_4 + 30 m_2^3, \\
k_7 &= m_7 - 21 m_2 m_5 - 35 m_3 m_4 + 210 m_2^2 m_3, \\
&\text{etc.}
\end{aligned} \quad \text{(M.A.8)}$$

M.A.5 The normal or Gaussian distribution

Let us apply the above consideration to a normal distribution.

A random variable x is normally distributed with parameters (m, σ) with $\sigma > 0$, if its probability density function is given by

$$f(x) = \frac{1}{\sigma\sqrt{2\pi}} \exp\left(-\frac{(x-m)^2}{2\sigma^2}\right). \tag{M.A.9}$$

The graph of (M.A.9) is symmetrical with respect to $x = m$, with maximum ordinate $1/\sigma\sqrt{2\pi}$ at $x = m$. The graph has inflection points at $x = m \pm \sigma$. It is thus apparent that

$$\langle x \rangle = m, \quad \langle (x - \langle x \rangle)^2 \rangle = \sigma^2.$$

The characteristic function is

$$C(u) = \langle \exp(iux) \rangle = \exp\left(miu - \frac{1}{2}\sigma^2 u^2\right).$$

The most convenient form of the normal distribution for tabulation is that corresponding to a *standardized random variable*, $x' = (x - m)/\sigma$. Its probability density function is

$$\phi(x) = \frac{1}{\sqrt{2\pi}} \exp\left(-\frac{x^2}{2}\right); \tag{M.A.10}$$

its cumulative distribution function is

$$\Phi(x) = \frac{1}{\sqrt{2\pi}} \int_{-\infty}^{x} \exp\left(-\frac{t^2}{2}\right) dt; \tag{M.A.11}$$

its characteristic function is

$$C(u) = \exp(-u^2/2).$$

It is easy to verify that

$$\langle x' \rangle = \left\langle \frac{x-m}{\sigma} \right\rangle = 0, \quad \langle (x' - \langle x' \rangle)^2 \rangle = 1.$$

If x is distributed according to (M.A.9), then

$$\langle x^n \rangle = \begin{cases} = 1 \times 3 \times \cdots \times (n-1)\sigma^n & \text{for even } n, \\ = 0 & \text{for odd } n; \end{cases} \tag{M.A.12}$$

$$\langle |x|^n \rangle = \begin{cases} = 1 \times 3 \times \cdots \times (n-1)\sigma^n & \text{for even } n, \\ = \sqrt{\frac{2}{\pi}} 2^n n! \sigma^{2n+1} & \text{for odd } n. \end{cases} \tag{M.A.13}$$

In fact, since

$$\int_{-\infty}^{+\infty} \exp(-\alpha x^2) dx = \sqrt{\frac{\pi}{\alpha}}, \tag{M.A.14}$$

derivatizing (M.A.14) k times with respect to α gives,

$$\int_{-\infty}^{+\infty} x^{2k} \exp(-\alpha x^2) dx = \frac{1 \times 3 \times \cdots \times (2k-1)}{2^k} \sqrt{\frac{\pi}{\alpha^{2k+1}}}. \tag{M.A.15}$$

If $a = 1/2\sigma^2$ is substituted in (M.A.15), we obtain (M.A.12) and the first of eqations (M.A.13). The second of equations (M.A.13) is proved if one considers that

$$\langle |x|^{2k+1} \rangle = \int_{-\infty}^{+\infty} |x|^{2k+1} f(x) dx = 2 \int_0^\infty x^{2k+1} f(x) dx$$

$$= \frac{2}{\sigma\sqrt{2\pi}} \int_0^\infty x^{2k+1} \exp\left(\frac{-x^2}{2\sigma^2}\right) dx.$$

If $y = x^2/2\sigma^2$, we obtain

$$\langle |x|^{2k+1} \rangle = \frac{2}{\sqrt{\pi}} \frac{(2\sigma^2)^{k+1}}{2\sigma} \int_0^\infty y^k \exp(-y) \, dy.$$

A property of the Γ function (see Appendix M.C) gives the second of equations (M.A.13).

M.A.6 The central limit theorem

Let $\xi_1, \xi_2, \ldots, \xi_n$ be a sequence of independent random variables: m_j and σ_j^2 are the mean and the variance value of ξ_j. No matter what the distribution of the independent variables ξ_j, subject to certain very general conditions, the sum $\xi = \xi_1 + \xi_2 + \cdots + \xi_n$ tends to the normal distribution as $n \to \infty$ and has parameters (m, σ), given by

$$m = m_1 + m_2 + \cdots + m_n, \quad \sigma^2 = \sigma_1^2 + \sigma_2^2 + \cdots + \sigma_n^2.$$

When all the random variables have the same distribution, we will say that Lindeberg–Lévy conditions are satisfied.

The central limit theorem does not hold with sufficient accuracy if n is not sufficiently large, or when the distributions are not the same and one or a few dominate the sums. Then, some supplementary terms should be taken into account giving the so-called Gram–Charlier and Edgeworth series (see Chapter 5).

The central limit theorem remains valid under certain conditions, even when the n variables are not independent. Bernstein (1927) showed that the central limit theorem would apply if the variables are not too closely related. A particular example occurs when ξ_i is correlated to a subset of the n variables; in crystallography, such an example reflects the correlation of one atom with its neighbours, while little correlation exists with atoms far apart. If Bernstein's conditions are satisfied one has to take care of the correlation between the random variables when m and σ^2 are calculated.

M.A.7 Multivariate distributions

The cumulative distribution function of n random variables, $\xi_1, \xi_2, \ldots, \xi_n$, is defined by

$$F(x_1, x_2, \ldots, x_n) = P\{\xi_1 \leq x_1, \xi_2 \leq x_2, \ldots, \xi_n \leq x_n\}.$$

F is clearly a single-valued, real and non-negative function of the x_j values. In the case of continuous random variables, the following properties hold:

1. The probability density function is given by

$$f(x_1,\ldots,x_n) = \frac{\delta^n f(x_1,\ldots,x_n)}{\delta x_1 \delta x_2 \cdots \delta x_n}.$$

2. *The marginal probability density* of $\xi_1, \xi_2, \ldots, \xi_k$ is given by

$$f(x_1, x_2, \ldots, x_k) = \int_{-\infty}^{+\infty} \cdots \int_{-\infty}^{+\infty} f(x_1, x_2, \ldots, x_n)\, dx_{k+1} \cdots dx_n, \quad \text{(M.A.16)}$$

and the *conditional probability density* of $\xi_1, \xi_2, \ldots, \xi_k$ when $\xi_{k+1} = a_{k+1}, \ldots, \xi_n = a_n$ is given by

$$f(x_1, \ldots, x_k | x_{k+1}, \ldots, x_n) = \frac{f(x_1, x_2, \ldots, x_n)}{\int_{-\infty}^{+\infty} \cdots \int_{-\infty}^{+\infty} f(x_1, \ldots, x_k, a_{k+1}, \ldots, a_n) dx_1 \cdots dx_k}. \quad \text{(M.A.17)}$$

3. The expected value of the function $g(\xi_1, \ldots, \xi_n)$ is

$$\langle g(\xi_1, \ldots, \xi_n) \rangle = \int_{-\infty}^{+\infty} \cdots \int_{-\infty}^{+\infty} g(x_1, \ldots, x_n) f(x_1, \ldots, x_n) dx_1 \cdots dx_n. \quad \text{(M.A.18)}$$

4. The moment of order $v_1 + v_2 + \cdots v_n$ is defined as

$$m_{v_1 - v_n} = \int_{-\infty}^{+\infty} \cdots \int_{-\infty}^{+\infty} x_1^{v_1} x_2^{v_2} \cdots x_n^{v_n} f(x_1, \ldots, x_n) dx_1 \cdots dx_n. \quad \text{(M.A.19)}$$

5. The characteristic function is

$$C(u_1, u_2, \ldots, u_n) = \langle \exp(i(u_1 x_1 + u_2 x_2 + \cdots + u_n x_n)) \rangle$$

$$= \int_{-\infty}^{+\infty} \cdots \int_{-\infty}^{+\infty} \exp(i(u_1 x_1 + \cdots u_n x_n)) \quad \text{(M.A.20)}$$

$$\times f(x_1, \ldots, x_n) dx_1 \cdots dx_n,$$

from which the *joint probability density* function is derived:

$$f(x_1, x_2, \ldots, x_n) = \frac{1}{(2\pi)^n} \int_{-\infty}^{+\infty} \cdots \int_{-\infty}^{+\infty} \exp[-i(u_1 x_1 + \cdots + u_n x_n)]$$

$$\times C(u_1, \ldots, u_n) du_1 \cdots du_n. \quad \text{(M.A.21)}$$

6. If every x_i is a single-valued function of the random variables $y_1, \ldots, y_n$ according to $x_i = h_i(y_1, \ldots, y_n)$, then

$$f(x_1, x_2 \ldots, x_n) dx_1 dx_2 \cdots dx_n = f[h_1(y_1, \ldots, y_n), \ldots, h_n(y_1, \ldots, y_n)]$$

$$\times |J| dy_1 \cdots dy_n, \quad \text{(M.A.22)}$$

where j is the *Jacobian* of the transformation given by

$$J = \frac{\delta(x_1, \ldots, x_n)}{\delta(y_1, \ldots, y_n)}.$$

7. The cumulant generating function $K(u_1, u_2, \ldots, u_n)$ is defined to be

$$K(u_1, u_2, \ldots, u_n) = \log C(u_1, u_2 \ldots, u_n).$$

We can now give, for bivariate, trivariate, and tetravariate distributions, the relation between cumulants and moments of lower order in the case in which $m_{100} = m_{010-} = 0$. For bivariate distribution:

$$\begin{aligned}
&k_{10} = k_{01} = 0, \\
&k_{20} = m_{20}, \quad k_{11} = m_{11}, \\
&k_{30} = m_{30}, \quad k_{21} = m_{21}, \\
&k_{40} = m_{40} - 3m_{20}^2, \quad k_{31} = m_{31} - 3m_{20}m_{11}, \\
&k_{22} = m_{22} - m_{20}m_{02} - 2m_{11}^2, \\
&k_{50} = m_{50} - 10m_{30}m_{20}, \\
&k_{41} = m_{41} - 4m_{30}m_{11} - 6m_{21}m_{20}, \\
&k_{32} = m_{32} - m_{30}m_{02} - 6m_{21}m_{11} - 3m_{12}m_{20}.
\end{aligned} \quad \text{(M.A.23)}$$

For trivariate distribution:

$$\begin{aligned}
&k_{111} = m_{111}, \\
&k_{211} = m_{211} - m_{200}m_{011} - 2m_{110}m_{101}, \\
&k_{311} = m_{311} - 3m_{210}m_{101} - m_{300}m_{011} - 3m_{111}m_{200} - 3m_{201}m_{110}, \\
&k_{221} = m_{221} - m_{200}m_{021} - 4m_{111}m_{110} - 2m_{120}m_{101} - 2m_{210}m_{011} - m_{020}m_{201}.
\end{aligned} \quad \text{(M.A.24)}$$

For tetravariate distribution:

$$k_{1111} = m_{1111} - m_{1100}m_{0011} - m_{1010}m_{0101} - m_{1001}m_{0110}, \quad \text{(M.A.25)}$$

etc.

In a number of distributions used in the text, the following relationships are true:

$$m_{10\cdots} = m_{010\cdots} = m_{30\cdots} = m_{50\cdots} = m_{110\cdots} = m_{0110\cdots} = 0. \quad \text{(M.A.26)}$$

M.A.8 Evaluation of the moments in structure factor distributions

It is normal procedure to expand the characteristic function of a joint probability distribution of s.f.s. (see Sections 4.A.2–4.A.5) in terms of cumulants and then, by means of formulas such as (M.A.23)–(M.A.26), to expand them in terms of joint moments of trigonometric structure factors. We will show here how such moments may be evaluated:

$$m_{pq-r} = \langle \gamma_1^p \gamma_2^q \cdots \gamma_n^r \rangle = \int_0^1 \int_0^1 \int_0^1 \gamma_1^p \gamma_2^q \cdots \gamma_n^r p \, \mathrm{d}x \mathrm{d}y \mathrm{d}z. \quad \text{(M.A.27)}$$

In (M.A.27), γ_i represents the trigonometric form ξ_i of the jth s.f. or its real or imaginary part: P is the probability density function of the independent random variables (e.g. x, y, and z). When no *prior* information is available, x, y, and z may be assumed to be independent random variables uniformly distributed over the range 0–1 so that $P \equiv 1$. The moment m_{pq-r} can then be evaluated from (M.A.27) for any space group by substituting the appropriate trigonometric expression for the γ_j values.

In any space group it is the symmetry operators which define the trigonometric expression of the s.f. Thus, there is no need to know the explicit trigonometric expression of the s.f. if one is able to handle symmetry operators.

Let us first consider monovariate distributions in an n.cs. space group of order m. then (see definitions in Section 4.A.3)

$$m_1 = \langle \xi(\boldsymbol{h}) \rangle = \left\langle p \sum_{s=1}^{m/p} \exp\left(2\pi i \boldsymbol{h} \boldsymbol{C}_s \boldsymbol{r}\right) \right\rangle = 0$$

$$m_2 = \langle \xi(\boldsymbol{h})\xi(-\boldsymbol{h}) \rangle$$

$$= \left\langle p^2 \sum_{s_1,s_2=1}^{m/p} \exp\left[2\pi i \boldsymbol{h} \left(\boldsymbol{C}_{s_1} - \boldsymbol{C}_{s_2}\right)\boldsymbol{r}\right] \right\rangle$$

$$= p^2 \left\langle \sum_{s_1 \neq s_2=1}^{m/p} \exp\left[2\pi i \boldsymbol{h} \left(\boldsymbol{C}_{s_1} - \boldsymbol{C}_{s_2}\right)\boldsymbol{r}\right] \right\rangle$$

$$+ p^2 \left\langle \sum_{s_1 \neq s_2=1}^{m/p} \exp\left[2\pi i \boldsymbol{h} \left(\boldsymbol{C}_{s_1} - \boldsymbol{C}_{s_2}\right)\boldsymbol{r}\right] \right\rangle.$$

Only the first of the two terms on the right-hand side provides non-vanishing contributions. Then

$$m_2 = p^2 \frac{m}{p} = pm.$$

Accordingly (see again the definition in Section 4.A.3)

$$\langle |\xi'(\boldsymbol{h})|^2 \rangle = \langle \xi'(\boldsymbol{h})\xi'(-\boldsymbol{h}) \rangle = m. \qquad \text{(M.A.28)}$$

For a cs. space group of order m,

$$\xi(\boldsymbol{h}) = p \sum_{s=1}^{m/p} \cos\left(2\pi \bar{\boldsymbol{h}} \boldsymbol{C}_s \boldsymbol{r}\right).$$

Simple calculations show that (M.A.28) is also valid in cs. space groups.

Let us now estimate, for a three-variate structure factor distribution, the mixed moment m_{111} when all three reflections are general ($p \equiv 1$). Then,

$$m_{111} = \langle \xi(\boldsymbol{h}_1)\xi(\boldsymbol{h}_2)\xi(-\boldsymbol{h}_1-\boldsymbol{h}_2) \rangle$$

$$= \left\langle \sum_{s_1,s_2,s_3=1}^{m} \exp\left\{2\pi \left[\bar{\boldsymbol{h}}_1\left(\boldsymbol{C}_{s_1} - \boldsymbol{C}_{s_3}\right) + \bar{\boldsymbol{h}}_2(\boldsymbol{C}_{s_2} - \boldsymbol{C}_{s_3})\right] \cdot \boldsymbol{r}\right\} \right\rangle.$$

Which gives a non-vanishing contribution only when $s_1 = s_2 = s_3$. This condition is satisfied m times, and at each time contribution to m_{111} is unitary. Therefore (if reflections are of general type)

$$m_{111} = m \qquad \text{(M.A.29)}$$

in any space group. It is worthwhile noting that (M.A.29) is not valid when reflections are of special type.

A further useful mixed moment is

$$m_{1111} = \langle \xi'(\mathbf{h}_1)\xi'(\mathbf{h}_2)\xi'(\mathbf{h}_3)\xi'(-\mathbf{h}_i - \mathbf{h}_2 - \mathbf{h}_3)\rangle.$$

Again, we calculate m_{1111} for general reflections for which $p \equiv 1$. Then,

$$m_{1111} = \left\langle \sum_{s_1,s_2,s_3,s_4=1}^{m} \exp\left\{2\pi \left[\overline{\mathbf{h}_1}(\mathbf{C}_{s_1} - \mathbf{C}_{s_4}) \right.\right.\right. \\ \left.\left.\left. + \overline{\mathbf{h}_2}(\mathbf{C}_{s_2} - \mathbf{C}_{s_4}) + \overline{\mathbf{h}_3}(\mathbf{C}_{s_3} - \mathbf{C}_{s_4})\right] \cdot \mathbf{r}\right\}\right\rangle.$$
(M.A.30)

The right-hand side of (M.A.30) provides non-vanishing contributions only when $s_1 = s_2 = s_3 = s_4$; therefore

$$m_{1111} = m \qquad \text{(M.A.31)}$$

for all the space groups. Again we notice that (M.A.31) is not valid when reflections are not of general type.

The reader will easily verify in $P\overline{1}$ the following moment values:

$$\begin{aligned} m_{20} &= \langle \xi^2(\mathbf{h}) \rangle = 2 & m_{111} &= \langle \xi(\mathbf{h}_1)\xi(\mathbf{h}_2)\xi(\mathbf{h}_1+\mathbf{h}_2) \rangle = 2 \\ m_{40} &= 6 & m_{113} &= \langle \xi(\mathbf{h}_1)\xi(\mathbf{h}_2)\xi^3(\mathbf{h}_1+\mathbf{h}_2) \rangle = 6. \end{aligned}$$
(M.A.32)

M.A.9 Joint probability distributions of the signs of the structure factors

In probabilistic theories for the determination of signs of s.f.s, the first aim is to find multivariate distributions of s.f.s with their magnitudes and signs. Since the magnitudes are obtained from measurements, it may be useful to find the joint probability distribution of signs under the conditions that the corresponding magnitudes of s.f.s are known. If we denote the jth s.f., its modulus, and its sign by E_j, R_j, and s_j, respectively, and the joint probability distribution function by P, then (Naya et al., 1964; Allegra, 1965),

$$P(R_1,\ldots,R_n) = \sum_{s_1=\pm 1}\cdots\sum_{s_n=\pm 1} P(R_1 s_1,\ldots,R_n s_n), \qquad \text{(M.A.33)}$$

$$P(s_1,\ldots,s_n) = P(s_1,\ldots,s_n|R_1,\ldots,R_n)$$
$$= \frac{P(R_1 s_1,\ldots,R_n s_n)}{\sum_{s_1=\pm 1}\cdots\sum_{s_n=\pm 1} P(R_1 s_1,\ldots,R_n s_n)}. \qquad \text{(M.A.34)}$$

(M.A.33), (M.A.34), and following formulas may be considered as particular applications of (M.A.16) and (M.A.17). If $P(s_1)$ is the marginal probability that the sign of E_1 is s_1 irrespective of the other signs $s_2,\ldots,s_n$, we have,

$$P(s_1) = \sum_{s_2=\pm 1}\cdots\sum_{s_n=\pm 1} P(s_1,\ldots,s_n). \qquad \text{(M.A.35)}$$

In general,

$$P(s_1,\ldots,s_r) = \sum_{s_{r+1}=\pm 1}\cdots\sum_{s_n=\pm 1} P(s_1,\ldots,s_n). \qquad \text{(M.A.36)}$$

When a sign product is considered, i.e. $s_1 s_2$, the probability of $s_1 s_2$ being $+1$ is

$$P^+(s_1 s_2) = P^{+,+}(s_1, s_2) + P^{-,-}(s_1, s_2). \qquad \text{(M.A.37)}$$

For a triple product $s_1 s_2 s_3$, then

$$P^+(s_1 s_2 s_3) = P^{+,+,+} + P^{+,-,-} + P^{-,+,-} + P^{-,-,+}. \qquad \text{(M.A.38)}$$

In general,

$$P^+(s_1 s_2 \ldots s_r) = \sum P^{s_1,\ldots,s_r}, \qquad \text{(M.A.39)}$$

where the summation goes over all the sets of signs for which $s_1 s_2 \ldots s_r = +$. The normalizing constants of the sign probabilities are readily found by means of

$$P^+ = P^+/(P^+ + P^+). \qquad \text{(M.A.40)}$$

In the end, the conditional probability for some signs given some others is (see equation (M.A.17))

$$P(s_1, s_2, \ldots, s_r | s_{r+1}, \ldots, s_n) = \frac{P(s_1, s_2, \ldots, s_n)}{\sum_{s_1,\ldots,s_r=\pm 1} P(s_1, s_2, \ldots, s_n)}. \qquad \text{(M.A.41)}$$

We are now able to calculate, from the distribution $P(s)$, the values $\langle s \rangle$ and the associated variance V:

$$\langle s \rangle = sP(s) + (-s)P(-s) = s(2P(s) - 1), \qquad \text{(M.A.40$'$)}$$

$$V = \langle s^2 \rangle - \langle s \rangle^2 = s - [2P(s) - 1]^2 = 4P(s)(1 - P(s)). \qquad \text{(M.A.40$''$)}$$

M.A.10 Some measures of location and dispersion in the statistics of directional data

The usual linear measures of location and dispersion can be inappropriate for circular distributions; e.g. the arithmetic mean gives absurd results if the observed angles in a sample of size 2 are 5° and 355°. The arithmetic mean, in fact, is 180°, whereas geometrical intuition indicates 0°. Suitable measures of location and dispersion have been developed in order to analyse directional data (see Mardia, 1972). We mention only two of them.

1. *Mean direction*

    Let P_j, $j = 1, \ldots, n$, be the points on the circumference of the unit circle corresponding to the angle Φ_j. If O is the centre of the circle, the mean direction $\langle \Phi \rangle_c$ is the direction of the resultant of the unit vectors $OP_1, \ldots, OP_n$.

    If

$$C = \frac{1}{n} \sum_{j=1}^{n} \cos \Phi_j = R \cos \langle \Phi \rangle_c, \qquad \text{(M.A.42)}$$

$$S = \frac{1}{n} \sum_{j=1}^{n} \sin \Phi_j = R \sin \langle \Phi \rangle_c, \qquad \text{(M.A.43)}$$

then nR is the length of the resultant and
$$\langle\Phi\rangle_c = \tan^{-1}(S/C). \tag{M.A.44}$$
It may easily be shown that $\langle\Phi\rangle_c$ has the following property:
$$\sum_{j=1}^{n} \sin\left(\Phi_j - \langle\Phi\rangle_c\right) = 0, \tag{M.A.45}$$
which corresponds in the linear case to
$$\sum_{j=1}^{n}\left(y_j - \langle y\rangle\right) = 0.$$
If Φ is a continuous variable with probability distribution $P(\Phi)$, then (M.A.42) and (M.A.43) are substituted by
$$C = \int \cos\Phi P(\Phi)d\Phi, \quad S = \int \sin\Phi P(\Phi)d\Phi.$$

2. *Circular variance*

A measure of angular dispersion of the values Φ_j about the angle θ is
$$D = \frac{1}{n}\sum_{j=1}^{n}\left[1 - \cos\left(\Phi_j - \theta\right)\right].$$
The dispersion is smallest about $\langle\Phi\rangle_c$. In fact, on equating the derivative with respect to θ to zero, (M.A.44) is obtained.

The circular variance is defined as
$$V_c = 1 - \frac{1}{n}\sum_{j=1}^{n}\cos\left(\Phi_j - \langle\Phi\rangle_c\right) = 1 - R.$$
We see immediately that $0 \leq V_c \leq 1$: in fact, if the angles Φ_j are very close to $\langle\Phi\rangle_c$ then $R \simeq 1$ and $V_c \simeq 0$. On the other hand, if the Φ_j are highly dispersed, then $R \simeq 0$ and $V_c \simeq 1$. It may easily be seen that V_c is invariant under a change of the zero direction.

An appropriate transformation of the variance is
$$S_c = [-2\ln(1 - V_c)]^{1/2},$$
which gives a measure somewhat analogous to the ordinary standard deviation on the line.

Let $l_j \exp(i\Phi_j), j = 1, \ldots, n$, be n vectors in the complex plane, and
$$C = \frac{1}{n}\sum_{j=1}^{n} l_j \cos\Phi_j = R\cos\langle\Phi\rangle_c,$$
$$S = \frac{1}{n}\sum_{j=1}^{n} l_j \sin\Phi_j = R\sin\langle\Phi\rangle_c,$$
$$R = \left(C^2 + S^2\right)^{1/2}.$$
Then nR is the length of the resultant vector and $\langle\Phi\rangle_c = \tan^{-1}(S/C)$ can be assumed to be the mean direction. $\langle\Phi\rangle_c$ has the following property:
$$\sum_{j=1}^{n} l_j \sin\left(\Phi_j - \langle\Phi\rangle_c\right) = 0.$$
The quantity $1 - R$ can again represent the variance.

APPENDIX M.B
MOMENTS OF THE P(Z) DISTRIBUTIONS

In Section 2.4, the probability $P(z)$ was obtained, where $z = R^2$ is the square of the normalized structure factor modulus. The variable z can be considered, in every respect, to be a *gamma variable* (Srinivasan and Subramanian, 1964); it satisfies a probability density function of type

$$\gamma_l(z) = \frac{\exp(-z)z^{l-1}}{\Gamma(l)}, \quad 0 \leq z \leq \infty$$

where Γ is the well-known *gamma function* defined by equation (M.C.1). In fact, if we consider the case $l = 1$ we obtain the acentric z-distribution (2.15) (z is then a γ_1 variable); if we choose $l = 1/2$ (then z is a $\gamma_{1/2}$ variable), we obtain (see (M.C.4)) the distribution

$$P(u)du = \frac{1}{\sqrt{\pi u}} \exp(-u)du,$$

which reduces to (2.16) when the new variable $z = u/2$ is introduced. The above properties may be used to easily calculate the higher moments for the centric and for the acentric distributions (see Section M.A.5):

$$<R^{2n}>_{\bar{1}} = \sqrt{\frac{2}{\pi}} \int_0^\infty R^{2n} \exp\left(-\frac{R^2}{2}\right) dR = \frac{(2n)!}{2^n n!}$$

$$<R^{2n+1}>_{\bar{1}} = \sqrt{\frac{2}{\pi}} \int_0^\infty R^{2n+1} \exp\left(-\frac{R^2}{2}\right) dR = \left(\frac{2}{\pi}\right)^{1/2} 2^n n!$$

$$<R^{2n}>_1 = 2 \int_0^\infty R^{2n+1} \exp(-R^2) dR = n!$$

$$<R^{2n+1}>_1 = 2 \int_0^\infty R^{2n+2} \exp(-R^2) dR = \Gamma(n + 3/2).$$

APPENDIX M.C
THE GAMMA FUNCTION

For real positive values of x, the gamma function $\Gamma(x)$ is defined by the definite integral

$$\Gamma(x) = \int_0^\infty t^{x-1} \exp(-t)\, dt. \qquad (M.C.1)$$

If we integrate by parts, it follows that

$$\Gamma(x+1) = x\Gamma(x), \qquad (M.C.2)$$

from which it is evident that, if x is a positive integer,

$$\Gamma(m+1) = m!. \qquad (M.C.3)$$

Table M.C.1

X	Γ(x)	(x)	Γ(x)
1.00	1.000	1.52	0.887
1.02	0.989	1.54	0.888
1.04	0.978	1.56	0.890
1.06	0.969	1.58	0.891
1.08	0.960	1.60	0.894
1.10	0.951	1.62	0.896
1.12	0.944	1.64	0.899
1.14	0.936	1.66	0.902
1.16	0.930	1.68	0.905
1.18	0.924	1.70	0.909
1.20	0.918	1.72	0.913
1.22	0.913	1.74	0.917
1.24	0.909	1.76	0.921
1.26	0.904	1.78	0.926
1.28	0.901	1.80	0.931
1.30	0.897	1.82	0.937
1.32	0.895	1.84	0.943
1.34	0.892	1.86	0.949
1.36	0.890	1.88	0.955
1.38	0.889	1.90	0.962
1.40	0.887	1.92	0.969
1.42	0.886	1.94	0.976
1.44	0.886	1.96	0.984
1.46	0.886	1.98	0.992
1.48	0.886	2.00	1.000
1.50	0.886		

Special values of the gamma function are

$$\Gamma\left(\frac{1}{2}\right) = \sqrt{\pi},$$

$$\Gamma\left(m + \frac{1}{2}\right) = \frac{1 \times 3 \times 5 \times \cdots \times (2m-1)}{2^m} \sqrt{\pi}. \quad \text{(M.C.4)}$$

The only minimum occurs at about $x_0 = 1.4616$. In Table M.C.1 the values of $\Gamma(x)$ for $1 < x < 2$ are shown.

APPENDIX M.D
THE HERMITE AND LAGUERRE POLYNOMIALS

The Hermite polynomials $H_v(x)$ are defined by the equation

$$H_v(x) = (-1)^v \exp\left(\frac{1}{2}x^2\right) \frac{d^v}{dx^v}\left[\exp\left(-\frac{1}{2}x^2\right)\right]. \quad \text{(M.D.1)}$$

The lowest-order polynomials are

$$\begin{aligned}
&H_0(x) = 1, \quad H_4(x) = x^4 - 6x^2 + 3,\\
&H_1(x) = x, \quad H_5(x) = x^5 - 10x^3 + 15x,\\
&H_2(x) = x^2 - 1, \quad H_6(x) = x^6 - 15x^4 + 45x^2 - 15,\\
&H_3(x) = x^3 - 3x, \quad H_7(x) = x^7 - 21x^5 + 105x^3 - 105x.
\end{aligned} \quad \text{(M.D.2)}$$

The orthogonality of the Hermite polynomials is easily seen by means of the two relations

$$\int_{-\infty}^{+\infty} \exp\left(-\frac{x^2}{2}\right) H_m(x) H_n(x) \, dx = 0, \quad m \neq n,$$

$$\int_{-\infty}^{+\infty} \exp\left(-\frac{x^2}{2}\right) [H_n(x)]^2 \, dx = \sqrt{2\pi} n!.$$

Thus a function $f(x)$ may be expanded in the orthogonal series

$$f(x) A_0 H_0(x) + A_1 H_1(x) + A_2 H_2(x) + \cdots,$$

where

$$A_\nu = \frac{1}{\sqrt{2\pi}\nu!} \int_{-\infty}^{+\infty} \exp\left(-\frac{x^2}{2}\right) f(x) H_\nu(x) \, dx.$$

Useful relations are

$$H_\nu(-x) = (-1)^\nu H_\nu(x), \quad H_{2\nu-1}(0) = 0, \quad \frac{d}{dx} H_\nu(x) = \nu H_{\nu-1}(x),$$

$$\sum_{\nu=0}^\infty \frac{H_\nu(x) H_\nu(y)}{\nu!} t^\nu = \frac{2}{\sqrt{1-t^2}} \exp\left(-\frac{1}{2(1-t^2)}\left(t^2 x^2 + t^2 y^2 - 2txy\right)\right) \quad (|t| < 1).$$
(M.D.3)

If

$$\phi(x) = \frac{1}{\sqrt{2\pi}} \exp\left(-\frac{x^2}{2}\right),$$

(M.D.1) may be rewritten as

$$\phi^{(\nu)}(x) = (-1)^\nu H_\nu(x) \phi(x).$$

Table (M.A.1) then proves useful in deriving the values of the Hermite polynomials up to sixth order. Polynomials of higher order may be derived by the recurrence relation

$$H_{\nu+1}(x) = x H_\nu(x) - \nu H_{\nu-1}(x). \quad \text{(M.D.4)}$$

The following formulas may be useful:

$$\int_{-\infty}^{+\infty} \exp(-x^2) x^m H_n(x) \, dx = \begin{cases} 0 & \text{if } n > m \quad \text{or} \quad n-m = \text{odd} \\ \dfrac{m!\sqrt{\pi}}{4^p p!}, & \text{if } m \geq n, \quad p = \dfrac{m-n}{2} \end{cases}$$
(M.D.5)

$$\int_0^{+x} \exp(-t^2) H_n(t)\, dt = (H_{n-1}(0) - H_{n-1}(x)) \exp(-x^2). \qquad \text{(M.D.6)}$$

Laguerre polynomials $L_\nu(x)$ are defined by the equation (Rodrigues' formula)

$$L_\nu(x) = \frac{1}{\nu!} \exp(x) \frac{d^\nu}{dx^\nu} \left[\exp(-x) x^\nu \right].$$

The lowest-order polynomials are

$L_0(x) = 1$
$L_1(x) = 1 - x$
$L_2(x) = 1 - 2x + \dfrac{x^2}{2}$
$L_3(x) = \left(6 - 18x + 9x^2 - x^3\right)/6$
$L_4(x) = \left(24 - 96x + 72x^2 - 16x^3 + x^4\right)/24$
$L_5(x) = \left(120 - 600x + 600x^2 - 200x^3 + 25x^4 - x^5\right)/120$
$L_6(x) = \left(720 - 4320x + 5400x^2 - 2400x^3 + 450x^4 - 36x^5 + x^6\right)/720.$

The orthogonality of the Laguerre polynomials is easily seen by means of the two relations

$$\int_0^\infty \exp(-x) L_n(x) L_m(x) = 0$$

$$\int_0^\infty \exp(-x) (L_n(x))^2 d(x) = 1.$$

Useful relations are

$$\frac{d}{dx}(L_n(x) - L_{n+1}(x)) = L_n(x)$$

$$(n+1) L_{n+1}(x) = (2n + 1 - x) L_n(x) - n L_{n-1}(x).$$

From Rodrigues' formula, may be derived:

$$\int_0^\infty \exp(-x) x^m L_n(x) dx = \begin{cases} 0 & n > m \\ \dfrac{(-1)^n (m!)^2}{n!(m-n)!} & m \geq n \end{cases} \qquad \text{(M.D.7)}$$

$$\int_0^x \exp(-t) L_n(t) dt = (L_{n-1}(x) - L_n(x)) \exp(-x). \qquad \text{(M.D.8)}$$

APPENDIX M.E
SOME RESULTS IN THE THEORY OF BESSEL FUNCTIONS

M.E.1 Bessel functions

Bessel functions of the first kind may be defined via the Laurent expansion of the generating function

$$\exp\left[\frac{1}{2} z \left(t - \frac{1}{t}\right)\right].$$

The coefficient of t^n in the expansion is the Bessel coefficient of argument z and order n and it is denoted by $J_n(z)$:

$$\exp\left[\frac{z}{2}\left(t - \frac{1}{t}\right)\right] = \sum_{n=-\infty}^{+\infty} t^n J_n(z), \tag{M.E.1}$$

where

$$\begin{aligned} J_n(z) &= \sum_{m=0}^{\infty} \frac{(z/2)^{n+m}}{(n+m)!} \frac{(-z/2)^m}{m!} \\ &= \sum_{m=0}^{\infty} \frac{(-1)^m (z/2)^{n+2m}}{m!(n+m)!}. \end{aligned} \tag{M.E.2}$$

In particular,

$$J_0(z) = 1 - \frac{z^2}{2^2} + \frac{z^4}{2^2 \times 4^2} - \frac{z^6}{2^2 \times 4^2 \times 6^2} + \cdots, \tag{M.E.3}$$

$$J_1(z) = \frac{z}{2} - \frac{z^3}{2^2 \times 4} + \frac{z^5}{2^2 \times 4^2 \times 6} - \frac{z^7}{2^2 \times 4^2 \times 6^2 \times 8} + \cdots. \tag{M.E.4}$$

$J_n(z)$ is an even function of z when n is even, and is an odd function when n is odd. Furthermore, $J_{-n}(z) = (-1)^n J_n(z)$.

In Fig. M.E.1, the function $J_n(z)$ is shown for $n = 0, 1, 2, 3$.

Equation (M.E.1) may be written in the form

$$\exp\left[\frac{z}{2}\left(t - \frac{1}{t}\right)\right] = J_0(z) + \sum_{n=1}^{\infty} \left\{t^n + (-1)^n t^{-n}\right\} J_n(z).$$

If $t = \pm \exp(i\theta)$, then the Jacobi expansion is obtained:

$$\begin{aligned} \exp(\pm i z \sin\theta) = J_0(z) &+ 2\sum_{n=1}^{\infty} J_{2n}(z) \cos 2n\theta \\ &\pm 2i \sum_{n=0}^{\infty} J_{2n+1}(z) \sin(2n+1)\theta, \end{aligned} \tag{M.E.5}$$

or

$$\exp(iz \cos\theta) = \sum_{n=-\infty}^{+\infty} i^n J_n(z) \cos n\theta. \tag{M.E.6}$$

Furthermore, if $t = \exp(-i\theta)$ and the right- and left-hand sides of (M.E.1) are multiplied by $\exp(in\theta)$, we obtain

$$\exp[i(n\theta - z \sin\theta)] = \sum_{m=-\infty}^{+\infty} \exp[i(n-m)\theta] J_m(z).$$

When this relation is integrated from α to $2\pi + \alpha$, the only term in the right-hand side which gives a non-zero contribution is that for which $m = n$. Then,

$$J_n(z) = \frac{1}{2\pi} \int_{\alpha}^{2\pi+\alpha} \exp[i(n\theta - z \sin\theta)] d\theta. \tag{M.E.7}$$

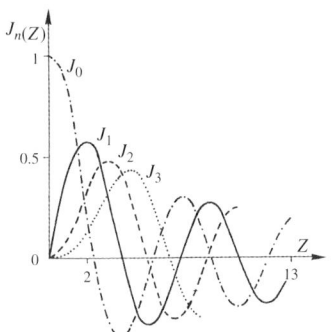

Fig. M.E.1
The functions $J_n(z)$ for $n = 0,1,2,3$.

When $n = 0$, (M.E.7) becomes

$$J_0(z) = \frac{1}{2\pi} \int_\alpha^{2\pi+\alpha} \exp(-iz\sin\theta)d\theta \qquad \text{(M.E.8)}$$

$$= \frac{1}{2\pi} \int_\alpha^{2\pi+\alpha} \exp(-iz\cos\theta)d\theta.$$

Further formulas which occur frequently in the main text are

$$\frac{i^n}{\pi} \int_0^\pi \exp(-iz\cos\phi)\cos n\phi\, d\phi = J_n(z) \qquad \text{(M.E.9)}$$

$$\int_0^\pi \exp(-iz\cos\phi)\sin n\phi\, d\phi = 0, \qquad \text{(M.E.10)}$$

$$J_0(Z^2 + z^2 - 2Zz\cos\phi)^{1/2} = \sum_{m=0}^\infty \varepsilon_m J_m(Z)J_m(z)\cos m\phi, \qquad \text{(M.E.11)}$$

where ε_m is the Newmann factor which is defined to be equal to 2 when n is not zero and to be equal to 1 when n is zero;

$$\int_0^\infty J_\nu(at)\exp(-p^2 t^2)t^{\mu-1}dt = \frac{\Gamma((\nu+\mu)/2)(a/2p)^\nu}{2p^\mu \Gamma(\nu+1)}$$

$$\times \exp\left(\frac{-a^2}{4p^2}\right) {}_1F_1\left(\frac{\nu-\mu}{2}+1; \nu+1; \frac{a^2}{4p^2}\right),$$

where $\mathrm{Re}(\nu+\mu) > 0$. Equation (M.E.12) is the Weber–Sonine integral formula, and ${}_1F_1$ is the confluent hypergeometric function defined in Section M.E.2. The integral is an expression in finite terms whenever $\mu - \nu$ is an even positive integer.

Particular cases of (M.E.12) are

$$\int_0^\infty J_n(at)\exp(-p^2 t^2)t^{n+1}\, dt = \frac{a^2}{(2p^2)^{n+1}}\exp\left(\frac{-a^2}{4p^2}\right), \qquad \text{(M.E.13)}$$

$$\int_0^\infty J_0(at)\exp\left(-\frac{1}{2}t^2\right)t^{n-1}\, dt = \frac{\Gamma(n/2)}{2^{1-n/2}}\exp\left(-\frac{a^2}{2}\right){}_1F_1\left(1-\frac{n}{2}; 1; \frac{a^2}{2}\right), \qquad \text{(M.E.14)}$$

$$\int_0^\infty J_1(at)\exp\left(-\frac{1}{2}t^2\right)t^{n-1}\, dt = a2^{(n-3)/2}\Gamma\left(\frac{1+n}{2}\right)\exp\left(-\frac{a^2}{2}\right)$$

$$\times {}_1F_1\left(\frac{3-n}{2}; 2; \frac{a^2}{2}\right)$$

(M.E.15)

(expression of ${}_1F_1$ for particular values of the parameters are given in Section M.E.2)

$$\frac{1}{2\pi}\int_0^\infty \int_0^{2\pi} \exp(-p^2 t^2 - iat\cos\phi)\, t\, dt\, d\phi = \frac{1}{2p^2}\exp\left(-\frac{a^2}{4p^2}\right). \qquad \text{(M.E.16)}$$

Equation (M.E.16) is readily delivered by combining (M.E.9) and (M.E.13). Let us now consider the generating function

$$\exp\left[i\frac{z}{2}\left(t - \frac{1}{t}\right)\right],$$

where z is a real number. If we expand $\exp(izt/2)$ and $\exp(-iz/2t)$ in a power series, we obtain

$$\exp\left[i\frac{z}{2}\left(t - \frac{1}{t}\right)\right] = \sum_{n=-\infty}^{+\infty} i^n t^n I_n(z),$$

where

$$I_n(z) = \sum_{m=0}^{\infty} \frac{(z/2)^{n+2m}}{m!(n+m)!} \quad (M.E.17)$$

are the modified Bessel functions of first kind. In particular, for $n = 1, 2$,

$$I_0(z) = 1 + \frac{z^2}{2^2} + \frac{z^4}{2^2 \times 4^2} + \frac{z^6}{2^2 \times 4^2 \times 6^2} + \cdots, \quad (M.E.18)$$

$$I_1(z) = \frac{z}{2} + \frac{z^3}{2^2 \times 4} + \frac{z^5}{2^2 \times 4^2 \times 6} + \frac{z^7}{2^2 \times 4^2 \times 6^2 \times 8} + \cdots. \quad (M.E.19)$$

From (M.E.2) and (M.E.17) the following formulas arise:

$$I_{-n}(z) = I_n(z), \quad \frac{d}{dz}I_0(z) = I_1(z), \quad I_n(-z) = (-1)^n I_n(z), \quad (M.E.20)$$

$$I_n(z) = i^{-n} J_n(iz), \quad I_{-n}(z) = i^n J_{-n}(iz), \quad (M.E.21)$$

$$I_{n+1}(z) = I_{n-1}(z) - \frac{2n}{z} I_n(z). \quad (M.E.22)$$

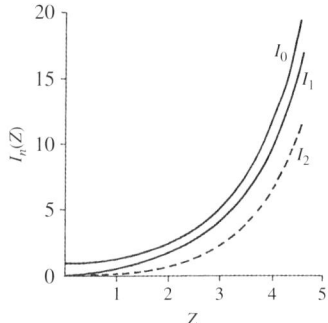

Fig. M.E.2
The functions $I_n(z)$ for $n = 0,1,2$.

I_0, I_1, and I_2 are shown in Fig. M.E.2.

If we write $z = i\alpha$ and $\theta = \pi/2 - \phi$ in (M.E.5), we obtain, because of equations (M.E.21)

$$\exp(\alpha \cos \phi) = I_0(\alpha) + 2\sum_{n=1}^{\infty} I_{2n}(\alpha) \cos 2n\phi + 2\sum_{n=0}^{\infty} I_{2n+1}(\alpha) \cos(2n+1)\phi, \quad (M.E.23)$$

$$I_n(z) = \frac{1}{2\pi}(i)^{-n} \int_{\alpha}^{2\pi+\alpha} \exp[in\theta + z \sin\theta] \, d\theta = \frac{1}{2\pi} \int_{\alpha}^{2\pi+\alpha} \exp[in\theta + z \cos\theta] \, d\theta. \quad (M.E.24)$$

In particular, when $n = 0$,

$$I_0(z) = \frac{1}{2\pi} \int_{\alpha}^{2\pi+\alpha} \exp(z \sin\theta) \, d\theta$$

$$= \frac{1}{2\pi} \int_{\alpha}^{2\pi+\alpha} \exp(z \cos\theta) \, d\theta. \quad (M.E.25)$$

Further formulas of frequent occurrence in the main text are

$$\frac{1}{\pi}\int_0^\pi \exp(z\cos\phi)\cos n\phi\, d\phi = I_n(z), \quad \text{(M.E.26)}$$

$$\int_0^\pi \exp(-z\cos\phi)\sin n\phi\, d\phi = 0, \quad \text{(M.E.27)}$$

$$I_0\left(Z^2 + z^2 + 2Zz\cos\phi\right)^{1/2} = \sum_{m=-\infty}^{+\infty} I_m(Z)I_m(z)\exp(im\phi)$$
$$= I_0(Z)I_0(z) + 2\sum_{m=1}^{\infty} I_m(Z)I_m(z)\cos m\phi, \quad \text{(M.E.28)}$$

$$\int_0^\infty I_n(at)\exp(-p^2 t^2) t^{n+1} dt = \frac{a^n}{(2p^2)^{n+1}}\exp(a^2/4p^2). \quad \text{(M.E.29)}$$

Successive differentiations of both sides of (M.E.29) with respect to p^2 yield expressions for

$$\int_0^\infty I_n(at)\exp(-p^2 t^2) t^{2m+n+1} dt, \quad m = 1, 2, \ldots.$$

A further useful formula is

$$\int_0^\infty I_m(at) I_m(bt)\exp(-p^2 t^2) t\, dt = \frac{1}{2p^2} I_m\left(\frac{ab}{2p^2}\right)\exp\left(\frac{a^2+b^2}{4p^2}\right). \quad \text{(M.E.30)}$$

Successive differentiations of both sides of (M.E.30) with respect to p^2 yield formulae for

$$\int_0^\infty I_m(at)I_m(bt)\exp(-p^2 t^2) t^{2n+1} dt, \quad n = 1, 2, \ldots.$$

Furthermore,

$$\int_0^\infty J_m(at)J_m(bt)\exp(-p^2 t^2) t\, dt = \frac{1}{2p^2} I_m\left(\frac{ab}{2p^2}\right)\exp\left(-\frac{a^2+b^2}{4p^2}\right). \quad \text{(M.E.31)}$$

M.E.2 Generalized hypergeometric functions

The *generalized hypergeometric function* with p numerator and q denominator parameters is defined as

$$_pF_q(a_1, a_2, \ldots, a_p; b_1, b_2, \ldots b_q; z) = \sum_{n=0}^{\infty} \frac{(a_1)_n (a_2)_n \cdots (a_p)_n}{(b_1)_n (b_2)_n \cdots (b_q)_n}\left(\frac{z^n}{n!}\right),$$

where

$$(e)_n = e(e+1)(e+2)\cdots(e+n-1), (e)_0 = 1.$$

In particular, the confluent hypergeometric function

$$_1F_1(a; b; z) = \sum_{n=0}^{\infty} \frac{(a)_n}{(b)_n}\left(\frac{z^n}{n!}\right)$$

is frequently used in the main text. If any one of the numerator parameters in $_pF_q$ is a negative integer then $_pF_q$ is a polynomial of finite order. Special cases of $_1F_1$ are

$$_1F_1\left(0;2;\frac{a^2}{2}\right)=1,$$

$$_1F_1\left(-1;2;\frac{a^2}{2}\right)=1-\frac{a^2}{4},$$

$$_1F_1\left(-2;2;\frac{a^2}{2}\right)=1-\frac{a^2}{2}+\frac{a^4}{24},$$

$$_1F_1\left(-3;2;\frac{a^2}{2}\right)=1-\frac{3}{4}a^2+\frac{a^4}{8}+\frac{a^6}{192},$$

$$_1F_1\left(0;1;\frac{a^2}{2}\right)=1,$$

$$_1F_1\left(-1;1;\frac{a^2}{2}\right)=1-\frac{a^2}{2},$$

$$_1F_1\left(-2;1;\frac{a^2}{2}\right)=1-a^2+\frac{a^4}{8},$$

$$_1F_1\left(-3;1;\frac{a^2}{2}\right)=1-\frac{3}{2}a^2+\frac{3}{8}a^4-\frac{1}{24}a^6.$$

APPENDIX M.F

SOME DEFINITE INTEGRALS AND FORMULAS OF FREQUENT APPLICATION

1. It is well known that

$$\int_{-\infty}^{+\infty}\exp(-u^2)du=\sqrt{\pi}.$$

If $u=x\sqrt{h/2}$, then

$$\int_{-\infty}^{+\infty}\exp\left(-\frac{1}{2}hu^2\right)du=\sqrt{2\pi/h}. \quad \text{(M.F.1)}$$

2. If the derivative of (M.F.1) is taken with respect to h, then

$$\int_{-\infty}^{+\infty}u^{2\nu}\exp\left(\frac{1}{2}hu^2\right)du=\frac{1\times 3\times 5\times\cdots\times(2\nu-1)}{h^{\nu+1/2}}\sqrt{2\pi}$$

$$=\frac{(2\nu)!}{2^\nu\nu!}\frac{\sqrt{2\pi}}{h^{\nu+1/2}}. \quad \text{(M.F.2)}$$

3. The identity can then be stated by means of a suitable expansion in a power series:

Some definite integrals and formulas of frequent application

$$\int_{-\infty}^{+\infty} \exp\left(itu - \frac{1}{2}hu^2\right) du = \int_{-\infty}^{+\infty} \sum_{v=0}^{\infty} \frac{(itu)^v}{v!} \exp\left(-\frac{1}{2}hu^2\right) du.$$

The contributions of the terms with odd values of v vanish; thus

$$\int_{-\infty}^{\infty} \exp\left(itu - \frac{1}{2}hu^2\right) du = \sum_{v=0}^{\infty} \frac{(it)^v}{v!} \int_{-\infty}^{+\infty} u^v \exp\left(-\frac{1}{2}hu^2\right) du$$

$$= \sum_{v=0}^{\infty} \frac{(it)^{2v}}{2v!} \frac{(2v)!}{2^v v!} \frac{\sqrt{2\pi}}{h^{v+1/2}}$$

$$= \sqrt{\frac{2\pi}{h}} \exp(-t^2/2h). \tag{M.F.3}$$

By analogy,

$$\int_{-\infty}^{+\infty} \exp\left(\pm tu - \frac{1}{2}hu^2\right) du = \sqrt{\frac{2\pi}{h}} \exp\left(\frac{t^2}{2h}\right). \tag{M.F.3'}$$

4. Let us consider the integral

$$\frac{1}{2\pi} \int_{-\infty}^{+\infty} (iu)^v \exp\left(-\frac{1}{2}u^2\right) \exp(-iut) du. \tag{M.F.4}$$

Since

$$(iu)^v \exp(-iut) = (-1)^v \frac{d^v}{dt^v} \exp(-iut),$$

(M.F.4) may be estimated:

$$\frac{(-1)^v}{2\pi} \frac{d^v}{dt^v} \int_{-\infty}^{+\infty} \exp\left(-\frac{1}{2}u^2\right) \exp(-itu) du = \frac{(-1)^v}{2\pi} \frac{d^v}{dt^v} \sqrt{2\pi} \exp\left(-t^2/2\right)$$

$$= \frac{(-1)^v}{\sqrt{2\pi}} \frac{d^v}{dt^v} \exp\left(-t^2/2\right).$$

Because of (M.D.1) we have

$$\frac{1}{2} \int_{-\infty}^{+\infty} (iu)^v \exp\left(-\frac{1}{2}u^2\right) \exp(-iut) du = H_v(t) \frac{1}{\sqrt{2\pi}} \exp\left(-\frac{1}{2}t^2\right). \tag{M.F.5}$$

We will prove that

$$\int_{-\infty}^{+\infty} \cdots \int_{-\infty}^{+\infty} \exp(i\tilde{\mathbf{T}}\mathbf{U}) \exp\left(-\frac{1}{2}\tilde{\mathbf{U}}\lambda\mathbf{U}\right) d\mathbf{U} = \frac{(2\pi)^{n/2}}{\sqrt{\lambda}} \exp\left(-\frac{1}{2}\tilde{\mathbf{T}}\lambda^{-1}\mathbf{T}\right), \tag{M.F.6}$$

where $\tilde{\mathbf{U}} = (u_1, u_2, \ldots, u_n)$ and $\tilde{\mathbf{T}} = (t_1, t_2, \ldots, t_n)$ are real vectors in n dimensions, λ is the determinant of the symmetric matrix λ, and λ^{-1} is the matrix's inverse. The tilde ($\sim$) indicates the transpose of a matrix.

In a more explicit form, (M.F.6) may be written as

$$\int_{-\infty}^{+\infty}\cdots\int_{-\infty}^{+\infty}\exp\left(i\sum_{j=1}^{n}t_j u_j\right)\exp\left[-\frac{1}{2}Q(u_1,u_2,\ldots u_n)\right]du_1 du_2\ldots du_n$$
$$=\frac{(2\pi)^{n/2}}{\sqrt{\lambda}}\exp\left[-\frac{1}{2}Q^{-1}(t_1,t_2,\ldots,t_n)\right],$$

(M.F.6′)

where Q is the quadratic positive form corresponding to λ and $Q^{-1} = \tilde{T}\lambda^{-1}T$ is its reciprocal form. When $n = 1$, (M.F.6) and (M.F.6') coincide with (M.F.3).

Since λ is a symmetric matrix, it is always possible to find an orthogonal matrix C such that $\tilde{C}\lambda C = K$ is a diagonal matrix with principal elements that are the eigenvalues λ_j of λ. Let us now introduce in (M.F.6) the two new variables

$$\tilde{x} = (x_1, x_2, \ldots, x_n) \quad \text{and} \quad \tilde{y} = (y_1, y_2, \ldots, y_n),$$

defined by $T = Cx$ and $y = \tilde{C}U$. Then $\tilde{T}U = \tilde{x}y$. Since $C = \pm 1$, the left-hand side of (M.F.6) becomes

$$\int_{-\infty}^{+\infty}\cdots\int_{-\infty}^{+\infty}\exp(i\tilde{x}y)\exp\left(-\frac{1}{2}\tilde{y}Ky\right)dy = \prod_{j=1}^{n}\int_{-\infty}^{+\infty}\exp(ix_j y_j)\exp\left(-\frac{1}{2}\lambda_j y_j^2\right)dy_j.$$

(M.F.7)

Applying (M.F.3) to every term of the product and remembering that the diagonal matrix with principal elements $1/\lambda_j$ coincides with K^{-1}, we have

$$\int_{-\infty}^{+\infty}\cdots\int_{-\infty}^{+\infty}\exp(i\tilde{T}U)\exp\left(-\frac{1}{2}\tilde{U}\lambda U\right)dU = \frac{(2\pi)^{n/2}}{\sqrt{\lambda_1\lambda_2\cdots\lambda_n}}\exp\left[-\frac{1}{2}\sum_{j=1}^{n}\left(\frac{x_j^2}{\lambda_j}\right)\right]$$
$$=\frac{(2\pi)^{n/2}}{\sqrt{\lambda}}\exp\left(-\frac{1}{2}\tilde{x}K^{-1}x\right).$$

and, since

$$\tilde{x}K^{-1}x = \tilde{x}C^{-1}\lambda^{-1}\tilde{C}^{-1}x = \tilde{x}\tilde{C}\lambda^{-1}Cx = \tilde{T}\lambda^{-1}T.$$

(M.F.6) is proved.

6. In the main text, the evaluation of integrals such as

$$\frac{1}{2\pi}\int_0^{2\pi}\exp\left(i\sum_{k}C_k\cos(\phi+\alpha_k)\right)d\phi$$

is frequently performed. For real C_k application of elementary trigonometric rules leads to

$$\sum_{k}C_k\cos(\phi+\alpha_k) = Z\cos(\phi+\xi),$$

(M.F.8)

where

$$Z^2 = \left(\sum_{k}C_k\cos\alpha_k\right)^2 + \left(\sum_{k}C_k\sin\alpha_k\right)^2 = \sum_{k,l}C_k C_l\cos(\alpha_k - \alpha_l),$$

$$Z\cos\xi = \sum_{k}C_k\cos\alpha_k, \qquad Z\sin\xi = \sum_{k}C_k\sin\alpha_k.$$

Some definite integrals and formulas of frequent application

Z and ξ are independent of ϕ, so that, according to (M.F.8)

$$\frac{1}{2\pi}\int_0^{2\pi}\exp\left[i\sum_k C_k\cos(\phi+\alpha_k)\right]d\phi = J_0(Z). \qquad \text{(M.F.9)}$$

7. $$\int_0^{2\pi}\cos(\tau-s)\exp\{-it[a\cos(q-\tau)-b\cos(r-\tau)]\}d\tau$$
    $$= (-2\pi i)\frac{J_1(tQ)}{Q}[a\cos(q-s)-b\cos(r-s)], \qquad \text{(M.F.10)}$$

    where
    $$Q^2 = a^2 + b^2 - 2ab\cos(q-r).$$

8. $$\int_0^\infty t^2 J_1(Qt)\exp\left(-\frac{p^2 t^2}{2}\right)dt = \frac{Q}{p^4}\exp\left(-\frac{Q^2}{2p^2}\right). \qquad \text{(M.F.11)}$$

9. $$\int_0^\infty t\exp\left(-\frac{1}{2}p^2 t^2\right)dt\int_0^{2\pi}\exp\{-it[a\cos(q-\tau)$$
    $$-b\cos(r-\tau)]\}d\tau = \frac{2\pi}{p^2}\exp\left(-\frac{Q^2}{2p^2}\right) \qquad \text{(M.F.12)}$$

10. $$\int_0^\infty t^2\exp\left(-\frac{1}{2}p^2 t^2\right)dt\int_0^{2\pi}\cos(s-\tau)$$
    $$\times\exp\{-it[a\cos(q-\tau)-b\cos(r-\tau)]\}d\tau = \left(-\frac{2\pi i}{p^4}\right)[a\cos(q-s)$$
    $$-b\cos(r-s)]\exp\left(-\frac{Q^2}{2p^2}\right). \qquad \text{(M.F.13)}$$

11. $$\int_0^\infty t^3\exp\left(-\frac{1}{2}p^2 t^2\right)dt\int_0^{2\pi}\exp\{-it[a\cos(q-\tau)-b\cos(r-\tau)]\}d\tau$$
    $$= \left(\frac{4\pi}{p^4}\right)\exp\left(-\frac{Q^2}{2p^2}\right)\left(1-\frac{Q2}{2p^2}\right). \qquad \text{(M.F.14)}$$

12. $$\int_0^\infty x^\mu \exp(-ax^2)I_0(qx)dx = \Gamma[(\mu+1)/2]/(2a^{(\mu+1)/2})$$
    $$\times {}_1F_1[(-\mu+1)/2; 1; -q^2/4a]\exp(q^2/4a) \qquad \text{(M.F.15)}$$

Where Γ is the gamma function and ${}_1F_1$ is the confluent hypergeometric function.

References

Abrahams, J. P. (1997). *Acta Crystallogr.*, D**53**, p. 371.
Abrahams, J. P. and Leslie, A. G. W. (1996). *Acta Crystallogr.*, D**52**, p. 30.
Abramowitz, M. and Stegun, I. A. (1972). *Handbook of Mathematical Functions*. Dover.
Adams, P. D., Grosse-Kunstleve, R. W., Hung, L. W. et al. (2002). *Acta Crystallogr.* D**58**, p. 1948.
Allegra, G. (1965). *Acta Crystallogr.*, **19**, p. 949.
Allegra, G. and Colombo, A. (1974). *Acta Crystallogr.*, A**30**, p. 727.
Altomare, A., Belviso, B. D., Burla, M. C., Campi, G., Cuocci, C., Giacovazzo, C., Gozzo, F., Moliterni, A., Polidori, G., and Rizzi, R. (2009e). *J. Appl. Crystallogr.*, **42**, p. 30.
Altomare, A., Burla, M. C., Cascarano, G., Giacovazzo, C., Guagliardi, A., Moliterni, A. G. G., and Polidori, G. (1996). *J. Appl. Crystallogr.*, **29**, p. 341.
Altomare, A., Burla, M. C., Cuocci, C., Giacovazzo, C., Gozzo, F., Moliterni, A., Polidori, G., and Rizzi, R. (2009d). *Acta Crystallogr.*, A**65**, p. 291.
Altomare, A., Caliandro, R., Camalli, M., Cuocci, C., da Silva, I., Giacovazzo, C., Moliterni, A. G. G., and Spagna, R. (2004). *J. Appl. Crystallogr.*, **37**, p. 957.
Altomare, A., Caliandro, R., Giacovazzo, C., Moliterni, A. G. G., and Rizzi, R. (2003). *J. Appl. Crystallogr.*, **36**, p. 230.
Altomare, A., Camalli, M., Cuocci, C., da Silva, I., Giacovazzo, C., Moliterni, A. G. G., and Rizzi, R. (2005). *J. Appl. Crystallogr.*, **38**, p. 760.
Altomare, A., Camalli, C., Cuocci, C., Giacovazzo, C., Moliterni, A., and Rizzi, R. (2009a). *J. Appl. Crystallogr.*, **42**, p. 1197.
Altomare, A., Camalli, M., Cuocci, C., Giacovazzo, C., Moliterni, A. G. G., and Rizzi, R. (2007). *J. Appl. Crystallogr.*, **40**, p. 743.
Altomare, A., Campi, G., Cuocci, C., Eriksson, L., Giacovazzo, C., Maggi, S., Moliterni, A., Rizzi, R., and Werner, P.-E. (2009b). *J. Appl. Crystallogr.*, **42**, p. 768.
Altomare, A., Cascarano, G., Giacovazzo, C., and Guagliardi, A. (1994a). *J. Appl. Crystallogr.*, **27**, p. 1045.
Altomare, A., Cascarano, G., Giacovazzo, C., Guagliardi, A., Moliterni, A. G. G., Burla, M. C., and Polidori, G. (1995). *J. Appl. Crystallogr.*, **28**, p. 738.
Altomare, A., Cascarano, G., Giacovazzo, C., and Viterbo, D. (1991). *Acta Crystallogr.*, A**47**, p. 744.
Altomare, A., Cuocci, C., Giacovazzo, C., Moliterni, A., and Rizzi, R. (20121). *J. Appl. Crystallogr.* **45**, p. 789.
Altomare, A., Cuocci, C., Giacovazzo, C., Kamel, G. S., Moliterni, A., and Rizzi, R. (2008a). *Acta Crystallogr.*, A**64**, p. 326.
Altomare, A., Cuocci, C., Giacovazzo, C., Moliterni, A., and Rizzi, R. (2008b). *J. Appl. Crystallogr.*, **41**, p. 592.
Altomare, A., Cuocci, C., Giacovazzo, C., Moliterni, A., and Rizzi, R. (2009c). *Acta Crystallogr.*, A**65**, p. 183.
Altomare, A., Cuocci, C., Giacovazzo, C., Moliterni, A., and Rizzi, R. (2010). *J. Appl. Crystallogr.*, **43**, p. 798.
Altomare, A., Cuocci, C., Giacovazzo, C., Moliterni, A. G. G., and Rizzi, R. (2006). *J. Appl. Crystallogr.*, **39**, p. 558.
Altomare, A., Cuocci, C., Giacovazzo, C., Moliterni, A., and Rizzi, R. (2012a). *Acta Crystallogr.*, A**68**, p. 244.

Altomare, A., Cuocci, C., Moliterni, A., and Rizzi, R. (2013). *J. Appl. Crystallogr.*, **46**, p. 476.
Altomare, A., Foadi, J., Giacovazzo, C., Moliterni, A. G. G., Burla, M. C., and Polidori, G. (1998). *J. Appl. Crystallogr.*, **31**, p. 74.
Altomare, A., Giacovazzo, C., Guagliardi, A., and Siliqi, D. (1994b). *Acta Crystallogr.*, A**50**, p. 311.
Altomare, A., Giacovazzo, C., Guagliardi, A., Moliterni, A. G. G., and Rizzi, R. (2000b). *J. Appl. Crystallogr.*, **33**, p. 1305.
Altomare, A., Giacovazzo, C., Guagliardi, A., Moliterni, A. G. G., Rizzi, R., and Werner, P.-E. (2000a). *J. Appl. Crystallogr.*, **33**, p. 1180.
Altschul, S. F., Madden, T. L., Schaffer, A. A., Zhang, J., Zhang, Z., Miller, W., and Lipman, D. J. (1997). *Nucleic Acids Res.*, **25**, p. 3389.
Andreev, Y. G., Lightfoot, P., and Bruce, P. G. (1997). *J. Appl. Crystallogr.*, **30**, p. 294.
Ardito, G., Cascarano, G., Giacovazzo, C., and Luic, M. (1985). *Z. Kristallogr.*, **172**, p. 25.
Baerlocher, C. (1982). *The X-Ray Rietveld System*. Institut für Kristallographie und Petrographie, Zürich.
Baerlocher, Ch., Gramm, F., Massüger, L., McCusker, L. B., He, Z., Hovmöller, S., and Zou, X. (2007a). *Science*, **315**, p. 1113,
Baerlocher, Ch., McCusker, L. B., and Palatinus, L. (2007b). *Z. Kristallogr.*, **222**, p. 47.
Baerlocher, Ch., McCusker, L. B., Prokic, S., and Wessels, T. (2004). *Z. Kristallog.*, **219**, p. 802.
Baggio, R., Woolfson, M. M., Declercq, J. P., and Germain, G. (1978). *Acta Crystallogr.*, A**34**, p. 883.
Baker, D., Bystroff, C., Fletterick, R. J., and Agard, D. A. (1993). *Acta Crystallogr.*, D**49**, p. 429.
Bart, J. C. J. and Busetti, A. (1976). *Acta Crystallogr.*, A**32**, p. 927.
Beevers, C. A. and Robertson, J. H. (1950). *Acta Crystallogr.*, **4**, p. 531.
Bentley, G. A. and Houdusse, A. (1992). *Acta Crystallogr.*, A**48**, p. 312.
Bergamaschi, A., Cervellino, A., Dinapoli, R., Gozzo, F., Henrich, B., Johnson, I., Kraft, P., Mozzanica, A., Schmitt, B., and Shi, X. (2010). *J. Synchrotron Rad.*, **17**, p. 653.
Bernstein, S. (1927). *Math. Ann.*, **97**, p. 1.
Bertaut, E. F. (1955a). *Acta Crystallogr.*, **8**, p. 537.
Bertaut, E. F. (1955b). *Acta Crystallogr.*, **8**, p. 544.
Bertaut, E. F. (1960a). *Acta Crystallogr.*, **13**, p. 546.
Bertaut, E. F. (1960b). *Acta Crystallogr.*, **13**, p. 643.
Beurskens, P. T., Gould, R. O., Bruins Slot, H. J., and Bosman, W. P. (1987). *Z. Kristallogr.*, **179**, p. 127.
Beurskens, P. T., Beurskens, G., Gelder, R., de Garcia-Granda, S., Gould, R. O., and Smits, J. M. M. (2008). *The DIRDIF2008 program system*, Crystallography Laboratory, University of Nijmegen, The Netherlands.
Beurskens, P. T. and Noordik, J. H. (1971). *Acta Crystallogr.*, A**27**, p. 187.
Beurskens, P. T., van der Hark, Th. E.M., and Beurskens, G. (1976). *Acta Crystallogr.*, A**32**, p. 821.
Bhat, T. N. (1990). *Acta Crystallogr.*, A**46**, p. 735.
Bhat, T. N. and Blow, D. M. (1982). *Acta Crystallogr.*, A**38**, p. 21.
Black, P. J. and Corby, R. N. (1975). In *Anomalous Scattering* (ed. S. Ramaseshan and S. C. Abrahams). Munksgaard, Copenhagen.
Blessing, R. H. (1997). *J. Appl. Crystallogr.*, **30**, p. 176.
Blessing, R. H. and Smith, G. D. (1999). *J. Appl. Cryst.*, **32**, p. 664.
Blow, D. M. (1957). X-ray analysis of haemoglobin: determination of phase angles by isomorphous substitution. PhD Thesis, Univ. of Cambridge.
Blow, D. M. (1958). *Proc. R. Soc.*, A**247**, p. 302.
Blow, D. M. (1985). In *Molecular Replacement* (ed. P. A. Machin). Proceedings of the Daresbury Study Weekend.
Blow, D. M. and Crick, F. H. C. (1959). *Acta Crystallogr.*, **12**, p. 794.

Blundell, T. L. and Johnson, L. N. (1976). *Protein Crystallography*. Academic, London.
Böhme, R. (1982). *Acta Crystallogr.*, A**38**, p. 316.
Böhme, R. (1983). *Z. Naturf.*, **38**, p. 304.
Booth, A. D. (1945). *Philos. Mag.*, **36**, p. 609.
Booth, A. D. (1946). *Proc. Roy. Soc.*, A**188**, p. 77.
Booth, A. D. (1947). *Proc. Roy. Soc.*, A**190**, p. 482.
Boultif, A. and Louër, D. (1991). *J. Appl. Crystallogr.*, **24**, p. 987.
Boutselakis, H. et al. (2003). *Nucleic Acids Res.*, **31**, p. 458.
Bragg, W. L. and Perutz, M. F. (1954). *Proc. R. Soc.*, **225**, p. 315.
Bragg, W. L. and West, J. (1930). *Phil Mag.*, **10**, p. 823.
Braun, P. B., Hornstra, J., and Leenhouts, J. I. (1969). *Philips Res. Rep.*, 24, p. 85.
Bricogne, G. (1974). *Acta Crystallogr.*, A**30**, p. 395.
Bricogne, G. (1976). *Acta Crystallogr.*, A**32**, p. 832.
Bricogne, G. (1984). *Acta Crystallogr.*, A**40**, p. 410.
Bricogne, G. (1988a). *Acta Crystallogr.*, A**44**, p. 517.
Bricogne, G. (1988b). In *Crystallographic Computing 4: Techniques and New Technologies* (ed. N. W. Isaacs and M. R. Taylor). Oxford University Press, New York.
Bricogne, G. (1991). *Acta Crystallogr.*, A**47**, p. 803.
Bricogne, G., Vonrhein, C., Flensburg, C., Schiltz, M., and Paciorek, W. (2003). *Acta Crystallogr.*, D**59**, p. 2023.
Brunelli, M., Wright, J. P., Vaughan, G. R. M., Mora, A. J., and Fitch, A. (2003). *Angew. Chem. Int. Ed.*, **42**, p. 2029.
Brünger, A. T., Adams, P. D., Clore, G. M., DeLano, W. L., Gros, P., Grosse-Kunstleve, R. W., Jiang, J.-S., Kuszewski, J., Nilges, M., Pannu, N. S., Read, R. J., Rice, L. M., Simonson, T., and Warren, G. L. (1998). *Acta Crystallogr.*, D**54**, p. 905.
Buerger, M. J. (1942). *X-Ray Crystallography*. Wiley, New York.
Buerger, M. J. (1946). *J. Appl. Phys.*, **17**, p. 579.
Buerger, M. J. (1951). *Acta Crystallogr.*, **4**, p. 531.
Buerger, M. J. (1953). *Proc. Natl. Acad. Sci. USA*, **39**, p. 674.
Buerger, M. J. (1956). *Proc. Natl. Acad. Sci. USA*, **42**, p. 776.
Buerger, M. J. (1959). Vector Space and its Application in Crystal Structure Investigation. Wiley, New York.
Buerger, M. J. (1969). In *Physics of the Solid State*. Academic, London.
Bullough, R. K. (1961). *Acta Crystallogr.*, **14**, p. 257.
Bullough, R. K. (1964). *Acta Crystallogr.*, **17**, p. 295.
Bunge, H. J., Dahms, M., and Brokmeier, H. G. (1989). *Crystallogr. Rev.*, **2**, p. 67.
Burdina, R. A. (1973). *Kristallografiya*, **18**, p. 694.
Burgi, H. B. and Dunitz, J. D. (1971). *Acta Crystallogr.*, A**27**, p. 117.
Burla, M. C., Caliandro, R., Camalli, M., Carrozzini, B., Cascarano, G. L., De Caro, L., Giacovazzo, C., Polidori, G., Siliqi, D., and Spagna, R. (2007c). *J. Appl. Crystallogr.*, **40**, p. 609.
Burla, M. C., Caliandro, R., Carrozzini, B., Cascarano, G. L., De Caro, L., Giacovazzo, C., Polidori, G., and Siliqi, D. (2007b). *J. Appl. Crystallogr.*, **40**, p. 211.
Burla, M. C., Caliandro, R., Carrozzini, B., Cascarano, G. L., De Caro, L., Giacovazzo, C., Polidori, G., and Siliqi, D. (2006a). *J. Appl. Crystallogr.*, **39**, p. 728.
Burla, M. C., Caliandro, R., Carrozzini, B., Cascarano, G. L., De Caro, L., Giacovazzo, C., Polidori, G., and Siliqi, D. (2006b). *J. Appl. Crystallogr.*, **39**, p. 527.
Burla, M. C., Caliandro, R., Carrozzini, B., Cascarano, G. L., De Caro, L., Giacovazzo, C., Polidori, G., and Siliqi, D. (2007a). *J. Appl. Crystallogr.*, **40**, p. 834.
Burla, M. C., Caliandro, R., Giacovazzo, C., and Polidori, G. (2010a). *Acta Crystallogr.*, A**66**, p. 347.
Burla, M. C., Caliandro, R., Camalli, M., Carrozzini, B., Cascarano, G. L., Giacovazzo, C., Mallamo, M., Mazzone, A., Polidori, G., and Spagna, R. (2012a). *J. Appl. Crystallogr.*, **45**, p. 357.
Burla, M. C., Camalli, M., Cascarano, G., Giacovazzo, C., Polidori, G., Spagna, R., and Viterbo, D. (1989a). *J. Appl. Crystallogr.*, **22**, p. 389.

References

Burla, M. C., Camalli, M., Carrozzini, B., Cascarano, G. L., Giacovazzo, C., Polidori, G., and Spagna, R. (2001). *J. Appl. Crystallogr.*, **34**, p. 523.
Burla, M. C., Carrozzini, B., Caliandro, R., Cascarano, G. L., De Caro, L., Giacovazzo, C., and Polidori, G. (2003a). *Acta Crystallogr.*, A**59**, p. 245.
Burla, M. C., Carrozzini, B., Cascarano, G. L., Giacovazzo, C., and Polidori, G. (2000). *J. Appl. Crystallogr.*, **33**, p. 307.
Burla, M. C., Carrozzini, B., Cascarano, G. L., Giacovazzo, C., and Polidori, G. (2011b). *Acta Crystallogr.*, A**44**, p. 1143.
Burla, M. C., Carrozzini, B., Cascarano, G. L., Giacovazzo, C., and Polidori, G. (2011c). *J. Appl. Crystallogr.*, A**67**, p. 447.
Burla, M. C., Carrozzini, B., Cascarano, G. L., Comunale, G., Giacovazzo, C., Mazzone, A., and Polidori, G. (2012b). *Acta Crystallogr.*, A**68**, p. 513.
Burla, M. C., Carrozzini, B., Cascarano, G. L., Giacovazzo, C., and Polidori, G. (2003b). *Acta Crystallogr.*, D**59**, p. 662.
Burla, M. C., Carrozzini, B., Cascarano, G. L., Giacovazzo, C., Moustiakimov, M., Polidori, G., and Siliqi, D. (2004). *Acta Crystallogr.*, D**60**, p. 1683.
Burla, M. C., Carrozzini, B., Cascarano, G. L., Giacovazzo, C., Polidori, G., and Siliqi, D. (2002). *Acta Crystallogr.*, D**58**, p. 928.
Burla, M. C., Cascarano, G., Fares, V., Giacovazzo, C., Polidori, G., and Spagna, R. (1989b). *Acta Crystallogr.*, A**45**, p. 781.
Burla, M. C., Cascarano, G., Giacovazzo, C., and Guagliardi, A. (1992). *Acta Crystallogr.*, A**48**, p. 906.
Burla, M. C., Giacovazzo, C., and Polidori, G. (2011a). *J. Appl. Crystallogr.*, **44**, p. 193.
Burla, M. C., Giacovazzo, C., and Polidori, G. (2010b). *J. Appl. Crystallogr.*, **43**, p. 825.
Burla, M. C., Polidori, G., Nunzi, A, Cascarano, G., and Giacovazzo, C. (1977). *Acta Crystallogr.*, A**33**, p. 949.
Bystroff, C., Baker, D., Fletterick, R. J., and Argard, D. A. (1993). *Acta Crystallogr.*, D**49**, p. 440.
Caliandro, R., Carrozzini, B., Cascarano, G. L., Comunale, G., and Giacovazzo, C. (2013). *Acta Crystallogr.*, A**69**, p. 98.
Caliandro, R., Carrozzini, B., Cascarano, G. L., De Caro, L., Giacovazzo, C., and Siliqi, D. (2008b). *Acta Crystallogr.*, A**64**, p. 519.
Caliandro, R., Carrozzini, B., Cascarano, G. L., De Caro, L., Giacovazzo, C., Mazzone, A. M., and Siliqi, D. (2006). *J. Appl. Crystallogr.*, **39**, p. 185.
Caliandro, R., Carrozzini, B., Cascarano, G. L., De Caro, L., Giacovazzo, C., Moustiakimov, M., and Siliqi, D. (2005c). *Acta Crystallogr.*, A**61**, p. 343.
Caliandro, R., Carrozzini, B., Cascarano, G. L., De Caro, L., Giacovazzo, C., Mazzone, A., and Siliqi, D. (2008a). *J. Appl. Crystallogr.*, **41**, p. 548.
Caliandro, R., Carrozzini, B., Cascarano, G. L., De Caro, L., Giacovazzo, C., and Siliqi, D. (2007a). *J. Appl. Crystallogr.*, **40**, p. 883.
Caliandro, R., Carrozzini, B., Cascarano, G. L., De Caro, L., Giacovazzo, C., and Siliqi, D. (2005a). *Acta Crystallogr.*, D**61**, p. 556.
Caliandro, R., Carrozzini, B., Cascarano, G. L., De Caro, L., Giacovazzo, C., and Siliqi, D. (2005b). *Acta Crystallogr.*, D**61**, p. 1080.
Caliandro, R., Carrozzini, B., Cascarano, G. L., De Caro, L., Giacovazzo, C., and Siliqi, D. (2007b). *J. Appl. Crystallogr.*, **40**, p. 931.
Caliandro, R., Carrozzini, B., Cascarano, G. L., Giacovazzo, C., Mazzone, A., and Siliqi, D. (2009). *Acta Crystallogr.*, A**65**, p. 512.
Camalli, M., Carrozzini, B., Cascarano, G. L. and Giacovazzo, C. (2012). *J. Appl. Crystallogr.*, **45**, p. 351.
Carrozzini, B., Cascarano, G. L., and Giacovazzo, C. (2010). *J. Appl. Crystallogr.*, **43**, p. 221.
Carrozzini, B., Cascarano, G. L., Comunale, G., Giacovazzo, C., and Mazzone, A. (2013a). *Acta Crystallogr.*, D**69**, p. 1038.
Carrozzini, B., Cascarano, G. L., Giacovazzo, C., and Mazzone, A. (2013b). *Acta Crystallogr.*, A**69**, p. 408.
Cascarano, G., Favia, L., and Giacovazzo, C. (1992c). *J. Appl. Crystallogr.*, **25**, p. 310.

Cascarano, G. and Giacovazzo, C. (1995). *Acta Crystallogr.*, A**51**, p. 820.
Cascarano, G., Giacovazzo, C., Camalli, M., Spagna, R., Burla, M. C., Nunzi, A., and Polidori, G. (1984). *Acta Crystallogr.*, A**40**, p. 278.
Cascarano, G., Giacovazzo, C., and Guagliardi, A. (1991). *Acta Crystallogr.*, A**47**, p. 698.
Cascarano, G., Giacovazzo, C., and Guagliardi, A. (1992a). *Acta Crystallogr.*, A**48**, p. 859.
Cascarano, G., Giacovazzo, C., and Guagliardi, A. (1992b). *Z. Kristallogr.*, **200**, p. 63.
Cascarano, G., Giacovazzo, C., and Luić, M. (1985a). In *Structure and Statistics in Crystallography*, (ed. A. J. C. Wilson). Adenine Press, New York.
Cascarano, G., Giacovazzo, C., and Luić, M. (1985b). *Acta Crystallogr.*, A**41**, p. 544.
Cascarano, G., Giacovazzo, C., and Luić, M. (1987b). *Acta Crystallogr.*, A**43**, p. 14.
Cascarano, G., Giacovazzo, C., and Luić, M. (1988a). *Acta Crystallogr.*, A**44**, p. 176.
Cascarano, G., Giacovazzo, C., and Luić, M. (1988b). *Acta Crystallogr.*, A**44**, p. 183.
Cascarano, G., Giacovazzo, C., Luić, M., Pifferi, G., and Spagna, R. (1987c). *Z. Kristallogr.*, **179**, p. 113.
Cascarano, G., Giacovazzo, C., and Viterbo, D. (1987a). *Acta Crystallogr.*, A**43**, p. 22.
Chang. G. and Lewis. M. (1997). *Acta Crystallogr.*, D**53**, p. 279.
Chapuis, G., Templeton, D. H., and Templeton, L. K. (1985). *Acta Crystallogr.*, A**41**, p. 274.
Cheary, R. W. and Coelho, A. (1992). *J. Appl. Crystallogr.*, **25**, p. 109.
Chen, Y. W., Dodson, E. J. and Kleiywegt, G. J. (2000). *Structure*, **8**, R213.
Chernishev, V. V. and Schenk, H. (1998). *Z. Kristallogr.*, **213**, p. 1.
Chothia, C. and Lesk, A. (1986). *EMBO J.*, 5(4), p. 823.
Cianci, M., Helliwell, J. R., and Suzuki, A. (2008). *Acta Crystallogr.*, D**64**, p. 1196.
Clastre, J. and Gray, R. (1950). *Compt. Rend.*, **230**, p. 1876.
Cochran, W. (1951). *Acta Crystallogr.*, **4**, p. 408.
Cochran, W. (1952). *Acta Crystallogr.*, **5**, p. 65.
Cochran, W. (1955). *Acta Crystallogr.*, **8**, p. 473.
Cochran, W. and Douglas, A. S. (1957). *Proc. R. Soc.*, A**243**, p. 281.
Cochran, W. and Woolfson, M. M. (1955). *Acta Crystallogr.*, **8**, p. 1.
Coelho, A. A. (2003). *TOPAS version 3.1*. Bruker AXS GmbH, Karlsruhe, Germany.
Collings, I., Watier, Y., Giffard, M., Dagogo, S., Kahn, R., Bonneté, F., Wright, J. P., Fitch, A. N., and Margiolaki, I. (2010). *Acta Crystallogr.*, D**66**, p. 539.
Collins, D. M., Konnert, J. H., and Stewart, J. M. (1996). *Proc. Macromolecular Crystallography Computing School*, Western Washington University (ed. P. Bourne and K. Watenpaugh).
Cooley, J. W. and Tukey, J. W. (1965). *Math. Comput.*, **19**, p. 297.
Coppens, P. and Hamilton, W. C. (1968). *Acta Crystallogr.*, B**24**, p. 925.
Corby, R. N. and Black, P. J. (1973). *Acta Crystallogr.*, B**29**, p. 2669.
Coulter, C. L. and Dewar, R. B. K. (1971). *Acta Crystallogr.*, B**27**, p. 1730.
Courbion, G. and Ferey, G. (1988). *J. Solid State Chem.*, **76**, p. 426.
Cowley, J. M. (1992). *Electron Diffraction Techniques*, Vols 1 and 2. Oxford University Press.
Cowley, J. M. (1953a). *Acta Crystallogr.*, **6**, p. 516.
Cowley, J. M. (1953b). *Acta Crystallogr.*, **6**, p. 522.
Cowley, J. M. (1953c). *Acta Crystallogr.*, **6**, p. 846.
Cowley, J. M. (1955). *Acta Crystallogr.*, **9**, p. 391.
Cowley, J. M. (1961). *Acta Crystallogr.*, **14**, p. 920.
Cowley, J. M. and Goswami, A. (1961). *Acta Crystallogr.*, **14**, p. 1071.
Cowley, J. M. and Moodie, A. F. (1957). *Acta Crystallogr.*, **10**, p. 609.
Cowtan, K. D. (1994). Joint CCP4 and ESF-EACBM Newsletter on Protein Crystallography, **31**, p. 34.
Cowtan, K. (1998). *Acta Crystallogr.*, D**54**, p. 750.
Cowtan, K. D. (1999). *Acta Crystallogr.*, D**55**, p. 1555.
Cowtan, K. D. (2001). *Acta Crystallogr.*, D**57**, p. 1435.
Cowtan, K. D. (2006). *Acta Crystallogr.*, D**62**, p. 1002.

Cowtan, K. (2008). *Acta Crystallogr.*, D**64**, p. 83.
Cowtan, K. (2012). *Acta Crystallogr.*, D**68**, p. 328.
Cramér, H. (1951). *Mathematical Methods of Statistics.* Princeton University Press.
Cranswick, M. D. (2008). Computer software for powder diffraction. In *Powder Diffraction: Theory and Practice*. Eds: R. E. Dinnebier and S. J. L. Billinge, RSC Publishing.
Crick, F. H. C. and Magdoff, B. S. (1956). *Acta Crystallogr.*, **9**, p. 901.
Crowther, R. A. (1969). *Acta Crystallogr.*, B**25**, p. 2571.
Crowther, R. A. (1972). In *The Molecular Replacement Method* (ed. M. G. Rossmann). Gordon & Breach, New York.
Cruickshank, D. W. J. (1949). *Acta Crystallogr.*, **2**, p. 65.
Cruickshank, D. W. J. and Rollett, J. S. (1953). *Acta Crystallogr.*, **6**, p. 705.
Cutfield, J. F., Dodson, E. J., Dodson, G. G., Hodgkin, D. C., Isaacs, N. W., Sakabe, K., and Sakabe, N. (1975). *Acta Crystallogr.*, A**31**, S21.
Dauter, Z., Dauter, M., and Dodson, E. D. (2002). *Acta Crystallogr.*, D**58**, p. 496.
Dauter, Z., Li, M., and Wlodawer, A. (2001). *Acta Crystallogr.*, D**57**, p. 239.
David, P. R. and Subbiah, S. (1994). *Acta Crystallogr.*, D**50**, p. 132.
David, W. I. F. (1987). *J. Appl. Crystallogr.*, **20**, p. 316.
David, W. I. F., Shankland, K., Cole, J., Maginn, S., Motherwell, W. D. S., and Taylor, R. (2001). *DASH User Manual*, Cambridge Crystallographic Data Centre, Cambridge, UK.
Davis, I. W., Leaver-Fay, A., Chen, V. B., Block, J. N., Kapral, G. J., Wang, X., Murray, L. W., Arendall, W. B. III, Snoeyink, J., Richardson, J. S., and Richardson, D. C. (2007), *Nucleic Acids Res.*, **35**, p. W375.
de la Fortelle, E. and Bricogne, G. (1997). *Methods Enzymol.*, **276**, p. 472.
Debaerdemaeker, T., Tate, C., and Woolfson, M. M. (1985). *Acta Crystallogr.*, A**41**, p. 286.
Debaerdemaeker, T., Tate, C., and Woolfson, M. M. (1988). *Acta Crystallogr.*, A**44**, p. 353.
Debaerdemaeker, T. and Woolfson, M. M. (1975). *Acta Crystallogr.*, A**31**, p. 401.
Debaerdemaeker, T. and Woolfson, M. M. (1983). *Acta Crystallogr.*, A**39**, p. 193.
Debaerdemaeker, T. and Woolfson, M. M. (1989). *Acta Crystallogr.*, A**45**, p. 349.
de Bruijn, N. G. (1970). *Asymptotic Methods in Analysis*. North-Holland, Amsterdam.
Debye, P. (1909). *Math. Ann.*, **67**, p. 535.
Declercq, J. P., Germain, G., Main, P., and Woolfson, M. M. (1973). *Acta Crystallogr.*, A**29**, p. 231.
Declercq, J. P., Germain, G., and Woolfson, M. M. (1975). *Acta Crystallogr.*, A**31**, p. 367.
Declercq, J. P., Germain, G., and Woolfson, M. M. (1979). *Acta Crystallogr.*, A**35**, p. 622.
de Gelder, R. (1992). *Thesis.* University of Leiden.
de Gelder, R., de Graaf, A. G., and Schenk, K. H. (1990). *Acta Crystallogr.*, A**46**, p. 688.
Delarue, M. (2008). *Acta Crystallogr.*, D**64**, p. 40.
de Rango, C., Tsoucaris, G., and Zelwar, C. (1969). *C.R. Acad. Sci., Paris*, **268**, p. 1090.
de Rango, C., Tsoucaris, G., and Zelwar, C. (1974). *Acta Crystallogr.*, A**30**, p. 342.
de Rosier, D. and Klug, A. (1968). *Nature, London*, **217**, p. 130.
de Titta, G. T., Edmonds, J. W., Langs, D. A., and Hauptman, H. A. (1975). *Acta Crystallogr.*, A**31**, p. 472.
de Titta, G. T., Weeks, C. M., Thuman, P., Miller, R., and Hauptman, H. A. (1994). *Acta Crystallogr.*, A**50**, p. 203.
de Wolff, P. M. (1961). *Acta Crystallogr.*, **14**, p. 579.
Destro, R. (1972). CECAM Workshop Reports.
Dickerson, R. F., Kendrew, J. C., and Standberg, B. E. (1961). *Acta Crystallogr.*, **14**, p. 1188.
Dickerson, R. E., Weinzierl, J. E., and Palmer, R. A. (1968). *Acta Crystallogr.*, B**24**, p. 997.

DiMaio, F., Terwilliger, T. C., Read, R. J., Wlodawer, A., Oberdorfer, G., Wagner, U., Valkov, E., Alon, A., Fass, D., Axelrod, H. L., Das, D., Vorobiev, S. M., Iwai, H., Pokkuluri, P. R. & Becker, D. (2011). *Nature*, **473**, 540.

Dodson, E. J. and Vijayan, M. (1971). *Acta Crystallogr.*, B**27**, 2402.

Dong, W., Baird, T., Fryer, J. R., Gilmore, C., McNicol, D. D., Bricogne, G., Smith, D. J., O'Keefe, M. A., and Hovmöller, S. (1992). *Nature*, **355**, p. 605.

Donnay, J. D. H. and Kennard, O. (1964). *Acta Crystallogr.*, **17**, p. 1337.

Dorset, D. L., Jap, B. K., Ho, M. S., and Glaeser, R. M. (1979). *Acta Crystallogr.*, A**35**, p. 1001.

Dorset, D. L. (1995). *Structural Electron Crystallography*. Plenum, New York.

Dorset, D. L. (1996). *Acta Crystallogr.*, A**52**, p. 480.

Doyle, P. A. and Turner, P. S. (1968). *Acta Crystallogr.*, A**24**, p. 390.

Drendel, W. B., Dave, R. D., and Jain, S. (1995). *Proc. Natl. Acad. Sci., USA*, **92**, p. 547.

Dumas, C. and van der Lee, A. (2008). *Acta Crystallogr.* D**64**, p. 864.

Dunbrack, R. L. and Cohen, F. E. (1997). *Protein Sci.*, **6**, p. 1661.

Elsenhans, O. (1990). *J. Appl. Crystallogr.*, **23**, p. 73.

Emsley, P. and Cowtan, K. (2004). *Acta Crystallogr.*, D**60**, p. 2126.

Emsley, P., Lohkamp, B., Scott, W. G., and Cowtan, K. (2010). *Acta Crystallogr.*, D**66**, p. 486.

Engel, G. E., Wilke, S., Konig, O., Harris, K. D. M., and Leusen, F. J. J. (1999). *J. Appl. Crystallogr.*, **32**, p. 1169.

Esnouf, R. M. (1997). *Acta Crystallogr.*, D**53**, p. 665.

Estermann, M. A. and Gramlich, V. (1993). *J. Appl. Crystallogr.*, **26**, p. 396.

Estermann, M. A., McCusker, L. B., and Baerlocher, C. (1992). *J. Appl. Crystallogr.*, **25**, p. 539.

Fan, H. F., Zhong, Z. Y., Zheng, C. D., and Li, F. H. (1985). *Acta Crystallogr.*, A**41**, p. 163.

Favre-Nicolin, V. and Černy, R. (2002). *J. Appl. Crystallogr.*, **35**, p. 734.

Fisher, J. E., Hancock, J. H., and Hauptman, H. A. (1970b). *NRL Report*, p. 7157.

Fisher, J. E., Hancock, J. H., and Hauptman, H. A. (l970a). *NRL Report*, p. 7132.

Fitzgerald, P. M. D. (1988). *J. Appl. Crystallogr.*, **21**, p. 273.

Florence, A. J., Shankland, N., Shankland, K., David, W. I. F., Pidcock, E., Xu, X., Johnston, A., Kennedy, A. R., Cox, P. J., Evans, J. S. O., Steele, G., Cosgrove, S. D., and Frampton, C. S. (2005). *J. Appl. Crystallogr.*, **38**, 249.

Foadi, J., Woolfson, M. M., Dodson, E., Wilson, K. S., Jia-xing, Y., and Chao-de, Z. (2000). *Acta Crystallogr.*, D**56**, p. 1137.

Fortier, S. and Hauptman, H. A. (l977a). *Acta Crystallogr.*, A**33**, p. 572.

Fortier, S. and Hauptman, H. A. (1977b). *Acta Crystallogr.*, A**33**, p. 829.

Fortier, S. and Hauptman, H. A. (1977c). *Acta Crystallogr.*, A**33**, p. 694.

Foster, F. and Hargreaves, A. (1963a). *Acta Crystallogr.*, **16**, p. 1124.

Foster, F. and Hargreaves, A. (1963b). *Acta Crystallogr.*, **16**, p. 1133.

French, S. and Wilson, K. (1978). *Acta Crystallogr.*, A**34**, p. 517.

Frigo, M. and Johnson, S. G. (2005) *Proceedings of the IEEE*, **93**(2), p. 216.

Fujinaga, M. and Read, R. J. (1987). *J. Appl. Crystallogr.*, 20, p. 517.

Furàsaki, A. (1979). *Acta Crystallogr.*, A**35**, p. 220.

Garrido, J. (1950). *Compt. Rend.*, **230**, p. 1878.

Gassmann, J. (1975). *Acta Crystallogr.*, A**31**, p. 825.

Gassmann, J. (1976). *Acta Crystallogr.*, A**32**, p. 274.

Gassmann, J. (1977). *Acta Crystallogr.*, A**33**, p. 474.

Gassmann, J. and Zechmeister, K. (1972). *Acta Crystallogr.*, A**28**, p. 270.

Germain, G. and Main, P. (1971). *Acta Crystallogr.*, B**27**, p. 368.

Germain, G., Main, P., and Woolfson, M. M. (1970). *Acta Crystallogr.*, B**26**, p. 274.

Germain, G., Main, P., and Woolfson, M. M. (1974). In *NATO Advanced Study Institute on Direct Methods in Crystallography*. NATO Advanced Study Institute, Erice, Italy.

Giacovazzo, C. (1974a). *Acta Crystallogr.*, A**30**, p. 626.

Giacovazzo, C. (1974b). *Acta Crystallogr.*, A**30**, p. 631.

Giacovazzo, C. (1975a). *Acta Crystallogr.*, A**31**, p. 252.

Giacovazzo, C. (1975b). *Acta Crystallogr.*, A**31**, p. 602.
Giacovazzo, C. (1976a). *Acta Crystallogr.*, A**32**, p. 74.
Giacovazzo, C. (1976b). *Acta Crystallogr.*, A**32**, p. 91.
Giacovazzo, C. (1976c). *Acta Crystallogr.*, A**32**, p. 100.
Giacovazzo, C. (l976d). *Acta Crystallogr.*, A**32**, p. 958.
Giacovazzo, C. (1977a). *Acta Crystallogr.*, A**33**, p. 933.
Giacovazzo, C. (1977b). *Acta Crystallogr.*, A**33**, p. 944.
Giacovazzo, C. (1979). *Acta Crystallogr.*, A**35**, p. 757.
Giacovazzo, C. (1980). *Direct Methods in Crystallography.* Academic, London.
Giacovazzo, C. (1983a). *Acta Crystallogr.*, A**39**, p. 685.
Giacovazzo, C. (1983b). *Acta Crystallogr.*, A**39**, p. 585.
Giacovazzo, C. (1988). *Acta Crystallogr.*, A**44**, p. 294.
Giacovazzo, C. (1989). *Acta Crystallogr.*, A**45**, p. 534.
Giacovazzo, C. (1991). *Acta Crystallogr.*, A**47**, p. 256.
Giacovazzo, C. (1993). *Z. Kristallogr.*, **206**, p. 161.
Giacovazzo, C., Altomare, A., Cuocci, C., Moliterni, A. G. G., and Rizzi, R. (2002). *J. Appl. Crystallogr.*, **35**, p. 422.
Giacovazzo, C., Burla, M. C., and Cascarano, G. (1992a). *Acta Crystallogr.*, A**48**, p. 901.
Giacovazzo, C., Guagliardi, A., Ravelli, R., and Siliqi, D. (1994a). *Z. Kristallogr.*, **209**, p. 136.
Giacovazzo, C., Ladisa, M., and Siliqi, D. (2002) *Acta Crystallogr.*, A**58**, p. 598.
Giacovazzo, C., Ladisa, M. and Siliqi, D. (2003). *Acta Crystallogr.*, A**59**, p. 569.
Giacovazzo, C. and Mazzone, A. (2011). *Acta Crystallogr.*, A**67**, p. 210.
Giacovazzo, C. and Mazzone, A. (2012). *Acta Crystallogr.*, A**68**, p. 464.
Giacovazzo, C. Mazzone, A., and Comunale, G. (2011). *Acta Crystallogr.*, A**67**, p. 368.
Giacovazzo, C., Moustiakimov, M., Siliqi, D., and Pifferi, A. (2004). *Acta Crystallogr.*, A**60**, p. 233.
Giacovazzo, C. and Siliqi, D. (1997). *Acta Crystallogr.*, A**53**, p. 789.
Giacovazzo, C. and Siliqi, D. (2002). *Acta Crystallogr.*, A**58**, p. 590.
Giacovazzo, C. and Siliqi, D. (2004). *Acta Crystallogr.*, D**60**, p. 73.
Gilmore, C. J., Nicholson, W. V., and Dorset, D. L. (1996). *Acta Crystallogr.*, A**52**, p. 937.
Gilmore, C. J., Shankland, K., and Fryer, J. R. (1993). *Ultramicroscopy*, **48**, p. 132.
Glykos, N. M. and Kokkinidis, M. (2000a). *Acta Crystallogr.*, D**56**, p. 1070.
Glykos, N. M. and Kokkinidis, M. (2000b). *Acta Crystallogr.*, D**56**, p. 169.
Goldberg, D. E. (1989). In Genetic Algorithms in Search, Optimization and Machine Learning, Addison-Wesley, New York.
González, A. (2003a). *Acta Crystallogr.*, D**59**, p. 3.
González, A. (2003b). *Acta Crystallogr.*, D**59**, p. 1935.
Gopal, K., McKee, E., Romo, T. D., Pai, R., Smith, J. N., Sacchettini, J. C., and Ioerger, T. R. (2007). *Bioinformatics*, 23(3), p. 375.
Gould, R. O., van der Hark, T. E. M., and Beurskens, P. T. (1975). *Acta Crystallogr.*, A**31**, p. 813.
Gozzo, F., De Caro, L., Giannini, C., Guagliardi, A., Schmitt, B., and Prodi, A. (2006). *J. Appl. Crystallogr.*, **39**, p. 347.
Gramlich, V. (1984). *Acta Crystallogr.*, A**40**, p. 610.
Green, D. W., Ingram, V. M., and Perutz, M. F. (1954). *Proc. R. Soc.*, A**225**, p. 287.
Green, E. A. and Hauptman, H. A. (1976). *Acta Crystallogr.*, A**32**, p. 43.
Greer, J. (1974). *J. Mol. Biol.*, **82**, p. 279.
Greer, J. (1985). *Methods Enzymol.*, **115**, p. 206.
Grosse-Kunstleve, R. W. and Adams, P. D. (2001). *Acta Crystallogr.*, D**57**, p. 1390.
Grosse-Kunstleve, R. W. and Brunger, A. T. (1999). *Acta Crystallogr.*, D**55**, p. 1568.
Gu, Y. X., Liu, Y. D., Hao, Q., and Fan, H. F. (2000). *Acta Crystallogr.*, A**56**, p. 592.
Gu, Y. X., Woolfson, M. M., and Yao, J. X. (1996). *Acta Crystallogr.*, D**52**, p. 1114.
Guinier, A. (1963). *X-Ray Diffraction*. Freeman, San Francisco.

Hall, S. R. (1970a). In *NATO Advanced Study Institute on Direct and Patterson Methods*. NATO Advanced Study Institute, Parma, Italy.
Hall, S. R. and Subramanian, W. (1982a). *Acta Crystallogr.*, A**38**, p. 590.
Hall, S. R. and Subramanian, W. (1982b). *Acta Crystallogr.*, A**38**, p. 598.
Han, F. and Langs, D. A. (1988). *Acta Crystallogr.*, A**44**, p. 563.
Hao, Q., Gu, Y. X., Zheng, C. D., and Fan, H. F. (2000). *J. Appl. Crystallogr.*, **33**, p. 980.
Harada, Y., Lifshitz, A., Berthou, J., and Jolles, P. (1981). *Acta Crystallogr.*, A**37**, p. 398.
Hargreaves, A. (1955). *Acta Crystallogr.*, **8**, p. 12.
Harker, D. (1936). *Chem. Phys.*, **4**, p. 381.
Harris, G. W. (1995). *Acta Crystallogr.*, D**51**, p. 695.
Harris, K. D. M., Tremaine, M., Lighfoot, P., and Bruce, P. G. (1994). *J. Am. Chem. Soc.*, **116**, p. 3543.
Harrison, R. W. (1989). *Acta Crystallogr.*, A**45**, p. 4.
Hašek, J. (1974). *Acta Crystallogr.*, A**30**, p. 576.
Hauptman, H. A (1970). New Orleans Meeting of the American Crystallographic Association.
Hauptman, H. A (1975a). *Acta Crystallogr.*, A**31**, p. 671.
Hauptman, H. A (1975b). *Acta Crystallogr.*, A**31**, p. 680.
Hauptman, H. A. (1976). *Acta Crystallogr.*, A**32**, p. 877.
Hauptman, H. A. (1982a). *Acta Crystallogr.*, A**38**, p. 289.
Hauptman, H. A. (1982b). *Acta Crystallogr.*, A**38**, p. 632.
Hauptman, H. A. (1995). *Acta Crystallogr.*, B**51**, p. 416.
Hauptman, H. A., Fisher, J., Hancock, H., and Norton, D. A (1969). *Acta Crystallogr.*, B**25**, p. 811.
Hauptman, H. A. and Fortier, S. (1977a). *Acta Crystallogr.*, A**33**, p. 575.
Hauptman, H. A. and Fortier, S. (1977b). *Acta Crystallogr.*, A**33**, p. 697.
Hauptman, H. A. and Green, E. A. (1976). *Acta Crystallogr.*, A**32**, p. 45.
Hauptman, H. A. and Karle, J. (1953). *The Solution of the Phase Problem I. The Centrosymmetric Crystal*, ACA Monograph No. 3. Polycrystal Book Service, New York.
Hauptman, H. A. and Karle, J. (1956). *Acta Crystallogr.*, **9**, p. 45.
Hauptman, H. A and Karle, J. (1959). *Acta Crystallogr.*, **12**, p. 93.
Hedel, R., Bunge, H. J., and Reck, G. (1994). In *Proc. 10th Int. Conf. Textures of Matter.*, H. J. Bunge (ed), Trans Tech Publ., Switzerland, pp. 2067–74.
Heinerman, J. J. L., Krabbendam, H., Kroon, J., and Spek, A. L. (1978). *Acta Crystallogr.*, A**34**, p. 447.
Helliwell, J. R., Helliwell, M., and Jones, H. R. (2005). *Acta Crystallogr.*, A**61**, p. 568.
Helliwell, M. (2000). *J. Synchrotron Rad.*, **7**, p. 139.
Hendersson, R. and Unwin, P. N. T. (1975). *Nature*, **257**, p. 28.
Hendrickson, W. A. (1975). *J. Mol. Biol.*, **91**, p. 226.
Hendrickson, W. A., Klippenstein, G. L., and Ward, K. B. (1975). *Proc. Natl. Acad. Sci., USA*, **72**, p. 2160.
Hendrickson, W. A., Love, W. E., and Karle, J. (1973). *J. Mol. Biol.*, **74**, p. 331.
Hendrickson, W. A., Pähler, A., Smith, J. L., Satow, Y., Merritt, E. A., and Phizackerley, R. P. (1989). *Proc. Natl. Acad. Sci., USA*, **86**, p. 2190.
Hendrickson, W. A. and Teeter, M. M. (1981). *Nature*, **290**, p. 107.
Hendrickson, W. A. and Ward, K. B. (1976). *Acta Crystallogr.*, A**32**, p. 778.
Hirshfeld, F. L. (1968). *Acta Crystallogr.*, A**24**, p. 301.
Hirshfeld, F. L. and Rabinovich, D. (1973). *Acta Crystallogr.*, A**29**, p. 510.
Hodgkin, D. C., Kamper, J., Lindsey, J., MacKay, M., Pickworth, J., Robertson, J. H., Shoemaker, C. B., White, J. G., Prosen, R. J., and Trueblood, K. N. (1957). *Proc. R. Soc. London Ser.*, A**242**, p. 228.
Hoppe, W. (1957). *Acta Crystallogr.*, **10**, p. 750.
Hoppe, W. (1962a). *Naturwissenschaften*, **49**, p. 536.
Hoppe, W. (1962b). *Acta Crystallogr.*, **15**, p. 13.
Hoppe, W. and Gassmann, J. (1968). *Acta Crystallogr.*, B**24**, p. 97.

Hoppe, W., Gassmann, J., and Zechmeister, K. (1970). In *Crystallographic Computing*. Munksgaard, Copenhagen.
Hoppe, W. and Jakubowski, U. (1975). In *Anomalous Scattering* (ed. S. Ramaseshan and S. C. Abrahams), p. 437. Munksgaard, Copenhagen.
Hosemann, R. and Bagchi, S. N. (1954). *Acta Crystallogr.*, **7**, p. 237.
Hovmöller, S., Sjögren, A., Farrants, G., Sundberg, M., and Marinder, B. O. (1984). *Nature*, **311**, p. 238.
Howells, E. R., Phillips, D. C., and Rogers, D. (1950). *Acta Crystallogr.*, **3**, p. 210.
Hu, J. J., Li, F. H., and Fan, H. F. (1992). *Ultramicroscopy*, **41**, p. 387.
Huang, T. C. (1988). *Australian Journal of Physics*, **41**, p. 201.
Huang, T. C. and Parrish, W. (1975). *Appl. Phys. Letters*, **27**, p. 123.
Huber, R. (1965). *Acta Crystallogr.*, **19**, p. 353.
Huber, R. and Schneider, M. (1985). *J. Appl. Crystallogr.*, 18(3), p. 165.
Huber, R. (1969). In *Crystallographic Computing Proceedings* (ed. F. R. Ahmed). Munksgaard, Copenhagen.
Hughes, E. W. (1949). *Acta Crystallogr.*, **2**, p. 34.
Hull, S. E., Viterbo, D., Woolfson, M. M., and Zhang, S.-H. (1981). *Acta Crystallogr.*, A**37**, p. 566.
Ibers, J. A. (1958). *Acta Crystallogr.*, **11**, p. 178.
Ida, T. and Toraya, H. (2002). *J. Appl. Crystallogr.*, **35**, p. 38.
Immirzi, A. (1980). *Acta Crystallogr.*, B**36**, p. 2378.
International Tables for Crystallography (1992). Vol. C, Ed.A. J. C.Wilson, Kluwer Academic Publishers, Dordrecht.
International Tables for Crystallography (1993). Vol. B, Ed. U. Shmueli, Kluwer Academic Publishers, Dordrecht.
International Tables for Crystallography (2005). Vol. A., Ed. T. Hahn, Springer, Dordrecht.
Izumi, F. (1989). *Rigaku J.*, **1**, p. 10.
James, R. W. (1962). The Optical Principles of the Diffraction of X-Rays. Bell and Sons, London.
Jamrog, D. C., Zhang, Y., and Phillips Jr., G. N. (2003). *Acta Crystallogr.*, D**59**, p. 304.
Jansen, I., Peschar, R., and Schenk, H. (1992). *J. Appl. Crystallogr.*, **25**, p. 231.
Järvinen, M. (1993). *J. Appl. Crystallogr.*, **26**, p. 525.
Jiang, J.-S. and Brunger, A. T. (1994). *J. Mol. Biol.*, **243**, p. 100.
Johnson, C. K. (1978). *Acta Crystallogr.* A**34**, S–353.
Jones, T. A and Thirup, S. (1986). *EMBO J.*, **5**, p. 819.
Jones, T. A. (2004). *Acta Crystallogr.*, D**60**, p. 2115.
Jones, T. A., Zou, J.-Y., Cowan, S. W., and Kjeldgaard, M. (1991). *Acta Crystallogr.*, A**47**, p. 110.
Joosten, K., Cohen, S. X., Emsley, P., Mooij, W., Lamzin, V. S., and Perrakis, A. (2008). *Acta Crystallogr.*, D**64**, p. 416.
Kariuki, B. M., Serrano-Gonzàlez, H., Johnston, R. L., and Harris, K. D. M. (1997). *Chem. Phys. Lett.*, **280**, p. 189.
Karle, J. (1970a). *Acta Crystallogr.*, B**26**, p. 1614.
Karle, J. (1970b). In *Crystallographic Computing*, (ed. F. R. Ahmed, S. R. Hall, and C. P. Huber). Munksgaard, Copenhagen.
Karle, J. (1976). In *Crystallographic Computing Techniques*. Munksgaard, Copenhagen.
Karle, J. (1980). Int. J. Quantum Chem.: Quantum Biol. Symp., **7**, p. 356.
Karle, J. (1983a). Proceedings of the School on Direct Methods and Macromolecular Crystallography. Medical Foundations of Buffalo, Buffalo.
Karle, J. (1983b). In *Methods and Applications in Crystallographic Computing* (ed. S. R. Hall and T. Ashida). Clarendon, Oxford.
Karle, J. (1984a). *Acta Crystallogr.*, A**40**, p. 1.
Karle, J. (1984b). *Acta Crystallogr.*, A**40**, p. 4.
Karle, J. (1989a). *Acta Crystallogr.*, A**45**, p. 303.

Karle, J. (1989b). *Phys. Today*, June, 22.
Karle, J. and Hauptman, H. A. (1950). *Acta Crystallogr.*, **3**, p. 181.
Karle, J. and Hauptman, H. A. (1957). *Acta Crystallogr.*, **10**, p. 515.
Karle, J. and Hauptman, H. A. (1958). *Acta Crystallogr.*, **11**, p. 264.
Karle, J. and Hauptman, H. A. (1961). *Acta Crystallogr.*, **14**, p. 217.
Karle, J. and Karle, I. L. (1966). *Acta Crystallogr.*, **21**, p. 849.
Kartha, G. and Parthasarathy, R. (1965). *Acta Crystallogr.*, **18**, p. 745.
Keegan, R. M. and Winn, M. D. (2008). *Acta Crystallogr.*, D**64**, p. 119.
Kennard, O., Wampler, D. L., Coppola, J. C., Motherwell, W. D. S., Watson, D. G., and Larson, A. C. (1971). *Acta Crystallogr.*, B**27**, p. 1116.
Kirkpatrick, S., Gelatt, C. D., and Vecchi, M. P. (1983). *Science*, **220**, p. 671.
Kissinger, C. R., Gelhaar, D. K., and Fogel, D. B. (1999). *Acta Crystallogr.*, D**55**, p. 484.
Kleywegt, G. J. (1997). *J. Mol. Biol.*, **273**, p. 371.
Kleywegt, G. J. and Jones, T. A. (1996). *Structure*, **4**, p. 1395.
Kleywegt, G. J. and Jones, T. A. (1997a). *Methods Enzymol.*, **277**, p. 208.
Kleywegt, G. J. and Jones, T. A. (1997b). *Acta Crystallogr.*, D**53**, p. 179.
Klug, A. (1958). *Acta Crystallogr.*, **11**, p. 515.
Klug, H. P. and Alexander, L. E. (1974). X-Ray Diffraction Procedures from Polycrystalline and Amorphous Materials. Wiley Interscience, New York.
Kluyver, J. C. (1906). *Proc. Sect. Sci. K. Ned. Akad. Wet.*, **8**, p. 341.
Koch, M. H. J. (1974). *Acta Crystallogr.*, A**30**, p. 67.
Kockelmann, W., Jansen, E., Schafer, W., and Will, G. (1985). *Report Forts-chungszentrum Julich*.
Kolb, U., Gorelik, T., Keble, C., Otten, M. T., and Hubert, D. (2007a). *Ultramicroscopy*, **107**, p. 507.
Kolb, U., Gorelik, T., and Otten, M. T. (2007b). *Ultramicroscopy*, **108**, p. 763.
Kraut, J. (1958). *Biochim. Biophys. Acta.*, **30**, p. 265.
Kroon, J., Spek, A. L., and Krabbendam, H. (1977). *Acta Crystallogr.*, A**33**, p. 382.
Kyrala, A. (1972). Applied Functions of a Complex Variable. Wiley, New York.
Lajzerowicz, J. and Lajzerowicz, J. (1966). *Acta Crystallogr.*, **21**, p. 8.
Lamzin, V. S. and Wilson, K. S. (1993). *Acta Crystallogr.*, D**49**, p. 129.
Larson, A. C. and von Dreele, R. B. (1987). *Los Alamos National Laboratory Report* LA-UR–86–748.
Lattman, E. E. and Love, W. E. (1970). *Acta Crystallogr.*, B**26**, p. 1854.
Le Bail, A. (2001). *Mater. Sci. Forum*, **378–381**, p. 65.
Le Bail, A., Duroy, H., and Fourquet, J. L. (1988). *Math. Res. Bull.*, **23**, p. 447.
Leonard, G. A., Sainz, G., de Backer, M. M. E., and McSweeney, S. (2005). *Acta Crystallogr.*, D**61**, p. 388.
Leslie, A. G. W. (1987). *Acta Crystallogr.*, A**43**, p. 134.
Lessinger, L. and Wondratschek, H. (1975). *Acta Crystallogr.*, A**31**, p. 521.
Levitt, D. G. (2001). *Acta Crystallogr.*, D**57**, p. 1013.
Li, D. X. and Hovmöller, S. (1988). *J. Solid State Chem.*, **73**, p. 5.
Lipson, H. and Woolfson, M. M. (1952). *Acta Crystallogr.*, **5**, p. 680.
Long, F., Vagin, A. A., Young, P., and Murshudov, G. N. (2008). *Acta Crystallogr.* D**64**, p. 125.
Lovell, S. C., Davis, I. W., Arendall, W. B., de Bakker, P. I., Word, J. M., Prisant, M. G., Richardson, J. S., and Richardson, D. C. (2003). *Proteins*, **50**, p. 437.
Lovell S. C., Word, J. M., Richardson, J. S., and Richardson, D. C. (2000). *Proteins: Struct. Funct. Genet.*, **40**, p. 389.
Lunin, V. Y. (1988). *Acta Crystallogr.*, A**44**, p. 144.
Lunin, V. Y. (1993). *Acta Crystallogr.*, D**49**, p. 90.
Lunin, V. Y., Afonine, P. V., and Urzhumtsev, A. G. (2002). *Acta Crystallogr.*, A**58**, p. 270.
Lunin, V. Y., Lunina, N. L., Petrova, T. E., Vernoslova, E. A., Urzhumtsev, A. G., and Podjarny, A. D. (1995). *Acta Crystallogr.*, D**51**, p. 896.
Lunin, V. Y. and Skovoroda, T. P. (1991). *Acta Crystallogr.*, A**47**, p. 45.

Lunin, V. Y. and Urzhumtsev, A. G. (1984). *Acta Crystallogr.*, A**40**, p. 269.
Lunin, V. Y., Urzhumtsev, A. G., and Skovoroda, T. P. (1990). *Acta Crystallogr.*, A**46**, p. 540.
Lutterotti, L. and Scardi, P. (1990). *J. Appl. Crystallogr.*, **23**, p. 246.
Luzzati, V. (1952). *Acta Crystallogr.*, **5**, p. 802.
Luzzati, V. (1953). *Acta Crystallogr.*, **6**, p. 142.
Luzzati, V., Mariani, P., and Delacroix, H. (1988). *Macromol. Chem. Macromol. Symp.*, **15**, p. 1.
MacGillavry, C. H. (1950). *Acta Crystallogr.*, **3**, p. 214.
Mackay, A. L. (1953). *Acta Crystallogr.*, A**30**, p. 607.
Magini, M., Licheri, G., Pischina, G., Piccaluga, G., and Pinna, G. (1988). *X-Ray Diffraction of Ions in Aqueous Solutions: Hydration and Complex Formation* (ed. M. Magini). CRC Press, Boca Raton, FL.
Main, P. (1967). *Acta Crystallogr.*, **23**, p. 50.
Main, P. (1977). *Acta Crystallogr.*, A**33**, p. 750.
Main, P. (1979). *Acta Crystallogr.*, A**35**, p. 779.
Main, P. (1990a). *Acta Crystallogr.*, A**46**, p. 372.
Main, P. (1990b). *Acta Crystallogr.*, A**46**, p. 507.
Main, P. (1993). *Z. Kristallogr.*, **206**, p. 201.
Main, P. and Hull, S. E. (1978). *Acta Crystallogr.*, A**34**, p. 353.
Mardia, K. V. (1972). *Statistics of Directional Data*. Academic, London.
Margiolaki, I., Wright, J. P., Fitch, A. N., Fox, G. C., and Von Dreele, R. B. (2005). *Acta Crystallogr.*, D**61**, p. 423.
Mariani, P., Luzzati, V., and Delacroix, H. (1988). *J. Mol. Biol.*, **204**, p. 165.
Markvardsen, A. J., David, W. I. F., Johnson, J. C., and Shankland, K. (2001). *Acta Crystallogr.*, A**57**, p. 47.
Masciocchi, N., Moret, M., Cairati, P., Sironi, A., Ardino, G. A., and La Monica, G. (1994). *J. Am. Chem. Soc.*, **116**, p. 7668.
Matthews, B. W. (1968). *J. Mol. Biol.*, **33**, p. 491.
Matthews, B. W. and Czerwinski, E. W. (1975). *Acta Crystallogr.*, A**31**, p. 480.
Matthies, S., Lutterotti, L., and Wenk, H. R. (1997). *J. Appl. Crystallogr.*, **30**, p. 31.
McCoy, A. J. (2004). *Acta Crystallogr.*, D**60**, p. 2169.
McCoy, A. J., Grosse-Kunstleve, R. W., Adams, P. D., Winn, M. D., Storoni, L. C., and Read, R. J. (2007). *J. Appl. Crystallogr.*, **40**, p. 658.
McCoy, A. J., Storoni, L. C., and Read, R. J. (2004). *Acta Crystallogr.*, D**60**, p. 1220.
McKenna, R., Xia, D., Willingmann, P., Ilag, L. L., Krishnaswamy, S., Rossmann, M. G., Olson, N. H., Baker, T. S., and Incardona, N. L. (1992). *Nature*, **355**, p. 137.
Metropolis, N., Rosenbluth, A. W., Rosenbluth, M. N., Teller, A. H., and Teller, E. (1953). *J. Chem. Phys.*, **21**, p. 1087.
Mighell, A. D. and Jacobson, R. A. (1963). *Acta Crystallogr.*, **16**, p. 443.
Miller, R., Gallo, S. M., Kholak, H. G., and Weeks, C. M. (1994). *J. Appl. Crystallogr.*, **27**, p. 613.
Mitchell, C. M. (1957). *Acta Crystallogr.*, **10**, p. 475.
Mö, F., Hjortas, J., and Svinning, T. (1973). *Acta Crystallogr.*, A**29**, p. 358.
Mooers, B. H. M. and Matthews, B. W. (2006). *Acta Crystallogr.*, D**62**, p. 165.
Moras, D., Lorber, B., Romby, P., Ebel, J.-P., Giegè, R., Lewit-Beutley, A., and Roth, M. (1983). *J. Biomol. Struct. Dynamics*, **1**, p. 209.
Morniroli, J. P., Redjaïmia, A., and Nicolopoulos, S. (2007). *Ultramicroscopy*, **107**, p. 514.
Morniroli, J. P. and Steeds, J. W. (1992). *Ultramicroscopy*, **45**, p. 219.
Morris, R. J., Blanc, E., and Bricogne, G. (2004). *Acta Crystallogr.*, D**60**, p. 227.
Morris, R. M., Perrakis, A., and Lamzin, V. S. (2002). *Acta Crystallogr.*, D**58**, p. 968.
Moss, D. S. (1985). *Acta Crystallogr.*, A**41**, p. 470.
Moss, B. and Dorset, D. L. (1983). *Acta Crystallogr.*, A**39**, p. 609.
Muirhead, H., Cox, J. M., Mazzarella, L., and Perutz, M. F. (1967). *J. Mol. Biol.*, **28**, p. 117.
Mukherjee, A. K., Helliwell, J. R., and Main, P. (1989). *Acta Crystallogr.*, A**45**, p. 715.

Murshudov, G. N., Dodson, E. S., and Vagin, A. A. (1996). In *Macromolecular Refinement. Proceedings of the CCP4 Study Weekend*, pp. 93–104. CLRC Daresbury Laboratory, Warrington, UK.
Murshudov, G. N., Skubák, P., Lebedev, A. A., Pannu, N. S., Steiner, R. A., Nicholls, R. A., Winn, M. D., Long, F., and Vagin, A. A. (2011). *Acta Crystallogr.*, D**67**, p. 355.
Murshudov, G. N., Vagin, A. A., and Dodson, E. J. (1997). *Acta Crystallogr.*, D**53**, p. 240.
Nagem, R. A. P., Ambrosio, A. L. B., Rojas, A. L., Navarro, M. V. A. S., Golubev, A. M., Garratt, R. C., and Polikarpov, I. (2005). *J. Synchrotron Rad.*, **12**, p. 420.
Nagem, R. A., Polikarpov, I., and Dauter, Z. (2003). *Methods Enzymol.*, **374**, p. 120.
Nan, J., Zhou, Y., Yang, C., Brostromer, E., Kristensen, O., and Su, X.-D. (2009). *Acta Crystallogr.*, D**65**, p. 440.
Navaza, J. (1987). *Acta Crystallogr.*, A**43**, p. 645.
Navaza, J. (1993). *Acta Crystallogr.*, D**49**, p. 588.
Navaza, J. (1994). *Acta Crystallogr.*, A**50**, p. 157.
Navaza, J. and Vernoslova, E. (1995). *Acta Crystallogr.*, A**51**, p. 445.
Naya, S., Nitta, I., and Oda, T. (1964). *Acta Crystallogr.*, **17**, p. 421.
Naya, S., Nitta, I., and Oda, T. (1965). *Acta Crystallogr.*, **19**, p. 734.
Needleman, S. B. and Wunsch, C. D. (1970). *J. Mol. Biol.*, **48**, p. 443.
Nelder, J. A. and Mead, R. (1965). *Comput. J.*, 7, p. 308.
Niederwanger, V., Gozzo, F., and Griesser, U. J. (2009). *J. Pharm. Sci.*, **98**, p. 1064.
Norby, P., Christensen, A. N., Fjellvag, H., Lehmann, M. S., and Nielsen, M. J. (1991). *Solid State Chem.*, **94**, 281.
Nordman, C. E. and Nakatsu, K. (1963). *J. Am. Chem. Soc.*, **85**, p. 353.
Nowacki, W. (1955). *J. Chem. Phys.*, **23**, p. 1172.
O'Keefe, M. A. (1973). *Acta Crystallogr.*, A**29**, p. 389.
Oldfield, T. (2002). *Acta Crystallogr.*, D**58**, p. 487.
Oldfield, T. J. and Hubbard, R. E. (1994). *Proteins: Struct. Funct. Genet.*, 18, p. 324.
Oszlányi, G. and Suto, A. (2004). *Acta Crystallogr.*, A**60**, p. 134.
Oszlányi, G. and Suto, A. (2005). *Acta Crystallogr.*, A**61**, p. 147.
Oszlányi, G. and Suto, A. (2008). *Acta Crystallogr.*, A**64**, p. 123.
Otwinowski, Z. (1991a). In *Isomorphous Replacement and Anomalous Scattering*, CCP4. Daresbury, SERC.
Otwinowski, Z. (1991b). In *Proceedings of the 1991 CCP4 Study Weekend*. CLRC Daresbury Laboratory, Warrington, UK.
Palatinus, L. and Chapuis, G. (2007). *J. Appl. Crystallogr.*, **40**, p. 786.
Palatinus, L., Dušek, M., Glaum, R., and El Bali, B. (2006). *Acta Crystallogr.*, B**62**, p. 556.
Pannu, N. S. and Read, R. J. (2004). *Acta Crystallogr.*, D**60**, p. 22.
Pannu, N. S., McCoy, A. J., and Read, R. J. (2003). *Acta Crystallogr.*, D**59**, p. 1801.
Parrish, W. (1965). In *X-ray Analysis Papers*. Centrex Publishing, Eindhoven.
Parthasarathi, V. (1975). *Ph.D. Thesis*. University of Madras.
Parthasarathy, S. and Parthasarathi, V. (1976). *Acta Crystallogr.*, A**32**, p. 57.
Patterson, A. L. (1934a). *Phys. Rev.*, **45**, p. 763.
Patterson, A. L. (1934b). *Phys. Rev.*, **46**, p. 372.
Patterson, A. L. (1939). *Nature*, **143**, p. 939.
Patterson, A. L. (1944). *Phys. Review*, **65**, p. 195.
Patterson, A. L. (1949). *Acta Crystallogr.*, **2**, p. 339.
Pauling, L. and Shapell, M. D. (1930). *Z. Kristallogr.*, **75**, p. 128.
Pavelcik, F. (2006). *J. Appl. Crystallogr.*, **39**, p. 483.
Pavelcik, F., Zelinka, J., and Otwinowski, Z. (2002). *Acta Crystallogr.*, D**58**, p. 275.
Pawley, G. S. (1981). *J. Appl. Crystallogr.*, **14**, p. 357.
Pearson, K. (1905). *Nature*, **72**, p. 294.
Peederman, A. F. and Bijvoet, J. M. (1956). *Proc. K. Ned. Akad. Wet.*, B**39**, p. 312.
Pepinsky, R. and Okaya, Y. (1956). *Proc. Nat. Acad. Sci.*, USA, **42**, p. 286.
Perrakis, A., Morris, R. M., and Lamzin, V. S. (1999). *Nature Struct. Biol.*, 6, p. 458.

Perutz, M. F. (1956). *Acta Crystallogr.*, **9**, p. 867.
Peschar, R., Schenk, H., and Capkova, P. (1995). *J. Appl. Crystallogr.*, **28**, p. 127.
Phillips, D. C. (1966). In *Advances in Structure Research by Diffraction Methods* (ed. R. Brill and R. Mason). Interscience, New York.
Podjarny, A. D. and Yonath, A. (1977). *Acta Crystallogr.*, A**33**, p. 655.
Ponder, J. V. and Richards, F. M. (1987). *J. Mol. Biol.*, **193**, p. 775.
Pontenagel, W. M. G. F. and Krabbendam, H. (1983). *Acta Crystallogr.*, A**39**, p. 333.
Pradhan, D., Ghosh, S., and Nigam, G. D. (1985). In *Structure and Statistics in Crystallography* (ed. A. J. C. Wilson). Adenine Press, New York.
Prandl, W. (1990). *Acta Crystallogr.*, A**46**, p. 988.
Prandl, W. (1994). *Acta Crystallogr.*, A**50**, p. 52.
Press, W. H., Tenkolsky, S. A., Vetterling, W. T., and Flannery, B. P. (1992). In *'Numerical Recipes': The Art of Scientific Computing*, pp. 436–48. Cambridge University Press.
Prince, E. (1993). *Acta Crystallogr.*, D**49**, p. 61.
Putz, H., Schon, J. C., and Jansen, M. (1999). *J. Appl. Crystallogr.*, **32**, p. 864.
Qian, B., Raman, S., Das, R., Bradley, P., McCoy, A. J., Read, R. J., and Baker, D. (2007). *Nature*, **450**, p. 259.
Rabinovich, D. and Shakked, Z. (1984). *Acta Crystallogr.*, A**40**, p. 195.
Rabinovich, D., Rozenberg, H., and Shakked, Z. (1998). *Acta Crystallogr.*, D**54**, p. 1336.
Ramachandran, G. N., Ramakrishnan, C., and Sasisekharan, V. (1963). *J. Mol. Biol.*, **7**, p. 95.
Ramachandran, G. N. and Raman, S. (1959). *Acta Crystallogr.*, **12**, p. 957.
Ramachandran, G. N. and Srinivasan, R. (1970). *Fourier Methods in Crystallography*. Wiley/Interscience, London.
Ramagopal, U. A., Dauter, M., and Dauter, Z. (2003). *Acta Crystallogr.*, D**59**, p. 1020.
Ramakrishnan, V. and Biou, V. (1997). *Methods Enzymol.*, **276**, p. 538.
Raman, S. and Lipscomb, W. N. (1961). *Z. Kristallogr.*, **116**, p. 314.
Rameseshan, S. (1963). In *Advanced Methods of Crystallography* (ed. G. N. Ramachandran). Academic, London.
Rameseshan, S. and Venkatesan, K. (1957). *Current Sci. India*, **26**, p. 352.
Rao, S. N., Jih, J. H., and Hartsuck, J. A. (1980). *Acta Crystallogr.*, A**36**, p. 878.
Ravelli, R. B., Theveneau, P., McSweeney, S., and Caffrey, M. (2002). *J. Synchrotron Rad.*, **9**, p. 355.
Rayleigh, Lord (1919a). *Phil. Mag.*, **37**, p. 321.
Rayleigh, Lord (1919b). *Phil. Mag.*, 37, p. 498.
Read, R. J. (1986). *Acta Crystallogr.*, A**42**, p. 140.
Read, R. J. (1999). *Acta Crystallogr.*, D**55**, p. 1759.
Read, R. J. (2001). *Acta Crystallogr.*, D**57**, p. 1373.
Read, R. J. (2003a). *Acta Crystallogr.*, D**59**, p. 1891.
Read, R. J. (2003b). *Crystallogr. Rev.*, **9**, p. 33.
Read, R. J. and Schierbeek, A. J. (1988). *J. Appl. Crystallogr.*, **21**, p. 490.
Refaat, L. S., Tate, C., and Woolfson, M. M. (1996a). *Acta Crystallogr.*, D**52**, p. 252.
Riche, C. (1973). *Acta Crystallogr.*, A**29**, p. 133.
Ridgen, D. J., Keegan, R. M., and Winn, M. D. (2008). *Acta Crystallogr.* D**64**, p. 1288.
Riello, P., Fagherazzi, G., Clemente, D., and Canton, P. (1995). *J. Appl. Crystallogr.*, **28**, p. 115.
Riemann, B. (1892). *Gesammelte Mathematische Werke*, 2nd edn., p. 424. Leipzig.
Rietveld, H. M. (1969). *J. Appl. Crystallogr.*, **2**, p. 65.
Rius, J. and Miravittles, C. (1989). *Acta Crystallogr.*, A**45**, p. 490.
Roberts, P. J., Petterson, R. C., Sheldrick, G. M., Isaacs, N. W., and Kennard, O. (1973). *J. Chem. Soc. Perkin Trans.*, 2, p. 1978.
Robertson, J. M. (1935). *J. Chem. Soc.*, p. 615.
Robertson, J. M. (1936). *J. Chem. Soc.*, p. 1195.
Robertson, J. M. and Woodward, I. (1937). *J. Chem. Soc.*, pp. 219–230.

Rodríguez, D. D., Grosse, C., Himmel, S., González, C., de Ilarduya, I. M., Becker, S., Sheldrick G. M., and Usón, I. (2009). *Nature Methods*, **6**, p. 651.
Rodríguez, D. D., Sammito, M., Meindl, K., de Ilarduya, I. M., Potratz, M., Sheldrick, G. M., and Usón, I. (2012). *Acta Crystallogr.*, D**68**, p. 336.
Rogers, D. (1965). In *Computing Methods in Crystallography*. Pergamon, London.
Rogers, D., Stanley, E., and Wilson, A. J. C. (1955). *Acta Crystallogr.*, **8**, p. 383.
Rogers, D. and Wilson, A. J. C. (1953). *Acta Crystallogr.*, **6**, p. 439.
Rossmann, M. G. (1960). *Acta Crystallogr.*, **13**, p. 221.
Rossmann, M. G. (1961). *Acta Crystallogr.*, **14**, p. 383.
Rossmann, M. G. (1990). *Acta Crystallogr.*, A**46**, p. 73.
Rossmann, M. G. and Blow, D. M. (1962). *Acta Crystallogr.*, **15**, p. 24.
Rossmann, M. G. and Blow, D. M. (1963). *Acta Crystallogr.*, **16**, p. 39.
Rowan, T. (1990). *Functional Stability Analysis of Numerical Algorithms*, Ph.D. thesis, Department of Computer Sciences, University of Texas at Austin, 1990.
Rozhdestvenskaya, I., Mugnaioli, E., Czank, M., Depmeier, W., Kolb, U., Reinholdt, A., and Weirich, T. (2010). *Mineral. Mag.*, **74**, p. 159–177.
Sayre, D. (1952). *Acta Crystallogr.*, **5**, p. 60.
Sayre, D. (1953). *Acta Crystallogr.*, **6**, p. 430.
Sayre, D. (1972). *Acta Crystallogr.*, A**28**, p. 210.
Sayre, D. (1973). *IBM Research Report* RC 4602.
Sayre, D. (1974a). *Acta Crystallogr.*, A**30**, p. 180.
Sayre, D. (1974b). In *NATO Advanced Study Institute on Direct Methods in Crystallography*. NATO Advanced Study Institute, Erice, Italy.
Scardi, P. (2008). Microstructural properties: lattice defects and domain size effects. In *Powder Diffraction: Theory and Practice* (eds R. E. Dinnebier and S. J. L. Billinge). RSC Publishing.
Schenk, H. (1971). *Acta Crystallogr.*, B**27**, p. 2037.
Schenk, H. (1972a). *Acta Crystallogr.*, A**28**, p. 412.
Schenk, H. (1972b). *Acta Crystallogr.*, A**28**, p. 661.
Schenk, H. (1973a). *Acta Crystallogr.*, A**29**, p. 480.
Schenk, H. (1973b). *Acta Crystallogr.*, A**29**, p. 77.
Schenk, H. (1974). *Acta Crystallogr.*, A**30**, p. 472.
Schevitz, R. W., Podjarny, A. D., Zwick, M., Hughes, J. J., and Sigler, P. B. (1981). *Acta Crystallogr.*, A**37**, p. 669.
Schiltz, M. and Bricogne, G. (2008). *Acta Crystallogr.*, D**64**, p. 711.
Schlünzen, F., Hansen, H. A. S., Thygesen, J., Bennett, W. S., Volkmann, N., Levin, I., Harms, J., Bartels, H., Zaytzev-Bashan, A., Berkovitch-Yellin, Z., Sagi, I., Franceschi, F., Krumbholz, S., Geva, M., Weinstein, S., Agmon, I., Boddeker, N., Morlang, S., Sharon, R., Dribin, A., Maltz, E., Peretz, M., Weinrich, V., and Yonath, A. (1995). *Biochem. Cell. Biol.*, **73**, p. 739.
Schneider, T. R. and Sheldrick, G. M. (2002). *Acta Crystallogr.*, D**58**, p. 1772.
Schwarzenbacher, R., Godzik, A., Grzechnig, S. K., and Jaroszewski, L. (2004). *Acta Crystallogr.*, D**60**, p. 1229.
Schwarzenbacher, R., Godzik, A., and Jaroszewski, L. (2008). *Acta Crystallogr.* D**64**, p. 133.
Shankland, K., David, W. I. F., Csoka, T., and McBride, L. (1998). *Int. J. Pharm.*, **165**, p. 117.
Shankland, K., David, W. I. F., and Sivia, D. S. (1997a). *J. Mater. Chem.*, **7**, p. 569.
Shankland, K., David, W. I. F., and Csoka, T. (1997b). *Z. Kristallogr.*, **212**, p. 550.
Sheldrick, G. M. (1982). In *Computational Crystallography* (ed. D. Sayre), pp. 506–14. Clarendon, Oxford.
Sheldrick, G. M. (1990). *Acta Crystallogr.*, A**46**, p. 467.
Sheldrick, G. M. (1997). In *Direct Methods for Solving Macromolecular Structures*. NATO Advanced Study Institute, Erice, Italy.
Sheldrick, G. M. (1998). In *Direct Methods for Solving Macromolecular Structures* (ed. S. Fortier). Dordrecht: Kluwer.
Sheldrick, G. M. (2002). *Z. Kristallogr.*, D**58**, p 1958.

Sheldrick, G. M. (2007). In *Evolving Methods for Macromolecular Crystallography* (ed. R. J. Read and J. L. Sussman). Springer, The Netherlands.
Sheldrick, G. M. (2010). *Acta Crystallogr.*, D**66**, p. 479.
Sheldrick, G. M. and Gould, R. O. (1995). *Acta Crystallogr.*, B**51**, p. 423.
Sheriff, S., Klei, H. E., and Davis, M. E. (1999). *J. Appl. Crystallogr.*, **32**, p. 98.
Shiono, M. and Woolfson, M. M. (1992). *Acta Crystallogr.*, A**48**, p. 451.
Shirley, R. (1980). *NBS Spec. Publ.*, **567**, p. 362.
Shmueli, U. and Weiss, G. H. (1986). *Acta Crystallogr.*, A**42**, p. 240.
Shmueli, U. and Weiss, G. H. (1992). *Acta Crystallogr.*, A**48**, p. 418.
Shmueli, U. and Weiss, G. H. (1995). In *Introduction to Crystallographic Statistics*. IUCr Oxford University Press.
Shmueli, U., Weiss, G. H., Kiefer, J. E., and Wilson, A. J. C. (1984). *Acta Crystallogr.*, A**40**, p. 651.
Silva, A, M. and Viterbo, D. (1980). *Acta Crystallogr.*, A**36**, p. 1065.
Sim, G. A. (1959a). *Acta Crystallogr.*, **12**, p. 813.
Sim, G. A. (1959b). *Acta Crystallogr.*, **12**, p. 511.
Simerska, M. (1956). *Czech. J. Phys.*, **6**, p. 1.
Simonov, V. I. (1976). In *Crystallographic Computing Techniques*, pp. 138–43. Munksgaard, Copenhagen.
Sisinni, L., Cendron, L., Favaro, G., and Zanotti, G. (2010). *FEBS J.*, **277**, p. 1896.
Sivia, D. S. and David, W. I. F. (1994). *Acta Crystallogr.*, A**50**, p. 703.
Sonneveld, E. J. and Visser, J. W. (1975). *J. Appl. Crystallogr.*, **8**, p. 1.
Srinivasan, R. (1961). *Acta Crystallogr.*, **14**, p. 607.
Srinivasan, R. and Subramanian, E. (1964). *Acta Crystallogr.*, **17**, p. 67.
Srinivasan, R. and Ramachandran, G. N. (1965a). *Acta Crystallogr.*, **19**, p. 1003.
Srinivasan, R. and Ramachandran, G. N. (1965b). *Acta Crystallogr.*, **19**, p. 1008.
Strahs, G. and Kraut, J. (1968). *J. Mol. Biol.*, **35**, p. 503.
Su, W.-P. (1995). *Acta Crystallogr.*, A**51**, p. 845.
Subbiah, S. (1991). *Science*, **252**, p. 128.
Subbiah, S. (1993). *Acta Crystallogr.*, D**59**, p. 108.
Subramanian, W. and Hall, S. R. (1982). *Acta Crystallogr.*, A**38**, p. 577.
Suortti, P. and Jennings, L. D. (1977). *Acta Crystallogr.*, A**33**, p. 1012.
Swanson, S. M. (1979). *J. Mol. Biol.*, **129**, p. 637.
Swanson, S. M. (1994). *Acta Crystallogr.*, D**50**, p. 695.
Takeuchi, Y. (1972). *Z. Kristallogr.*, **135**, p. 120.
Tang, D., Jansen, J., Zandbergen, H. W., and Shenk, H. (1995). *Acta Crystallogr.*, A**51**, p. 188.
Taylor, W. J. (1953). *J. Appl. Phys.*, **24**, p. 662.
Taylor, D. J. and Woolfson, M. M. (1975). In Tenth International Congress of Crystallography, Amsterdam, Contribution 02. p. 2–12.
Taylor, D. J., Woolfson, M. M., and Main, P. (1979). *Acta Crystallogr.*, A**35**, p. 870.
Templeton, D. H. and Templeton, L. K. (1988). *Acta Crystallogr.*, A**44**, p. 1045.
Templeton, D. H., Templeton, L. K., Philips, J. C., and Hodgson, K. O. (1980). *Acta Crystallogr.*, A**36**, p. 436.
Templeton, L. K., Templeton, D. M., Phizackerley, R. P., and Hodgson, K. O. (1982). *Acta Crystallogr.*, A**38**, p. 74.
Ten Eyck, L. F. (1973). *Acta Crystallogr.*, A**29**, p. 183.
Terwilliger, T. C. (1997). *Methods Enzymol.*, **276**, p. 530.
Terwilliger, T. C. (2001). *Acta Crystallogr.*, D**57**, p. 1755.
Terwilliger, T. C. (2003). *Acta Crystallogr.*, D**59**, p. 38.
Terwilliger, T. C. (2004). *J. Synchrotron Radiation*, **11**(1), p. 121.
Terwilliger, T. C., Adams, P. D., Read, R. J., McCoy, A. J., Moriarty, N. W., Grosse-Kunstleve, R. W., Afonine, P. V., Zwart, P. H., and Hung, L.-W. (2009). *Acta Crystallogr.*, D**65**, p. 582.
Terwilliger, T. C. and Berendzen, J. (1999). *Acta Crystallogr.*, D**55**, p. 849.
Terwilliger, T. C. and Berendsen, J. (2001). *International Tables for Crystallography F.*, pp. 303–309.

Terwilliger, T. C. and Eisenberg, D. (1983). *Acta Crystallogr.*, A**39**, p. 813.
Terwilliger, T. C. and Eisenberg, D. (1987). *Acta Crystallogr.*, A**43**, p. 6.
Terwilliger, T. C., Read, R. J., Adams, P. D., Brunger, A. T., Afonine, P. V., Grosse-Kunstleve, R. W., and Hung, L-W. (2012). *Acta Crystallogr.*, D**68**, p. 861.
Thygesen, J., Weinstein, S., Franceschi, F., and Yonath, A. (1996). *Structure*, **4**, p. 513.
Tivol, W. F., Dorset, D. L., McCourt, M. P., and Turner, J. N. (1993). *Microsc. Soc. Am. Bull.*, **23**(1), p. 91.
Tollin, P., Main, P., and Rossmann, M. G. (1966). *Acta Crystallogr.*, **20**, p. 404.
Tollin, P. and Rossmann, M. G. (1966). *Acta Crystallogr.*, **21**, p. 872.
Tong, L. (1993). *J. Appl. Crystallogr.*, **26**, p. 748.
Tong, L. (2001). *Acta Crystallogr.*, D**57**, p. 1383.
Tong, L. and Rossmann, M. G. (1990). *Acta Crystallogr.*, A**46**, p. 783.
Toraya, H. (1986). *J. Appl. Crystallogr.*, **19**, p. 440.
Tremaine, M., Kariuki, B. M., and Harris, K. D. M. (1997). *Angew. Chem. Int. Ed. Engl.*, **36**, p. 770.
Tsoucaris, G. (1970). *Acta Crystallogr.*, A**26**, p. 492.
Turner, P. S. and Cowley, J. M. (1969). *Acta Crystallogr.*, A**25**, p. 475.
Unwin, P. N. T. and Hendersson, R. (1975). *J. Mol. Biol.*, **94**, p. 425.
Ursby, T. and Bourgeois, D. (1997). *Acta Crystallogr.*, A**53**, p. 564.
Urzhumtsev, A. G., Vernoslova, E. A., and Podjarny, A. D. (1996). *Acta Crystallogr.*, D**52**, p. 1092.
Vagin, A. and Teplyakov, A. (1997). *J. Appl. Crystallogr.*, **30**, p. 1022.
Vagin, A. and Teplyakov, A. (1998). *Acta Crystallogr.*, D**54**, p. 400.
Vagin, A. and Teplyakov, A. (2000). *Acta Crystallogr.*, D**56**, p. 1622.
Vainshtein, B. K. (1964). *Structure Analysis By Electron Diffraction*. Pergamon Press, Oxford.
Vainshtein, B. K. and Kayushina, R. L. (1967). *Sov. Phys. Crystallogr.*, **11**, p. 468.
Vainshtein, B. K. and Pinsker, Z. G. (1949). *J. Phys. Chem.* SSSR, **23**, p. 1058.
van der Hark, T. E. M., Prick, P., and Beurskens, P. T. (1976). *Acta Crystallogr.*, A**32**, p. 816.
van der Putten, N. and Schenk, H. (1977). *Acta Crystallogr.*, A**33**, p. 856.
Vaughan, P. A. (1958). *Acta Crystallogr.*, **11**, p. 111.
Veilleux, F. M. D. A. P., Hunt, J. F., Roy, S., and Read, R. J. (1995). *J. Appl. Crystallogr.*, **28**, p. 347.
Vermin, W. J. and de Graaf, R. A. G. (1978). *Acta Crystallogr.*, A**34**, p. 892.
Vicković, I. and Viterbo, D. (1979). *Acta Crystallogr.*, A**35**, p. 500.
Vijayan, M. (1980). *Acta Crystallogr.*, A**36**, p. 295.
Vincent, R. and Midgley, P. A. (1994). *Ultramicroscopy*, **53**, p. 271.
Visser, J. W. (1969). *J. Appl. Crystallogr.*, **2**, p. 89.
Voigt-Martin, I. G., Yan, D. H., Yakimansky, A., Schollmeyer, D., Gilmore, C. J., and Bricogne, G. (1995). *Acta Crystallogr.*, A**51**, p. 849.
von Eller, G. (1973). *Acta Crystallogr.*, A**29**, p. 63.
Wang, B. C. (1985). *Methods Enzymol.*, **115**, p. 90.
Wang, D. N., Hovmöller, S., Kihlborg, L., and Sundberg, M. (1988). *Ultramicroscopy*, **25**, p. 303.
Weeks, C. M., de Titta, G. T., Hauptman, H. A., Thuman, P., and Miller, R. (1994). *Acta Crystallogr.*, A**50**, p. 210.
Weeks, C. M. and Miller, R. (1999). *Acta Crystallogr.*, D**55**, p. 492.
Weinzierl, J. E., Eisenberg, D., and Dickerson, R. E. (1969). *Acta Crystallogr.*, A**25**, p. 380.
Weiss, M. S., Sicker, T., and Hilgenfeld, R. (2001). *Structure*, **9**, p. 771.
Wenk, H. R., Downing, K. H., Hu, Meisheng, and O'Keefe, M. A. (1992). *Acta Crystallogr.*, A**48**, p. 700.
Werner, P.-E., Eriksson, L., and Westdahl, M. (1985). *J. Appl. Crystallogr.*, **18**, p. 367.
Wessels, T., Baerlocher, Ch., and McCusker, L. B. (1999). *Science*, **284**, p. 477.
West, C. D. (1954). *J. Chem. Phys.*, **22**, p. 150.
Weyl, H. (1915–16). *Math. Ann.*, **77**, p. 313.

References

White, P. and Woolfson, M. M. (1975). *Acta Crystallogr.*, A**31**, p. 53.
Will, G. (1979). *J. Appl. Crystallogr.*, **12**, p. 483.
Williams, T. V. (1982). *Ph.D. Thesis*. University of North Carolina, Chapel Hill.
Wilson, A. J. C. (1942). *Nature*, **150**, p. 152.
Wilson, A. J. C. (1949). *Acta Crystallogr.*, **2**, p. 318.
Wilson, A. J. C. (1963). *Mathematical Theory of X-ray Powder Diffraction*. Philips Technical Library.
Wilson, A. J. C. (1978). *Acta Crystallogr.*, A**34**, p. 474.
Wilson, C. and Agard, D. A. (1993). *Acta Crystallogr.*, A**49**, p. 97.
Winn, M. D. et al. (2011). *Acta Crystallogr.*, D**67**, p. 235.
Woolfson, M. M. (1977). *Acta Crystallogr.*, A**33**, p. 219.
Woolfson, M. M. (1978). In *Direct Methods of Solving Crystal Structures*, International School of Crystallography, Erice, Lectures 4 and 13.
Wrinch, D. (1939). *Phil. Mag.*, **27**(7), p. 98.
Wu, H. and Hendrickson, W. A. (1996). *Acta Crystallogr.*, S**52**, C–55.
Wu, J. S., Spence, J. C. H., O'Keeffe, M., and Groy, T. L. (2004). *Acta Crystallogr.* A**60**, p. 326.
Yang, C. and Pflugrath, J. W. (2001). *Acta Crystallogr.*, D**57**, p. 1480.
Yang, C., Pflugrath, J. W., Courville, D. A., Stence, C. N., and Ferrara, J. D. (2003). *Acta Crystallogr.*, D**59**, p. 1943.
Yao, J. X. (1981). *Acta Crystallogr.*, A**37**, p. 642.
Yao, J.-X. (2002). *Acta Crystallogr.*, D**58**, p. 1941.
Yao, J. X., Dodson, E. J., Wilson, K. S., and Woolfson, M. M. (2006). *Acta Crystallogr.*, D**62**, p. 901.
Young, R. A. (1993). *The Rietveld Method*. Oxford University Press.
Zachariasen, W. H. (1945). *Theory of X-Ray Diffraction in Crystals*. Wiley, New York.
Zachariasen, W. H. (1952). *Acta Crystallogr.*, **5**, p. 68.
Zhang, K. Y. J. and Main, P. (1990a). *Acta Crystallogr.*, A**46**, p. 41.
Zhang, K. Y. J. and Main, P. (1990b). *Acta Crystallogr.*, A**46**, p. 377.
Zhang, D., Oleynikov, P., Hovmöller, S., and Zou, X. (2010). *Z. Kristallogr.*, **225**, p. 94.
Zhang, X. J. and Woolfson, M. M. (1982). *Acta Crystallogr.*, A**38**, p. 683.
Zou, X. D., Hovmöller, S., and Oleynikov, P. (2011). *Electron Crystallography*. Oxford University Press.
Zou, X. D., Mo, Z. M., Hovmöller, S., Li, X. Z., and Kuo, K. H. (2003). *Acta Crystallogr*, A**59**, p. 526.
Zou, X., Sukharev, Y., and Hovmöller, S. (1993). *Ultramicroscopy*, **52**, p. 436.
Zuev, A. D. (2006). *J. Appl. Cryst.*, **39**, p. 304.

Index

A
Absolute scaling of the intensities 43
Allowed origin
  definition of 65, 66, 76, 81
  tabulation of 70–72, 77, 79
Anomalous
  difference 340, 349
  scattering 336
  scattering applied to powder data 363
  scattering substructure 345, 350
ARCIMBOLDO 289
Arp/warp 196
Atomic scattering factor 8
Atomicity postulate 108
Automatic model building 184

B
Banks (Crystallographic) 24–26
Basic postulate of structural crystallography 17–24
Bessel functions 385

C
Central limit theorem
  in Wilson distributions 28, 50, 52
  statement of 375
Characteristic function 94, 98, 371
Charge flipping 199–201
C-map 225
Cochran formula 105
Cochran integral criterion 231
Co-crystallization 315
Convergence procedure 135
Convolution, definition of 2
Crystal lattice 2
Crystal system 4
Cumulant generating function 95, 98, 373
Cumulants of a distribution 95, 98, 373
Cumulative distribution 35

D
Debye formula 59
Decomposition of powder patterns 259
Difference Fourier synthesis 164, 201–203, 206–211

Difference Patterson synthesis 318
Directional data 380
Double Patterson 232
Dynamic scattering 237

E
EDM procedures 178
Electron density
  covariance in 168, 174
  difference 164, 201–203, 206–211
  hybrid 166, 205, 211–212
  interpretation 138, 184
  map calculation 10
  modification (EDM) procedures 178
  observed 162
  properties 156
  resolution bias in 158–162
  variance of 168, 174
Electron diffraction
  precession and rotation cameras 244
  traditional techniques 239, 241
Electron scattering
  non-kinematical 237
  properties of 235
Envelopes 190–191
Equiphasic surface 62
Ewald sphere 9
Excitation error 251

F
Figures of merit 132, 137, 138
Fourier syntheses, see Electron density
Free lunch 184
Friedel law
  definition 10
  violation 337

G
Gamma function 382
Gaussian distribution, see Normal distribution
Generalized hypergeometric function 389

Index

H
Half-bake approach 147
Harker sections 218
Hauptman–Karle family
  definition of 69
  tabulation of 70, 71, 77, 79
Heavy atom
  derivative 314
  method 219
  substructure 219, 318, 325, 333
Hermite polynomials 383
Histogram matching 180, 193
Hybrid Fourier syntheses 166, 205, 211–212

I
Implication transformation method 221
Isomorphous
  data scaling 322
  derivative 314
  difference 318
  difference Patterson synthesis 318
Isomorphous replacement techniques 315
  and direct methods 331

J
Joint probability distributions, see Probability distribution functions

L
Lack of closure error 327
Laguerre polynomials 383
Le Bail method 261

M
MAD
  algebraic bases of 352
  data refinement 365
  probabilistic approach to 354
  wavelength definition 340
Magic integers approach 135
MIR
  algebraic bases of 320
  method 315
  probabilistic approach to 327
MIRAS techniques 360
Molecular averaging 181
Molecular replacement
  six-dimensional search 279
  stochastic approaches 289
  techniques 275
Molecular scattering factor 8
Moments of a distribution 373, 378
Multisolution procedures 134

N
Neighbourhoods 87
Neutron scattering 245
Non-crystallographic symmetry
  definition of 181
  of translational type 308
  operators 304, 305
Normal distribution 374
Normalized structure factor
  definition of 31
  distribution of, in Wilson statistics 31

O
Observed Fourier synthesis 162
One-phase structure seminvariant 75, 233
Origin definition 61
Origin translation 62

P
P_{10} formula for triplet invariants 111, 121
Patterson
  deconvolution of 218
  function 215–217
  function of second kind 233
  superposition methods 223
  symmetry 217
Pawley method 261
Peak shape modelling 259
Permissible origin, see Allowed origin
Phase extension
  in direct space 177
  in reciprocal space 140
Phasing magnitudes 88
Phasing shells 87
Point group symmetry 4
Positivity postulate
  definition of 108
  violation of 247
Powder data
  ab initio phasing methods for 267
  full pattern decomposition 258
  indexing of 264
  non-ab initio phasing methods for 270
  peak overlapping in 254
  peak resolution 255
Probability distribution functions
  conditional 85–87
  cumulants of 95, 373
  joint 83, 93–103, 375
  mathematical bases of 370
  method of 93–103
  moments of 371
  of signs 379
  when a model is available 152

Profile shape function 259
Psi-zero triplets 138

Q

Quadrant permutation technique 135
Quartet invariant
    algebra of 116
    finding 149
    probabilistic estimation in any space group 123–124
    probabilistic estimation in P1 and P$\bar{1}$ 112–116

R

Random phase approach 136
Reciprocal space 5
Reciprocal lattice 7
Relax algorithm 212–213
Representations
    for isomorphous structures 91
    of a structure invariant 88
    of a structure seminvariant 89
Resolution bias 158, 183
Rietveld approach 258
Rotation function
    definition 280, 282, 284
    in Cartesian space 294
    symmetry of 299

S

SAD
    algebraic approach 344
    case ambiguities 346
    method 340
    probabilistic approach to 354, 360
Sayre–Hughes equation 230
Selenoproteins 341
Sequence
    docking 185
    identity 277
Shake & Bake 144
Sigma-A 154, 173
SIR method
    algebraic bases 317
    and Direct Methods 331
    for centric reflections 330
    probabilistic approach to 360, 368
SIRAS
    algebraic bases of 347
    probabilistic approach to 360, 368
    techniques 347
Skeletonization techniques 182
Soaking 315
Solvent
    content 190
    flattening 180
    model of 192
Space group
    definition of 5
    identification of 42, 36–41, 266
Starting set of phases 134
Statistical weight
    in Wilson distributions 32
    tabulation of 33–34
Structure factor
    algebra 53
    definition of 8–9, 11
    statistics of 56–58
Structure invariant
    definition of 63
    two-phase 152
Structure seminvariant
    definition of 69
    representations of 89
Symmetry operators 3, 12–13
Systematic absences 15–16

T

Tangent formula
    derivation 128–130
    limits 141
    weighted 136
Texture effects 272
Translation functions 286
Triplet invariant
    estimation in any space group 120
    estimation in P1 104, 108, 110, 117, 120–121
    estimation in P$\bar{1}$ 107
    for isomorphous structures 170
    via P_{10} formula 111, 121

U

Unit cell
    chemical content 49
    definition of 1, 2

V

VLD (vive la difference)
    difference Fourier synthesis 201–203, 206–211
    hybrid Fourier syntheses 205, 211–212
    phasing approach 201

W

Wilson
    plot 43–49
    statistics 28